Advances in

POWER STATION
CONSTRUCTION

Pergamon Titles of Related Interest

BULLARD
Trends in Electric Utility Research

COMMISSION OF THE EUROPEAN COMMUNITIES
Fusion Technology 1984

DAVIES
Protection of Industrial Power Systems

FROST
Nuclear Fuel Elements

HERBST
Automatic Control in Power Generation, Distribution and Protection

HEWITT & WALLEY
Gas-Liquid Flow and Heat Transfer

INSTITUTION OF CHEMICAL ENGINEERS
Energy: Money, Materials and Engineering

KUFFEL & ZAENGL
High Voltage Engineering

MARKS *et al.*
Aspects of Civil Engineering Contract Procedure, 3rd Edition

PENNER
New Sources of Oil and Gas

SIMPSON
Fracture Problems and Solutions in the Energy Industry

SUBRAMANYAM
Computer Applications in Large-Scale Power Systems

TAHER
Energy: A Global Outlook, 2nd Edition

Pergamon Related Journals

Annals of Nuclear Energy
Civil Engineering for Practising and Design Engineers
Energy
Energy Conversion and Management
Energy Developments in Japan
International Journal of Heat and Mass Transfer
Journal of Heat Recovery Systems
Progress in Energy and Combustion Science
Progress in Nuclear Energy

Sample copy gladly sent on request

Advances in
POWER STATION
CONSTRUCTION

Generation Development and Construction Division,
Central Electricity Generating Board,
Barnwood, Gloucester, UK

PERGAMON PRESS

OXFORD · NEW YORK · TORONTO · SYDNEY · FRANKFURT

U.K.	Pergamon Press Ltd., Headington Hill Hall, Oxford OX3 0BW, England
U.S.A.	Pergamon Press Inc., Maxwell House, Fairview Park, Elmsford, New York 10523, U.S.A.
CANADA	Pergamon Press Canada Ltd., Suite 104, 150 Consumers Road, Willowdale, Ontario M2J 1P9, Canada
AUSTRALIA	Pergamon Press (Aust.) Pty. Ltd., P.O. Box 544, Potts Point, N.S.W. 2011, Australia
FEDERAL REPUBLIC OF GERMANY	Pergamon Press GmbH, Hammerweg 6, D-6242 Kronberg, Federal Republic of Germany
JAPAN	Pergamon Press Ltd., 8th Floor, Matsuoka Central Building, 1-7-1 Nishishinjuku, Shinjuku-ku, Tokyo 160, Japan
BRAZIL	Pergamon Editora Ltda., Rua Eça de Queiros, 346, CEP 04011, São Paulo, Brazil
PEOPLE'S REPUBLIC OF CHINA	Pergamon Press, Qianmen Hotel, Beijing, People's Republic of China

Copyright © 1986 Central Electricity Generating Board

First edition 1986

Library of Congress Cataloging-in-Publication Data
Main entry under title:

Advances in power station construction.

.1. Electric power-plants — Great Britain. I. Central Electricity Generating Board. Generation Development and Construction Division.
TK57.A73 1985 621.31′213′0941 85–19113

British Library Cataloguing in Publication Data
Advances in power station construction.
1. Electric power-plants — Design and construction — Great Britain — History
I. Central Electricity Generating Board. Generation Development and Construction Division
621.31′21′0941 TK1193.G7

ISBN 0–08–031677–8 (Hardcover)
ISBN 0–08–031678–6 (Flexicover)

Printed in Great Britain by A. Wheaton & Co. Ltd., Exeter

Foreword

THIS book is about power stations — specifically about the construction of modern power stations by the Central Electricity Generating Board in England and Wales over the past decade. It describes the work of the CEGB's Generation Development and Construction Division, perhaps better known throughout the world as simply 'Barnwood' where it has its Headquarters in Gloucester, UK. Barnwood was formed in the early 1970s to concentrate the CEGB's then dispersed engineering construction resources to cope with the smaller number but greatly increased size and complexity of modern power station projects. Perhaps uniquely over the ten years since its formation Barnwood has managed the construction of all types of station; coal-fired, oil-fired, nuclear, pumped storage and hydro. This book tells the story of these various projects and gives detailed descriptions of the respective stations. However, it is not intended as a comprehensive description of power station technology — that is dealt with in a separate text 'Modern Power Station Practice' which was first published in 1959 and currently being updated. Rather it is intended to convey the scale of such projects and the many decisions and compromises which have to be made in the course of managing their construction.

The considerations, evaluations and factors leading up to the decision as to when, where and what type of power station to build are described in Chapter 1. The processes leading up to the selection of appropriate sites are outlined in terms of systems engineering and systems planning with attendant architectural and civil engineering considerations. Station layout is described for the various types of plant and fuel, as well as the principal auxiliary plant items which influence the best utilisation of available space. Developments in boiler, turbine and gas turbine plant design are presented as well as the general developments in nuclear reactor types and the options they present. Station main electrical power systems and control and instrumentation systems are discussed in terms of state-of-the-art equipment and needs. The design of the primary auxiliary systems such as cooling water, fire protection, air services, heating and ventilating and water treatment are also discussed.

Chapters 2, 3, 4 and 5 are devoted respectively to Littlebrook D oil-fired station, Drax coal-fired station, Dinorwig pumped storage station and Heysham 2 advanced gas-cooled reactor nuclear station. In each case, the chapter describes the systems and plant installed, mentioning improvements over earlier designs, and giving design parameters and performance details of main plant items, auxiliary systems and the overall station.

The extent of plant described in these chapters is selective because of the enormous range of equipment installed in a modern power station and the equally vast extent of information required to operate the station safely and efficiently.

Chapter 6 describes the intended design of the Sizewell B pressurised water reactor nuclear station, the construction of which at the time of publication is subject to the outcome of a Public Inquiry.

Chapter 7, entitled 'Project Management', explains the organisation, procedures and practices relating to the management of power station projects, particularly in relation to the most recent stations. The principal aspects discussed include management and control of design resources, planning and programming, contracting procedures, cost control, quality assurance, site organisation, commissioning and industrial relations, and the use of computer techniques in project management.

Like the construction of a power station, the preparation of this book is the work of a great many people primarily at Barnwood but using a great deal of technical information supplied by the CEGB's many contractors. Whilst it is impracticable to name all the individuals who have contributed to the volume it is appropriate to note the effort of Mr. H. E. Johnson at GDCD Barnwood who compiled and edited the many contributions.

Marshall of Goring

THE LORD MARSHALL OF GORING KT, CBE, FRS
Chairman — Central Electricity Generating Board

Contents

3. Drax Coal-fired Power Station

4. Dinorwig Pumped Storage Power Station

5. Heysham 2 — AGR Nuclear Power Station

6. Sizewell B — PWR Nuclear Power Station

7. Project Management

Comprehensive List of Contents

Index

Editorial Panel

J. G. COLLIER, B.SC. (ENG.), F.ENG., F.I.CHEM.E., F.I.MECH.E., HON. F.I.NUC.E.

B. POWELL, O.B.E., M.ENG., C.ENG., F.I.E.E., F.I.MECH.E.

P. M. BILLAM, M.A.

R. N. G. BURBRIDGE, O.B.E., C.ENG., F.I.MECH.E., F.I.E.E.

B. V. GEORGE, B.TECH., C.ENG., M.I.MECH.E.

J. F. C. JEBSON, F.I.P.M., F.B.I.M.

J. M. KAY, D.M.A., A.C.I.S.

Editorial Consultants

J. LAWRENCE, C.ENG., M.INST.F.

H. MASDING, C.ENG., M.I.E.E.

This book has been compiled and edited from information provided by several engineer-authors and departments within the Generation Development and Construction Division of the CEGB at Barnwood. It contains information, both textual and graphical, which has been supplied to CEGB by contractors, manufacturers and plant suppliers for design, construction, operation and maintenance purposes, and for which GDCD wishes to make due acknowledgement.

Abbreviations and Units in Common Use in this Book

AC	alternating current
ACS	average cold spell
ACW	auxiliary cooling water
AGR	advanced gas-cooled reactor
ALARP	as-low-as-reasonably-practicable
AOD	above ordnance datum
APC	Atomic Power Construction
ASSASSIN	agricultural system for the storage and subsequent selection of information
AUX	auxiliary
AVR	automatic voltage regulator
BCD	burst cartridge detector
BCF	bromochlorofluoromethane
BD	board
BEA	British Electricity Authority
BFP	boiler feed pump
BFPT	boiler feed pump turbine
BNDC	British Nuclear Design & Construction
BNFL	British Nuclear Fuels Ltd.
BR	British Rail
BS	British Standard
BTM	bromotrifluoromethane
BTRS	boron thermal regeneration system
BWR	boiling water reactor
CAD	computer-aided design
C & I	control and instrumentation
CCR	central control room
CCWS	component cooling water system
CEGB	Central Electricity Generating Board
CISD	Computing Information Systems Department
CMR	continuous maximum rating
CO_2	carbon dioxide
CPA	contract price adjustment
CPU	central processing unit
CSS	containment spray system

CVCS	chemical and volume control system
CT	current transformer
CW	cooling water
Cr/Mo/V	chromium/molybdenum/vanadium
DA	deaerator
DBE	design basis earthquake
DC	direct current
DDC	direct digital control
DHB	decay heat boiler
DIA	diameter
DIM	design intent memorandum
DORIS	drawing office records system
DPS	data processing system
DTV	draught tube valve
ECC	emergency communications centre
ECON	economiser
EHWL	extreme high water level
EIC	emergency indication centre
ELWL	extreme low water level
ESB	essential supplies building
ESD	Engineering Services Department
ESFAS	engineered safety features actuation system
ESV	emergency stop valve
FD	forced draught
FM	fuelling machine
FRF	fire resistant fluid
GEN	generator
GDCD	Generation Development and Construction Division
GIS	guaranteed instrument supplies
GRP	glass reinforced plastic
GSW	general service water
GT	gas turbine
HADV	high activity debris void
HDLC	high integrity high speed data link
HP	high pressure
HT	high tension
HTR	high temperature reactor
H & V	heating and ventilating
HVAC	heating, ventilating and air conditioning
ID	induced draught
IFDF	irradiated fuel disposal facility
IGV	inlet guide vanes
IP	intermediate pressure
ISI	in-service inspection
LER	liquid earthing resistor
LH	left hand
LOCA	loss of coolant accident

LP	low pressure
LWR	light water reactor
MMC	Monopolies and Mergers Commission
MSLB	main steam line break
MTBF	mean time before failure
N-16	nitrogen 16
NAECI	National Agreement for the Engineering Construction Industry
NAC	net avoidable cost
NCB	National Coal Board
NDT	non-destructive testing
NEC	net effective cost
NEDO	National Economic Development Office
NII	Nuclear Installations Inspectorate
NNC	National Nuclear Corporation
NJC	National Joint Council
NPC	Nuclear Power Company
Ni/Nb	nickel/niobium
NSSS	nuclear steam supply system
OD	ordnance datum
od	outside diameter
PA	primary air
PAX	private automatic exchange
PCPV	prestressed concrete pressure vessel
PERM	programmable equipment for relaying and measurement
PF	pulverised fuel
PJC	project joint council
PLC	programmable logic controller
PMB	project management board
PMP	project master programme
PMT	project management team
PPS	primary protection system
PRES	pressure
PVC	poly-vinyl-chloride
PWR	pressurised water reactor
RACS	reactor auxiliaries cooling system
RCC	rod cluster control
RCDT	reactor coolant drains tank
RCS	reactor coolant system
RFW	reserve feed water
RH	right hand
RHRS	residual heat removal system
RPS	reactor protection system
RPV	reactor pressure vessel
RUHS	reserve ultimate heat sink
SD	site datum
SF_6	sodium hexafluoride
SGHWR	steam generating heavy water reactor

SIS	safety injection system
SMART	scheduling manpower and resource technique
SNAP	system of network analysis programs
SNUPPS	standardised nuclear unit power plant system
SPS	secondary protection system
SSD	secondary shutdown
SSE	safe shutdown earthquake
SSS	synchronous self shifting
TEMP	temperature
TG	turbine generator
TNPG	The Nuclear Power Group
TRANS	transformer
TSV	turbine stop valve
Th	thermal
UHF	ultra high frequency
UK	United Kingdom
UKAEA	United Kingdom Atomic Energy Authority
US	United States
USA	United States of America
USB	upper stabiliser brush
UO_2	uranium oxide (enriched)
UPS	uninterruptable power supplies
UPVC	unplasticised poly-vinyl-chloride
VDU	visual display unit
VHF	very high frequency
VFSE	variable frequency starting equipment
WT	water treatment
ZVI	zone of visual influence

Annotations on illustrations are generally in upper case letters and therefore abbreviations which appear as lower case in the text are printed in upper case on illustrations.

UNITS

mm	— millimetre	μ	— micro (10^{-6})
m	— metre	m	— milli (10^{-3})
m^2	— square metre	k	— kilo (10^3)
m^3	— cubic metre	M	— mega (10^6)
ha	— hectare	G	— giga (10^9)
		T	— tera (10^{12})
s	— second	A	— ampere
min	— minute	V	— volt
h	— hour	W	— watt
d	— day	J	— joule
y	— year	N	— newton
pa	— per annum	H	— henry
rev/min	— revolutions per minute	e	— electrical
		so	— sent out
g	— gram		
t	— tonne	Th	— thermal
tU	— tonne of uranium	Hz	— hertz
			(cycles per second)

% wt — percent by weight
% vol — percent by volume
vpm — volumetric parts per million
°C — degrees centigrade

bar — pressures quoted are absolute unless otherwise stated
(g) — gauge
cSt — centistokes (kinematic viscosity)
sec — seconds (viscosity — Redwood 1)
Sv — sievert

1.

Construction History and Development

1. INTRODUCTION

At the time of nationalisation in 1948, the electricity supply industry in the UK consisted of some 560 separate supply undertakings both private and municipal, each being a local monopoly and all committed to the cheapest service and a modest profit. The industry was generally overseen, from a financial point of view, by the government Electricity Commissioners, and from an operational point of view by the national Central Electricity Board which maintained a form of grid system.

There was at that time a wide diversity in the range and quality of supplies available to the public. These consisted of 100V, 200V, 240V and 2000V 50Hz single phase supplies, 415V and 6600V 50Hz 3-phase supplies, and in some cases, 250V and 500V dc supplies. Standardisation in the voltages available and in the design of industrial and domestic appliances which had been started before 1939 was being vigorously pursued.

There was equally a wide diversity in the size and the design of power stations and the boiler and turbine plant which generated the electricity. They were, in the main, small stations of less than 50MW, with the boilers connected to a common steam 'range' system which fed a variety of turbine generators that had been added to the local system over the previous 20 to 30 years.

The advancement of the technology was modest; steam temperatures of 454°C and steam pressures of 45 bar were common. Auxiliary power supplies were for the first time taken from unit transformers or separate works generators driven by the main turbine generator shafts. The overall thermal efficiencies were of the order 20-23%.

Automatic control of the 'range' steam pressure was normal practice together with automatic control of water level, combustion, draught and voltage. There was little or no telemetry and the control and instrumentation was based on the direct and single measurement of the properties of the working steam. Coal, both the high quality steam coal and mine refuse, was the only significant fuel available for power electricity generation. There were a number of chain grate and retort type stokers being built, but the main boiler development was directed towards pulverised fuel systems.

The 132kV grid system at that time provided a limited interconnection between the different generating authorities, and did provide a system of mutual standby. The total capacity of all the utilities was in the order of 12,000MW and the annual units supplied were around 40TWh.

The 1947 Electricity Act which was implemented in 1948, created the British Electricity Authority (BEA) which consisted of 14 Area Boards including two Scottish Boards and 12 Generating Divisions. These were arranged geographically, with each Area Board responsible to the central authority for the distribution and consumer services and each Generating Division responsible to the BEA for the generation and transmission. The area of UK covered by the new Authority comprised England, Wales and South of Scotland.

The construction of new power stations was under the technical direction of the BEA Chief Engineer at the London Headquarters, who maintained and controlled departments which covered the design of power plant, transmission system design, future system planning and siting requirements. The station design in terms of detailed engineering and contract management was, in most cases, delegated to the Divisions on a geographical basis, with some design work being done by the Headquarters organisation and consulting engineers.

The Divisions initiated a degree of standardisation of 30MW and 60MW plants which were individual boiler/turbine units instead of the common steam range systems. Between 1948 and 1952, some 5800MW of plant comprising 150 units of 30MW and 60MW was installed in 66 power stations. Typical of the stations of that era is Marchwood oil-fired station comprising eight 60MW units, see Fig 1.1.

The Divisions were responsible for the generation, supply and economic and efficient operation of the stations to the grid requirements, as determined by the 'merit' order, as well as being required to maintain a new construction programme within their areas.

The next major development was the design and construction of the first 100MW non-reheat, single shaft turbine generator fed by a single directly connected 100MW boiler. Six of these units were ordered for the power station at Castle Donnington in the early 1950s.

These units were followed by similar 120MW units but including reheat at several stations, leading to 200MW units at High Marnham in the mid 1950s.

Along with the development of the coal-fired, or conventional, stations, civil nuclear power became a reality with eight Magnox stations being built following the successful prototype development at Calder Hall.

In 1958, a reorganisation of the electricity supply industry included changes within the generation organisation and the formation of the Central Electricity Generating Board (CEGB). The 12 Divisions were regrouped as five operating regions which were made responsible for the generation and supply from the stations within their respective areas.

Design and construction of new stations was, at this time, vested in three Project Groups which, between them, covered the area of England and Wales on a geographically related basis.

In 1960, following technical appraisals with the principal main plant manufacturers, it was decided to standardise on single shaft machines at 3000rev/min 500MW output with single reheat. The steam conditions at the turbine stop valves were 160 bar, with superheat and reheat temperatures of 565°C for coal-fired units and 538°C for oil-fired units. Contracts were placed in 1961 for four units at West Burton and also at Ferrybridge C. These were followed throughout that decade with further orders which in all totalled 49 units.

In 1964 the first advanced gas-cooled reactor (AGR) was ordered. This is a uniquely designed British reactor based on developments of the gas cooling system used in the earlier Magnox stations. The steam conditions are the same as for conventional stations. Since this first order, a further six stations are in operation or under construction.

The operational development of the 500MW unit led to the first 660MW units being ordered in 1966, using the same steam conditions as the 500MW units.

In 1971, the CEGB revised its organisation to give a greater degree of delegated authority to the five operating regions. At the same time, the design, development and construction of new plant was consolidated by the formation of the Generation Development and Construction Division (GDCD) for new power stations.

MARCHWOOD OIL-FIRED POWER STATION

MARCHWOOD TURBINE HALL SHOWING EIGHT 60 MW TURBINE GENERATORS

FIG. 1.1   A TYPICAL 60MW UNIT POWER STATION

Also in 1971, the production of nuclear fuels was vested in a single company, British Nuclear Fuels Ltd (BNFL), under the auspices of the government's Department of Energy and the United Kingdom Atomic Energy Authority (UKAEA).

In the early 1970s, the industry was set to exploit the cheap residual fuel oil then available, with the construction of oil-fired units at Grain, Littlebrook D and Ince B. Due to the 1974 and 1979 international crises of fuel oil supplies and inflation of fuel costs, all the expected benefits of oil-fired units have not materialised. These stations consequently became operationally low in the merit order and would only be capable of realising their expected economic returns if costs and availability of coal were adversely altered. However, they do provide a diversity of capability which can take advantage of major fluctuations in the economic generation and availability of other types of plant.

The formation of the GDCD at its Barnwood, Gloucester location occurred over the period 1971-74 in the bringing together of the three existing project groups and the London Headquarters specialist departments.

The fundamental approach to the design, construction and commissioning of new power plant remained unchanged over the following decade. Basic design and main plant parameters were established and developed with plant manufacturers. Front line control of contract management and site construction was vested in the station project team which consisted of multidiscipline engineering groups, contracts, finance, programming, commissioning and quality assurance. Specialist plant departments gave design support to the teams and, in the case of nuclear projects, the additional nuclear safety support.

During this period, GDCD was concerned with the design and construction of all types of power stations, coal, oil, nuclear, pumped storage and hydro.

Finally, in this brief history, the CEGB finds itself in the mid 1980s having a planning margin of 28% generating capacity, which is considered to be the optimum. The overall thermal efficiency is reaching the practical maximum for a system which is mostly fossil-fired thermal plant with a low load factor. Nuclear power is likely to provide a maximum of 20% of the total demand in the next decade.

Fig 1.2 shows the generating capacity commissioned for each year 1946-1984 in terms of the generating plant size and numbers of units.

The following sections and chapters of this book are devoted to the development of power station technology, broadly during the period 1973 to 1983 which was the first decade of the Generation Development and Construction Division.

From the stations that feature during this period, the most recent of each type has been selected for description. Littlebrook D represents oil-fired stations, Drax represents coal-fired stations, Dinorwig represents pumped storage stations, Heysham 2 represents AGR nuclear stations and Sizewell B represents PWR nuclear stations. In the case of Drax, its two stages of construction (First Half and Completion) are discussed as a continuous development.

Within the allocation of space devoted to each chapter, it has not been possible to discuss all aspects of the plant at the selected stations. The choice and extent of information included was determined by the individual authors without the constraint of a rigid synopsis common to all chapters.

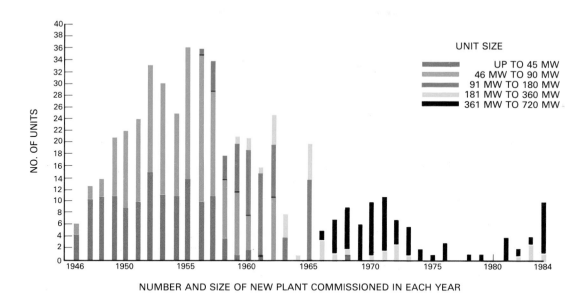

NUMBER AND SIZE OF NEW PLANT COMMISSIONED IN EACH YEAR

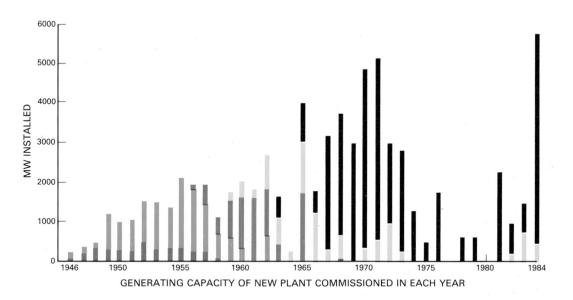

GENERATING CAPACITY OF NEW PLANT COMMISSIONED IN EACH YEAR

FIG. 1.2   GENERATING PLANT SIZE AND CAPACITY COMMISSIONED 1946-1984

The major developments and innovations mentioned are, of necessity, selective and are intended to give the reader as broad a coverage as possible of the principal types of station, systems and plant, as well as the organisational processes which affect the options and decisions within a changing consumer environment.

2. SYSTEMS ENGINEERING

2.1 Selection of Alternative Generating Plants

The power stations described in Chapters 2 to 6 were constructed primarily because at the time of their conception it was demonstrated that when added to the existing electricity supply system to meet increasing capacity needs, they would show a benefit by reducing overall system cost. The method of analysis adopted is to establish, from previous experience or by preliminary design studies, the capital cost, fuel costs, manpower and other operating costs, for each alternative type of plant. The annual costs of owning and operating the plant are then:

(1) Annual charges on capital, including repayment of the capital and interest. This charge may include the costs of raising finance during the construction phase, any government taxes related to capital, and allowances for the cost of decommissioning and dismantling the station.

(2) Annual costs of fuel for the expected hours of use (load factor), including any costs associated with disposal of waste (ash, nuclear waste).

(3) Annual costs of maintenance, manpower, etc.

When the requirement is for a single generating unit to supply a defined load, all alternatives will, of course, be required to operate at the same load factor and the selection procedure will readily show which alternative has the minimum total of the above costs.

However, it is far more usual for the plant to be required as an addition to an existing integrated system to meet load growth or to replace obsolescent plant. In this situation the selection procedure begins by deciding how the new plant will be absorbed into the existing load pattern and at this stage it must be recognised that, irrespective of the initial capital cost, any plant on a large system will be allocated a loading pattern strictly related to its operating cost, which is related primarily to its fuel costs. Thus, apart from considerations of system security, the plants having lowest generating costs are loaded first in meeting an increasing demand, and will have the highest annual load factor. This means that in comparing alternative plants to meet a given power demand, the economic calculations described here may justifiably result in the use of different load factors for the various types of plant.

Moreover, it is necessary, in a large system, to take account of the fact that during the lifetime of a generating unit its annual load factor may change from year to year due to growth of the demand or due to the addition, in later years, of plants having lower operating costs. The economic selection thus begins with an estimate, for each alternative, of its probable lifetime loading pattern and the savings in system cost which would result if the particular plant reduces the duration of operation of existing plants having higher running costs. The variable costs and savings throughout the lifetime are averaged using well-established accountancy methods of discounted cash flow.

This type of procedure formed the basis on which the plants here described were selected. The nuclear plant, having low fuel and operating costs was predicted to have a high load factor throughout its lifetime, and these low operating costs offset the high capital charges. The implications on plant design are that the nuclear units will run, as far as they are able, at virtually constant full load output.

The coal-fired unit having a higher fuel and operating cost, was predicted to have a high load factor in the early years of its life, but to operate intermittently for increasing periods in its later years. The uncertainty in predicting load patterns many years in advance results in the need to specify for these plants a combination of operating regimes, which is onerous in terms of load-cycling capability, including the ability to respond reliably to demands for rapid loading and deloading.

At the time when the oil-fired units were selected, the lifetime cost of oil was expected to be lower than that of coal, and the plants were specified to have high initial load factors, but with the provision to operate intermittently later in the lifetime. However, the abrupt rise in oil cost has meant that oil-fired units have been allocated a low lifetime load factor since they were commissioned, and they have been generally called upon to operate in a flexible peaking role. Steps are being taken to monitor the effect of this type of operation on the expected life of the major plant components.

The pumped storage plant can be regarded in system operational terms as complementing both the nuclear and the coal-fired plant. When, in the future, surplus nuclear capacity becomes available at night-time, the storage of water for regeneration of electricity during the daytime is itself economic. In addition, the very rapid loading which has been possible is of considerable value in relieving some of the coal-fired plant of the need to respond to rapid changes in system demand, with a resulting saving in system operating cost. In this context, the pumped storage plant is also overcoming inherent limitations in the capability of the peaking gas turbine plant in the CEGB system. The response from shutdown to full load for the gas turbines is of the order of minutes, whereas the Dinorwig pumped storage plant responds in seconds, which is of considerable value in meeting system needs.

2.2 Application of Systems Engineering in Plant Design

The selection process described in Section 2.1 is based on the known capital and operating costs of the generic types of plant. Having chosen a particular plant for a nominated site, further systems engineering studies are carried out.

2.2.1 *Economic Optimisations*

The starting point for these is the preparation of economic data based on the early predictions of the economic plant load factor. These data give an indication of the extent to which it is worthwhile spending additional money at the construction stage to increase station efficiency, reduce power consumption of auxiliary plant, and improve reliability.

In the case of advanced gas-cooled reactors, it is the practice to order an essentially standard nuclear steam supply system. Consequently any improvement in the efficiency of the turbine generator is seen as an increase in station output. This increase is valued at a high level because of the effective reduction in specific capital cost and because the additional output displaces less-efficient generating plant on the system. The final feed temperature is lower than in a conventional fossil-fired plant, because this parameter has an influence on the design of the boilers within the reactor, and on the associated power consumed by the gas circulators. The system benefits resulting from the low operating cost of nuclear plant mean that great emphasis is placed on plant reliability to achieve a high load factor. This has a considerable influence when considering alternative refuelling proposals (on-load, off-load). An emphasis on high efficiency for fossil-fired boilers and turbine generators arises because of the relatively high cost of fuel. It has been CEGB practice predominantly to standardise the operating steam pressure (166.5 bar at the boiler stop valve) and to use the highest feasible superheat and reheat temperature. In the case of coal-fired plant this is 568°C (at the boiler stop valve) but materials problems in oil-fired plant resulted in the need to reduce this to 541°C. Bled steam feed heating is employed to the maximum extent possible within the constraint that the highest temperature corresponds to the saturation temperature of the exhaust from the high pressure steam turbine. In addition to the cooling water system optimisations mentioned in Section 13, optimisations are carried out on condenser vacuum, pressure drops and similar parameters. Additionally, the manufacturer is encouraged to carry out further detailed optimisations using the economic data derived from system studies. It is worth noting that the choice of four low pressure exhausts for the steam turbines at Littlebrook D oil-fired power station was based on economic optimisation which, at that time, predicted the cost of additional oil fuel to be lower than the capital savings achieved by adopting four instead of six low pressure exhausts, which is the normal CEGB practice. The value of achieving good plant reliability and annual availability remains high for fossil-fired plant. Extensive monitoring of plant on the CEGB system has resulted in the establishment of a data bank of causes of plant failure and the associated outage times for repair. At the time of preparation of the plant specification, fault analyses are carried out to determine policy on auxiliary plant provision, such as the need for standby boiler feed pumps.

Similar economic criteria relating to the efficiency of plant components apply to the pumped storage system, with additional emphasis on the penalty of hydraulic losses in relation to the civil engineering costs. Numerical values have been derived for the benefit in system operating cost of being able to respond to load changes in a few seconds. These values can be directly compared with costs incurred in providing the robust hydraulic components and control gear which are necessary to achieve the high rates of load change.

2.2.2 *Plant System Studies*

It has been CEGB practice, especially for fossil-fired plant, to develop the overall station concept and then to place contracts individually with the manufacturers of the main plant items and certain of the auxiliary plant components. In order that each supplier has freedom to optimise his design, continual monitoring during the design stage is necessary to ensure compatibility at the interfaces between contractors and to reach decisions where conflicts may arise.

Where possible and economically justified, it has been the practice to specify the boiler steaming capacity to be some 5% greater than that required by the turbine generator to give the specified output. This margin has been assessed as adequate to cover changes which the turbine generator manufacturer may make as a result of his own optimisations and known changes in the turbine generator/condenser performance during its lifetime.

Technical audits carried out on operating plant have indicated areas where deficiencies in performance of parts of the main or auxiliary plants have together resulted in a station performance lower than the design value. This feedback of experience reinforces future systems engineering work in preparing specifications.

2.2.3 *Investigations of Operational Capability*

As a direct result of its low fuel and operating costs, nuclear plant is expected to operate on the CEGB system at base load, ie constant full load power output, until early into the next century. However, as the proportion of nuclear plant in the generation plant mix increases it will inevitably be required to part-load in order to match the generation to the demand. No nuclear plant currently being designed is expected to have to reduce load on a regular basis below 50% power. When load following, it will be expected to carry its share of grid frequency control. Its design requirements in respect of grid disconnection and other disturbances is similar to that of conventional plant. For the purposes of cumulative damage to plant components eg fatigue, creep and corrosion, plant operational fault and test histories over design life are predicted and used for the purposes of structural analysis. To this end, and in order to design operational procedures and control systems and for the purpose of fault analysis, it has been necessary to develop large and detailed mathematical models of individual plants. These models are used to resolve any commissioning difficulties, are validated against actual plant transient data and then used in the longer term for ongoing plant support.

Coal-fired plant is subjected to load excursions throughout its life, and it has been necessary to develop similar mathematical modelling techniques both to assess the safe life of components at high pressure and temperature, and to develop the most suitable overall control philosophy. The same mathematical techniques are applied to the development of plant simulators for operator training.

During the design phase of the Dinorwig pumped storage project, considerable design effort was devoted to the investigation of the dynamic forces on plant and control gear and their ability to respond to short-term load changes. This work was incorporated in a mathematical model of the whole system to investigate the hydraulic and electrical transients.

2.3 Studies on Future Plant Alternatives

The foregoing review of the plant selection and the systems engineering functions starts with the premise that the choice is being made from a number of alternatives of proven cost and performance. To a large extent this is the case, because CEGB recognises the benefits of standardisation which allows interchangeability of spares and replication of proven plant, so that experience from operation of earlier plant can be incorporated into later designs. The electricity supply industry must also keep in mind the need to develop future alternatives which may better suit the needs in the longer term. For stations based on conventional coal-fired boilers, the options under continuing study are the increase of steam temperatures and pressures to improve efficiency, and increase in unit size to gain additional benefits of reduced generating costs.

An increase in the proportion of nuclear plant on a system improves the economic case for energy storage but the number of sites available for pumped storage may be limited. The alternative storage of compressed air in large caverns for later discharge through a fired gas turbine may be less onerous in terms of site selection. The duty would be for a regular generating pattern, with response times similar to those of conventional gas turbines.

The renewable energy sources such as wave power and wind turbine generators, although having zero fuel cost, cannot be credited with ability to provide full power on demand, and may have a relatively low annual load factor. In estimating their value in an existing system, account must be taken of the need to provide other plant to ensure firm supplies, and of the operational demands made on other plant to offset the variability of the wind turbine generator output. Fig 1.3 shows the 200kW wind turbine which was installed at Carmarthen Bay power station in 1982. It has a three-bladed, 25m diameter rotor.

A third category of new plant development arises from the possible need for future controls on emissions of oxides of sulphur and nitrogen from coal-fired plant. The feasibility studies and research on this topic have the object of reaching conclusions on the most cost-effective solution to the problem. The alternatives broadly are flue gas treatment, production of coal gas for use in a gas turbine combined cycle, and combustion in a fluidised bed combustor with a gas turbine combined cycle.

In addition, there has been an incentive to study the methods and comparative costs for conversion of (now uneconomic) oil-fired plant to coal firing. It is generally found that to burn coal in a boiler designed for oil firing results in an appreciable fall in output. The possibilities of substituting oil/coal or water/coal mixtures are under continuous review.

3. SYSTEM PLANNING

3.1 Introduction

The construction of a new power station takes typically about seven to eight years from the decision to initiate the process of procuring the station to the commissioning of the first unit. The CEGB's annual plans therefore include the provision for specific new generating stations that are planned for commissioning in the period seven to nine years ahead (referred to as the 'planning years'). Before it can embark on the ordering of a new power station the CEGB must have received the Secretary of State's consent

FIG. 1.3   200kW WIND TURBINE INSTALLED AT CARMARTHEN BAY POWER STATION

required under Section 2 of the Electric Lighting Act 1909, together with any related consent and licences, and must separately have received financial sanction from the Government.

The CEGB has to evaluate the need for power stations in the light of its statutory duties. It considers whether there is a need for new capacity in order to maintain an adequate security of supply, or to give greater economy or to improve the security of fuel supply by allowing the types and sources of fuel or primary energy to be diversified. In addition it may be justifiable to build a new form of generating capacity in order to prepare the ground for a possible future benefit.

3.2 Capacity Considerations

Capacity requirement is determined by the need to meet the peak demand of the year. The first step in estimating generating capacity requirement is therefore to forecast the peak demand for each future winter up to the planning years. The forecast presumes that the peak is most likely to occur on working weekdays in December to February during a spell of cold weather of average severity and is thus described as the 'average cold spell' (ACS) winter peak demand. ACS conditions are determined by a statistical analysis of past weather data and the variation in demand caused by weather variations.

Each year the Electricity Supply Industry prepares new estimates of the unrestricted ACS winter peak demand, the corresponding values of restricted peak demand after allowing for the expected reduction in peak demand by load management, and the total number of units of electricity (unit requirements) to be produced by the CEGB or purchased from external supplies. The unit requirement therefore equals the sum of the CEGB's sales of electricity to Area Boards and to its direct consumers and the transmission losses on the CEGB system.

After consideration of the various forecasts, recommendations are made to the Electricity Council as to estimates of demand and unit requirement up to the planning years. The Electricity Council then formally adopts these forecasts, together with provisional estimates for the subsequent two years, on behalf of the Electricity Supply Industry in England and Wales.

In order to meet the statutory requirement to provide a continuous supply of electricity except in cases of emergency, the industry has over many years aimed to provide sufficient generating capacity to meet the future demand with a high degree of security. Since it is impracticable to ensure 100% security of supply there will, on occasions, be insufficient generating capacity to meet demand even after the application of load management. In such circumstances, the first action would be to reduce the voltage and/or frequency within permissible statutory limits. This has the effect of reducing the magnitude of demands which are sensitive to voltage or frequency while maintaining continuity of supply to all consumers. In this way the overall demand can be reduced by up to 7.5%; but if the remaining demand still exceeds the generation available then some consumers must be disconnected.

It is the CEGB's function to ensure that sufficient generating capacity is provided to meet the generation standard and it achieves this by planning a reserve margin of generating capacity called the 'planning margin'. This is defined as the percentage margin of additional generating plant planned to be in service in the planning years over and above that needed to meet the peak demand to be met by the CEGB.

The CEGB and Electricity Council makes estimates of the expected average availability and of the expected magnitude of variabilities of availability and forecast demand. A simple statistical calculation then gives the size of planning margin that meets, or approximately meets, the security standard. With the present estimates of parameter values, the planning margin is set at 28%. The greater part of the 28% margin is required to cover the forecast loss of availability of generating plant at the time of peak demand.

3.3 Economic Considerations

The provision of new capacity to meet the forecast demand is not the only reason which might justify the construction of new generating plant. New construction might also be justified on economic grounds and might allow the retirement of some existing capacity. In principle, a plant is retained in service until it becomes more economic to replace it with new capacity. Evaluations are made for certain economic indicators for existing stations and for the potential new stations that might be built:

(1) For existing stations, the annual avoidable cost is evaluated, on a year-by-year basis, of retaining certain stations or parts of stations in an operable condition. This cost is called the 'net avoidable cost' (NAC) expressed in units of £/kWpa.

(2) For new generating station options for commissioning by the planning years, the CEGB calculates the net effect on total system costs of building and operating the station over its lifetime and converts this into an average annual cost, in units of £/kWpa, called the 'net effective cost' (NEC).

These indicators allow two economic comparisons to be made. Firstly, the comparison of NEC for alternative new generating plant options allows, for given assumptions of input parameter values, the identification of the most economic option, namely the one with the lowest NEC. Secondly, for that option, it is possible to test whether it is economic to install the new plant and decommission existing capacity.

When making an economic appraisal of alternative new generating station options, it is necessary to assess the probable cost of installing and running each station and its impact on other system operating costs, and to ensure that there is likely to be sufficient fuel available at an acceptable price throughout its expected operating life. Some generating plant options may have a relatively short construction time and have the potential of being economic after a short period of generation. However, the planner must consider all options, including ones with a construction lead-time of seven to eight years and operating life of up to 40 years. Hence the planner must take a view of electricity demand, fuel availability and fuel price up to about 50 years ahead. Indeed for some projects it may be necessary to look further.

15

3.4 Future Requirement Predictions

The inter-relationship between estimates of economic activity, fuel prices, energy supply and demand, electricity demand and the implications for electricity supply have been more fully examined through the development of economic scenarios (ie imagined sequence of future events).

The scenarios set out a spectrum of possible future developments which can be used in a variety of planning studies. More specifically:

(1) They form a valuable aid to the judgement of the range of plausible outcomes that should be allowed for in planning, especially as regards the future extent and composition of economic activity, energy supply and demand, energy conservation, fuel prices and availabilities, and electricity demand.

(2) The relative economic merit of alternative generating plant types for each scenario are evaluated.

(3) The implications for economic operation and security of fuel supplies of alternative generating plant development options within each scenario are considered.

The scenario approach does not require the CEGB to estimate specific probabilities of occurrence of the alternative scenarios but provides a background against which planning judgements can be made for a highly uncertain future. However, each scenario is considered with care when it has been fully developed and is judged whether or not it still appears to be plausible and with a significant prospect of occurrence in real life. Provided the scenarios individually pass this test, the CEGB aims ideally to plan so as to be able to respond to any one of these plausible outcomes. In practice, some reasonable latitude would be acceptable; for example, in the case of a scenario entailing rapid growth of electricity demand, it might be practically necessary to accept a moderately lower standard of security of supply for some interim period before generating capacity could be fully adjusted to the requirement. However, it is planned to avoid a really serious failure to achieve a secure and economic supply for any plausible scenario.

The scenarios allow the CEGB to examine the risks attached to alternative generating plant options which arise in particular from variations in future electricity demand or fuel prices. In addition, the analysis of risk covers uncertainties which attach to the alternative options, in particular as regards capital cost, performance factors, lifetime and construction time. The wider strategic aspects are also considered, of which two are worth particular mention, namely the security of fuel supplies and making provision for future investment options.

Fig 1.4 shows the effect of particular scenarios on the estimation of system demand up to the end of the century, using 1979/80 as the base year.

In order to assess the economic merits of different types of generating plant it is necessary to make estimates of capital cost, construction period, lifetime and availability in service, all of which are relevant to the overall value of the plant. The construction period and the incidence of expenditure over that period are important in relation to the total capital investment and the time when a return may be expected on that investment; the lifetime and availability (together with the estimated fuel and running costs) determine what that return will be. Endeavours are made to ensure that,

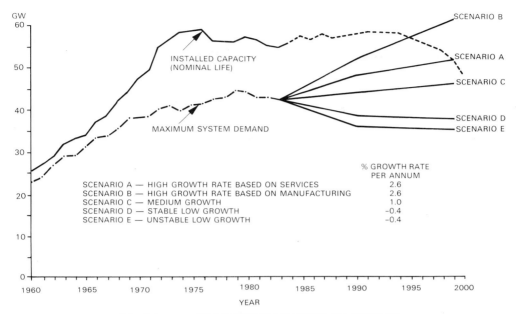

FIG. 1.4   EFFECT OF SCENARIOS ON DEMAND

as far as possible, these estimates are central ones (ie ones which are as likely to be high as they are to be low) and the sensitivity of the results of the economic appraisals to changes either way in the estimated values is examined.

As the economic appraisal must represent the performance of a new station over its lifetime, it must take account of the other generating plant which may be on the system over that period. It is therefore necessary to make assumptions about the types of generating plant which might be installed in the future and their cost and availability. For this purpose it has been assumed that the values of capital cost, construction period, lifetime and availability for later stations would be the same as for the stations being appraised unless there is justification for doing otherwise.

4.   SITE SELECTION

4.1 System Planning Studies

Early planning work begins with the examination of system load flows and the identification of future generation and transmission needs. This process shows regional requirements and notional locations of generation sites. One important factor which is taken into account is a CEGB policy to develop existing sites wherever possible, if this satisfies the system requirements. By developing such sites to their full capacity, as determined by any technical and environmental limits, advantage can be taken of existing facilities, such as transmission outlet, improved local roads and minimising the amount of new works. At an early stage, the main plant and overall

17

station design and construction features are selected. This applies particularly to matters such as station efficiency, availability, capital cost, siting areas and construction programme.

At the completion of this review, a list of alternative generation sites will have been compiled. A ranking of the sites is then carried out based on an initial technical appraisal, capital cost and construction programme for each site.

4.2 Authority to Build a New Power Station

Station design and siting studies are carried out, to the point where an application is made for government consent to develop a site. This procedure, a statutory obligation, is a request to the Secretary of State, under the provision of Section 2 of the 1909 Electric Lighting Act, to build a power station.

In addition to Section 2 Consent, the CEGB requires planning permission under the Town and Country Planning Act of 1971. Part of this Act empowers the Secretary of State to direct that planning permission is granted at the same time as Section 2 Consent. However, the Secretary of State may attach conditions, as he thinks appropriate, in regard to the planning approval.

Following receipt of Section 2 Consent, the CEGB proceeds with the design and construction of the project.

An important part of the investigation programme is consultation with Ministerial and Local Authorities. As part of the procedure for ensuring that all parties are fully aware of agreements which have been negotiated and which must be observed during the station design and construction period, a document called Station Development Particulars is issued, which records all discussions and agreements with parties and also contains a schedule of statutory consents which must be obtained.

The Station Development Particulars also contain a technical section dealing with the transmission connections and parameters of the main plant, particularly the generator transformer, so that they are properly matched to the transmission system. The details cover matters such as power factor, synchronous impedance, frequency regulation, the dynamic response of the unit to changes in load demand and guidelines in the electrical auxiliary system to ensure that this is a reliable network.

4.3 Site Investigation Programme

Fig 1.5 shows a programme of preliminary site investigations carried out to assist in the evaluation of site feasibility or the ranking of multiple sites. The figure shows details of investigations for a nuclear project.

4.3.1 *Investigations Common to Fossil-Fuelled and Nuclear Projects*

Following the preparation of notional site layouts, initial site work begins with boreholes to confirm that the location is suitable for heavy plant loads, foundation design and the site hydrology. At coastal and estuarine sites, tests are done off-shore to look at the movement of cooling water and settle the location of intake and outfall structures. Fig 1.6 shows examples of the use of infra red imagery taken by satellite which provides data of thermal discharges from existing stations.

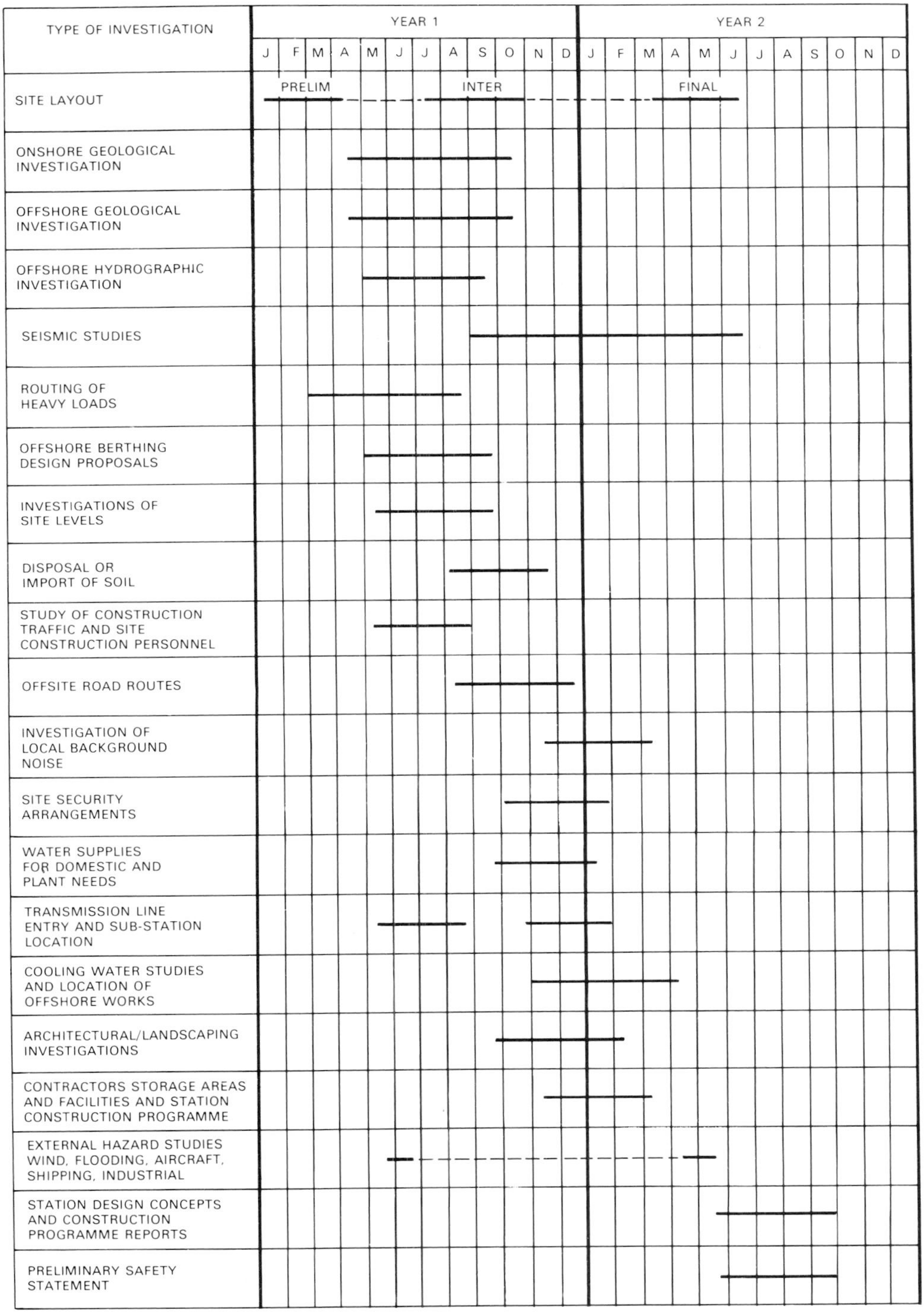

TYPE OF INVESTIGATION	YEAR 1												YEAR 2											
	J	F	M	A	M	J	J	A	S	O	N	D	J	F	M	A	M	J	J	A	S	O	N	D
SITE LAYOUT		PRELIM						INTER								FINAL								
ONSHORE GEOLOGICAL INVESTIGATION																								
OFFSHORE GEOLOGICAL INVESTIGATION																								
OFFSHORE HYDROGRAPHIC INVESTIGATION																								
SEISMIC STUDIES																								
ROUTING OF HEAVY LOADS																								
OFFSHORE BERTHING DESIGN PROPOSALS																								
INVESTIGATIONS OF SITE LEVELS																								
DISPOSAL OR IMPORT OF SOIL																								
STUDY OF CONSTRUCTION TRAFFIC AND SITE CONSTRUCTION PERSONNEL																								
OFFSITE ROAD ROUTES																								
INVESTIGATION OF LOCAL BACKGROUND NOISE																								
SITE SECURITY ARRANGEMENTS																								
WATER SUPPLIES FOR DOMESTIC AND PLANT NEEDS																								
TRANSMISSION LINE ENTRY AND SUB-STATION LOCATION																								
COOLING WATER STUDIES AND LOCATION OF OFFSHORE WORKS																								
ARCHITECTURAL/LANDSCAPING INVESTIGATIONS																								
CONTRACTORS STORAGE AREAS AND FACILITIES AND STATION CONSTRUCTION PROGRAMME																								
EXTERNAL HAZARD STUDIES WIND, FLOODING, AIRCRAFT, SHIPPING, INDUSTRIAL																								
STATION DESIGN CONCEPTS AND CONSTRUCTION PROGRAMME REPORTS																								
PRELIMINARY SAFETY STATEMENT																								

FIG. 1.5   SITE INVESTIGATION PROGRAMME FOR A NUCLEAR POWER STATION

Off-site investigations concern any road improvements needed for the delivery of heavy loads and the ability of local roads to handle the construction traffic. Another important matter is the whereabouts of civil engineering aggregate. Estimates are made of the aggregate needed and the volume of construction and operational traffic up to the stage where the station is fully operational.

A parallel exercise looks at the station construction. This work includes the determination of the finished site level and the site levelling methods, the disposition of construction contractors' plant and storage areas, early on-site road construction, the timing and phasing of deep site excavations and the preparation of a preliminary construction schedule.

An on-site location of the transmission high voltage substation has important consequences on the general site layout. Water supplies for domestic and plant purposes must be investigated. A suitable water source is selected and then an appropriate treatment process is determined.

Visual environmental issues are an important part of this initial work. Independent architects and landscape consultants are engaged to give professional advice on the site layouts and station building designs. Plant environmental influences are studied, such as the dispersion of chimney gas and particulate matter, noise and the discharge of liquid waste.

4.3.2 Nuclear Stations

The special investigations for nuclear projects are mostly concerned with safety matters and early work concentrates on population density and any hazards which affect the selection of the site. This category of hazards comprises seismicity of the site, the risk from aircraft activity, toxic or explosive gases from industrial plant or nearby shipping at coastal sites. Coastal features are studied to ensure that they will not result in a hazard to the site during its operational life up to the decommissioning process.

The reliability of the grid system connections to the site influences the design of the station essential electrical supplies system. This needs an investigation of the grid system up to the high voltage connections to the station auxiliary electrical system.

Station Design and Preliminary Safety Statements are prepared to support the selection of the site. If a public inquiry is necessary, specialist resources are made available to prepare evidence and witnesses at the inquiry.

4.3.3 Coal-Fired Stations

Preliminary work specific to the fuel consists of the preparation of layouts, with the coal handling system having a major influence on the disposition of buildings. Discussions take place, where the coal is rail-borne, with the National Coal Board and British Rail to establish a daily delivery schedule.

Ash disposal needs long term planning to select and purchase disposal areas. Various types of off-site disposal processes are studied consisting of ash transport in a dry, semi dry or wet condition and provision is made on the site for ash sales.

4.3.4 Pumped Storage Stations

Initial surveys for prospective sites are carried out to establish an order of preference based on technical and economic considerations.

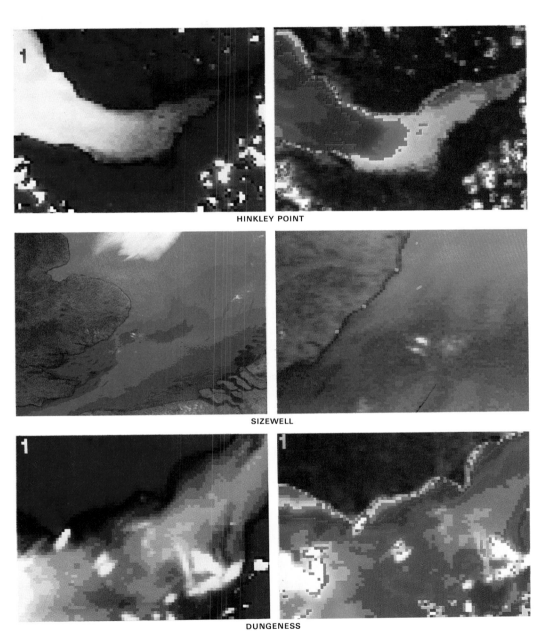

HINKLEY POINT

SIZEWELL

DUNGENESS

FIG. 1.6   SATELLITE IMAGERY OF THERMAL DISCHARGES FROM POWER STATIONS

The process of determining the order of preference consists of preparing notional station hydraulic schemes between high and low level reservoirs, the consideration of alternative pumping turbine capacities and the construction process together with the effect on the environment.

4.4 Capital Costs

The estimation of capital costs is a major factor in the site selection process. At the stage of investigating and comparing the merits of various sites, the major variation is normally the civil engineering content. This reflects the site specific ground conditions and consequences for the main building foundations, the extent of cooling water works, site levelling, roads, coastal protection, architectural and landscaping requirements.

Two other cost elements are important, one is a start-to-finish allowance which is a sum added to an estimate before a project is sanctioned, reflecting a measure of general uncertainty in the station design and because no tenders have been sought at that stage. The second element concerns the financial consequences of the length of the construction period. When a proposal for constructing a new station is presented for financial sanction, a formal target programme is used. The target programme is as short as is economically attainable, since it is in the consumer's interest that it should be so. However, for planning and investment appraisal, longer periods are used which contain a long range judgement of alternative construction strategies.

4.5 Site Selection

In parallel with the investigation processes described the transmission technical needs, capital costs and any wayleaves problems for routes to the various sites are evaluated.

A further step in the economic study is the estimate of total system capital and revenue costs on a year-by-year basis. This procedure takes account of the type of station, the mix of existing and new plant, system demand and plant availability.

The completion of the investigation and reporting processes described, together with these economic indices, are the final steps in the site selection process. They enable recommendations to be made on the selection of a specific site based on the widest possible technical and economic assessment.

5. ARCHITECTURE AND CIVIL ENGINEERING

5.1 Introduction

In financial terms civil engineering represents about 25%-30% of the capital cost of a modern power station, whether nuclear or fossil-fired. For hydro stations and pumped storage this proportion may well be 70%.

Essentially the civil engineering function provides buildings, structures and foundations to support and protect mechanical and electrical plant. Architects use aesthetic and logistic considerations to strike a balance between the bare structural requirements of power station buildings and over-elaboration, whilst the quantity surveyors strive to achieve 'value for money'.

In most respects power station buildings and structures are specified, planned, designed and constructed in a very similar manner to those throughout industry generally. Perhaps the main distinguishing feature lies in the combination of many such large structures being built in relatively close physical proximity, to a complex programme that is dictated largely by plant considerations.

To avoid unnecessary obsolescence in the selection of mechanical and electrical plant there is advantage in deferring final choices to await new developments or to prove modifications and manufacturers' improvements. Such delay would automatically cause commensurate delays in the supply of civil design layout and loading data and intensify the design effort needed to permit the civil construction to start the entire power station building process. To mitigate this unavoidable programme pressure, design contracts are placed with the suppliers of the principal mechanical components ahead of the hardware contracts, thereby obtaining advance civil loadings and layout information.

For recent stations the contract strategy has been to break down the total civil works into its natural sub-divisions, ie steelwork, foundations, piling, structures. Such breakdown not only eases the designer's task, but also allows fullest advantage to be taken of the healthy competition throughout the construction industries. Both nuclear and fossil-fired station civil works, when so sub-divided, offer opportunities not only for the major national contractors, but for the numerous medium-to-large companies and for the specialists.

5.2 Site Selection

5.2.1 *Architectural Considerations*

The 1947 Electricity Act imposes two stringent, but somewhat conflicting statutory duties upon the CEGB:
(1) Section 2 - "to develop and maintain an efficient co-ordinated and economical system of electricity supply".
(2) Section 37 - "to take into account any effect which its proposals would have on the natural beauty of the countryside and on flora, fauna, features, buildings and objects of special interest".

The second of these duties, which may be summed up as 'showing concern for the environment', was unique in its time as a statutory goal for a major industry. Both duties must be regarded as of equal importance; the CEGB is neither required to develop the cheapest possible electricity system nor to protect the environment at all costs; a balanced judgement for each development must be reached.

England and Wales is one of the most heavily populated areas in the world, wherein some 40% of the land is protected to form national parks, green belts, nature reserves, etc; 60% of the coastline is protected and 10% of the total land area is already built upon. The remaining areas are predominantly upland regions and agricultural land. Opposing claims on land-use rightly impose considerable constraints on industrial developments such as a power station.

Assessing the merits and disadvantages of alternative power station sites is a lengthy, complex and contentious process involving several engineering disciplines together with a wide range of socio-environmental expertise.

From the initial site selection stage, environmental design, architecture and landscape architecture contribute to the layout and siting, helping to establish a form which will reconcile the station development within its setting. Extensive visual analysis is undertaken, including photo montage studies using computer draughting techniques and zone of visual influence (ZVI) studies, to confirm the local inter-visibilities objectively. Such ZVI studies are normally limited to an 8km radius from the site, experience having established that the impact of any station beyond that distance is minimal. These studies take into account not only the topography of the existing land form, but also the 'mantle' of trees, shrubs, hedgerows and buildings. Thus, it is possible to establish exactly the importance of each element of the mantle and the extent to which it should be preserved or protected.

Whilst it is impossible to hide such large engineering structures as power stations within the small scale United Kingdom landscape, the longer range views are normally tolerable. Particular care is taken to simplify the form of the upper part of the station as this is always dominant from long distances. Under most conditions cooling tower and chimney plumes regrettably act as long range pointers to the location of any station sited in flat country. Fig 1.7 shows Littlebrook D power station sited on the flat terrain of the Thames estuary.

Paradoxically such enormous structures can contribute to the visual scene by establishing a kind of order, but this can be readily spoilt by any clutter of improvised buildings and stuctures such as switchgear, transformers, coal plant buildings or even lighting columns, which provide a reference against which the scale of the main structures can be assessed. Generally power stations become simpler and cleaner in outline at higher levels, emphasising the importance of carefully considered modelling of the ground to help conceal or soften the jarring effects of the lower portions. Shaping such elements is not only important visually, but may facilitate long term maintenance. Fig 1.8 shows some of the external cladding features of the Littlebrook D buildings. Here and elsewhere, landscape design must aim towards the greatest possible simplification — to make the result appear not contrived, but inevitable.

Good landscape design endeavours to retain and use the character of the original landscape. It is rarely appropriate to create an entirely new landscape on completion of the structures. The planting of trees both on and off site at visually strategic points is often highly beneficial and there is a need to co-operate with local planning authorities and adjoining land owners to achieve acceptable schemes.

However, it will be realised that such massive building complexes cannot be concealed behind belts of trees. There is also an essential difference of timescale between the buildings and the landscape, whereby the buildings reach an immediate maturity in contrast to the much slower maturity dictated by nature of the landscape features. Fig 1.9 shows the effect of careful landscaping at the Didcot coal-fired station in Oxfordshire.

FIG. 1.7   LITTLEBROOK D POWER STATION

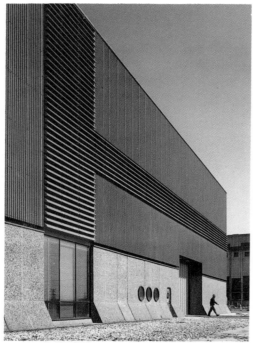

FIG. 1.8   LITTLEBROOK D POWER STATION BUILDINGS

FIG. 1.9   DIDCOT POWER STATION LANDSCAPING

5.2.2 *Geotechnical and Civil Engineering Considerations*

In parallel with the architectural and landscape architectural comparison between potential sites, it is equally important to assess many other characteristics, such as the need to prove that sites offer adequate foundation conditions, without excessive foundation costs being incurred.

Geotechnical work for each selected site ranges from conceptual studies to construction, it being not unusual to have four or five investigations on an adopted site. In recent years much of the activity has centred on the ever more complex process of finding and proving new power station sites, particularly ones for nuclear plant.

Initial desk studies locate probably suitable foundation areas. For each site initially identified as promising, it is normally also possible to undertake a desk study of the likely foundation conditions and thereby make initial site comparisons. Such comparisons provide a short list at which stage preliminary site investigation studies, each comprising a few relatively shallow bore holes with standard soil mechanics laboratory testing, would be undertaken to confirm the local geological and superficial deposit information.

At the conclusion of this second stage, a much more detailed assessment may well be needed to prove the foundations viability of the chosen site, or perhaps a few sites. Here the scope of the work will call on most of the up-to-date techniques in geotechnics.

Nuclear plant site investigations call for more rigorous examination than conventional stations, as seismic loadings are now considered to be of special significance for the key safety-related structures and systems. For both types of stations however, geotechnical investigations have been required on a much more intensive basis in recent years, perhaps because the knowledge and equipment in this field have become much more searching, complex and specialised as the science of geotechnics has rapidly advanced. To illustrate this development, it was not unusual to build stations 20 years ago on the basis of a few 25m deep bore holes, with associated tests. Nowadays a modern nuclear station could well call for 30 or more bore holes of up to 150m in depth supported by a much wider range of associated testing both on site and in the laboratory.

International and European experience has led to the need to assess many natural and man-made hazards as potential power station loadings to standards beyond that normally required in UK structural engineering. Wind, sea level, flooding and seismic risks are extensively studied for all future nuclear station sites.

Seismic hazards present exceptionally difficult assessment problems in an area of low seismicity like UK since, by definition, an extremely long time extrapolation must be based on a minimum of reliable data from earlier events. Such uncertainties can readily lead to unexpectedly high magnitude predictions.

Geological investigations need to be exceptionally rigorous to reach meaningful estimates of seismic hazard. In addition to deep boreholes, structural geological interpretation and laboratory testing, historical researches, micro-seismic arrays and extensive geophysical techniques have been introduced. Interestingly, the power and definition of recently-developed geophysical investigation techniques suggests that the traditional order of proceeding from exploratory boreholes to geophysics could profitably be reversed in future investigations.

Seismic hazards have forced a major change in the preferred type of nuclear station foundations, massive rafts and large scale contact foundations taking over from the pattern piling that has long been used extensively for fossil-fired stations.

In 1985 a site investigation covering only a single chosen site could well cost £1.5M. It is also likely that for any coastal location the hydrographic work would need to be supplemented with an offshore site investigation, to prove the adequacy of formation beneath the selected positions for the cooling water intake and outlet and the tunnelling, at a further cost of £1M.

Investigations of wind, sea level, and flood extreme loadings are normally less intensive than those for the seismic hazard, since much reference data is normally available for UK sites from the Meteorological Office, local Coast Guards and Admiralty records. Nevertheless the emphasis placed on proving claims for extreme levels leads to a great extension in the 'traditional' work content.

Beyond the determination of loadings for a new site, the investigation phase must also consider the influence of the location and the preferred layout on associated civil engineering structures. Such work typically includes feasibility assessments and cost comparisons of direct cooling and cooling towers, novel methods of constructing cooling water intake and outfall tunnels and culverts, and alternative forms for the off-take structures themselves. Much comparative preliminary design and cost is required.

Having examined the architectural, geotechnical and extreme loading characteristics of a particular site, many other considerations require to be assessed in producing the global assessment, such as, the effect of the proposal on the local road system, infra-structure, water and sewerage.

Building such major engineering structures as power stations within the small scale features of England and Wales rightly induces great public interest. The need for public inquiry is virtually obligatory for all such proposals. Added safety considerations in the case of nuclear stations intensify the duration and interest of that public examination of projects.

5.3 Structural Engineering

In considering structural forms, a further contrast becomes apparent between modern nuclear and fossil-fired units. Hanging of main boilers has traditionally dictated the use of structural steelwork on a massive scale, (30,000t to 40,000t per 2000MW) whereas the principal structures of UK nuclear stations are predominantly built in prestressed and reinforced concrete. This trend has been carried into the main pressure vessel area of gas-cooled reactors.

In the early 1960s, massive new pressure retaining structures, beyond the capability of welded steel, were needed to shield and contain the later Magnox reactors and second generation (AGR) of gas-cooled reactors. These prestressed concrete pressure vessels (PCPV) were designed and built by the nuclear consortia of the day to operate at pressures and temperatures well above anything previously experienced in conventional prestressed concrete design. Because PCPVs were novel in both concept and execution, the Nuclear Installations Inspectorate required specialist civil engineers to act independently as the 'Appointed Engineer' during the construction phase and 'Appointed Examiner' during the operational phase of the PCPVs - a very early application of modern quality assurance methodology. Their functions included ensuring that each design was adequate, that the PCPV was built precisely in accordance with that design and that the vessel remained fully capable of meeting its function throughout its operational life.

The main safety element in any PCPV is the prestressing system itself. In conventional prestressed concrete practice, these tendons are tensioned and held on their anchorages only long enough to enable the tendon duct to be fully grouted and the load transferred into the tendon/grout interface. In PCPVs however, because the vessel operates continuously at an elevated temperature, creep in the concrete and relaxation in the tendons themselves cause the overall loss of prestress to be considerably greater than in typical prestressed concrete structures. This makes grouting inadvisable in that it prevents any possible periodic restoration of the tendon tension at appropriate stages during the operating life of the structure. Ungrouted tendons are universally adopted in UK PCPVs.

Two problems result. Firstly the tensioned tendons themselves have to be held permanently by their anchorage components, making it necessary to test and prove the adequacy of the entire tendon/anchorage system to a much enhanced standard of quality control beyond the normal. This includes extensive testing of all the individual components, preliminary design testing of the entire system and a proportionate series of production control tests of components and system as manufacture proceeds. Secondly the absence of grouting exposes the steel tendons to a more corrosive environment than is normal in any fully grouted beam. Greases have been developed and used to minimise the corrosion hazard with varying degrees of success. In devising such protection it has been found necessary to distinguish between the damp cold construction conditions, when condensation onto the steel exacerbates the exposure problems, and where the possibility of stray currents from nearby welding returns may also accelerate corrosive attack. The later warm operating conditions are naturally less demanding, despite elevated temperatures theoretically increasing the rate of chemical corrosion. In general, some difficulties have been experienced with tendon corrosion during the construction stage, but very little during operation.

As concrete cannot be made impermeable to high pressure gas it has always been necessary to provide a mild steel liner for each PCPV. To maintain a reasonable temperature regime in the resulting composite structure, extensive insulation and water cooling are provided. The integrity necessary to maintain insulation to operate effectively within the reactor gas pressure and temperature regime and to maintain the water cooling immediately outside the liner, but fully embedded in concrete, again calls for exceptionally high engineering standards of design and fabrication. A total of 14 PCPVs have now been successfully built, tested and commissioned, all of which have been cleared at each critical stage to standards well above those normally adopted in structural engineering.

Fig 1.10 shows an early stage of AGR base and support walls construction.

5.3.1 *Cooling Towers*

To date, high wind loading has presented many more structural problems in fossil-fired stations than in nuclear, since the latter have avoided the use of natural draught cooling towers through being sited on the coast. Towers and chimneys are unusual structures in having little other superimposed loading except wind and are

FIG. 1.10   AGR BASE AND SUPPORT WALLS

consequently the more vulnerable to error in the estimation of extreme loadings. Hyperbolic towers compound the problem for designers through their complex geometry of shape and by the unique thinness of shell (125mm-175mm) required to resist the dominant membrane stresses.

Successful design by extrapolation between 1920 and 1960 was brought into question in 1965 when three Ferrybridge C power station towers collapsed in very high winds. Since then, considerable research and design activity has added immeasurably to the knowledge of both structural and thermal performance.

Prior to the Ferrybridge incident, wind loading had been assumed to act in quasi-static manner as a horizontal pressure and indeed this approach remains the only practical way of designing a new shell. However extensive wind tunnel testing on detailed models of some eight CEGB station multiple tower layouts and on layouts for overseas clients has demonstrated this to be a gross oversimplification. Such tests showed unequivocally that the wind loading included a substantial resonant element which, above a threshold value, tended to increase its contribution by the fourth power of the wind velocity, whereas the quasi-static element is only dependent on the square law. In magnitude this resonant element is dependent on the location of other large bluff structures up-stream, most notably other towers. Hence wind direction as well as wind velocities is now recognised as being very critical in determining the critical loading on any tower built within a group. As it is impractical to contemplate building towers far enough apart to avoid any wind induced structural effects, the spacing is a chosen minimum to avoid thermal performance interaction.

Another feature whose significance has only been appreciated in recent years is the accuracy of shell construction. Due to the collapse of a cooling tower elsewhere in the UK, it was shown that the simple uplift tension value on the windward meridian that forms the basis of tower design and assessment generally and was the initial failure element at Ferrybridge C, is not necessarily the only criterion in these membrane structures. Deviations in shape from that presumed in design can induce horizontal tensions of considerable magnitude, which, in the ultimate, can serve as the 'trigger' to cause the membrane to move into a total collapse mode. Again, recent work suggests that the overall deviations of the shell centre line from its true design position are less important in this regard than relatively sharp changes in profile. Smoothness in construction in this respect may well be more valuable than absolute accuracy. Equally it has been shown important to maintain a considerably higher proportion of horizontal reinforcement in tower shells than simple membrane analysis suggests to be necessary.

The vast majority of cooling towers in the CEGB system are natural draught towers, using timber or asbestos cement sheet packs as the heat transfer surface. They are known as 'wet' towers. From an environmental viewpoint their heat transfer is predominantly evaporative thereby causing massive plumes of water vapour under high humidity conditions. This plume is frequently more visually obstructive than the station itself, acting as a reverse arrow when seen from many miles away. To avoid the plume completely involves the use of 'dry' cooling, a technique adopted on one station, Rugeley A, both for this reason and because water supplies for cooling appeared limited. The dry system depends on radiator type cooling elements with a water-to-air heat transfer. Being much less efficient than the evaporative cooling, the specific size of a 'dry' tower will always be considerably larger than the equivalent 'wet' one. Further, because of the high costs of the heat exchanger grids and the extra size, the overall cost of 'dry' cooling has been found to be several times that of 'wet' cooling. As such it is not a preferred choice under UK conditions.

Attempts have also been made to reduce the visual impact of multiple tower stations by enhancing the cooling duties of a single natural draught tower shell by a factor of about three or four. At Ince B a standard 114.5m high shell has been surrounded by a large podium structure which houses 35 fans of 9m diameter to assist the air flow through the packing. See Fig 1.11. From extensive pre-testing, it was decided that induced air conditions worked better than forced draught fans and in consequence the packings are arranged around the periphery of the podium, beneath annular water distribution trays. An annular return flume runs round the foundations beneath the packing. The Ince B tower has now met its design specification requirements.

In capital cost terms the assisted draught cooling tower is not dissimilar from the total of the three to four alternative normal draught towers it replaces, but the auxiliary power costs needed to drive the 35 fans, amounting at peak to some 6MW (relative to the 1000MW station output) is clearly a price to be paid for the aesthetic advantage offered by the single shell.

A more recent development in the field of natural draught towers has been forced by the concern surrounding asbestos. Fortunately the consequential change to purpose-made fabrications of thin sheet plastic has proved to offer a substantial heat transfer advantage over the asbestos cement, without incurring any overall cost penalty. It is therefore anticipated that towers over the next few years will be re-packed with plastic rather than the original timber splash packing laths, as there is an even larger heat transfer advantage in such an exchange. Asbestos packing is naturally long lived and is unlikely to be replaced so profitably. The very large extended surface offered by such plastic packings gives this heat transfer advantage and also considerably lessens the pack weight compared with the asbestos cement alternative. However, as so often occurs, this advantage has a corresponding disadvantage in that any deposition of scale from the circulating cooling water represents a significant percentage increase in the weight of the pack itself and, if untreated, can threaten its structural stability. To avoid this risk, with certain waters, it is necessary to maintain a suitable pH level to avoid or limit scale build-up. The cost of this acid dosing equipment together with the acid itself has to be offset against the financial advantages that the enhanced heat transfer given by plastic represents.

5.3.2 *Chimneys*

From a structural viewpoint, chimneys are not dissimilar from cooling towers in being largely wind loaded. They are, however, subject to different aero-dynamic effects, the possibility of forced transverse oscillations due to the establishment of Von Karman vortex generation being a feasible risk. Again, since the Ferrybridge C cooling tower collapse, there has been considerable research effort into the wind loading of chimneys, including full scale field tests to measure damping characteristics, surface pressures and basic wind speed variations with height. Laboratory work in wind tunnels and analytical support have also expanded design appreciation greatly. This work coupled with that of other investigators has markedly changed the basis of design of chimneys in recent years.

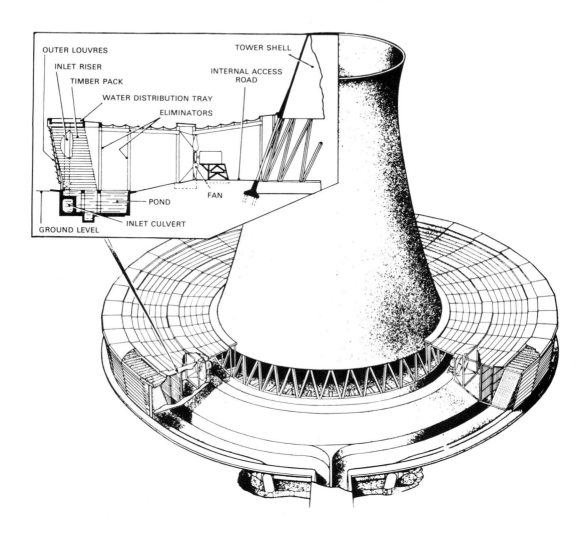

FIG. 1.11   INCE B POWER STATION FAN ASSISTED DRAUGHT COOLING TOWER

Chimney design has also been changed fundamentally by environmental pressures to increase the number of flues within a single wind shielding, in order to achieve a higher efflux velocity and to avoid a separate visual intrusion being associated with each individual unit. When a wind shield contains two, three or four flues a circular shape readily accommodates them without undue waste of space. Above that number however, the geometry suggests that advantage could be derived by using elliptical shaped flues for better 'packing' within the circular wind shield.

Unfortunately this change tends to negate the very successful experience of acid-resisting brick linings, which has been experienced for both coal and oil-fired units over many years. After some disappointments however, the concrete flues of the Drax chimney, which are substantial elipses of about 15m × 10m axes, have been most successfully protected over the last 15 years by a sprayed application of fluoro-elastomer. Although originally expected to last only about five years, this material now appears to be capable of giving a 25 year life under high merit coal burning conditions. Subsequent formulation improvements may even add to this success in later applications.

On conventional brick-lined chimneys on oil-fired stations, some difficulties have been encountered due to acid smuts agglomerating high in the chimney under changing load conditions and subsequently being ejected to cause acidic environmental damage. A heated section of flue has been introduced at Littlebrook D with a view to preventing the coalescence of the oil smuts during reduced load conditions.

5.4 The Rival Merits of Reinforced Concrete and Steel

As in civil engineering generally, the rival merits of steel and reinforced concrete clash in many parts of the design of power stations. Apart from the normal structural beam, column and slab alternatives, three examples deserve reference. Firstly to support turbine generator sets it was formerly the usual policy to use reinforced concrete blocks, believing that the greater damping characteristics of that medium would tend to reduce the overall vibration of the sets. However from some 15 years ago the virtues of steel turbine blocks led to a tranche of stations in which this structural form eased the congestion around and beneath the turbine and in general has been found no less satisfactory than the reinforced concrete which preceded it. An alternative which has not yet been adopted is the combination of steel on purpose tuned springs which is claimed to be a further technical advance. It is however noteworthy that both the steel and the spring loaded alternatives do not offer any economic advantage over reinforced concrete and consequently the decisions remain to be taken on a station-by-station basis. On balance, the steel alternative is preferred against the apparent advantage of reinforced concrete as a more malleable alternative. Fig 1.12 shows the foundation steelwork of a 660MW five cylinder turbine generator.

An area in which concrete has largely displaced steel is in the vortex casings of main cooling water circulation pumps. There the high cost of fabricating or casting metal in three-dimensional convolutions is considerably more expensive than the concrete alternative, where the admittedly complex initial formwork can be used several times

FIG. 1.12   660MW TURBINE GENERATOR STEEL FOUNDATIONS

for a multi-unit station. Some quite exceptionally fine concrete volutes have been made on most of the recent CEGB stations. Despite initial fears, evidence of subsequent cavitation in operation has not arisen.

A third and even more finely balanced comparison between steel and concrete concerns the main columns of conventional fossil-fired units. With the hanging boiler it has become standard policy to use very heavy steel box columns to carry loads from the boiler sling deck down to the foundations. In some other countries this duty has been taken by massive reinforced concrete columns with considerable success and for a future fossil-fired station it would be a very serious contender within the CEGB.

Simple cost comparisons between steel and concrete columns for this key application are not entirely meaningful because the choice also influences boiler house layout, construction programme and the optimum form of adjacent structures. Although substantially larger in section than its equivalent steel column, concrete offers an inherent additional lateral stiffness that may be used to relieve partially the wind loading on adjacent roof members. Such extra stiffness also encourages the adoption of low level roof steelwork erection, with the completed structure being subsequently lifted up the concrete columns.

Slip forming of concrete columns offers programme advantages during the early stages of station construction. This technique was used effectively on Hartlepool and Heysham 1 nuclear stations to provide weathertight main buildings at a relatively early stage.

Apart from size, the major disadvantage of concrete columns in fossil-fired boiler houses is the relative difficulty of making attachments to the column unless such fixings can be pre-planned in detail at the design stage.

6. PLANT LAYOUT

6.1 Introduction

Plant layout is concerned with the optimised location of equipment to provide efficient use of space and to establish correct relationships between plant items and their related systems. To a large extent, design specifications affect the efficiency of the station and layout influences the capital costs and, to some extent, the running costs.

Plant layout objectives should achieve minimum construction costs, minimise energy flow losses, give access for operation and adequate space and facilities for maintenance. The solutions are influenced by the individual plant purchased and, in the overall layout assessment, conflicts arise between the ideals for each plant item in its relationship with associated plant. The best compromise is sought.

The major items, boiler, reactor, turbine and generator transformer, together with the number of units in the power station, determine the initial overall dimensions of the main buildings. These major plant items must be co-ordinated with their interconnected flow process (main steam and feed system) and their own supportive systems (fuel supply, waste products, cooling water, power supplies) and the control and instrumentation related to the control room. The boiler house and turbine house must be linked to give the best overall arrangement, minimising overall building volumes consistent with adequate maintenance and operation access.

6.2 Main Plant Orientation

6.2.1 *Turbine Generators*

With fossil-fired plant, the initial determination is between transverse, longitudinal, or angled layout of the turbine generator. With 4-flow low pressure (LP) turbines, the overall dimensions of the turbine with the condensing, feed heating and general turbine auxiliary plant alongside produces a plan area approximately square. With 6-flow LP turbines, siting the auxiliary plant alongside produces a rectangular area. These areas must clearly be covered by main cranes or have auxiliary hooks mounted, so the plan area of the turbine hall follows from the alignment of the appropriate number of such areas, but includes the requirement for access via loading bays and the provision of adequate laydown areas for machine parts during overhaul.

The square plan area shows that with 4-flow LP turbines, the turbine hall dimensions are little affected whether a longitudinal or transverse arrangement is adopted. Generator rotor withdrawal space may marginally increase the width required for a transverse arrangement and so requires a slightly larger crane span, but the advantage of the shortest possible symmetrical main steam and reheat pipework offsets this. See Fig 1.13.

An angled turbine generator layout can produce a compact 'square' for a 6-flow LP exhaust turbine by having the longest item of plant as a diagonal feature, eg Drax. Complications arise in utilising the remaining triangular areas to give a good layout of auxiliary plant. See Fig 1.14.

The 6-flow LP turbine arranged longitudinally produces a narrower turbine hall with short span main cranes. Against this is the disadvantage of longer, asymmetrical, main steam and reheat pipework. Also the spacing between boilers is increased to provide room for a loading bay.

With 6-flow LP turbines arranged transversely requiring a wider turbine hall, the gains in main steam pipework are offset by cranage problems, see Fig 1.15. The long crane span can be split by providing two cranes on separate parallel longitudinal rails, but the complication of the supporting columns splitting the turbine hall and the consequent associated blind spots, not approachable by the cranes, must be taken into account.

The same disadvantage applies if, in a transverse arrangement of turbine generator sets, separate cranes are provided longitudinally over each set. Erection and maintenance requirements are then likely to demand a total of four independent cranes.

6.2.2 *Boilers*

It is necessary to determine which way the boilers will face relative to the turbine hall; whether the 'firing wall' (or simply the furnace in the case of corner fired boilers) is to be on the turbine side or the remote side. If on the turbine hall side, front-firing indicates the positioning of coal bunkers, feeders and pulverised fuel mills between the boiler and turbine for economic pulverised fuel pipe routing. This arrangement minimises flue duct lengths to the precipitators and chimney, but the disadvantages

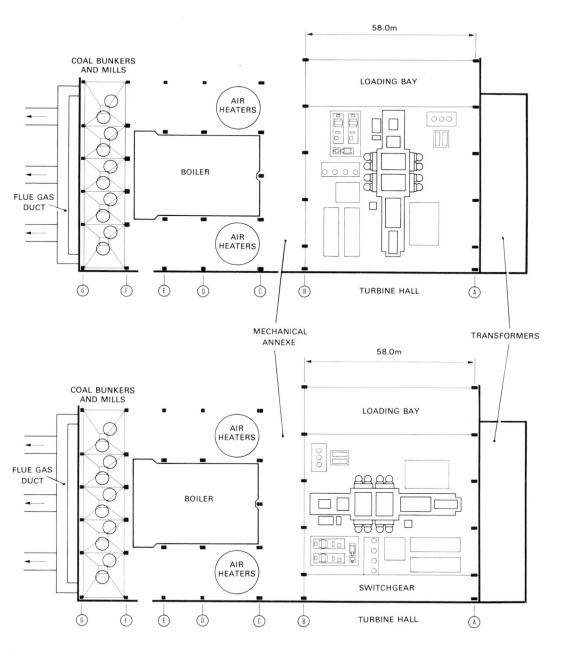

FIG. 1.13   LONGITUDINAL AND TRANSVERSE ARRANGEMENTS OF 4-FLOW LP TURBINE GENERATORS

are the bringing of dirt and noise from the mills into the centre of the station, which is undesirable, and lengthening of main steam and reheat pipes through having to cross the fuel handling annexe. See Fig 1.16.

Turning the boiler through 180° produces a better arrangement for coal plant at the outer side of the boiler house, but the gas flow is then remote from the approach to the precipitators. Extra ducting is required to return the gases around the boilers and the boiler width overall is then increased. Main and reheat steam pipes are shortened by the removal of the fuel handling annexe.

With oil-fired boilers, a longitudinal arrangement of 6-flow LP turbines requires the boilers to be opened out from minimum centres. Instead of coal plant, it may be appropriate to position other plant between boilers, such as auxiliary boilers or standby generators. Alternatively, with 4-flow LP turbines, the transverse arrangement with boilers on near minimum centres gives a very compact arrangement.

6.2.3 *Nuclear Plant*

Fewer options are presented by nuclear plant. In the case of advanced gas-cooled reactors (AGR) with one turbine generator per reactor, placing two AGRs on their minimum centre distance, as defined by construction space and access for operation, with transverse turbine generators on coincident or similar centre lines, gives a compact arrangement. However, it follows that the fuel handling and active components services block will be located at one end of the reactor hall, and this gives operational disadvantages, since all fuelling machine operations involve traversing over the nearer reactor. More attractive from the operational point of view, is a central fuel and active components services block.

This opening out of reactor centres may permit various economic options in the turbine house arrangement. In the case of pressurised water reactors (PWR) with two turbine generators per reactor, being without an intermediate pressure cylinder, the 6-flow LP design is shorter overall; the square plane is required for auxiliaries as already described. Consequently the advantages of shorter pipework can be achieved with a transverse arrangement set about the reactor centre line. AGR and PWR arrangements are shown in Fig 1.17.

The layout of nuclear plant is influenced particularly by the need to ensure that at all times sufficient plant is available to maintain essential cooling of the reactor. In the event of a plant failure causing the depressurisation of an AGR, it is essential that reactor cooling can be maintained by the gas circulators and the spread of fire or hot gas release be confined to a quadrant of the reactor building by segregation and fire barriers. Segregation is also required of essential cooling water supplies, pipework and power and control cables associated with the safety-related plant within and between adjacent quadrants.

Although plant is designed such that the consequences of failure of any single pipe are acceptable, the consequential damage to adjacent pipes and equipment due to pipe whip or missiles from the failed pipe, requires the segregation of pipework which performs essential cooling or other safety-related functions.

The possibility of earthquakes applying forces to plant and pipework and causing large scale damage, with consequential nuclear safety risks, requires seismic supports and special foundations for plant, pipework and building structures.

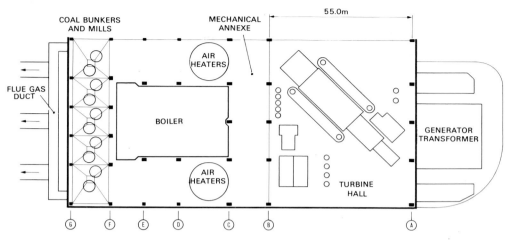

FIG. 1.14   DIAGONAL LAYOUT ARRANGEMENT OF 6-FLOW LP TURBINE GENERATORS

6.3 Auxiliary Plant Layout

The feedwater pumping and heating plant is logically placed between the turbine generator and boiler. To provide the required net positive suction head under all conditions, including load transients for the feed pumps, a high level deaerator also provides a buffer feed store. To minimise head losses in the interconnecting feed suction pipework, the feed pumps are best situated, as near as possible, directly beneath the deaerator. A mechanical annexe between the boiler and turbine hall can contain a deaerator, feed pumps and the vertical runs of main steam and reheat pipework. Pumps in this location will be out of reach of the turbine hall cranes and will require additional cranes for erection and maintenance.

6.4 High Pressure Pipework

As stated earlier, the main steam, reheat and feed pipework layouts are important factors in the overall station layout. This follows from the very high cost and inherent stiffness. Reheat is normally employed and, in addition to cost considerations, lengths must be kept to a minimum to avoid incurring too great a pressure drop between the boiler and the turbine. The layout must provide the necessary degree of flexibility to ensure that maximum permissible thrusts from expansion are not exceeded on the boiler, turbine or feed pumps; this may include additional loops and/or bends. Computer aided design techniques are available for these thrust and stress calculations. Anchors for seismic qualification are required on modern nuclear stations.

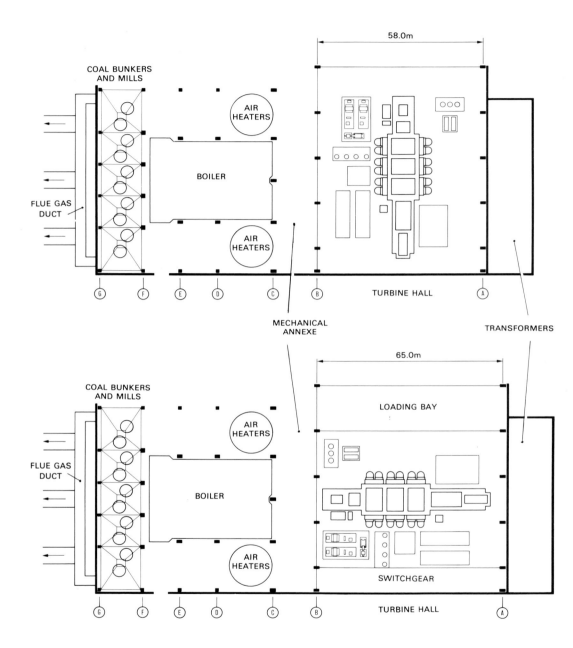

FIG. 1.15   LONGITUDINAL AND TRANSVERSE ARRANGEMENTS OF 6-FLOW LP TURBINE GENERATORS

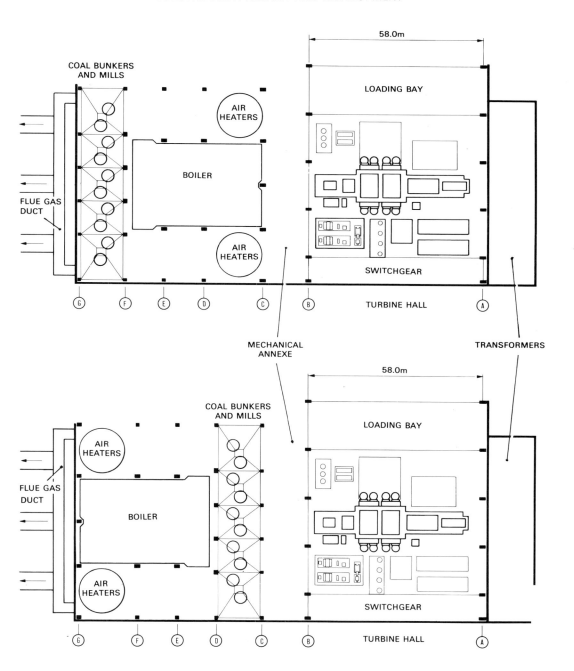

FIG. 1.16   LAYOUTS SHOWING ALTERNATIVE BOILER ORIENTATIONS

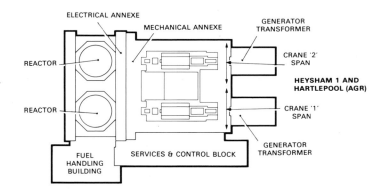

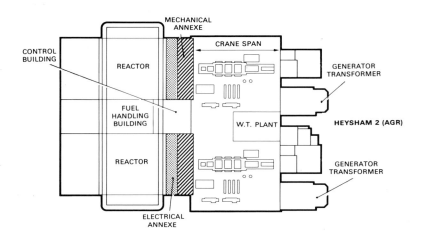

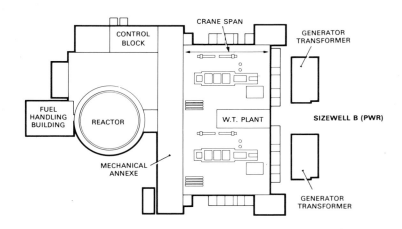

FIG. 1.17   LAYOUTS FOR NUCLEAR STATIONS

6.5   General Services - Low Pressure Pipework

Space is required for the layout of typical services, ie towns water, river water for auxiliary cooling, reserve and make-up demineralised feed water, hypochlorite dosing solution, blowdowns recovery, fire hydrant system, fixed water spray protection, ash sluicing supplies, potable water, compressed air for sootblowing or general services, and hydrogen, nitrogen and propane services from central storage centres. Many of these lend themselves to trunk main routing along the length of the station with branch supplies to each unit and/or ring main distribution systems.

This is an area where the use of appropriate models can provide visualisation and greatly facilitate error free routing. Routes and pressure drops must be optimised and access for maintenance provided to pipework and valves, including appropriate provisions for drains, air releases and drainage falls.

Water storage tanks represent a large space requirement, for instance, some $6820m^3$ reserve of towns water would be appropriate for a 2000MW station. This, and reserve feed water, is not economically stored in the main building. External tanks at ground level avoid hazards to electrical equipment, which are borne in mind in the positioning of all header tanks at high levels.

6.6   Site and Station Levels

The main factors governing the choice of station and site levels, which are an important aspect of the overall layout, are as follows:

(1) The need to protect the station against the risk of flooding.

(2) The capital cost of civil works.

(3) Cooling water pumping costs.

(4) Construction requirements.

For economy and general convenience on the site, it is desirable to construct the station basement with roads and rail sidings etc, at the existing ground level. Some minor levelling is usually required, but the aim is to minimise excavation and spoil removal or the importation of filling material. If a site is below the predicted flood level, there are two main methods of affording protection. The first and surest way is to lift the station basement and all other installations essential for the safe operation of the station above flood level. This method can be adopted where the cost of required filling material is cheap and the cost of pumping cooling water is not significantly increased. This is only likely to pertain for a cooling tower system. The second method is to rely upon flood banks, where the filling material and pumping costs are expensive. However, if a ground level exists well above predicted flood level, then to minimise pumping costs, it can be economic to excavate to lower the turbine house basement.

With closed cooling tower systems, the circulating water static head is purely between the pump suction culverts and the distribution headers inside the cooling towers. The level of a station with this system is not dependent on the absolute sea or river level for determination of major pumping costs, although flood banks may be required for protection from an adjacent river and make-up water pumping costs are affected.

6.7 Fuel Supplies and Storage

6.7.1 *Coal Plant*

The costs of coal deliveries by British Rail (BR) from the National Coal Board (NCB) are dependent not only on the distances involved, but also on demurrage rates for locos and rolling stock and the efficient use of rail capacity by high speed permanently-coupled wagons. These costs can be minimised by providing the most rapid and efficient turnround at the unloading point, ie the power station. For this reason, the favoured arrangement for coal unloading at any power station site is the merry-go-round system, whereby bottom-opening hopper wagons unload the coal into underground hoppers, with the train running on to leave the site without stopping. For the train to turn round and return to the loading colliery, a loop is required, with a 181m minimum radius of track, and having the appropriate standing room for signal delays, means that a considerable area is required for such a loop arrangement. The land within the loop provides a convenient coal stock-out area.

Where insufficient area is available, or where access problems exist, a compromise solution can be adopted with sidings before and after the unloading track hopper and with provision for the loco to run round the train prior to exit from the site.

The track hoppers are situated as close to the boiler house as possible to minimise conveyor lengths, but still providing sufficient distance for the rise from underground hoppers to boiler house bunker tops to be achieved at a suitable inclination angle, allowing for junction towers as required.

Another factor in the coal plant layout derives from the NCB working arrangements requiring a week's coal burn to be delivered in five working days. Thus, on average, two sevenths of each day's delivery must be stocked out for reclamation at the weekend. Consequently, stock-out and reclaim on a regular basis must be facilitated, and large travelling bucket wheel machines on rail tracks are required for access to and from the appropriate parts of the total fuel stocks. Longer term strategic stocks are held as part of the total stock, but transport to and from these more remote areas of the coal stock area is more economically achieved by bowl scraper mobile equipment.

Similar layout considerations may apply to coastal stations with sea-borne coal deliveries with short-term stocks as a 'buffer' between ship arrivals and longer term strategic stocks.

6.7.2 *Fuel Oil Plant*

No attempt has been made to provide road or rail-borne transport of heavy fuel oil for the main boilers of modern oil-fired power stations. Stations have been sited either close to oil refineries where direct piped fuel is available, or on coasts/estuaries where deliveries from sea-going tankers can be received. Quantities stored depend on a judgement of the security of supplies according to the proximity or otherwise of the source, and factors such as whether import and export to other nearby consumers is required. At least two, and possibly up to five, large storage tanks are required. Ideal situations would be close to the main buildings on the 'fuel delivery side', but leaving adequate distances to minimise fire hazard to the stations (and from other plant and equipment to the tanks themselves); also adequately firm ground conditions are required and a suitable area large enough for a bund to contain the contents of one tank in case of a tank fracture.

On coal-fired stations, the need for boiler lighting-up oil requires delivery and storage arrangements. The quantities of the lighter grade of oil needed are relatively small and so delivery is normally by road tanker. Storage is in tanks within a bund located as close to the main boiler house as other layout considerations allow.

6.8 Ash and Dust Handling

The layout must provide means of disposal for furnace bottom ash and for the large quantities of pulverised fuel dust produced as waste products. Although purchasers can be found at times for certain quantities of these waste products, in the construction industry, in concretes or simply as land-fill, long term dumping provisions are required. These can be close to the site or some distance away involving the pumping of dust as slurry, for example, to local natural or artificial lagoons, or transport by rail in a dry condition or by road in a wet condition. Market opportunities vary over the life of the station; some dumping grounds may become full or otherwise unavailable and disposal economics vary. Consequently, the layout is likely to require several disposal options to be kept open in the longer term, whatever the immediate or initial short-term disposal may be.

6.9 Electrical Equipment Layout

The type of power station to be constructed affects the layout of the electrical plant. The electrical load requirements vary considerably from one type of station to another. In a coal-fired station the high electrical loads for the milling plant are located at the rear of the boiler house, whereas in an oil-fired station, the electrical load for the fuel handling is considerably less and is at the oil pumping and heating location. The major oil heating is generally achieved by utilising the heat from a steam supply.

The configuration of the river and load-bearing properties of the ground affect the location of the main building and the positioning of outlying buildings, ie the cooling water pumphouse, the make-up water pumphouse and the coal handling plant.

The design of the electrical system, is determined by the size and number of generating units and the electrical drives which are required. An outline of the system (a power system diagram) must be determined before the layout can proceed.

Having designed the electrical auxiliary system, the following major equipment must be located:
(1) Transformers - generator, station, unit and auxiliary.
(2) Auxiliary switchgear - 11kV, 3.3kV and 415V.
(3) Generator and main connections.
(4) Power and control cables.
(5) Control room.

The layout of the power station electrical equipment must take note of the following design considerations:
(1) All switchgear and electrical control equipment should be housed in satisfactory clean and dry conditions.

47

(2) The electrical switchgear and distribution equipment should be located as near as possible to the electrical load centre.

(3) The location of the central control room should be positioned to suit the best man-movement arrangement, together with the best position for cabling facilities.

(4) Cable and electrical plant fire risk precautions. The principal consideration is that one incident on any unit should not affect any of the other units.

(5) Oil-filled transformers must be located externally to the main buildings.

(6) The provision of air/oil-cooled transformers for various locations.

6.9.1 *Main Switchgear*

The introduction at Littlebrook D and Dinorwig of SF6 (sulphur hexafluoride) high voltage, gas-filled, metal clad switchgear for 400kV substations, provides advantages of protected insulation, minimal fire risk and low noise levels, as well as having a compact design which allows a smaller site area to be needed.

6.9.2 *Transformers*

The large oil-filled power transformers have to be located externally to the building for fire considerations, with a preference that the generator transformer be located as close as possible to the generator. This ensures that the main connections are as short as possible. These connections for large 660MW units carry high currents in the order of 19,200A. Therefore, the generator transformer is normally located externally to the turbine house.

The other large transformers to be accommodated are the station, unit and 11kV/3.3kV auxiliary transformers. These are generally oil-cooled transformers and must located externally to the buildings. A convenient location for these transformers is adjacent to the generator transformer area, particularly as the unit transformer is normally teed-off the main connections near to the generator transformer terminals to save on conductor material.

When using interstage cooling, the same low pressure cooling water circuit can be utilised and run in parallel for both generator and unit transformers. The siting of the coolers presents a layout problem, since they must be placed near the unit in the case of a shunt arrangement, with the coolers across the condenser, and within an area close to the transformer for security of the oil lines. Lifting beams must be provided in the cooler cell for handling the heat exchanger tube nests.

The coolant is normally drawn from the turbine generator unit auxiliary cooling water system and on high head cooling water systems (associated with cooling towers), individual valveless atmospheric discharges are arranged on the water/oil coolers because of the water/oil pressure differential.

A most important transformer to be accommodated near the main buildings is the station transformer. There are usually at least two of these and they step down the grid substation voltage to the highest plant auxiliary voltage required in the station. This means that these transformers are wound for operating at 400kV/11kV or 275kV/11kV on modern power stations. These transformers must be sited so that they are available for commissioning the first plant in a new power station.

The remaining transformers, ie turbine unit, boiler unit, coal plant, ash plant, etc should each be located as near as possible to the associated lower voltage switchgear to reduce conductor costs and losses. A new design of low voltage switchgear (415V) incorporates a transformer built into the suite of panels, but problems associated with size limitation and heat dissipation can arise.

Oil-filled transformers are a fire hazard and each must be separately protected with an emulsifying water system initiated by temperature rise in the transformer cells, hence transformers must be separated by fire walls. Large volumes of water and oil can be released into the compound and adequate arrangements must be made for drainage to a large sump fitted with oil separators (in case there is a failure on the oil circuit of the transformer). The water can be pumped away to the station drains but the oil must be retained and disposed of by road tanker or other means.

Cooling of oil-filled auxiliary transformers is usually by radiators, but the generator and unit transformers are now so large and the heat generation is such that the heat in the oil is dissipated by cooling water in the heat exchanger. The limiting factor on transformer cooling, by other than natural means, is the hydraulic strength of the transformer tank. The head on the tank is limited to the level of the conservator and these are kept as low as possible to minimise tank pressures.

6.9.3 *Auxiliary Switchgear*

Conventional Stations

It is preferable to locate auxiliary switchgear at the electrical load centre. This means allocating a large area in the centre of the station for accommodation of all electrical plant. A good location is in the annexe between the turbine house and the boiler house, but this means having a much wider annexe than is normally required for mechanical plant. In allocating all electrical equipment in this central location there are practical difficulties which are as follows:

(1) Short 11kV connections are preferable for the station and unit transformer connections.
(2) The 11kV/3.3kV transformers are also generally oil-filled and must be located externally to the building. Therefore long cable runs would be necessary.
(3) The space in this central location is expensive and a wider central annexe would increase the overall cost of the station.
(4) When all equipment is centralised, this concentrates cabling in a very small area and produces congestion.

Various alternative locations have been considered as follows:

(1) Dispersed switchgear and equipment rooms — this provides multi-locations for various plant items and in general takes more ground area and is therefore considered to be an unsatisfactory layout. Dispersed switchgear is generally used where outlying buildings are a distance away from the main turbine and boiler house buildings.
(2) Central electrical equipment annexe — only one annexe is provided for accommodating all electrical equipment located between the turbine house and boiler house. This has been used on at least one station but creates problems as previously mentioned and is considered to be an undesireable layout.

(3) Electrical equipment annexes ('A' line and central) — this provides two annexes. One major 'A' line annexe is provided to accommodate the majority of equipment, eg 11kV switchgear, 3.3kV switchgear, turbine house 415V switchgear, batteries, instrument supplies and miscellaneous equipment. A smaller central annexe is provided to accommodate boiler house 415V switchgear, cable marshalling facilities and miscellaneous equipment. 'A' line refers to the first row of columns on the station site, see Fig 1.13.

(4) Electrical equipment annexes ('A' line, central and rear of boiler house — this provides three annexes. A minor electrical annexe in the 'A' line, adjacent to the turbine house, accommodates items of equipment, eg 3.3kV switchgear, batteries, instrument supplies and miscellaneous equipment.

The central annexe accommodates 11kV and 415V switchgear for turbine auxiliaries and other miscellaneous equipment.

The electrical annexe at the rear of the boiler house accommodates 11kV, 3.3kV and 415V switchgear associated with boiler auxiliaries, together with other miscellaneous equipment.

This layout of equipment has been utilised in the past and would be considered for use on future coal-fired power stations.

Nuclear Stations

The location of electrical equipment and switchgear items in nuclear power stations depends on the type of nuclear reactor installation being considered. The nuclear island is generally detached from the main turbine house building and it is convenient to provide a centralised annexe between the turbine house and reactor buildings for the majority of electrical equipment.

Special requirements are necessary for the essential auxiliary supplies, which must be maintained under various conditions, including the possibility of an earthquake. To cater for this, a number of safety loops are provided, eg four separate loops with completely separate emergency electrical supply systems. These must be located at different positions to ensure that under all fault conditions, at least one emergency loop is always available.

6.9.4 Generator Main Connections

The generator design has the generator line and neutral connections brought out below the machine. Care needs to be taken in the design of the connections to allow adequate heat dissipation. The design of the associated neutral current transformers must allow for mounting below the machine, supported from the basement floor and independent of the generator. The use of steel foundation blocks may reduce the available space in the generator terminal area. However, it is a requirement to install air-insulated main connections up to the machine terminals and to provide an air-insulated flexible connection there, onto the line terminal adaptors. The stator cooling water supply pipes, resistance columns, etc must be accommodated in this area and due regard must be paid to this in the design. Although water-cooled connections throughout are technically feasible, the higher losses associated with these connections make the arrangement uneconomic.

The generator is connected to the generator transformer by aluminium busbars, air-insulated and carried in trunking. Each phase is suspended by insulators in individual circular ducts of about 1.2m outside diameter. This design saves space and provides better control for cooling the busbars.

Apart from designing the busbar system to a minimum length, changes of direction must also be minimised since, under short-circuit conditions, high stresses are set up in the bars and mechanical failure could occur at the weakest points, ie bends and sets, etc. All busbars are flexibly connected at or near the generator terminals to prevent transmission of vibration and work-hardening of the aluminium. Busbar arrangements are determined to a certain extent by the layout of the turbine generator.

With a transverse layout of machine, a simpler, probably shorter, route can be obtained. The busbars on this layout should run below the generator, since an overhead arrangement requires dismantling of the copper work each time the generator rotor is withdrawn. The neutral point is made and earthed under the generator and incorporates generator protection current transformers. The unit transformer is teed-off the main busbars just before the connection to the low voltage side of the generator transformer. With this arrangement of busbars, the winding connections are quite simple, since the three phases are arranged across the generator frame.

A further point in favour of the transverse machine layout with respect to busbars, is that usually very little plant is located at the end of the machine at basement level. However with the longitudinal arrangement the basement area is sterilised to a certain extent, since the busbars pass overhead.

Busbar arrangements for a longitudinal arrangement of machine are complicated when the connections are brought from under the machine. With top connected busbars, the arrangement is much simpler although the star point is still brought out under the generator, thus requiring two sets of 3-phase terminals. With this latter arrangement the busbars and enclosure can be vulnerable to damage from loads carried by the overhead crane.

To take the connections from a side of a longitudinally arranged machine would produce difficulties in arranging the stator hydrogen cooling system, since the tubes carrying the coolant are generally arranged down the sides of the stator, thus blanking-off the windings. This arrangement would also block access alongside the generator on the one side where the connections are taken out, creating unnecessary access problems around the machine.

6.9.5 *Power and Control Cables*

The cabling required for a modern unit with fully computerised automatic and remote control systems may be regarded as the spinal cord of the unit.

The rules regarding cable segregation have been revised following power station fires which caused major restrictions in station output. The cables on one unit have to be segregated entirely from that of any other unit to ensure that in a fire incident only one generating unit output is lost. This considerably affects the layout of the station, as separate routes for all cables from each unit to the control room have to be established. To assist this very complex task, a computer data system called CADMEC (computer aided design and management of electrical contracts) has been developed. This handles all aspects of cabling, such as design, contracts, planning, control and evaluation, and provides a ready feedback to designers and management. Information is collected at all stages of design, construction and commissioning, as well as throughout the operating life of the station.

Cabling facilities have been provided in the following ways:

(1) Sub-basements - some power stations have been provided with sub-basements under the turbine, boiler house and switchgear areas, eg Drax. These basements are used for both cabling and mechanical pipe connections. This provides very good cabling access for installation, but presents problems when fire segregation is considered. It is also a fairly costly method as space is wasted in a number of areas.

(2) Subways - some power stations have been provided with no cable basement but have cable subways interconnecting the electrical plant items with local basements or trenches under the switchgear areas, eg Ince B.

(3) Above ground level - some power stations provide no cable basements or cable subways and all cabling connections are above ground level in the turbine house/boiler house areas, eg Grain and Littlebrook D.

The present philosophy is to adopt a combination of (2) and (3), utilising cable subways for interconnection of plant areas, and providing overhead cableways in the turbine house and boiler house.

Depending on the control room position, the cables can terminate at either end with relay equipment from which telephone-type cables are run to the control room. Since the control room is best situated off-centre on the electrical annexe side, the relay equipment should be placed at this end and cross cabling between units run in the electrical annexe.

This arrangement reduces the lengths of connections from the main cable runs to switchgear and plant to a minimum, provided the switchgear units are logically placed in relation to the plant they serve.

All cabling must be protected from mechanical damage, initially by the sheathing selected, but also by the method of installation. It should not be possible to walk on a cable, run a truck over it, or for it to be hit by a load swinging from a crane.

There is a special hazard associated with hydrogen-cooled generators and where cables are in close proximity to the magnetic fields. Cables are routed to avoid the induction of currents in hydrogen pipework.

Cables are not routed in the vicinity of the generator main connections, except for those connected to electrical terminal points necessarily located in this area, to eliminate the effect of magnetic fields on the cables.

6.9.6 *Location of Control Room*

The present practice is to provide a central control room for the station which includes all the facilities for controlling each of the generating units provided, together with controls for both main and auxiliary switchgear equipment.

A considerable amount of equipment is involved in controlling the various items in the power station and this is now centralised in one location and covers the provision of computers, automatic control systems, control panels, remote control equipment, telecommunications equipment and alarms. It is essential to provide this type of equipment with clean and controlled environmental conditions. Therefore a centralised heating and ventilating system is installed.

The location of the control room is determined by both man-movement relationships and cabling connection requirements. It is essential to reduce the length of cable runs from each unit to the control room to a minimum. For a 4-unit station layout, the control room would be located between units 2 and 3 in a separate control room building as close to the turbine house as possible.

Power stations at present are being built with three 660MW units, which is not an ideal arrangement for centralising auxiliaries and control rooms. With a 3-unit arrangement, the control room has been located at one end of the station, either in line with the central annexe or adjacent to the 'A' line annexe. The preferred location is adjacent to the 'A' line annexe as this improves the co-ordination of cable routes from the switchgear to the control room. The control room is generally a multi-storey block which is necessary to house all the equipment involved.

Clearly there are many ways of laying out equipment internally in a power station and the selection of the best method is governed by a number of factors including cost, man-movement relationships, best technical electrical system arrangements, space limitations within the confines of the mechanical plant and fire risk precautions.

7. BOILERS

7.1 Introduction

The two decades prior to the formation of GDCD saw a rapid increase in unit size and as a result, the 500MW and 660MW boilers constructed in the 1960s and early 1970s were major extrapolations of experience obtained on much smaller units. 1970-80 has been a period of consolidation rather than major innovation. It has seen modest extrapolation from secure foundations, extensive development programmes for new components and processes and focus on the details of arrangement, fabrication processes and mechanical features. The result has been boilers with higher initial availability than hitherto and achievement of design parameters and efficiency without the need for major modifications or acceptance of reduced component life times.

7.2 Coal-Fired Boilers

For many years CEGB has employed design coal specifications which allow flexibility in usage of coals mined within economic transport distance of its power stations. This has led to relatively wide specifications of gross calorific value, grindability index and the properties of ash in the fuel, with repercussions on the inventory of milling plant and the sizing of the furnace. Milling plants have been sized on the basis of the limits of 95% supplies, with one spare mill if verticle spindle mills are provided and no spare mill with tube ball mills, it being accepted that for 100% supplies some loss of output will result. Furnaces have been sized on the basis of ash initial deformation temperature for 95% supplies, again some reduction of output being accepted for 100% supplies.

As operating pressure and unit size increased, furnace exit gas temperatures related to initial ash deformation temperature could not be achieved without incorporation of superheater or reheater surfaces within the furnace. The 200MW, 300MW and 350MW boilers had various arrangements of furnace wall superheaters and sometimes

53

separate superheat and reheat furnaces. However, the incorporation of specific requirements for 2-shift operation, that is frequent and rapid starts after overnight shutdown, led to the majority of the 500MW designs having platen superheaters in the furnace with reheaters remote from the furnace to allow startup without reheater cooling until synchronisation.

Satisfactory operation of low pressure boilers with superheater outlet steam temperature of 568°C led to the specification of this temperature for both superheater and reheater in all the 500MW coal-fired boilers, with the exception of the Kingsnorth coal and oil-fired boilers where the outlet steam temperatures were limited to 541°C to avoid the use of austenitic steels which were susceptible to oil ash corrosion. A progressive increase in the chlorine content of coal delivered has resulted in higher corrosion rates than were anticipated, but reconsideration of the choice of steam temperatures prior to specification of the Drax Completion 660 MW boilers showed that, even when appropriate corrosion allowances were incorporated, an economic return was obtained from the higher steam temperature.

On the 500MW boilers designed for bituminous coals (volatile content about 30%), two firing systems predominated, front wall firing and corner firing. Both achieved high combustion efficiency and stable conditions under extremes of operation. For low volatile (anthracitic) coals, down-shot firing had been developed for smaller boilers, the fuel being injected downwards from an arch below which was a refractory belt to maintain temperatures in the ignition zone. This was developed for the 500MW boilers into a double down-shot system with arches on front and rear walls, the combustion gases merging before passing to the convection surface.

The security of design of the 500MW coal-fired boilers varied significantly from manufacturer to manufacturer. In some cases the boilermakers had had a continuity of involvement right through 60MW, 100MW, 120MW, 200MW, 300MW and 350MW units before commencing the 500MW designs, while others had missed significant points in this sequence. Even where there was continuity, the rapid progression in unit size resulted in the 500MW boilers being designed before operating experience had been obtained on the 300MW and 350MW boilers. If steps in unit size had been missed, the 500MW boilers were being designed before significant operating experience had been obtained on units as small as 120MW. In neither case therefore could it be claimed that the designs were secure, security clearly being lower when the reference design was of smaller capacity.

Realising this, there was a major programme to obtain information on the performance of the 500MW boilers to provide:
(1) Design data for future plants.
(2) Identification of weaknesses which required rectification.
(3) Development of operating techniques which minimised life usage.

These trials which required the provision of up to 700 instruments on a boiler were carried out on the first of each basic boiler type constructed and more restricted tests on some of the variants of the basic type. To ensure that the information was fed forward into subsequent designs the manufacturers had equal access to the test data.

The first results from these trials became available during the early design stage of the Drax 660MW boilers. While this information led to the specification of more realistic draught plant margins, provision of larger temperature margins on superheater and reheater tubes and headers and improved thermal performance and sealing of flue gas airheaters, utilisation of the same mill and burner system, as in the 500MW boilers, required a change from front wall firing to opposed firing. While there was considerable experience overseas with opposed firing, early operation of Drax showed that furnace heat absorption was greater than predicted, leading to difficulty in achieving design superheater and reheater outlet steam temperatures.

In all other respects the boilers were highly successful and when installation of a further three units at Drax was required, it was decided to replicate boilers 1 − 3 except where there were known deficiencies, components were no longer available, metrication impacted on the design or the design was non-compliant with new safety and environmental legislation.

Design contracts were let with the manufacturer well in advance of tendering to examine the design options available within the broad replication concept. This was to achieve design superheater and reheater outlet steam temperatures, provide realistic corrosion allowances for the high chlorine coal to be supplied to the station, reduce the incidence of furnace ash hopper bridging, improve the distribution of fuel to the burners, improve platen superheater tube alignment and avoid fretting damage in tube assemblies. These studies led to designs which promised improvement both in performance and ease of construction, while taking full advantage of the operating information being accumulated from boilers 1 − 3 and the improved manufacturing equipment then being installed at the manufacturer's works. They confirmed that despite the highly corrosive coal to be supplied to the station it remained feasible to continue to employ superheater and reheater outlet temperatures of 568°C. This maintained the higher cycle efficiency that had hitherto been achieved on coal-fired units, compared with oil-fired units which were limited to outlet steam temperatures of 541°C by oil ash corrosion considerations.

Even with design contracts completed well in advance of acceptance of the formal tender, a tender in many respects remains an outline and there is a significant period early in the contract, during which the detail is developed and the interactions between the boiler plant, the turbine, the auxiliary systems, structures and civil works are identified and where appropriate, resolved. To ensure that this early design work is carried out within the constraints of ordering of long delivery items and letting of related contracts (boiler house foundations, structural steelwork, main steam and feed piping, etc) design networks have been developed which are translated into schedules of design submissions and approvals. Adherence to the timescale of these activities has ensured that the construction programme is not put at risk by late changes to design details or late material orders, etc.

In addition to progressing the contract design work, there remained some development and proving of systems during the early part of the contract. Three major areas of development were carried out in this period:

(1) Flow distribution tests of scale models of the final ductwork layout to ensure acceptable uniform gas flow into the dust extraction plant.

(2) Development of pulverised fuel (PF) pipework runs, using a technique employing PF/air mixtures in small scale pipework to obtain a geometry which was not susceptible to PF blockage and maintained equal distribution between pipes.

(3) Tests of the oil light-up burners to establish the reliability of operation and durability of components. For these tests, six light-up burners were installed on a mill group of one of the operating units and were subjected to real operating cycles over a prolonged period.

In addition to the development of design details, resolution of interactions, system and component development to be carried out in the early stages of the contract, manufacturing procedures covering welding, inspection and heat treatment, are submitted for approval during almost the whole contract period. While much of the approval process is verification that well-developed processes are being employed in appropriate conditions, when new processes are introduced the results of pre-production tests and early production runs have to be examined before final approval can be given. During the Drax Completion contract, the proportion of welding carried out by automatic processes was progressively increased. As experience with these processes accumulated and inspection records were correlated with the continuously recorded welding parameters, it became possible to substitute process control as the primary quality control procedure with inspection on a sampling basis. With the achievement of lower weld repair rates and the rationalisation of inspection requirements, the cost benefits of these processes are now being realised and a further increase in their use can be anticipated on future contracts.

While the Drax First Half and Completion boilers are the only 660MW coal-fired boilers purchased by CEGB, it has been the CEGB's policy to ensure that at any time each of the UK boilermakers has a well-substantiated design available for its potential requirements. While Babcock Power Ltd has designed and constructed all the CEGB's 660MW coal-fired boilers, NEI Parsons Ltd has developed 660MW designs initially based on its successful 500MW boilers but latterly on designs developed by its licensor Combustion Engineering of the USA. The successful operation of these latter designs at unit sizes up to 800MW, coupled with continuing analysis of performance of the CEGB's 500MW units provides designs of similar security to that based on Drax Completion.

Fig 1.18 and 1.19 show respectively the construction of a coal-fired burner wall and the block erection of boiler heat exchange surfaces.

7.3 Oil-Fired Boilers

The 500MW programme included only one oil-fired boiler design, the Fawley design, which was used at Fawley and Pembroke. These boilers were designed when the largest boilers the contractor had in operation were the 120MW coal-fired Richborough units and the 60MW oil-fired conversions at Marchwood. The only boiler specifically designed for oil firing which was commissioned during the contract stage of the Fawley and Pembroke boilers was the 120MW Bankside B unit which had a firing system completely different from the Fawley design.

Oil-fired boilers differ markedly from coal-fired boilers in three main respects. Firstly, the rate of combustion of fuel oil is greater than coal, allowing smaller furnace volume. Secondly, the low ash content allows more closely pitched superheater and reheater surfaces to be employed and the restrictions of ash softening temperature do not apply. Thirdly, the absence of alkaline ash which can absorb the acidity of the products of combustion require low excess air combustion to inhibit the oxidation of

FIG. 1.18   CONSTRUCTION OF BOILER BURNER WALL

FIG. 1.19   BLOCK ERECTION OF BOILER HEAT EXCHANGE SURFACES

SO_2 to SO_3 coupled with maintenance of air heater and duct temperatures above acid dew point to avoid agglomeration and subsequent emission of acid smuts. Because of the susceptability of austenitic steels to oil ash corrosion, superheater and reheater outlet temperatures have to be restricted to 541°C to allow the superheaters and reheaters to be designed using ferritic steels only.

While the Fawley design boilers were of similar layout to coal-fired boilers, with radiant superheaters installed in the furnace and parallel primary superheater and reheater in the rear pass, with gas biasing dampers to provide reheater temperature control, they are approximately two thirds the size of coal-fired units of equivalent output. See Fig 1.20. The small furnace coupled with rapid combustion generated high heat fluxes and the commissioning tests showed that circulation margins were only just adequate and subsequently water side corrosion occured in the furnace floor tubes. Inadequate allowance for corrosion by the boiler ash coupled with some excess temperatures, limited the life of some superheater and reheater surfaces. Nevertheless the availability of the boilers was high and they achieved their full output with combustion efficiency and better than specified.

During the early commissioning of the Fawley 500MW boilers, CEGB issued enquiries for 660MW oil-fired boilers for operation with turbines developed for the coal-fired and nuclear programmes. The successful tender, initially for three units and subsequently for five units on the Grain site, marked a radical departure from coal-fired practice with all the superheater and reheater surfaces protected from direct furnace radiation by a water-cooled screen. This arrangement avoided many of the metal temperature problems encountered on the Fawley design. This was particularly important as, for the first time, a specified requirement was for the plant to designed on combined corrosion and creep, and by employing effectively two furnaces, size extrapolation uncertainties were reduced.

The contract development of this design showed that its initial promise was being fulfilled and when it became clear that there was a requirement for further 660MW oil-fired boilers, a contract was awarded to the designer of the Fawley boilers for a design study for a 660MW boiler which used to the maximum the detailed operating and test information then being accumulated on the Fawley boilers. This led to a design which was in some respects, markedly different from the 500MW boilers but was securely based on a modest extrapolation from the earlier design. Compared with the Fawley design, the design which later became the Littlebrook D boilers had no radiant superheater in the furnace; the whole superheater and reheater was protected from direct radiation from the furnace by a water cooled screen and reheater temperature control was by gas recirculation instead of gas biasing. The boiler had greater margins in the furnace circulation system, the superheaters and reheaters were designed with allowances for oil ash corrosion in the choice of materials and scantlings and the burners had a 4 : 1 turn down capability (compared with 1.5 : 1 for the Fawley design). This latter was specified to improve burner life (because deterioration is greater when burners are off load) and improve boiler flexibility, particularly the achievement of loading rates during hot starts.

An important aspect of the study was the development of a design which minimised site work, particularly pressure part welding. The design developed was a major advance on Fawley/Pembroke in this respect. The superheaters and reheaters were delivered to site in blocks incorporating inlet and outlet headers which after lifting into

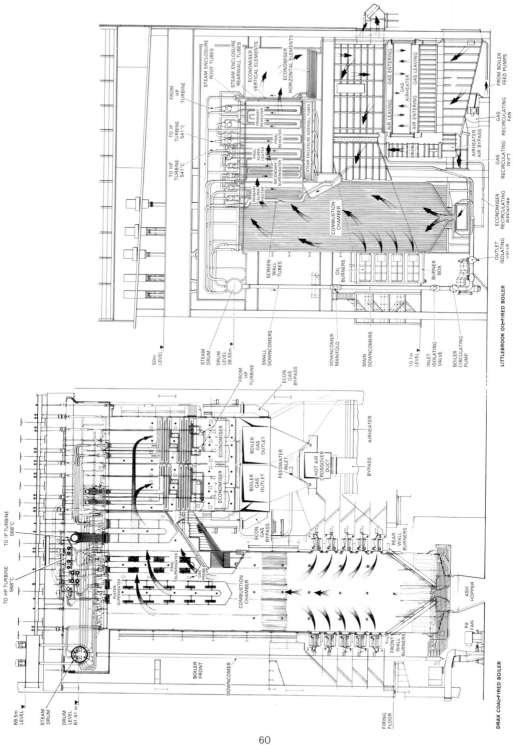

FIG. 1.20   COMPARISON OF SIZES OF COAL-FIRED AND OIL-FIRED BOILERS

position, only required header-to-header welding. Similarly, the furnace and enclosure walls were delivered as large panels, with, where appropriate, their headers already installed. For works manufacture the maximum use of automatic welding equipment was encouraged. The automatic processes were introduced progressively, manual production capability being maintained in parallel to ensure that any production development problems that might be encountered did not put the overall manufacturing programme at risk. The automatic processes immediately showed lower repair rates than manual processes, easing the achievement of manufacturing programmes and allowing rationalisation of inspection procedures.

As with Drax Completion, design networks were developed to ensure that the post tender design work was carried out within the constraints of ordering of long delivery items and letting of related contracts. In addition to progressing the contract design work, there remained some development and proving of systems during the early part of the contract. Three major areas of development were carried out in this period:
(1) Development of the burners in a full size test rig.
(2) Installation of a prototype burner complete with management equipment in a boiler at Fawley for environmental and endurance trials.
(3) Flow distribution tests on a scale model of the airheater to obtain gas and air flow distributions which avoided low temperature areas in the heat exchange surface which could be potential sites for acid smut production.

In addition to the care taken to achieve low excess air combustion and avoid low temperature areas in the airheater and duct work, each of the three chimney flues has a heated outlet section to reduce further the potential for acid smut generation. The heat input required was determined from tests on electrically-heated panels installed at Fawley. While this system was appropriate for test purposes, once installed the panels were virtually inaccessible. To overcome this difficulty the Littlebrook D scheme uses a hot air recirculation system with external electrical heaters and fans easily accessible from the chimney windshield.

8. TURBINE GENERATORS

8.1 General Considerations on Choice of Large Machines

The cost of turbine generator sets shows a marked reduction with increase of size. Roughly, doubling the size reduces the specific capital cost by 20%. However, savings in the supply and installation cost could be nullified by shortcomings in reliability, availability or efficiency of the larger machines and it is necessary to judge the most attractive propositions accordingly. The physical arrangement of a high pressure (HP), intermediate pressure (IP), three double flow low pressure(LP) cylinders and hydrogen cooled generator remained standard for 500MW and the first generation 660MW sets.

The 500MW sets had been designed and developed by four different manufacturers when 300MW sets were being commissioned or not yet commissioned. Experience had mainly been related to 120MW sets and design effort committed on a very wide front, with a variety of ratings at home and overseas. Although some 500MW sets suffered design and quality problems when commissioned, a wide study of loss of availability on a component-by-component basis showed no common output related cause, but rather served to illustrate the paramount importance of soundly engineered and proven component design.

Moving blading on 500MW turbines had been substantially better than that of 120MW, 200MW and 300MW turbines with no repetition of failures due to vibration and overstressing of blades and shrouds found on the smaller turbines. 660MW was similar to 500MW blading and well-supported by wheel chamber testing.

For sets smaller than 660MW it has to be recognised that there had been an evolution of different or partly different designs from those created 15 years before. Mergers and relocation of manufacturers had resulted in destruction of patterns, jigs and fixtures. Even if these were re-created, major redesign of castings and forging scantlings would be necessary to avoid delay by rectification and weld repairs during the programme stage. Present day foundries would decline to accept some existing casting designs. Known weaknesses demanding continued maintenance and operator effort had to be eliminated.

Sets smaller than 500MW would, of necessity, embody changes and prototype components; they could not be expected to have the same degree of security as offered by the large machines. This demonstrated the need for concentration of the design and development efforts of turbine manufacturers, identification of the appropriate skills and emphasis on component design using a modular approach to ensure reliability and availability, independent of nameplate rating.

In the late 1960s development of 660MW sets by GEC and C.A. Parsons began in a more concentrated way with a high quality approach to design of major components. The soundness of this approach has quickly been demonstrated because it has been possible to apply existing 660MW design features successfully as a solution to some of the failures on 500MW sets.

This output rating was initially chosen as being the modest extrapolation of the later 500MW sets which took advantage of the identified margins. The second tranche of 660MW plant has been developed from these machines, in particular using the modular approach to design.

In summary, the major considerations have been:

(1) The high cost and prototype nature of small size sets, eg 120MW or 200MW built to contemporary standards and quality.
(2) The technical attractiveness of 660MW designs compared with 500MW sets.
(3) The possibility of obtaining high availability and the flexibility to meet system requirements with any set size.
(4) The adoption of the modular approach and the use of proven components in all major plant items.
(5) The international trends towards large sets and large and strong manufacturing enterprises.

8.2 Steam Conditions and Operating Modes

Variations in steam pressure and temperature do not have a major impact on the choice of unit size in fossil-fired stations. Virtually identical conditions have been chosen for the 200MW and larger units with only a tentative excursion into super-critical pressures, which were not economic at typical CEGB load factors. For the original coal-fired stations a superheat and reheat steam temperature of 565°C at the turbine stop valve was selected. Subsequently there have been minor variations according to the corrosion properties of different fuels, but the lower temperature of 538°C instead of 565°C has been adopted for all oil-fired stations. There is no reason to envisage changing the temperatures in the foreseeable future. 538°C was also adopted as being more economic for all AGR stations after Dungeness B.

The mechanical solutions adopted to accommodate the high inlet temperature differ from one unit size to another and are associated with varying degrees of complication and, from time to time, innovation. In the case of 660MW turbines, a proper balance has been obtained for all critical components between the 30 year life from creep considerations and adequate thermal fatigue resistance to withstanding defined 2-shift and load cycling duties.

The modular design concept permits the size and number of LP turbines to be selected to give the optimum turbine exhaust area and exhaust pressure to suit the prevailing cost and efficiency criteria. The first station in the second tranche of 660MW plant (Littlebrook D) had two double-flow LP turbines, this being marginally more economic than three. For subsequent stations ordered since the rapid fuel price rises of the 1970s (whether nuclear or coal or oil-fired) three double-flow LP turbines have proved more economic, in common with other countries with relatively high fuel costs and relatively low cooling water temperatures.

Fig 1.21 shows the Littlebrook 660MW turbine generator under construction.

8.2.1 *Operating Flexibility*

The 500MW units had a specified requirement for 2-shift duty. The stipulated performance was that after the plant was off-loaded in 30 minutes and a shutdown period of 6 hours, it should be run up to speed in about 10 minutes and subsequently loaded to full rated output in 20 minutes. This was demonstrated with tests at West Burton, Ferrybridge C and Fawley in 1972. A number of hot starts were successfully performed and there were no restrictions on the turbine generator. The agreed procedure for starting the turbine had required only slight modification and had proved a sound basis for all subsequent flexible operation. Because of better casing symmetry and rotor profile, 660MW turbines would be, by definition, superior in this respect and this has been demonstrated in normal commercial operation at Grain, Drax and Littlebrook D for which the number of hot, warm and cold starts were additionally specified.

The thermal efficiency of the 660MW turbine generator sets equals or improves on that of the better 500MW sets, and some detailed improvements have been introduced where this is not precluded by conditions of interchangeability. 660MW sets were inherently sounder and more reliable than earlier ones; replicate designs of existing 500MW sets would retain obsolete components in some cases, eg shrunk-on turbine discs and flexible rotors.

8.3 Key Principles of 660MW Turbine Generator Components and Systems

The main features of 660MW turbine generators affecting reliability, availability and performance are as follows:
(1) High pressure and intermediate pressure rotors.
(2) Low pressure rotors.
(3) Generator rotors.
(4) High pressure casings.
(5) Intermediate pressure casings.
(6) Low pressure casings.
(7) Generator stators.
(8) Blading.
(9) High pressure and intermediate pressure steam chests.
(10) Boiler feed pump turbine.
(11) General structure.
These features are described as follows:

8.3.1 *High Pressure and Intermediate Pressure Rotors*

The HP and IP rotors are of solid, single piece construction, which provides greater thermal stability during steady and abnormal transient conditions. The rotors are substantially stiffer than on earlier turbines and have the inherent ability to minimise the transmission of large amplitude vibrations from one cylinder to another in the event of disturbances, as well as offering greater stability in the event of rubbing.

8.3.2 *Low Pressure Rotors*

The LP rotors are all simple one-piece construction and contrast with built-up rotors having shrunk-on discs on earlier turbines. They are immune to fatigue and corrosion cracking mechanisms in highly stressed regions and have a much lower susceptability to brittle fracture from a given defect. Current steelmaking and non-destructive testing practices assure the delivery of adequate solid forgings from several sources.

8.3.3 *Generator Rotors*

Monobloc gas-cooled generator rotors incorporate all the lessons from 500MW experience and have as a consequence achieved excellent settled availability very quickly. Rotor dynamic and thermal behaviour at full output has been demonstrated to be secure. The 660MW programme has been sufficiently large to permit a national spares holding of interchangeable rotors in the event of any problem arising in a particular family.

8.3.4 *High Pressure Casings*

The 660MW turbines have HP casings with pressure sleeves, which permit significant reductions in the flange and bolt dimensions, thus minimising thermal strain and distortions. This must inherently improve the performance in both 2-

shifting and load-following service. It should also safeguard reliability, by the avoidance of rubs and improve availability by reducing the likelihood of thermal cracking of casings in the long term.

8.3.5 *Intermediate Pressure Casings*

The 660MW turbines have IP casings of much simpler and more robust shape than earlier sets, particularly with regard to static deflection.

8.3.6 *Low Pressure Casings*

The LP casings are similar for both manufacturers' machines. Advantage has been taken of the experience with earlier turbines to improve the resistance to thermal distortion of the 660MW casings.

8.3.7 *Generator Stators*

Improvements introduced to insulation, bracing and winding support have obviated conductor fatigue cracking experienced on some 500MW generators. Simple hose arrangements for stator bar water cooling have proved effective and there has been no repetition of core overheating experienced earlier. Higher manufacturing cleanliness and quality control standards allied to stringent testing have secured high availability on 660MW generators.

8.3.8 *Blading*

The record of the 500MW turbine blading is good. The 660MW turbine blading is of similar type; emphasis has been placed on its freedom from vibration and additional testing has been applied to this end. The rating of the blades is within well established limits in both the 500MW and 660MW types.

8.3.9 *High Pressure and Intermediate Pressure Steam Chests*

No novel feature is incorporated in the 660MW turbines other than the improvements of detail which have been established by experience and trouble shooting on the earlier turbines. In the 660MW design, the steam chests are inherently more suitable for meeting load variations than the earlier designs.

8.3.10 *Boiler Feed Pump Turbine*

The normal practice on large units is to drive the boiler feed pump by an auxiliary turbine. The internal efficiency of an auxiliary turbine can be sufficiently high to be economically competitive at full load with electric motor drives. At part-load, a variable speed capability is necessary to avoid throttling losses on feed regulator valves. This can be simply provided by the steam turbine where an electric drive would require a hydraulic coupling, slip-ring motor or other solution.

In the UK, the decision was taken in the mid 1960s to develop advanced-class feed pumps of only two or three stages, running at relatively high speed, about 6500rev/min. These pumps, with their stiff shafts and generous running clearances, have the capability to accept thermal shock and to run vapour locked without damage, and they have now demonstrated their expected higher reliability. To improve availability, cartridge construction features were adopted to permit quick replacement (one shift) of the rotating assembly if operational needs required a pump replacement.

Fig 1.22 shows the Heysham 2 main boiler feed pump and its associated turbine.

8.3.11 *General Structure*

The general structure of the 660MW sets is superficially similar to that of 500MW sets, both being installed on steel foundations. The use of steel foundations rather than concrete offers both technical and constructional advantages due to the greater predictability and stability of its properties, and the more controlled conditions of works fabrication. However, a great deal of work has been done to improve the provisions for guiding expansion and contraction, accurate alignment, access for maintenance and other features. These are expected to contribute in no small measure to both availability and flexibility.

8.4 Plant Maintainability

The following provisions for reducing planned outage and maintenance times have been included in the latest designs exclusively for 660MW sets:
(1) Sleeved HP casing and improved scantlings to minimise thermal distortion.
(2) Spring-backed glands throughout to proven designs, to ensure ease of replacement and rotor security.
(3) Checking of internal clearances without dismantling cylinders on the later 660MW turbines.
(4) Redundancy built into the design of electronic governors (mainly 2-out-of-3 type) which permits, in the unlikely event of a fault, the location and complete replacement of any control channel while the unit is on load. Removal of the centrifugal governor and oil pumps from the front pedestal has simplified a complex maintenance area.
(5) Limited 'ad hoc' trials with forced air cooling on 500MW turbines has encouraged incorporation of this feature into later 660MW sets. Barring times can be reduced from about 70 to 25 hours giving quick access to rotors and casings for maintenance.

8.5 Further 660MW Development

A shortened version of the 660MW turbine was being offered in the mid 1970s having four LP flows instead of six. It incorporated a 2-module LP cylinder with underslung condensers already developed in some detail for home and export markets. The HP and IP cylinders and generator already existed. The underslung condenser and its associated double-flow LP turbine became a standard module and can be used in varying numbers of 4, 6 or even more flows, thus giving the possibility of common LP design for a variety of cooling water temperatures, unit ratings and superheat and wet steam cycles.

The modular approach concentrated design resources and skills and offered the prospect of standard sets for all conventional and nuclear applications in the UK.

The existing modular designs of plant are capable of being utilised for a unit output of 750MW while still retaining substantial interchangeability with 660MW components. This might involve an increase of initial steam pressure from 160 to 176 bar for enhanced efficiency. Developments are either completed or in hand by the plant manufacturers which will permit the output of a 3000rev/min single-shaft

FIG. 1.21   660MW TURBINE GENERATOR UNDER CONSTRUCTION

FIG. 1.22   MAIN BOILER FEED PUMP AND TURBINE

superheat machine to be raised to around 900MW. Increased thermal efficiency, possibly at the expense of some flexibility, could be achieved by an increase in supercritical initial pressure to about 240 bar. An increase in steam temperature above 565°C is considered more appropriate to a later phase of development.

9. NUCLEAR PLANT

9.1   The Magnox Reactor Programme

The commercial Magnox stations evolved steadily from the successful basic Calder Hall design of the 1950s. This was based on a graphite-moderated, carbon dioxide (CO_2) cooled core with natural uranium metal fuel clad in a low absorption cross section material whose major constituent was magnesium, with a gas pressure of 7.8 bar and bulk gas inlet and outlet temperature of 140°C and 340°C respectively.

The first two stations for CEGB, at Berkeley and Bradwell, came to commercial power in mid 1962. The Berkeley design was based very much on Calder Hall with small increases in parameters, eg gas pressure 9.65 bar, bulk gas inlet and outlet temperatures of 160°C and 345°C but with a net electrical output of 137.5MW per reactor compared with some 45MW from each Calder Hall reactor. The Bradwell design, however, was significantly more advanced, with a more compact layout, a gas pressure of 10.14 bar and a gas temperature at inlet and outlet of 180C° and 390°C. Net electrical output was 150MW per reactor. Both reactor cores were placed inside steel pressure vessels, cylindrical at Berkeley (like Calder Hall) but spherical at Bradwell, each designed to be refuelled at full power (unlike Calder Hall, which is refuelled off-load) and each with boilers raised above the level of the reactor core in order to promote natural circulation of the CO_2. The gas was pumped around the circuit by circulators, each situated beneath a boiler; eight at Berkeley, each driven by an induction motor through a hydraulic coupling, and six at Bradwell, driven by induction motors fed from variable frequency sets in the turbine hall. It can be seen that a need for part-load operation had been envisaged, although in the event the stations seem almost certain to operate at base load throughout their lives. Refuelling at full-load was always specified; although an important reason originally for this requirement was the ability to remove failed fuel on-load, operational statistics now show that, apart from a small number early in life, the incidence of failed fuel has been very low. However, on-load refuelling has enabled one or two reactors to operate continuously without shutdown between biennial overhauls, thus contributing greatly to high availability.

Four further Magnox stations were designed and built with their cores inside spherical steel pressure vessels. There was economic pressure to increase reactor output from that of Berkeley and Bradwell. As the welding techniques improved, so thicker plate could be used which in turn allowed higher gas pressures typified by 76.2mm thick plate at Bradwell (10.14 bar) to 104.8mm plate at Sizewell A to permit a pressure of 19.23 bar. The higher gas pressures allowed increased fuel rating and despite the much higher output, vessel diameters remained almost the same as at Bradwell. These reactors were constructed at Hinkley Point A (at power in early 1965) Trawsfynydd (also early 1965) Dungeness A (late 1965) and Sizewell A (early 1966).

The design electrical output from each 2-reactor station had virtually stabilised, being 500MW for Hinkley Point A and Trawsfynydd, 550MW for Dungeness A and 580MW for Sizewell A. The number of boiler/circulator units per reactor also decreased from six at the 500MW stations to four at the other two.

A feature of the motor/circulator drive for the first four stations (up to Trawsfynydd) was the inclusion of a flywheel, which gave a slow rundown under the fault condition of loss of all power to the circulator motors. Typically, the time to half speed was about 30 seconds and temperature transient increases arising from this fault in a pressurised reactor were minimised and therefore acceptable. The Hinkley Point A gas circulator drive system was very similar to that at Bradwell, but the Trawsfynydd system was different, with induction motors receiving power from the main station electrical system and gas mass flow control by inlet guide vanes (IGV) and a flow bypass. The starting current for these fairly large motors, including flywheels, was however high. The Dungeness A drive system was different again, using a back pressure steam turbine driving the gas circulator directly. This unitised system made an all circulator failure virtually impossible and the stored steam was as effective as a flywheel. The Sizewell A drive system reverted to IGV control and large induction motors with no flywheel. Two lines of prompt detection of loss of circulator power had to be fitted (low voltage and underspeed) which ensured a sufficiently rapid shutdown that the 5 second halfspeed time on rundown still gave acceptable transients.

Gas temperatures on these four stations had also begun to equalise. At Hinkley Point A, the bulk gas inlet/outlet temperatures were 180°C/373°C; at Trawsfynydd 200°C/390°C; at Dungeness A 245C°/410C°; and at Sizewell A 215C°/410C°. Mean fuel ratings had also increased, from about 0.65MWe/t at the first three stations to about 0.9MWe/t at the other three stations. The peak can surface temperature remained virtually unchanged, being strictly limited by specification, and better heat transfer surfaces were developed to allow optimisation within this limit.

Further developments seemed to lie in the direction of increased gas pressure, and perhaps by eliminating the most onerous fault condition, which was defined as the sudden failure of the 'cool' ductwork between gas circulator and reactor. The way forward lay in the development of the pre-stressed concrete pressure vessel (PCPV) concept, inside which core and boilers were contained so that there was no 'cool' duct. The gas pressure was raised in this design to 24.13 bar in a cylindrical vessel, using the developed Dungeness form of circulator drive with the turbine outside the PCPV boundary. Gas temperatures were 250°C/412°C and mean fuel rating increased to 0.96MWe/t. The PCPV concept involved the development of insulation on the inside of the steel gas-tight liner to the vessel cavity; together with a system of cooling water pipes on the other side of the liner, the concrete was kept acceptably cool. It was also necessary to develop once-through boilers to minimise the number of substantially sized penetrations through the vessel wall. Oldbury includes these developments and has a design output of 600MWe from its two reactors; it came on line at the end of 1967. See Fig 1.23.

The last of the Magnox reactors was at Wylfa, which was based on a doubled, PCPV version of Sizewell A, with circulators driven by four large induction motors and flow control by IGV. The PCPV had, however a spherical cavity with the boilers fitted in

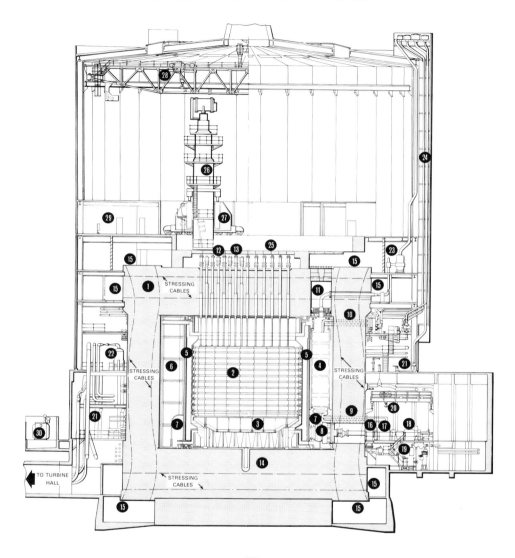

KEY

1	PRE-STRESSED CONCRETE PRESSURE VESSEL	16	GAS CIRCULATOR SHIELD DOORS
2	GRAPHITE CORE	17	GAS CIRCULATOR PONY MOTOR
3	CORE SUPPORT GRID	18	GAS CIRCULATOR TURBINE
4	BOILER	19	GAS CIRCULATOR AUXILIARIES
5	BOILER SHIELD WALL	20	GAS CIRCULATOR CRANE 25TON
6	BOILER END PIECE	21	STEAM AND FEED PIPEWORK
7	GAS CIRCULATOR OUTLET DUCT	22	BOILER START-UP VESSELS
8	GAS CIRCULATOR	23	REACTOR SAFETY VALVES AND FILTERS
9	BOILER FEED PENETRATIONS	24	RELIEF VALVE PIPES TO ATMOSPHERE
10	H P AND L P STEAM PENETRATIONS	25	CHARGE FLOOR
11	BOILER LOADING SLOT	26	CHARGE/DISCHARGE MACHINE
12	CHARGE STANDPIPE	27	CHARGE/DISCHARGE MACHINE GANTRY
13	CONTROL STANDPIPE	28	CHARGE HALL CRANE 25TON
14	DEBRIS MORTUARY TUBE	29	B C D ROOM
15	PRESSURE VESSEL STRESSING GALLERIES	30	TRANSFORMERS

FIG. 1.23   OLDBURY POWER STATION MAGNOX REACTOR

the space between the cylindrical core and shield, and the vessel wall. Gas pressure was 27.57 bar, temperatures 247°C/414°C and design output 1180MWe from its two reactors. It came to power early in 1971.

The aforementioned values are design figures. However, evidence became available in the mid 1960s of much higher mild steel oxidation rates in CO_2 than had been expected. This affected mild steel components operating at core outlet gas temperatures and, at these design values, would have reduced the station life. A careful appraisal of all possible options was made, as a result of which it was decided to limit bulk gas outlet temperature on all reactors to 360°C, with the proviso that this could be increased to 380°C for up to one month in every year. Every reactor except Berkeley was affected, and at each other station, parameters were re-optimised within this limitation to give maximum possible output. The reactors which suffered the greater loss of output were Dungeness A and Oldbury, where HP steam from the boilers was used to drive the circulator turbine. There was also a limitation on decreasing the core inlet temperature (in an attempt to restore the temperature rise up the fuel channels) because of stored energy considerations in the graphite moderator. This reduced output has, naturally, affected the lifetime load factors of all stations (except Berkeley) based on design output, although load factors based on declared (de-rated) output have been high.

The Magnox/uranium bar fuel has behaved well. The original target burn-up was a mean channel discharge irradiation of 3000MWd/t, and this has been steadily increased until it is now almost twice that figure (the actual values varying between stations). Magnox cladding failures have been rare. Some stations have adopted finned heat transfer surfaces on the fuel elements of a 'herringbone' form in place of the original polyzonal, helically-finned cans; both types have performed well.

9.2   The Advanced Gas-Cooled Reactor Programme

Construction at Wylfa had started in 1963, at about which time CEGB was giving serious consideration to the type of reactor upon which to base the second nuclear power programme. It was decided to open the competition to water reactors as well as to commercial reactor designs based on the advanced gas-cooled reactor (AGR) concept of which the small unit at Windscale could be regarded as a demonstration reactor. Enquiry specifications were issued in 1964, as a result of which tenders were received from the nuclear consortia for three AGRs, three pressurised water reactors (PWR) and one boiling water reactor (BWR). Assessment covered many facets, such as the state of technical development, safety, economics and development potential. The choice was eventually narrowed to that between a BWR offered by The Nuclear Power Group (TNPG) and AGR tendered by Atomic Power Constructions (APC). At that time, it appeared that the technological comparisons overall gave little advantage to either system. The BWR capital cost was lower, fuel costs were approximately the same, but the AGR was found to benefit markedly from its ability to refuel on load. Because of that, there was no need to shutdown the reactor between biennial overhauls whereas the BWR had to shutdown every year for refuelling. Assuming that refuelling would be made to coincide with major plant overhaul every second year, then there would be about a three weeks loss of generation every alternate year and this affected the economics sufficiently to make the AGR the preferred system. Thus the contract for construction of the first AGR at Dungeness B was placed in 1965 and work on site began in 1966.

Three more AGR contracts each also for a design gross electrical output of 1320MW, were placed fairly quickly after that. Construction began on Hinkley Point B in 1967, on Hartlepool in 1968 and on Heysham 1 in 1970, and South of Scotland Electricity Board (SSEB) had also ordered Hunterston B of a closely similar design to that for Hinkley Point B, later in 1967 (Fig 1.24). These later two stations began commercial operation in 1976, but the others were being commissioned in the power-raising phase at the end of 1983.

The design philosophy for the commercial AGRs was to make use of the technical knowledge obtained from the design and operation of magnox reactors; thus the PCPV concept, with core and once-through boilers inside the vessel cavity, was adopted. Capital costs would be reduced because of a higher core power density (5MWe/tU) compared with 1MWe/tU for Wylfa) and an increased steam cycle efficiency (41.8% compared with Oldbury, the best Magnox, with a design figure of 33.6%) resulting from higher gas temperatures. The mean rating of the fuel in AGRs varies from about 15MW/tU at Hinkley Point B to just over 12MW/tU at Hartlepool/Heysham 1 at full design output. The fuel had to be changed from natural uranium metal to slightly enriched uranium oxide (UO_2) and the cladding from Magnox to a 20%Cr/25%Ni/Nb stainless steel which had to operate at temperatures up to 825C° for the whole fuel dwell time. The cladding had low height ribs to promote heat transfer, and the fuel element consisted of 36 'pins' in cluster form held inside a double graphite sleeve. The pin length was just over 1m, the UO_2 pellets 14.5mm dia, and there were eight elements per channel, except in Dungeness B which had seven. A long tie-bar through the centre of these elements was coupled to the fuel plug unit structure above, by means of which the complete fuel assembly (from standpipe closure to bottom elements retaining feature) could be lifted into the single charge machine which serves both reactors. Thus refuelling was by single channel access, whereas multiple channels were refuelled from a single standpipe on Magnox reactors. These features are common to all AGR designs, fuel elements being identical.

Because of the gas temperature difference from Magnox, the gas flow pattern was different. The graphite temperatures had to be above about 250°C to avoid Wigner energy accumulation and below about 550°C to avoid significant thermal oxidation. The upper limit had to be reduced even further to 450°C-500°C to enable designers to take account of irradiation-induced dimensional changes in graphite. A re-entrant core flow path was adopted in which roughly half the cool gas from the circulators was directed to the top of the core structure, from where it flowed downwards through passages in and between keyed graphite columns before mixing with the other half at entry to the fuel channels. A mild steel gas baffle separated the cool re-entrant gas from the hot gas coming from the fuel channels, steel guide tubes being provided to contain the hot gas when crossing the cool gas plenum above the core. The re-entrant flow features in all AGR designs.

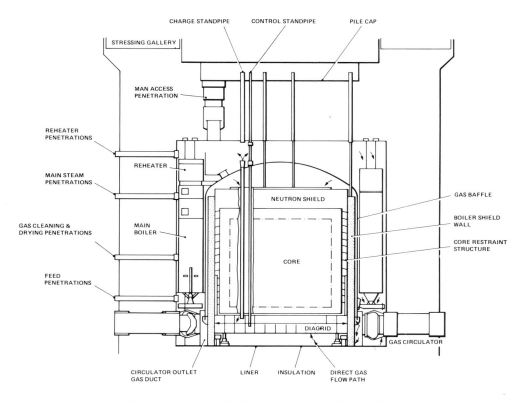

FIG. 1.24   ADVANCED GAS-COOLED REACTOR

Radiolytic oxidation of the graphite was minimised by a careful control of minor constituents in the coolant gas, eg 1.5vol% carbon monoxide, 300vpm methane, 400vpm water. This composition was expected to give full moderator design life of 30 years at 85% load factor without forming any carbonaceous deposits on the pins. Coolant gas compositions close to this are needed on all AGRs.

The boilers were all of once-through design giving superheat and reheat steam conditions compatible with those of fossil-fired 660MW turbine generator sets. The boiler designs differed however, those at Hartlepool and Heysham 1 being helically-coiled in cylindrical casings fitted into eight holes in the PCPV surrounding the core, to give a 'telephone dial' cross-section. Boilers at the other stations were of serpentine design in casings in an annulus around the core, shield and restraint structure and outside the gas baffle. The particular arrangement at Hartlepool and Heysham 1 allowed feed and steam penetrations to be situated in a closure unit at the top of the boiler with the circulator and motor on a vertical shaft below the boiler, thus permitting the vessel to be pre-stressed by circumferential wire winding. These penetrations on the other designs passed through the vessel wall and the circulator/motor shaft was horizontal. Except for Dungeness B, the circulator and motor were encapsulated inside vessel penetrations, so that no rotating shaft had to pass through the penetration closure.

Bulk gas inlet/outlet temperatures varied slightly between designs at roughly 320°C/650°C, so that extensive use of stainless steel was necessary wherever the hot gases impinged. The liner and gas baffle had to be insulated to protect against temperatures excessive for the underlying material and the insulation had to withstand high noise fields produced by the circulators.

9.2.1 *AGR Boilers*

During the 1970s power demand stagnated and no more nuclear power stations were ordered until 1978, when the CEGB and SSEB placed contracts for a further two 660MW AGR stations at Heysham 2 and Torness respectively.

For the boilers at Heysham 2 there has been a continuity of development. While the concept of serpentine monotube boilers at Hinkley Point B has been retained for Heysham 2, the material for the economiser has been changed from carbon steel to 1% Cr to provide greater resistance to the combined effect of erosion and corrosion. An improved transition joint constructed from 9% Cr and type 316 austenitic steel has been developed and the header arrangements have been modified to reduce the penalty of tube isolation following a single tube leak. The decay heat boiler has been increased in size to allow simplification of the shutdown and trip procedures. Fabrication development has resulted in the introduction of automatic butt welding, numerically controlled machine bending of platens and robotic welding of tube spacers.

Unlike fossil-fired boilers, an AGR boiler is largely inaccessible after installation. With this in mind, quality assurance procedures have been developed to obtain the highest integrity and also provide identification for individual material batches within each boiler unit to aid diagnosis of operational problems.

9.3 Safety Assessment and Licensing Procedure

The granting of licences for the construction and operation of nuclear power stations by the Nuclear Installations Inspectorate (NII) requires safety assessment of the reactor in the form of formal documentation as follows:

(1) Design safety criteria are laid down for any new reactor design, including details of external hazards of which account must be taken, permissible radiation doses to the public and to operators, permissible activity release, and the required integrity of components, engineered safeguards and protective devices whose failure could lead to undesirable releases of radioactivity. For well-known reactor types, more detailed criteria are given to help the assessment of the safety of the proposed design.

(2) A preliminary safety report (PSR) is produced which describes the station including outline drawings of the main and auxiliary plant, and diagrams of all important systems. It shows that the proposed design would be capable of meeting the criteria in item (1), provide preliminary safety analyses of critical fault conditions, and a preliminary assesement of the standard intended for the proposed safeguards. Research and development work, as then foreseen, would be listed.

(3) With the acceptance of the PSR by the Licensing Authority, detailed design proceeds on a secure basis. Work is then directed firmly towards the production of a pre-construction safety report (PCSR), which would be adequately detailed, and

not leave until the construction stage any matters of safety importance. Development work which has justified the design is described, also any such work still to be completed (eg long-term metallurgical tests); quality assurance procedures are described to give confidence that the plant will perform as predicted. Sufficient safety-related design will have been done to show that the criteria and standards required can actually be met, and for assessors to be able to reach fairly firm conclusions. Based on this report, the Licensing Authority (NII) is requested to give approval for the commencement of permanent construction at site and the issue of a Site License.

(4) Despite best endeavours, there could be parts of the safety case as presented in the PSR, for which further development or substantiation would be needed during the construction phase. It is now common practice for any safety-related changes which may be necessary because development or substantiation did not accord with expectations, and which therefore required changes to the text of the PCSR, to be communicated to the NII for their information or approval. It has also been found advantageous to send a steady flow of documents sustantiating the safety case eg integrity documents, or stress analyses, to the NII during the construction period, so that they are aware of the way which the detailed safety case is being built.

(5) Towards the end of the construction period, and not less than 6 months before the date at which it would be planned to load fuel into the reactor (taking account of any constructional programme delays which may have occured), a station safety report is produced. This would present an up-dated version of the PSR, taking account of changes that have arisen, making due reference to necessary supporting reports, giving a comprehensive fault and activity release analysis, and giving the evidence upon which the safety claims are based. The NII are asked to use this documentation in support of requests for consent and approval to load fuel.

(6) Commissioning tests of plant items and systems are carried out both before and after fuel loading. Test results are assembled into report form to show that the design intent and requirements had been met, and these reports would be used to obtain consent and approval from NII for power raising, culminating in attaining full power.

Vital to these procedures, and often to variation of operating conditions when the units are commercially operational, is the scope and depth of the safety analyses carried out. Complex 3-dimensional transient codes have been produced, the use of which shows temperature changes over the core volume caused by local reactivity perturbations, and from these computer results can be demonstrated the effectiveness of the protection systems. All credible faults are modelled, mainly one or two-dimensionally, which show that, as long as the reactor protection systems work, the safety criteria are met. The reliability of protection systems and engineered safeguards, particularly those connected with post trip heat removal, is derived quantitavely both with and without plant undergoing maintenance. Detailed reports are produced on the integrity of all safety-related structures, especially all parts of the main pressure envelope and of plant items whose failure might lead to damage to the main pressure circuit. Seismic analysis is performed to show that, under earthquake conditions, the reactor can be safely shutdown and the post trip heat removed to an acceptable reliability. Estimates of radiation doses to operators are made, as well as doses to the public in the event of extreme fault conditions.

9.4 Choice of Future Reactor Systems

During the 1970s, attention was directed towards establishing an additional nuclear power station option in the UK in addition to the gas-cooled reactors. In particular, effort was devoted to the steam generating heavy water reactor (SGHWR), see Fig 1.25, and the pressurised water reactor (PWR), see Fig 1.26. Ultimately in January 1978, the UK Government decided to have carried out all the necessary work to firmly establish the option of the PWR, in addition to giving the go-ahead for the design of two more AGR stations at Heysham 2 and Torness (SSEB).

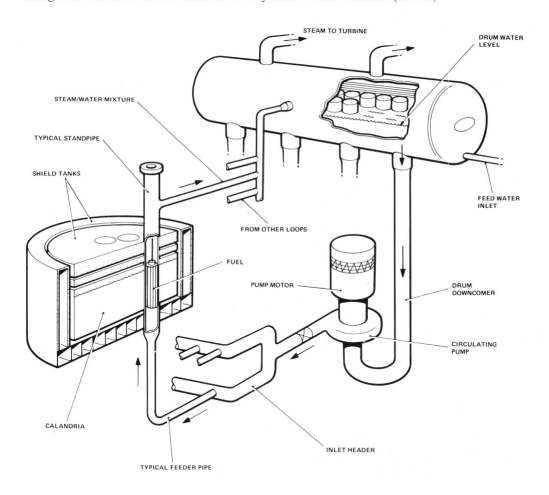

FIG. 1.25   STEAM GENERATING HEAVY WATER REACTOR

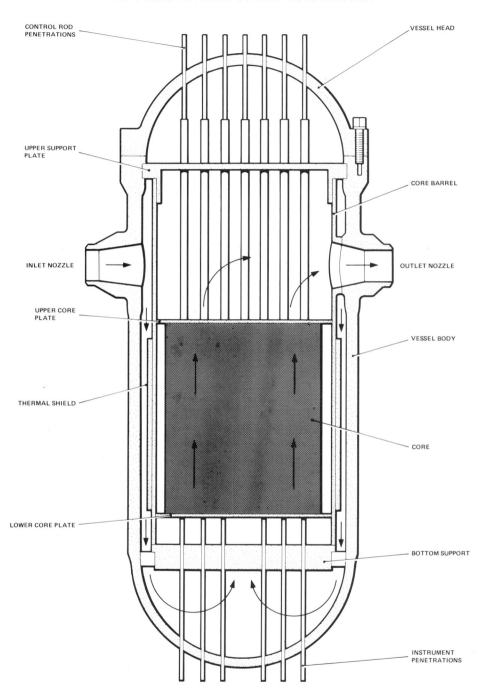

CONTROL ROD
PENETRATIONS

VESSEL HEAD

UPPER SUPPORT
PLATE

CORE BARREL

INLET NOZZLE

OUTLET NOZZLE

UPPER CORE
PLATE

VESSEL BODY

THERMAL SHIELD

CORE

LOWER CORE PLATE

BOTTOM SUPPORT

INSTRUMENT
PENETRATIONS

FIG. 1.26   PRESSURISED WATER REACTOR

Two years later, the Government agreed to the selection of the Westinghouse PWR for development in the UK and in April 1980 the CEGB issued a letter of intent to the NNC authorising the design and, subject to approval, the manufacture of a PWR. On 1 October 1980, CEGB made a public announcement of its proposal to construct the first PWR at Sizewell. Applications for Section 2 Consent was submitted to the Secretary of State for Energy on 30 Jan 1981. Before granting consent, he ordered a Public Enquiry to be held.

10. GAS TURBINES

10.1 Reason for Installation of Emergency Gas Turbines

The employment of gas turbines by the CEGB dates from 1961 when it was recognised that disconnection of a section of the grid could result in a severe lowering of grid voltage and frequency.

In such circumstances, station outputs are progressively reduced through the lowered output of their frequency-conscious auxiliaries, and ultimately cascade tripping of stations occurs. Once disconnected from the grid, the loss of the auxiliaries means that the station cannot build up to power again, even though it is otherwise fully operational.

It was therefore decided to arrange for the auxiliaries of the 500MW stations then being built, to be fed under lowered frequency conditions by an independent supply. Gas turbines, in which the high pressure exhaust from aircraft jet engines is ducted to mechanically separate power turbines, were chosen for this purpose, since no other prime mover was available which was of the right size and able to reach full load within 2 minutes of a start. The power turbines were directly coupled to ac generators whose size was fixed by the auxiliary needs of the station, in the range 17.5MW to 29MW.

Such gas turbines would also provide a capability for starting the station from cold when disconnected from the grid (called a 'black start') as well as adding to its output for meeting peak load demands. Such gas turbine generator units were also of the right size to be installed on nuclear power station sites to provide independent emergency and essential supplies.

In all, 82 emergency and essential supply gas turbines with a total capacity of 1788MW were installed in 23 power stations.

10.2 Power Turbines

Power turbines to drive the ac generators were designed and manufactured by several companies. The greater numbers of machines were to the English Electric/GEC and Rolls Royce designs and both employed a 2-stage design of turbine, the disc being overhung in two journal bearings. These machines were designed for a 30 years life and this necessitated the use of heavy section casings and components which were subjected, during startup, to a near step temperature rise of up to 450°C. In the early machines, certain fabrications such as inlet and exhaust ducts experienced short-term thermal fatigue cracking, and considerable redesign work was necessary to overcome this on later machines.

With the 17.5MW Rolls Royce design, the hot gas from an Olympus gas generator is fed axially into the power turbine and discharges to a stack sideways through a volute exhaust.

With the English Electric/GEC design providing 17.5MW or up to 35MW output, two Avon or Olympus generators are required for higher outputs, each supplying one half of the turbine annulus, the exhaust passing axially to the stack.

10.3   Gas Generators

The supply of hot gas under pressure necessary for the operation of the power turbine was provided by an aircraft jet engine. At the time they were installed there were two British engines of suitable power available:
(1) The Avon capable of producing sufficient gas for 14.5MW of power.
(2) The Olympus capable of producing sufficient gas for 17.5MW of power.

Thus in the double configuration employed by English Electric/GEC up to 35MW could be obtained, whereas the Rolls Royce design of power turbine employing a single Olympus was limited to 17.5MW.

The Avon gas generator was a single shaft machine having a single, 17-stage compressor driven by a 3-stage turbine. Some 79.4kg of air per second was compressed to 9.2 atmospheres, raised to a maximum cycle temperature of 900°C and, after expansion in the compressor drive turbine, was available for use in the power turbine at an inlet pressure of some 2.6 atmospheres and 640°C.

The Olympus was a 2-shaft machine, the compression to nearly 10 atmospheres being performed by two compressors, each being driven at optimum speed by a separate turbine stage. It delivered a gas flow of 104.4kg/s at conditions very similar to those of the Avon.

10.4   Advantages of Aero Engine Based Gas Generators

These were seen as follows:
(1) Immediate availability with most of the development charges met.
(2) Repair by replacement of the gas generator was possible. The relatively light weight of some 2t was also of advantage in this context and the set could be rapidly returned to operation with refurbishment of the gas generator at central stores or works.
(3) Returning gas generators to works gave the prospect of a continuously improved standard of machine through incorporation of the latest modifications.

10.5   Gas Generator Experience

Many design modifications to the gas generators were found to be necessary, over a number of years, in order to achieve a satisfactory overhaul life under the conditions imposed by power station operation as compared with aircraft usage. Amongst these may be cited:
(1) Continuous operation at sea level pressure imposed higher static and dynamic loads on bearings, blades and their fixings, duct assemblies, etc, necessitating strengthened designs. To avoid casing cracking, flexibility had to be built into the mounting points to simulate the compliance of an aircraft wing as compared with

the rigidity of power station foundations. Whereas aero engines can be run in strictly controlled regimes thus avoiding operation in a reasonant speed zone, the CEGB requirements demanded an ability to operate continuously at any power over the working range. For this reason redesign of compressor and turbine blades was necessary.

(2) The use of diesel fuel as opposed to aviation kerosene introduced problems both of fuel management and combustion chamber design. Good filtration was necessary especially on early machines where the fuel was used also as a control medium and the altered volatility, viscosity and flame radiation necessitated much redesign of the flame tubes and methods of location to achieve a satisfactory life.

In the early machines control was through the gas generator-mounted hydraulic fuel control unit, as designed for the aircraft application. In that role it did not need to ensure that the two gas generators could run-up from start to idle in step with one another and in particular with gas delivery pressures in synchonism. However where two gas generators fed a single power turbine as in the English Electric/GEC machines, it was necessary to shutdown the machine if any pressure imbalance occurred, since this could result in rapid fatigue of the power turbine blading.

This was the cause of much starting unreliability on these machines until the development of new systems which replaced the aircraft control unit by a system which scheduled the fuel to maintain the gas generator within required parameters.

(3) The presence of sulphur in the fuel also gave an added complication by combining with sodium chloride either from the air or fuel to form products which attacked the high temperature turbine blades. Good fuel management systems were therefore necessary, together with efficient air filtration. Since also the compressor required cleaning to restore performance, ways of doing this without transferring any salt adhering to the compressor blades had to be evolved.

Development of the gas generators ultimately made them extremely reliable in operation and brought them to the point where an overhaul life of some five to ten years at the anticipated load factor could be expected.

10.6 Peak Load Stations

With the launching of the emergency sets, the CEGB was aware of a need for turbines of larger capacity which could be placed on sites of old, inefficient steam stations and also be used for reinforcement of the system in certain power consuming areas.

The result was stations designed to meet peak load demands (peak lopping) containing sets with outputs of between 55MW and 70MW capacity which could easily be incorporated into the 132kV grid and readily developed from the existing gas generators. Six such stations, each equipped with two sets, were built. Each set was made up with four gas generators which were combined together in various ways to produce the required output. As opposed to the air cooling employed for the emergency machines the ac generators were all hydrogen cooled as was the practice for that size of ac generator at the time.

In accordance with a CEGB need for power factor compensation under certain conditions of system operation, all the machines were able to operate as synchronous compensators, a clutch being interposed between the turbine and ac generator for this purpose. This automatically engaged the turbine when generating and disengaged it from the ac generator when reactive compensation operation was selected.

10.7 Original Peak Load Set Design

The 56MW of the English Electric machine was obtained by placing a double Avon unit of 28MW at each end of the ac generator.

Rolls Royce drove the 70MW ac generator by placing four power turbines of 17.5MW each in tandem with the ac generator shaft. This required that the Olympus gas generators be placed at right angles to this axis with a special design of inlet and exhaust volute.

A total installed capacity of these early designs was 724MW.

10.8 Later Peak Load Set Design

During the 1970s a further construction programme of peak load sets was embarked upon. It consisted of ten 70MW sets installed by GEC Gas Turbines and eight 70MW sets by Rolls Royce; a total of 1260MW. All employed the Olympus as a gas generator.

The GEC sets are similar in layout to the early installations, the 70MW ac generator being driven however by two 35MW turbines (Fig 1.27).

The Rolls Royce design was changed to overcome problems of their earlier machines. The same arrangement of gas generator at right angles to its corresponding power turbine was employed, but these were arranged in pairs on a common shaft to cancel out end thrusts. A new design of inlet and exhaust volute was also employed (Fig 1.28).

In both cases, manufacture of the power turbines themselves was also rationalised. The earlier machines were factory assembled, tested, dismantled, shipped to site and re-assembled at site. However, there was a greater move towards modules which could be works assembled, piped and wired under controlled shop conditions, and delivered to site complete. The sets themselves were shipped on their own baseplates and skidded into position on site. This, coupled with the reduced foundations which are in consequence required, provided a saving of some 14 weeks in construction time.

10.9 Industrial Gas Turbines

Also during the 1970s it was decided to install industrial gas turbines. The incentive for doing so was a practical determination of the comparative economics of these and aero-based units. Such evidence as there was favoured the industrial unit for reasons of longer design life and the absence of any weight limitations, both of which offered the possibility of a design free of the dominating influence of transient and small forces.

There was also the possibility that greater cost savings with scale would be shown by the industrial over the aero-based machine where improved output is obtained through a multiplicity of gas generators.

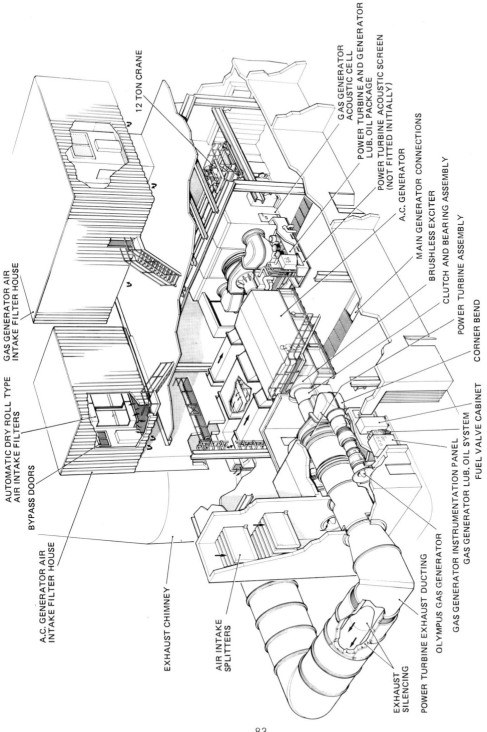

FIG. 1.27   GEC PEAK LOAD SET LAYOUT

12 TON CRANE

GAS GENERATOR
ACOUSTIC CELL

POWER TURBINE AND GENERATOR
LUB. OIL PACKAGE

POWER TURBINE ACOUSTIC SCREEN
(NOT FITTED INITIALLY)

A.C. GENERATOR

MAIN GENERATOR CONNECTIONS

BRUSHLESS EXCITER

CLUTCH AND BEARING ASSEMBLY

POWER TURBINE ASSEMBLY

CORNER BEND

FUEL VALVE CABINET

GAS GENERATOR LUB. OIL SYSTEM

GAS GENERATOR INSTRUMENTATION PANEL

OLYMPUS GAS GENERATOR

POWER TURBINE EXHAUST DUCTING

EXHAUST
SILENCING

AIR INTAKE
SPLITTERS

EXHAUST CHIMNEY

A.C. GENERATOR AIR
INTAKE FILTER HOUSE

BYPASS DOORS

AUTOMATIC DRY ROLL TYPE
AIR INTAKE FILTERS

GAS GENERATOR AIR
INTAKE FILTER HOUSE

83

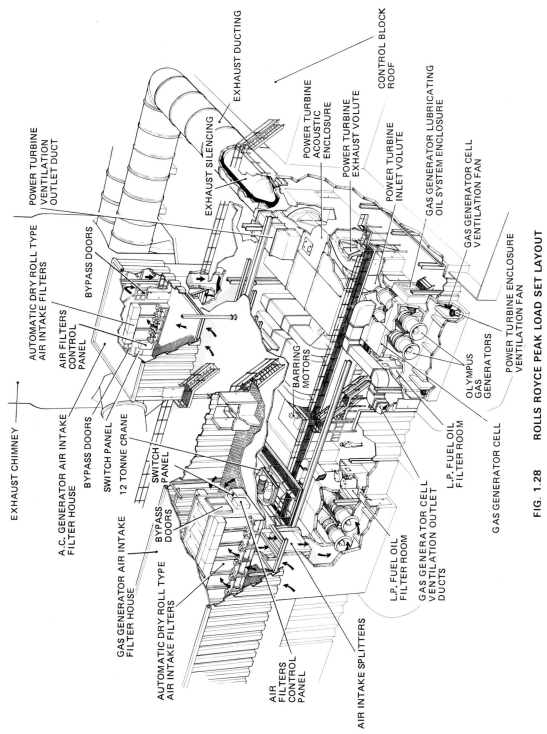

EXHAUST DUCTING

EXHAUST SILENCING

POWER TURBINE VENTILATION OUTLET DUCT

POWER TURBINE ACOUSTIC ENCLOSURE

POWER TURBINE EXHAUST VOLUTE

POWER TURBINE INLET VOLUTE

CONTROL BLOCK ROOF

GAS GENERATOR LUBRICATING OIL SYSTEM ENCLOSURE

GAS GENERATOR CELL VENTILATION FAN

AUTOMATIC DRY ROLL TYPE AIR INTAKE FILTERS

AIR FILTERS CONTROL PANEL

BYPASS DOORS

BARRING MOTORS

POWER TURBINE ENCLOSURE VENTILATION FAN

OLYMPUS GAS GENERATORS

EXHAUST CHIMNEY

A.C. GENERATOR AIR INTAKE FILTER HOUSE

BYPASS DOORS

SWITCH PANEL

12 TONNE CRANE

SWITCH PANEL

BYPASS DOORS

L.P. FUEL OIL FILTER ROOM

GAS GENERATOR CELL

GAS GENERATOR AIR INTAKE FILTER HOUSE

AUTOMATIC DRY ROLL TYPE AIR INTAKE FILTERS

AIR FILTERS CONTROL PANEL

AIR INTAKE SPLITTERS

L.P. FUEL OIL FILTER ROOM

GAS GENERATOR CELL VENTILATION OUTLET DUCTS

FIG. 1.28   ROLLS ROYCE PEAK LOAD SET LAYOUT

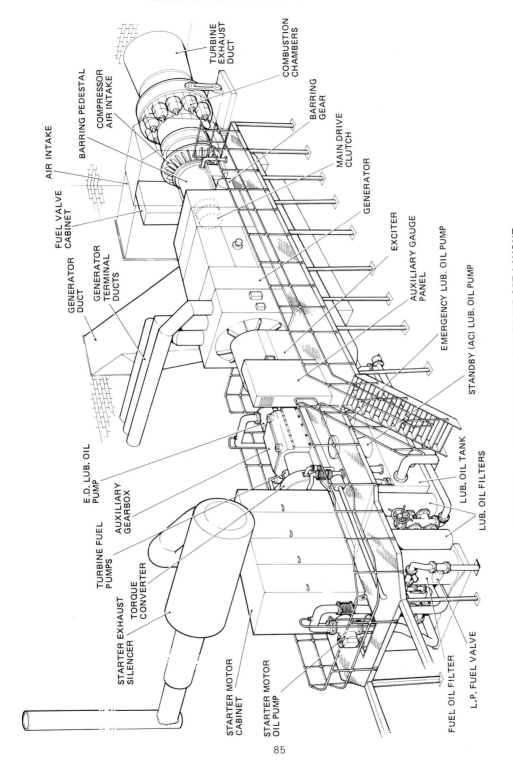

TURBINE EXHAUST DUCT

COMBUSTION CHAMBERS

BARRING PEDESTAL

COMPRESSOR AIR INTAKE

BARRING GEAR

AIR INTAKE

MAIN DRIVE CLUTCH

GENERATOR

FUEL VALVE CABINET

GENERATOR TERMINAL DUCTS

GENERATOR DUCT

EXCITER

AUXILIARY GAUGE PANEL

EMERGENCY LUB. OIL PUMP

STANDBY (AC) LUB. OIL PUMP

LUB. OIL TANK

LUB. OIL FILTERS

E.D. LUB. OIL PUMP

AUXILIARY GEARBOX

TURBINE FUEL PUMPS

TORQUE CONVERTER

STARTER EXHAUST SILENCER

STARTER MOTOR CABINET

STARTER MOTOR OIL PUMP

FUEL OIL FILTER

L.P. FUEL VALVE

FIG. 1.29   INDUSTRIAL GAS TURBINE SET LAYOUT

85

Two prototype GEC sets were installed using EM610B machines each of 51MW/58MW output. Each machine consists of a 13-stage compressor of 8:1 pressure ratio mounted on the same shaft as the 2-stage turbine and supported on two bearings. A thrust bearing is incorporated in the forward pedestal.

The airflow of 290kg/s is raised to a temperature of 900°C in the 10 combustion chambers and, after expanding through the turbine, passes to the stack via the axial exhaust duct and silencer.

The drive is taken from the air inlet end of the machine and connects with the ac generator via a synchronous self shifting (SSS) clutch. Next in the line of shafting is the exciter and finally a starting and auxiliaries pack which includes all lubricating and fuel pumps, filters and the 150kW starting diesel. The general arrangement of the machine is shown in Fig 1.29.

Initial running showed prototype problems which necessitated modifications in some areas including compressor blade fixing, combustion system, bearing operation and off-setting of end thrust. These problems were however overcome and the first set went on to commercial load in 1975.

10.10 Conclusion

The aero-based gas turbine became available to the CEGB at a crucial time, and it could reasonably be stated that the concepts of design and application required an imaginative step to employ them in this way. Whether it was employed in the station emergency role, essential supply at the nuclear stations, or for peak lopping it provided a valuable contribution to system security through its short (2-3 minutes) start time to full load, and to system planning by its short construction time. It finally became a highly developed and reliable machine but most of the plant including the industrial machine now has an extremely low load factor. The high cost of diesel fuel has made their operation uneconomic but it is in any case probable that much of their peak lopping role has been displaced by the Dinorwig pumped storage scheme.

11. MAIN ELECTRICAL POWER SYSTEMS

11.1 Introduction

Power stations incorporate a number of electrical systems to provide the power supplies and control and instrumentation facilities which make possible the safe and efficient operation of the station. These systems comprise a large number of different equipment types and technologies, which have to be interfaced correctly with each other and the plant they power and control. In common with all other plant these systems must meet all the station and system operating needs with the necessary levels of reliability. They must ensure personnel and plant safety and make possible the orderly and successful response of operators to all normal and abnormal plant states, with optimum plant availability and efficiency. They must in themselves represent a minimum cost of ownership, taking account of first (initial) cost, maintenance and replacement cost, and at the same time they must be adaptable to changing station and system operating needs.

The systems involved comprise those for 11kV, 3.3kV and 415V ac power, generator/generator transformer primary connections, dc power, guaranteed instrument supply, communications, modulating control, sequence startup and shutdown, remote manual control, data and alarm acquisition and display, plant operational and maintenance interlocking, plant protection, and the complete operator interface with all plant systems (including central and emergency control rooms, their desks and panels).

11.1.1 *Station Requirements*

The starting point for the design of such systems must be the identification of all detailed requirements for each system. These relate to both the operation and control of the station plant and the functioning of the grid system. They include special issues such as operator work load, plant protection and environmental factors, eg chimney emissions and nuclear hazard. The requirements are initially derived from the station design particulars as well as CEGB operational safety procedures and policies. An extensive list of quantified requirements is established at the outset of station design.

The station electrical systems must allow the station to meet the following principal requirements:

(1) Maintain rated output over a stated frequency range and operate continuously at any lower rating, usually down to half load.
(2) Meet a defined rate of load change and withstand specified step increases in demand as well as sudden load loss.
(3) Be suitable for 2-shift operation.
(4) The unit systems must remain stable in a specified fashion under a variety of system and station fault situations.
(5) Power and control systems must have defined levels of redundancy and fault tolerance and meet a variety of abnormal situations.

11.2 Electrical System Design

The selection of the best system design for providing power to auxiliary plant must first consider the cost of all schemes meeting the operational criteria. Usually there will be several alternatives for which initial cost differences are small. Other considerations then become important, such as operational convenience, conservative plant rating, likely failure modes and consequences, the use of proven plant as far as possible and the lifetime cost of ownership. The latter is heavily dependent on the frequency and convenience of maintenance and the ease of plant replacement and modification.

The different types of power station designed during the past 10 years have brought a corresponding variation in the electrical systems needs. Site-related factors such as system connections have also influenced auxiliary system designs.

The systems selected for the three units at the oil-fired Littlebrook D Station followed the established unit and station transformer scheme after considering four possible alternatives. The economics of the choice were dictated by the existence of a 132kV substation on the site, which was used to derive the station and unit startup supplies. Had there not been a 132kV substation on the site, or if its life was likely to be limited by system considerations, the station transformer scheme adopted would have been economically less attractive.

Dinorwig, being a pumped storage scheme, had a fundamental requirement that the main machines must function as both generators and pump-motors. This required phase-reversal facilities which could be arranged, most economically, at the machine terminals. Therefore, the electrical system incorporated generator voltage switches, suitable for an operational voltage of 18kV and a rated current of 11.5kA. A starting equipment busbar system for starting the machines as pump-motors had to be provided for use by all six units. The station supplies were derived from the generator side of the generator-motor transformers, since there was no 132kV supply available and no economic justification for providing this.

The coal-fired Drax Completion project broadly followed the initial outline design proposed at the time of building the first three units. Four 132kV/11kV station transformers were installed, with the First Half of the station to cater for the eventual total of six units. By changes to the 11kV interconnections and updated equipment, the electrical system was designed to meet a change of operational emphasis from the original base load station intent, to a 2-shifting regime likely to be required in the early life of the new plant. The rather complicated and extensive system which resulted, brought with it the need for operational flexibility, but with safeguards. To meet this, a microprocessor-based interlock system was introduced on the 11kV system to prevent permissible fault levels being exceeded. It included facilities for the control room staff to simulate system configurations before attempting switching. This appeared to be a cheaper and more acceptable alternative to the provision of additional system-fed transformers and busbars which would have offered simpler operating regimes.

The Heysham 2 AGR project, as all nuclear stations, required an electrical system to meet safety considerations, especially post trip, as well as normal operation. The electrical system had to achieve the same level of integrity as the 4-quadrant reactor plant concept. This was achieved at 11kV by providing each quadrant with a dedicated switchboard which supplied two duplicated, independent and segregated supply trains (called X and Y) at 3.3kV and 415V. The 11kV supplies were derived from the 400kV busbars via the generator and unit transformers, and from the 132kV busbars via the station transformers. Generator voltage switches, justified on safety grounds, were used in this station, as they enable 11kV supplies to be provided via the generator transformer when the main units are not in service. To meet the safety considerations, the X and Y systems incorporate eight diesel generators connected at the 3.3kV level and uninterruptible power supplies derived from battery-backed inverters.

The electrical systems being proposed for the Sizewell B PWR project, meet the power requirements of one reactor and two turbine generator units. The requirement for secure supplies to the reactor cooling pumps is met by operating the 11kV unit and station boards in parallel, via an inductor, which solves the conflict between the competing requirements of supply security and system fault level. The generator and station transformers are all connected at the 400kV level.

The increasing need for high security and quality of the electrical supplies to station control equipment and computers, has been met by the development of inverter systems ranging from 10kVA to 50kVA single-phase and up to 100kVA 3-phase. The systems are battery-backed and used in conjunction with main supplies via static switches, capable of changeover within 10ms. These modern designs have been developed to overcome a range of problems identified from unsatisfactory experience in the past, with such systems using both motor-generator sets and inverters.

Looking to the future, the development of electrical power systems will have regard to advancing technology in many equipment areas, drawing heavily on microprocessor technology and incorporating suitable developments, including vacuum switchgear, static protection and new cable types.

11.2.1 *Electrical System Analysis*

Before the choice of the optimum design of electrical system can be confirmed, each of the envisaged alternatives is subjected to detailed analysis of its performance and reliability.

System analysis was formerly confined to steady state performance calculations, supported by limited dynamic studies using batch system computer programs.

However, fully interactive computer analysis programs are now available in GDCD which provide the capability to analyse all aspects of system dynamic performance. Batch and interactive computer programs are now available for evaluation of the reliability of electrical power systems.

The computer programs have been validated by calculation and, where appropriate, field tests are used to model all large integrated power system networks and also isolated systems supplied from gas turbines and diesel generators. The models include the representation of prime movers, governors and excitation systems.

These interactive system analysis programs allow studies to be carried out on all aspects of system performance including load flow, both steady and transient, symmetrical and asymmetrical faults, dynamic performance of motors, prime movers, governors and automatic voltage regulators, motor starting stability, slow and fast transients, quick switching, reliability and protection.

The design of electrical systems also involves the design of the necessary protection schemes for all the station electrical systems from the station grid terminals. These include the unit overall protection, auxiliary generator protection and all busbar, transformer and motor protection schemes. Computer programs are also used in this work and in the determination of relay and fuse settings.

11.2.2 *Electrical Plant*

The design of successful electrical systems must include the specification and selection of suitable plant of various types to equip the systems. The first objective will always be, as far as possible, to use equipment of proven performance and reliability. However system needs change and operational experience often requires specifications to be varied while account must be taken of technological advances.

The CEGB always attempts to buy plant to national standards covering all aspects of design, manufacture and testing. However, these standards have to be augmented by features special to the CEGB, relating to the very high levels of reliability required, as well as the implications of the CEGB operating and safety rules. To illustrate the impact of this approach, some examples of special design features of electrical plant for power stations are reviewed in the following sub-sections.

Transformers

The generator transformer is a vital element of the generating unit whose unreliability will cause loss of generation for the several weeks it takes to replace a failed transformer. The service history of these transformers was unacceptable in the

1960s, with a number of failures and consequently heavy operating loss. The designs for the 800MVA units were examined and detailed designs, which had sound test and service experience supporting them, were agreed with the makers and registered. Only transformers to these registered designs are now purchased and no changes to these designs are accepted, unless consequent higher reliability can be demonstrated or changes are unavoidable. Several other features of transformer design have contributed significantly to higher system reliability and performance. The elimination of water cooling as far as possible (Dinorwig being a notable exception) is seen as an important step in improved reliability following several serious failures due to water leakage into the transformer.

Fig 1.30 shows a 23.5kV/432kV 800MVA generator transformer installed at Drax power station. It consists of three single-phase units inside a noise enclosure, with air/oil coolers on the right hand side and air insulated 23.5kV phase isolated delta connections above the noise enclosure.

Recent trends in auxiliary plant design have led to the almost exclusive use of direct on-line starter motors even for the largest drives, eg feed pumps. This has required the 11kV and lower voltage systems to be stiffened, ie designed with as low an impedance as permitted by fault level considerations. Stability margins for such systems can be improved if their voltage can be automatically increased with load increase and this has led to the use of on-load tap changers on some auxiliary transformers.

The need to eliminate oil as a fire risk has prompted the development of air insulated transformers of adequate reliability for 3.3kV and 415V systems.

High impedance generator neutral earthing via transformers requires oil-less transformers to limit fire risks if they are to be located in the turbine hall. Cast resin transformers have been developed because air insulated transformers need continuous energisation to maintain winding insulation.

Generator Main Connections

The heavy current busbar connections between the generator and transformer have been the subject of considerable investigation and development. A programme of full-scale short circuit testing has led to the evaluation of connections able to withstand the high short circuit forces and thermal effects. To assist access while observing the CEGB safety rules, special 3-phase motor driven earthing switches have been developed and extensive electrical and coded key interlocking has been designed. Dinorwig, with its extremely complicated busbar system, provided special problems in this area. Extensive computer analysis of all operating and maintenance states and procedures was employed to validate the interlocking system designed for the Dinorwig main connections.

Switchgear

In the switchgear field, while the well-proven air break designs are still used for all auxiliary voltages, various detailed modifications have been introduced to meet service experience. However considerable development work is proceeding on alternative designs, since air break technology does not match up to modern station operating needs, in particular, in the matter of frequency of operation and consequent wear of operating mechanisms. The increasing flexibility of operating regimes is subjecting switchgear to many more operations than present designs were developed for and frequency of maintenance has increased. The vacuum switch offers the prospect of many more operations between overhauls due to the much smaller operating forces required from its mechanism. It is envisaged that once vacuum bottles are available covering the complete range of requirements, switchgear of this type will progressively replace air break designs.

Fig 1.31 shows a switchboard of air circuit breakers used to control 11kV auxiliary plant, with one circuit breaker withdrawn for inspection.

The most significant development in the area of switchgear is the increasing use of generator circuit breakers. Wherever feasible, station transformers are connected to the 132kV system. Some sites provide only 275kV and 400kV systems, and the high cost of transformers and switchgear for these systems encouraged consideration of arrangement, which could reduce the number of these plant items necessary. One approach is to abandon conventional station transformers, and to allocate the duty to the unit/generator transformer combination associated with each main generator. For this concept, the generator switch must be connected at a point between the generator terminals and the unit transformer tee-off; one advantage is that the electrical supplies to essential auxiliaries would be more reliable. The switch performance requirement is governed by the operational requirements of the station; it should have the ability to handle the synchronising of the generator and the interruption of full-load current. However, the use of a fault-rated circuit breaker is preferred, in order to permit clearance of generator faults without disturbance of power supplies to the essential auxiliaries.

The Hartlepool and Heysham 1 nuclear power stations provided the first applications, within the CEGB, of the concept of switching large generators at generator voltage. The philosophy was later adopted for the pumped storage project at Dinorwig where, as previously mentioned, the switching of the generator at its terminals in conjunction with isolators is the chosen method of phase reversal. The generator switches at Dinorwig are also an essential part of both the auxiliary supply and pump starting system.

The switchgear at Hartlepool, Heysham 1 and Dinorwig is of the air blast type, designed for connection directly into the phase isolated generator main connections, and has a fault-interrupting capability at each station. Each assembly comprises three single-phase units, preserving the phase isolation of the connections, the three phases being controlled pneumatically to operate in unison. The main current carrying parts of the designs at Hartlepool and Heysham 1 are water cooled, the water circulating in a closed circuit through a water cooled heat exchanger. The switchgear at Dinorwig is similar in design, but has a lower current rating and therefore forced air cooling, as opposed to water cooling, is possible.

Fig 1.32 shows one pole of a 3-phase generator circuit breaker and its connection into the phase-isolated 18kV generator main connection at Dinorwig.

At Heysham 2, the generator voltage circuit breakers are water cooled and of similar design to the stations already mentioned.

The electrical system design for Sizewell B is currently based on the use of generator voltage switchgear, with the consequent improvement in the reliability of 400kV supergrid supplies to the power station auxiliary system.

FIG. 1.30   GENERATOR TRANSFORMER

FIG. 1.31   11kV AIR CIRCUIT BREAKER SWITCHBOARD

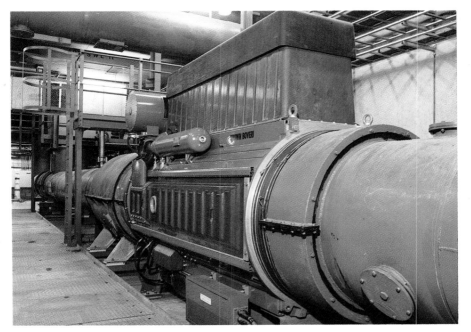

FIG. 1.32   GENERATOR CIRCUIT BREAKER AND MAIN CONNECTIONS

FIG. 1.33   SWITCHGEAR TERMINAL BOX

Cables

Power station cabling has developed considerably, resulting from service and installation experience. In particular the problem of the speed of fire spread by cables has been tackled vigorously.

Specifications have been written for testing and approval of a range of cables and accessories suitable for use in modern power stations, covering control and instrumentation, multipair and multicore cables and power cables up to 11kV. For the latter, solid extruded insulation has replaced paper insulated metal sheathed cables to meet fire retardance requirements. They are lighter, easier to install and terminate and eliminate compound filled boxes. Cables system design has to recognise both technical and economic requirements. Typical examples of economic factors are the selection of aluminium conductors for large power cables and the adoption of a rationalised range of cables to improve economics of supply. Until now, PVC has been the basic insulating material for use on voltages up to 3.3kV and for sheathing all cables. It has the advantage of being readily treated to give reduced propagation to fire, whilst being a very practical and economic material for manufacture of cables. A disadvantage of PVC is that it produces large volumes of black, toxic, corrosive smoke when it burns. However, advances in cable insulating compound material have been so rapid over the past few years that PVC cables are likely to be replaced by cables using polymeric material, which has low smoke, acid and toxic emissions when it burns and is fire retardent. This development will free station layouts from the need to provide dedicated and enclosed areas for cable installation and will minimise the risks which have occurred, to plant and personnel, during several serious power station fires involving cable systems.

Fig 1.33 shows the installation of 11kV reduced propagation cables with polymeric insulation and PVC oversheath in a switchgear terminal box. Air clearances between the cables eliminate the need for compound filling. The termination comprises insulated gland, cable screen stress relief core and conductor compression lug.

Computers are used extensively in both the design and installation of power station cable systems. The work ranges from the use of computer programmes for cable sizing and current balance in circuits employing several single core cables per phase, to the design of cable systems for control and instrumentation signals. One computer programme used is named CADMEC which initially supports the cable system design by handling items such as cable sizing, tray allocation, cable routing and signal core location, and is then used to monitor progress throughout the installation of the cables.

Instrument Supplies

Some of the instrument and control systems within a power station are essential for safe plant operation and shutdown and require uninterruptable or 'no-break' power supplies. Various battery-backed systems employing both motor-generator sets and inverters were used in the past, but operating experience was poor. Therefore a programme of investigation and research has been carried out to identify and solve all the problems with such systems. The problems which affect the design of such systems include the nature of the load (eg computers, which produce harmonic distortion up to 50%), transformer in-rush currents, high temperatures affecting component life, inadequate specification, selection and testing of electronic components and restraints on battery performance.

A detailed system specification was developed to deal with these problems once their sources had been identified, and satisfactory systems have been manufactured to this specification and installed at new stations following Littlebrook D. Also, existing stations are being refitted with systems to meet the new specification. The technical solutions to the problems include the specification of low impedance, low in-rush current transformers and the development of special point-on wave static switches.

11.3 Summary

From the foregoing, it is clear that most electrical equipment used within power stations for particular requirements has to be specified in detail. Therefore, ESI and CEGB Standards and Specifications are produced which identify all requirements additional to national standards. The ability of products to meet these requirements is assessed and acceptable products are approved.

The special requirements of power station equipment led to particular test regimes being developed. In the case of Heysham 2, the requirement was to demonstrate the ability of the reactor shutdown system to operate successfully after an earthquake. The electrical plant constitutes a major proportion of the shutdown system and has therefore been the subject of extensive seismic qualification analysis and testing.

A similar requirement applies to Sizewell B but in this case there is an additional need to demonstrate that certain electrical plant can withstand the effects of a loss of coolant accident (LOCA) or main steam line break (MSLB) and then continue to operate until the reactor reaches a safe condition. All of these requirements have had a very significant effect on equipment specifications and have required extensive development and testing programmes.

12. CONTROL AND INSTRUMENTATION

12.1 Introduction

Control and instrumentation (C&I) has a major part to play in the operation of power stations and it affects their design. Efficient C&I systems reduce the capital cost of the plant by reducing the margins that have to be allowed for uncertainties and hence reduce operating costs by:
(1) Reducing plant damage by better control.
(2) Increasing efficiency by identifying losses and controlling them.
(3) Increasing availability by reducing spurious trips and permitting faster return to power by improved operator displays and post incident diagnosis.

During the 1960s, C & I systems design and associated hardware developed rapidly and new technology was applied in a controlled manner to reap the benefits of these developments, without taking risks that could prejudice construction programmes or early operation. Improvements in control equipment enabled the range of automatic control to be extended from the 100% to 70% load range in the early 1970s, to a range from 100% to 50%. Simulation has been employed in the design stage to represent the plant and test the control hardware before and during commissioning. Control loops are often brought into service from the first time the plant is synchronised. Generating plant, particularly prototype, does not always perform as the supplier expects and therefore control loops are modified as necessary.

All these requirements call for great sophistication in the C&I equipment. Computers have been increasingly employed to provide the facilities required, economically and with the flexibility to adapt to changed operational requirements. The design of systems has to take into account many factors, eg safety, ergonomics, equipment reliability and operating procedures. Full advantage is taken of research and development, both in the CEGB and elsewhere in the UK and overseas.

12.2 Basic Requirements of Control and Instrumentation Systems

12.2.1 *Closed Loop Control*

The control systems enable the plant to meet the demands of the national grid network whilst having the ability to operate efficiently and to meet cumulative plant damage and stack emission requirements. The control systems for a typical large coal-fired station are indicated in Fig 1.34 and comprise:

(1) Unit load control system that accepts a megawatt demand from grid control, via the operator, and automatically regulates the generated load to meet the demand. This incorporates control of the pressure of the steam delivered to the turbine.

(2) Control loops associated with each mill to regulate the production and flow of pulverised fuel to meet the load demand.

(3) A control loop which regulates the flow of combustion air to match the flow of fuel.

(4) A control loop which regulates the induced draught fans to control the gas pressure in the combustion chamber.

(5) A control loop which regulates the flow of feedwater to maintain a desired level of water in the steam separator drum.

(6) Control loops which regulate the flow of attemperator spray water to control the temperature of the steam delivered to the turbine.

On earlier stations these control loops were implemented using electronic analogue controllers, but on the Drax Completion project direct digital control was employed for the reasons given in Section 12.3.

The control loops are allocated to control centres employing microprocessors and analogue and digital input and output equipment. The distribution of control loops is chosen on the basis of the effect of failures and the extent to which manual control can be maintained. Data and fault diagnosis information is passed from the control centres to the central processor by data links which are also used for down-line loading when first setting up the equipment on reinstatement after maintenance.

12.2.2 *Manual Control and Direct Indications*

Direct manual control of plant actuators is provided to allow regulation of the plant in the event of the automatic control not being available, eg because of equipment failure. These controls are located in the unit control desk and panels. At Drax Completion the desk design uses a modular type system which embodies the indications and facia alarms associated with these controls.

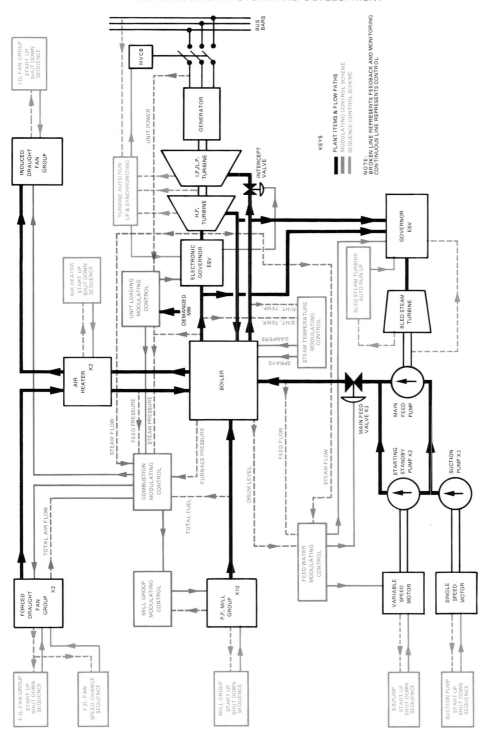

FIG. 1.34   BASIC CONTROL SYSTEM FOR TYPICAL LARGE COAL-FIRED STATION

12.2.3 *Monitoring and Computer-Based Displays*

The large amount of data made available to the operator can most effectively be displayed by the use of computer-driven colour visual display units (VDUs). Typically, five of these are fitted in the unit desk for data and alarms and display selection; alarm handling is at keyboards. See Fig 1.35.

A computer system scans the analogue and digital inputs. Some of the data collection points are mounted near to the plant, thus economising on cabling. After storage and processing, the information is displayed on the VDU. The analogue values are compared with alarm limits and, if they pass outside these limits, alarms are initiated. The system provides permanent records in the form of printed logs for subsequent analysis.

12.2.4 *Sequence Control*

In order to reduce the load on the operator, speed plant startup and make it more repeatable and reliable, a system of sequence control is applied to all main rotating plant, ash/dust plant, chemical monitoring and cooling water.

On earlier stations these sequence control systems were implemented using electromechanical relay or semiconductor logic, but for Drax Completion, computer-based systems have been installed. The complete sequence control system is implemented in seven control centres, each having its own microprocessor and digital input and output equipment. This provides signals to indicate, on the central control room unit control desk, the sequence progress and draws the attention of the operator if it is held up.

The systems use the CUTLASS sequence language in which the logic is defined in algorithm state machine notation with a system of automatic documentation. See Section 12.4.

12.3 **System Design**

Control systems initially were designed and supplied by plant contractors. The plant characteristics were not always fully understood and the C&I equipment available had limited performance and reliability. Problems arose particularly at the interface between plant supplied by different contractors.

As a result, many systems were modified and commissioned by CEGB staff and sometimes subjected to early partial or extensive refit. This unsatisfactory situation led to the current approach in which the C&I systems are designed by the CEGB. Furthermore, the new software-based technologies make it easier to support the systems over station life, including modification as a result of obsolete equipment or changed operating needs. The ability to make changes easily is made more important by the increasing need to operate the newer large units in a 2-shifting or more flexible mode, with wide and frequent variation in output. During such variations in output there is a risk of steam temperature excursions that reduce the life of the pressure parts and poor combustion control, giving rise to acid corrosion in oil-fired plant and unsatisfactory chimney emissions. These conditions demand more flexible and wide ranging automatic control systems and better plant monitoring.

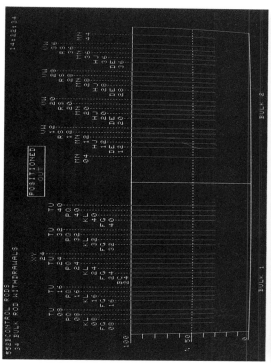
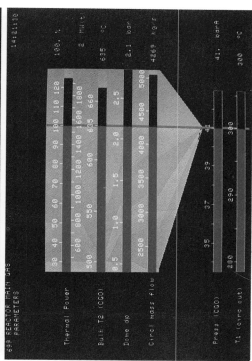
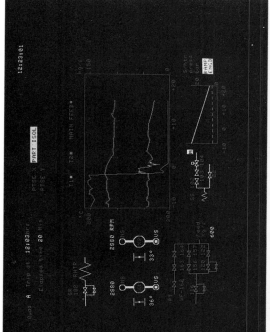
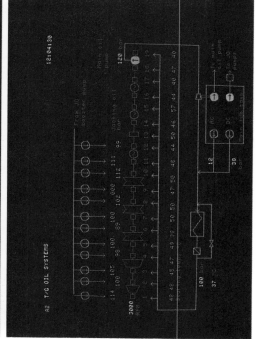

FIG. 1.35   EXAMPLES OF VDU DATA DISPLAYS

The approach adopted on stations designed during the period 1974 to 1984 emphasised the need for 'front end design' and co-operation with the main plant contractors to establish the control concepts in detail and develop more advanced control strategies. In parallel, and with the collaboration of suppliers of control hardware, CEGB establishes performance specifications, type tests for equipment and gives formal approval to ranges of modular hardware that can be configured to form a complete control system.

This policy resulted in the control systems for Grain and Littlebrook D being commissioned in record time with systems operational over a wide power range.

Although the analogue controllers at Littlebrook D have proved very satisfactory, direct digital control (DDC) offers significant advantages as follows:

(1) Greater flexibility, to meet the different plant requirements at different parts of the load range and the need for modification when the plant characteristics differed from those for which the system was developed, or new operating modes were required.

(2) Greater development potential in implementing advanced control strategies, particularly in respect of problems during power changes associated with the 2-shifting arrangements.

(3) More effective handling of the large number of digital inputs and outputs associated with the operator's controls and indications.

(4) A reduction in cost and maintenance and the prospect of improved reliability.

In view of these advantages there has been a change from analogue control to DDC techniques. The computer hardware can be procured from the computer industry on a 'hardware only' contract basis but the software requires a different approach.

12.4 Software

New projects and stations requiring refurbishment involve DDC and on-line data processing systems which needs many hundreds of man years work on software production if each project is dealt with on an individual basis. This has been significantly reduced by a CEGB policy of standardisation and re-usability of software. No suitable proprietary software system existed at the time the policy was adopted and agreement was obtained to develop a new software system specifically intended for power station use called CUTLASS (computer users technical language and software system) to be used on all appropriate applications.

The development requirements and software to meet these are now issued and supported to cover general purpose, DDC, sequence control and display applications. The DDC and sequence control workshops have enabled users to exchange views and establish good practice in CUTLASS programming.

CUTLASS has so far been used for Drax Completion, Heysham 2 and in over 50 refurbishment projects.

The CUTLASS policy enables application software to be produced by engineers who are familiar with station plant and its operation, but with a minimal knowledge of computing. These engineers can take the software production from the initial system design through writing and testing of the code installation at site and final documentation.

The trend towards software-based control systems includes high integrity systems used for interlocks in nuclear station fuel route machinery and for post trip cooling systems.

While the failure modes and reliability of hardware is relatively amenable to analysis, the use of software raises new problems. Therefore, considerable effort is devoted to establishing good practice in the writing of software for high integrity applications, particularly to meet high levels of quality assurance. Rigorous testing and validation is required using special techniques, including diverse software and hardware to avoid common mode faults.

12.5 Hardware

The CEGB uses equipment selected from the best of the proprietary hardware that has been evaluated and preferably type tested and approved for both new stations and refits. This policy increases confidence in the adequacy of the performance of the equipment, avoids development costs and minimises the variety of equipment and associated documentation costs, station spares holding and staff training.

The evaluation very often reveals deficiencies in the equipment, whether UK or foreign made, and these are corrected before an approval is issued, so raising the standard of equipment available. The results of evaluation and approval are documented as formal approvals that are distributed throughout CEGB and among contractors. Originally the assessed and approved equipment ranged from small electrical and electronic devices to large electromechanical actuators. This process is being extended to computer-based hardware that is compatible with the use of CUTLASS.

12.6 The Operator Interface

The design of control rooms and the layout of control desks are established by analysis of the requirements of plant operation and of the operator's task in relation to the operating procedures. Use is made of full scale static mock-ups of the desk, computer-aided and simple simulations of the computer-based displays and alarm systems. (Fig 1.36). A particularly important issue is the integration of computer-based and conventional alarm systems with other displays and controls. Extensive research in this area is being carried out.

Design documents for VDU displays including colour coding have been produced for the guidance of engineers designing and coding data and alarm displays. The controls and indicators required on control desks have been analysed and rationalised to a set of standard modules. These modules are fitted in a matrix and their type and position can be decided at a relatively late stage when front end design is well advanced.

This arrangement enables a formalised approach to be adopted, with computerised handling of the associated data to check the proper association of controls and indications into functional groups.

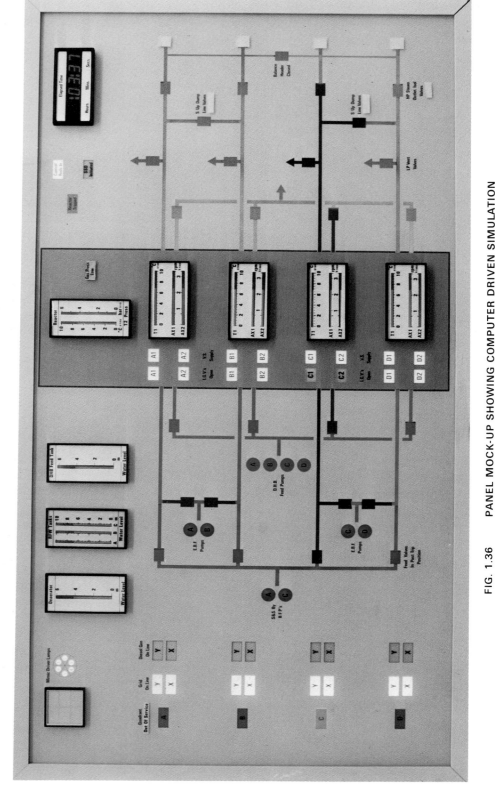

FIG. 1.36   PANEL MOCK-UP SHOWING COMPUTER DRIVEN SIMULATION

Computer-aided design (CAD) techniques are used in conjunction with the mock-up to experiment with different control room layouts and module locations, with a view to optimising their position and producing the desk drawings and associated schedules. Fig 1.37 shows an example of an AGR station control room layout produced by the CAD technique.

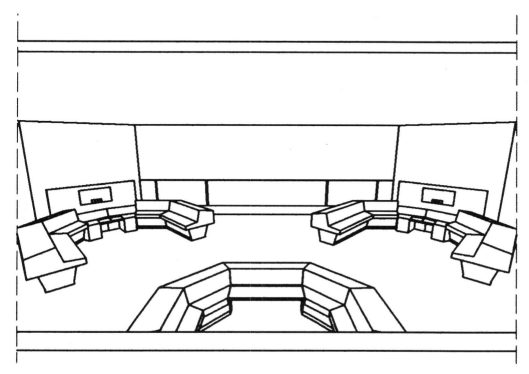

FIG. 1.37   EXAMPLE OF CAD PRODUCED DRAWING SHOWING AGR STATION CONTROL ROOM LAYOUT

Standards have been established for the lighting and air conditioning requirements of control rooms.

Ergonomics and human factors generally are important in many aspects in power station design and operation. Special attention has been given to these matters in the design of control rooms, desks and panels, particularly since the events of the US Three Mile Island Nuclear Power Station incident.

Although a policy is followed of using British and International Standards, these are often found to be insufficiently explicit. It is necessary to write and issue Standards and Specifications to reflect the precise and special requirements of instrumentation and electronic equipment in the power station environment.

CEGB engineers have co-operated closely with the control and instrumentation industries trade association to write and issue Electrical Supply Industry and CEGB Specifications. Modifications to existing equipment on operating nuclear stations are strictly controlled by the plant modification procedure.

13. COOLING WATER SYSTEMS

13.1 Introduction

This major system, supplying cooling water for the turbine condenser, requires consideration at a very early stage in the planning studies. In the case of a direct cooling system, it is necessary to decide on acceptable locations for offshore works, to consider any marine life environmental problems and to carry out negotiations with Local Authorities and other bodies concerned with offshore interests such as the Water Authorities.

Environmental issues which have come to the fore mostly concern the effect on marine life of offshore warm water discharges and the draw-in of massive quantities of water. Much research work on this has been implemented by CEGB over the last 12 years. Together with information pooled from statutory bodies such as the Ministry of Agriculture and Fisheries, this is now beginning to establish some guidelines for providing some protection for the offshore marine life.

A modern steam turbine of 660MW capacity requires up to $24m^3/s$ of water to cool the condensers, and up to six of these turbines might be used in a modern station, such as Drax. The cooling water flow required is greater than most inland sources can supply and often leads to the use of indirect, tower cooled designs. Where the water quantity is sufficient, in estuarine and coastal locations, extensive civil engineering works are required to minimise recirculation and protect the environment.

This Section is mainly concerned with the design and construction of direct cooled systems which do not require the assistance of cooling towers, but the implications of the latter on station design are briefly mentioned in Section 13.12.

13.2 The System

A basic 4-pump and debris drum screen cooling water system for a 2 to 4-unit fossil-fired power station is shown in Fig 1.38. The condenser cooling water arrangements for AGR and PWR nuclear stations are similar, but with the additional facility of cooling water for reactor systems, as described in Section 13.11.

Table 1.1 shows typical cooling water quantities and overall temperature increase of the water on its progress through the system.

TABLE 1.1. COOLING WATER FLOW AND TEMPERATURE INCREASE PER UNIT

Type of Station	Unit Size (MW)	Condenser Water m^3/s	Station Services m^3/s	Reactor Auxiliaries m^3/s	Temperature Increase °C
Fossil	660	15–17	1.5–1.7	—	10–12
AGR	660	15–17	1.5–1.7	1.5	10–12
PWR	660	24	2.0	1.8	11–13

The arrangement in Fig 1.38 shows a unit flow path for each screen and pump without the elaborate interconnection at the pump suction chamber which was a feature of older stations. This unitisation has been made possible by more reliable screens.

The system comprising four condenser cooling water pumps has been established from a series of reliability studies. These show the frequency of pump and screen breakdowns, the deterioration of the system flow following loss of a pump, the provision of interconnection on the pump discharge pipework and the selection of the seasonal water inlet temperature used for the optimisation of the condenser cooling system and water flow. This process, which is briefly described in Section 13.5, provides adequate cooling capacity over a large period of the year with three pumps in service whilst the fourth pump caters for unscheduled pump outages.

The arrangement of valves in Fig 1.38 shows the provision for a 3-unit station, such as Littlebrook D. The pump discharge valves are the quick-closing butterfly type, having adjustable closure time features to minimise pressure surges following pump trips and to reduce the amount of water backflow when one pump is lost with others still in operation. The valves on the interconnecting manifold are of the standard slow-closure butterfly design. The valves on the interconnecting manifold have been arranged to enable one condenser culvert to be taken out of commission along with the associated condenser and retain the other units in service. This arrangement also enables maintenance to be carried out on one interconnecting valve with the loss of only one turbine generator unit.

A single intake and discharge tunnel for the station is shown in Fig 1.38. The choice of single or twin tunnels from the standpoint of availability has been examined in detail from a survey of existing stations. It is concluded that a single flow path gives satisfactory availability and the choice is accordingly taken after considering the arrangement of offshore structures, the type of construction, depth of the water, geology and any future development of the site.

Water supplies for the auxiliary turbine plant cooling can utilise the pressure drop across the condenser. However, an alternative is to use a separate auxiliary supply pump. This has the advantage of providing auxiliary cooling water to a number of heat exchangers when the main cooling pumps are not needed. This scheme also gives an improved syphonic head and freedom from the loss of system priming which is a problem with large vertical auxiliary heat exchangers.

13.3 Location of Offshore Works

The effective dispersion of cooling water thermal discharge for a 2000MW station to the river, estuary or sea is a very complex process, and the following criteria must be taken into account in locating intakes and discharges:

(1) Adequate separation of the structures to minimise recirculation of the hot discharge into the intake and to obtain acceptable maximum temperatures.

(2) In considering (1) an economic balance must be achieved between costly lengths of underground offshore tunnelling work to obtain adequate separation of intake and outfall and the penalty of recirculation. The latter raises intake temperatures and lowers the cycle efficiency.

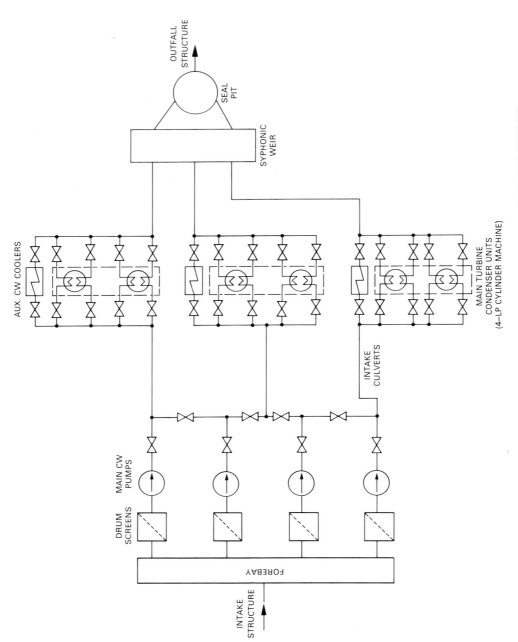

FIG. 1.38   FOSSIL-FIRED POWER STATION BASIC COOLING WATER SYSTEM

(3) Temperature limits are imposed by the authorities responsible for the cooling medium. The Thames Water Authority, for example, currently set a 25°C maximum mixed water temperature and a 12°C rise between the intake and outfall. In addition the Authority needs to know about the effects of any outfall flows on surface disturbance, thermal mixing characteristics and oxygen exchange.

Investigations for the location of offshore structures at a potential site begin 8 to 10 years before the project completion. The initial work comprises hydrographic surveys of the area to establish surface to depth flow patterns, change of tide conditions and any transport of silt. Records of temperature and flow history are examined together with a study of any changes in the sea or river bed profiles.

At the same time it is frequently necessary to examine the problem further by using a scale model of the area. A typical model scale is 1:100. Unfortunately, water depths at this scale are only a few centimetres and this is not sufficient for acceptably similar flow conditions to develop. For this reason, depths are modelled to a larger scale than the horizontal dimensions. In the past, this has been done using alternative locations of model intake and outfall structures and the latter have been designed to inject heated water into the model to examine the plume trajectory and growth. However, the vertical/horizontal distortion for models of large areas (typical values are about 1:10) make it impossible to accurately scale the thermal plume. Most models show a plume which is too wide and has a lower peak temperature than the full scale system. For this reason it has been necessary to apply mathematical techniques and construct effective models. This work is being done against the background of full scale tests to validate the model predictions, and to gain some idea of the actual plumes. A comparison of thermal plume widths and vertical mixing rates from a discharge of $57m^3/s$ to $24m^3/s$ is shown in Fig 1.39. This illustrates that the smaller discharge has a broader plume width, which shows the wide variations encountered and the need for full theoretical treatment of the subject.

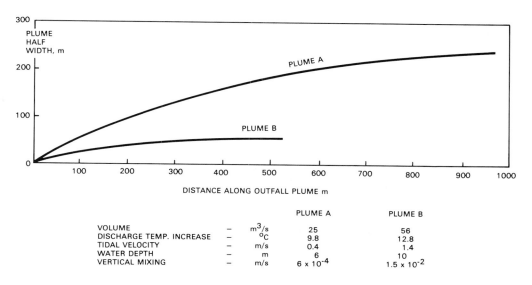

			PLUME A	PLUME B
VOLUME	–	m^3/s	25	56
DISCHARGE TEMP. INCREASE	–	°C	9.8	12.8
TIDAL VELOCITY	–	m/s	0.4	1.4
WATER DEPTH	–	m	6	10
VERTICAL MIXING	–	m/s	6×10^{-4}	1.5×10^{-2}

FIG. 1.39   COMPARISON OF THERMAL PLUME WIDTHS AND VERTICAL MIXING RATES

For tidal water, the slack tide growth and change of direction make analysis even more difficult and some degree of recirculation is usually inevitable at all the UK tidal sites.

In addition to the formation of the main plumes, which have similarity with the development of chimney emissions, random eddies through the mechanism of eddy-diffusion, transport heat from the primary plume to the overall flow area. This causes a resultant rise in the overall temperature of the surrounding estuary or sea water. The lateral dispersion mode is an important parameter for most conditions and values vary from $0.1m^2/s$ to $29m^2/s$ for the UK coast. On parts of the East Anglian coast there is a north/south tidal flow and a further phenomenon known as 'residual drift' which means that there is a gradual northerly or southerly movement of water over each tide. Offshore measurements have shown a southerly bias for central Suffolk of $0.09m^2/s$ to $0.06m^2/s$ representing a continuous input of water from the north.

For a project sited in this area and with typical lateral diffusion of $0.1m^2/s$, a 2000MW project would result in a plume width of 600m to 700m superimposed on an overall background temperature rise of approximately 1.5°C. Since the background rise can affect offshore water to a distance of 1km or more, economic studies usually dictate that this increase in ambient temperature must be accepted and siting of the offshore works must be such as to minimise direct plume entrainment by the intake.

At the present time, the most promising method for solving this type of problem appears to be the study of the offshore flow regime by full scale surveys or hydraulic model tests and the use of this flow as input data for a grid pattern in the mathematical model. The latter can also vary major parameters to see how these affect plume characteristics and the background temperature rise.

For offshore hydrographic surveys, fixed current meters are employed supplemented by thermistor stringers which record temperatures from the water surface to the seabed.

Where there is an existing station alongside a new project, other techniques are used. One is infra-red imagery using aircraft flying over the station area. Either full size or model aircraft can be employed to take photographs showing warm water surface levels.

Another technique which shows potential is the use of overhead satellites (see Fig 1.6). The satellites used for weather forecasting can show the changes of sea temperature and at present these are accurate to 0.2°C for a 1km square. In the future it is planned to obtain better resolution to squares of 120m. This method can provide long term studies of flow and temperature changes.

13.4 Environmental Considerations

Extensive research has been, and continues to be, carried out throughout the world on the behaviour of fish in the vicinity of power station cooling water intakes. The work has been aimed at reducing fish entrainment, particularly high quality species, and has taken several forms which are as follows:
(1) Underwater lighting, which is switched on and off to disturb fish from the illuminated source.
(2) Compressed air bubbles to form a buoyant screen, which inhibits fish movement across the bubbles.

(3) Electrical impulses generated in the water adjacent to the intake, which repel fish by a small electric shock.
(4) Study of fish reaction to changes in the direction or velocity of water.

This international work has been reviewed and it has been concluded that the most beneficial results arise from understanding fish reaction to water velocity and direction changes. Briefly, it is a general finding amongst researchers that where intake velocities can be reduced to a low value, in the order of 0.5m/s, fish can select their direction of movement without being diverted or entrained into a local high velocity aperture. For offshore intakes, this low velocity can be achieved by a concrete cap which overhangs the intake vertical downshaft, giving a peripheral velocity in the order of 0.5m/s. In addition to minimising the horizontal velocity, the cap also reduces the possibility of drawing fish downwards. This is an important feature, since it has been established that fish are less able to swim against upward or downward flows.

Cooling tower plumes and water carry-over are also a source of environmental concern. Theoretical studies, mathematical and physical models are used to examine these issues. It can now be shown that large concentrations of towers do not affect the local meteorology and carry-over of water droplets from cooling towers, using the most up-to-date design of eliminators, can be reduced to 0.001% of the total water throughput.

13.5 System Optimisation

One optimisation concerns the selection of the turbine exhaust pressure, condenser size, water throughput and condenser temperature rise. A computer program is used which varies all these parameters including the size of the circulating water structures, pumps and screens to estimate capital and annual operating costs. A set of results for the condenser and culverts is shown in Fig 1.40 and 1.41.

Another optimisation is the selection of the turbine house basement level for a direct cooled station. This study has been mathematically modelled to show how total capital and operating costs change as the basement is moved over a range of several metres. Capital cost changes are brought about by basement and culvert excavation costs and the size of pumps as the condenser shifts in line with the basement. Operating costs effected by the cooling water pump energy consumption change over the range of basement movement. However, it is necessary to exercise judgement over the optimisation results to ensure that the basement and operating floor positions are suitable for operational access, plant headroom and the level of the adjoining boiler house or reactor building. The result of a study for Littlebrook D is shown in Fig 1.42.

13.6 Hydraulic Gradient

Each project presents the problem of determining a system hydraulic gradient giving minimum pumping power, and at the same time taking into account uncertainties in the calculation of hydraulic losses for complicated system shapes. Arising from discrepancies between calculated and actual system losses, the British Hydro-Mechanics Research Association was commissioned to produce a design code for hydraulic calculations. This investigation showed the need for caution about accepting published data without knowledge of the experimental technique and the definition of the losses. Both of these aspects can lead to grossly different

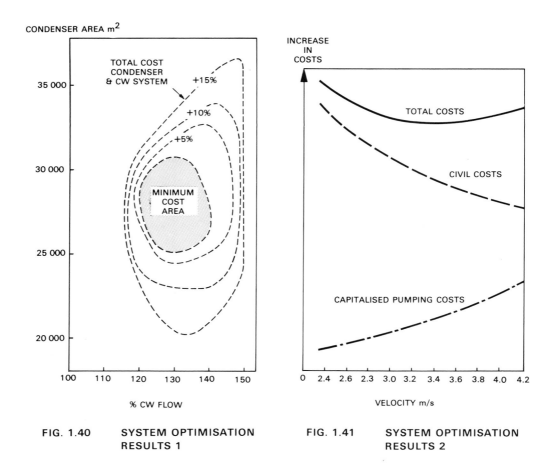

FIG. 1.40   SYSTEM OPTIMISATION
RESULTS 1

FIG. 1.41   SYSTEM OPTIMISATION
RESULTS 2

interpretations of the data. It must be stressed, however, that complicated manifolds and interactions of different shapes and fittings still require hydraulic modelling, both to calculate and optimise the hydraulic efficiency.

A gradient for a direct cooled tidal system is shown in Fig 1.43. The features which are of most concern to the designer are the syphon height, syphon seal and that part of the system at sub-atmospheric pressure. Current practice is to adopt a syphon height of approximately 9m for the lowest barometric pressure in the area of a site, minimum operating flow, low tide and maximum fluid temperatures. The latter requires discussion with the condenser designer, since higher temperatures are frequently experienced in the upper condenser tube rows in the order of 6°C to 7°C above the datum design temperature.

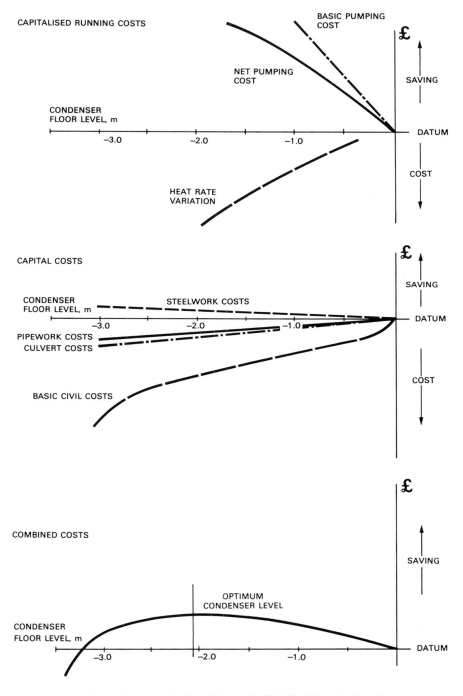

FIG. 1.42   LITTLEBROOK D OPTIMISATION RESULTS

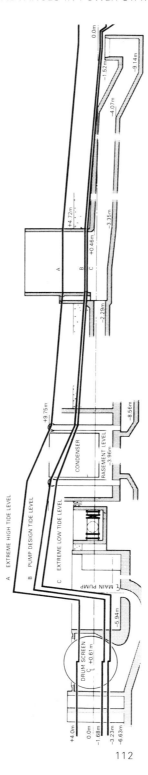

FIG. 1.43   DIRECT COOLED TIDAL SYSTEM GRADIENT

The system in Fig 1.43 shows a station positioned at a high level owing to difficult ground conditions. In order to minimise pumping costs, the maximum condenser syphon is maintained by the use of a syphonic seal design subjected to modelling tests. This feature also ensures that the culvert downstream of the condenser discharge remains full during unit shutdown.

For a complicated hydraulic network involving parallel and series circuits, a program has been developed which will rapidly compute circuit flows and system resistance so that a large number of schemes can be economically assessed.

13.7 Pumps, Screens and Valves

A mixed flow pump design has been established in consultation with the pump manufacturers. Also, during the late 1960s, this included development of the vertical spindle concrete volute casing pump, which has proved very successful in operation. Investigations have been made into the merits of variable speed pumps, but these have shown that any operational savings in running at reduced speed and power for certain flow conditions are outweighed by the cost of the variable speed equipment.

Experience of screening plant in direct cooled stations has been confined to the band and the drum types of plant positioned before the main pumps. A comparison of pump house dimensions for either type of arrangement has shown similar costs, but, after a close study of both designs along with operational experience, this has resulted in a preference for drum design on account of its simplicity, fewer rotating parts and lower maintenance costs.

In recent years the pressure type of strainer on the pump discharge has become feasible for large flows and has been installed at Grain power station. A comparison of drum and pressure screens for a 4-pump arrangement shows that overall costs, including capitalised value of additional pressure drop across the latter, are almost identical. The main disadvantage with the pressure screen is the lack of protection to the pump for certain sites with debris problems.

Considerable progress has been made in the design of butterfly valves and these are now acceptable for most isolation purposes, without the use of complicated mechanical sealing devices. It is apparent however, in certain locations, ie manifolds and right-angle connections to culverts, that hydraulic modelling of the valve and its immediate upstream and downstream approaches, is necessary to obtain minimum blade excitation forces.

The design of variable closure rate valves at the pump discharge to control the pressure changes following pump failure has also undergone considerable changes, particularly in the hydraulic actuating mechanism and reliable designs are now available.

Some consideration has been given to large butterfly valves fabricated in glass reinforced plastic (GRP). Trial valves of 1.6m diameter were fabricated and installed at Fawley power station and operational experience has been satisfactory. A comparison of life time costs shows that the GRP type is competitive with conventional materials. In the future, improved fabrication processes will show a considerable economic advantage for this type of material.

13.8 Pressure Surge Analysis

The methods of analysing pressure transients arising from pump failures have progressed considerably with the availability of computing techniques. Calculation of system behaviour, particularly those using syphon recovery, is now outside the range of any hand methods owing to the system complexity.

The analysis is normally carried out for a number of pump trip combinations and tidal conditions as follows:

(1) Open surface surge conditions between the offshore intake and the pumphouse screen chamber and also between the syphon seal pit discharge and the offshore outfall. This ensures that there is no likelihood of station flooding causing hazards to personnel and plant.

(2) Pressure transient changes in the closed section of the system between the pumps and syphon seal leading to possible water hammer.

The open surface section is regarded as a mass oscillation system, so the inertia in the system is considered to be of greater importance than the elasticity.

Several methods have been employed for the closed circuit section. One method adopts the characteristics for wave action processes in the culverts and the Kutta Merson integration for the end conditions, such as pumps and discharge valves, condenser and surge chamber.

Difficulties are still experienced with uncertainties in the pump quadrature data, since manufacturers are unable to provide performance tests in all operating modes.

The timescales over which water hammer and mass oscillation effects are found, are radically different and this is usually dealt with by using separate methods of calculation for the two cases. A notable example of this was the Dinorwig pumped storage station analysis where the tunnels are very long but the operational characteristics of the hydraulic machines are of comparatively short timescale.

Many researchers have shown the importance of air in attenuating closed system transients and how it changes the pressure wave velocity. Tests are being carried out on operating stations to see what boundary limits can be established for surge studies. Preliminary work has shown variations between 810m/s and 450m/s acoustic velocity, during the high and low tide states at one direct cooled station.

A hydraulically actuated valve on the pump discharge is used for control of pressure surges. It is designed for the initial 90% fast closure in not less than 1s and the remaining 10% in not less than 10s. Surge calculations then determine the closure rate which gives acceptable down and up-surge conditions. Since many stations are designed for syphonic assistance, the sub-atmospheric operating region after the condenser gives pressures approaching the vapour limit with certain pump trip and tide combinations. The example shown in Fig 1.44, requires air admission valves to fill the cavities and minimise the subsequent up-surge. It will be seen that valves must open or close in approximately 2s and designs are being discussed with manufacturers to ensure fast response and high reliability.

A considerable number of site tests has shown that the 2-speed valve principle is a satisfactory feature for surge control and normal pressure increases can be contained within the zero flow pump head. There is also a tendency for actual pressures to be lower than the calculated value and it is thought that the presence of air is responsible

114

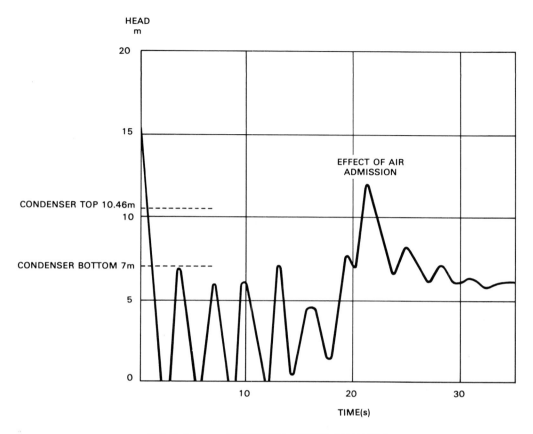

FIG. 1.44   PRESSURE SURGE ANALYSIS

for this attenuation. The investigations to determine system air contents will help to understand these discrepancies. The surges which occur at the pump house forebay following a trip of four pumps range from +2.5m to 2m.

13.9 Hydraulic Design of Structures, Plant and Modelling Techniques

In arriving at the most efficient and economic hydraulic and structural design, it is necessary to use a combination of calculation and modelling techniques and the following briefly describes the major points of concern.

In order to obtain the most economic design it is usual to construct a hydraulic model of the complete structure to a scale of about 1:20 to 1:30. Tests are carried out on the notional layout for all combinations of pumps, screens and tidal conditions. The main purpose of the tests is to:

(1) Determine head losses through the entire structure.
(2) Avoid any air entrainment vortices.
(3) Ascertain the degree of pre-rotation or flow instability before the pump intake.

Modifications and re-tests are usually necessary and when the pumphouse design has been established, it is current practice to arrange for the pump contractor to construct a further hydraulic model of a single intake and pump entry using a 1:15 scale model. This model is used to examine the flow approach to the pump in detail and also to assess pump performance characteristics.

Research into vortex action indicates the need for model testing. Currently tests over a range of unity Froude to 2.5 Froude velocities are being carried out. A large number of these tests have now been done and these have confirmed that this approach is satisfactory. Some researchers have advocated the use of identical velocities in the full scale and model simulation, but this procedure does not accurately model surface flows.

Intake structures and complicated valve manifolds also require modelling to minimise losses and, in particular, to ensure that valve blades will not be damaged by hydraulic instability.

Air models have advantages in closed pipe flow situations to obtain reasonable Reynolds number similarity. The outfall design for a direct cooled station does not always need head loss studies, since a surplus is frequently available after the syphon seal pit.

One area of importance which gives rise to difficult modelling problems is the syphon seal pit and any subsequent discharge chamber. The object of modelling is to ensure that the fluid does not break away from the culvert roof and to determine a satisfactory weir crest design.

A further phenomenon which has recently been experienced is flow and level instability between the down shaft after the syphon seal and the offshore discharge structure. This is caused by air entrainment in the down shaft, which changes the density of this column and initiates a regular oscillation. It is well known that air bubble modelling is very uncertain and later station designs are based on minimising the free fall of water to an open surface.

13.10 Cooling Water Pumphouse Layout

A pumphouse designed and constructed for a 2000MW fossil-fired station is shown in Fig 1.45. The hydraulic arrangement of the distribution forebay, screens and pumps is the outcome of hydraulic model tests.

The dimensions of this building are governed by the following factors:

(1) The overall width is determined by the hydraulic conditions in the common forebay, with a requirement for good distribution to each screen opening and on the face of the screen panel to give a uniform flow through the screen mesh. The length is determined by the width of the screen chambers and the height by the submergence of the screens below extreme low water level and the height of the crane used for maintenance.

(2) A uniform velocity distribution to the pump to avoid pre-rotation of water at the pump suction entry.

(3) The curvature of the pump section pipework.

(4) A separation between the pump and discharge valve to eliminate valve blade vibration caused by flow eddies at the pump outlet.

(5) The interconnecting manifold.

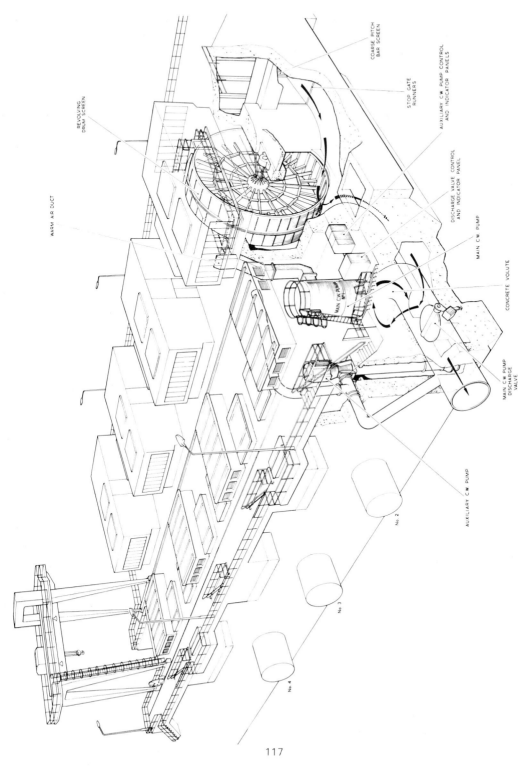

FIG. 1.45   2000MW FOSSIL-FIRED STATION PUMPHOUSE

This layout incorporates a division wall between the two pairs of pumps; this feature is included to ensure that any unexpected flooding of the pumphouse does not affect more than two pumps.

In this design the interconnecting manifold at the pump discharge is built into the mass concrete of the pumphouse. This is done to obtain an economic pumphouse dimension within the cofferdam used to build the structure. A hydraulic analysis was done to ensure balanced flow in the manifold culverts for all combinations of pump operation.

13.11 Reactor Essential Cooling Water System

This system, which serves reactor ancillary plant, has a safety role and must be designed so that total loss of the system will not occur more than once in 10,000 operations per year. A schematic arrangement shown in Fig 1.46 uses four pumps, physically segregated in the main cooling water pump house, and drawing water from the clean side of the main screens. The pump suctions are covered with water at extreme low water level and draw sufficient water for all cooling purposes through the screens, even when they are substantially blocked.

To perform the safety role, only one pump and its associated half system is required. The system operates satisfactorily with loss of all off-site power, supplies being obtained from the essential diesel generators.

13.12 Cooling Tower System

Indirect systems are usually used inland and utilise cooling towers to reject the condenser heat load to atmosphere.

Cooling towers normally operate by cascading water through a rising air flow and the cooling is by a combination of convection and evaporation. About 1% of the circulating flow is lost by evaporation and to avoid the build-up of salts in the system, about 2% of the flow is continuously purged from the system. The total loss of water is made up by extracting water from a natural water source such as a river.

For an indirect cooled system, a further optimisation is needed to obtain the best match of the towers, turbine condenser, pumping plant and civil works. The first step in the process is to compute a range of tower thermodynamic performance data, which for a natural draught tower design would comprise annual capital and operating costs, re-cooled water temperature and tower dimensions. The costs and re-cooled temperature information are then fed into the condenser optimisation programme to obtain a total system cost, which includes the effect of the changing turbine exhaust pressure. The final cost information shows the optimum low cost range and the sensitivity of the cost variation to changes in all the parameters.

13.13 Material Protection

Experience of impressed current cathodic protection has been disappointing. Arising from this, research work has been carried out which has led to a better understanding of the theoretical basis of protection for power station plant. This incorporates more efficient anode spacing and length, anodes of more robust construction which withstand maintenance activity and more reliable control systems.

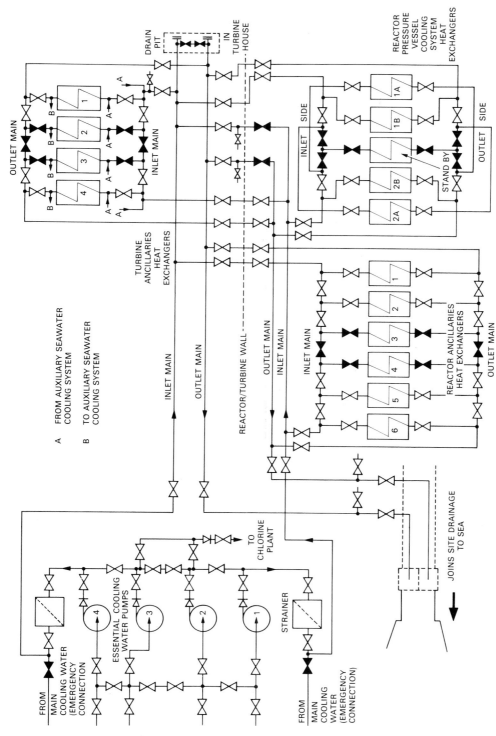

REACTOR ESSENTIAL COOLING WATER SYSTEM

FIG. 1.46

Trials carried out at Pembroke power station have been successful and this technique now shows promise for the future. In the interim period protective internal linings are being used.

13.14 Chlorination

Several methods of mussel control have been examined, consisting of flow reversal in intake tunnels to provide a heated water source and the design of the system to give a high velocity shearing flow along culvert walls, which will unsettle mussels from the surface. None of these has been very satisfactory and chlorine treatment remains the simplest and most efficient method. Water Authorities have been concerned about the effects of residual chlorine in the water discharged from the station and some research has been done to ascertain acceptable residuals. It is now general practice to adopt a figure of 0.5mg/litre. In accordance with Government policy to reduce the amount of bulk chlorine transport, all of the later sea water cooled stations use electrochlorination equipment and inland stations use sodium hypochlorite dosing.

13.15 Cooling Water System Design and the Interaction of Design Departments

The design of the entire cooling system involving the civil, mechanical, electrical and research disciplines, requires a well organised flow of information to meet a project programme. The early activities for siting begin some 10 to 12 years ahead of project commissioning and this work is mainly concerned with evaluating the full potential of an offshore area.

The network diagram in Fig 1.47 shows how the cooling water system design evolves and describes the flow of information passing between the various design departments.

14. STATION SERVICES

14.1 Introduction

This Section describes the main mechanical systems which serve a number of boiler, reactor and turbine units. Because these systems are linked to all units they are termed 'station services'. Also included are descriptions of systems dealing with environmental control, such as heating, ventilating and air conditioning, fire protection and smoke removal.

14.2 Auxiliary Steam

A steam supply which can be available when the main boilers are not operational is required for a number of services. For an oil-fired station, this consists of atomising steam for oil burners, deaerator heating, steam airheaters, airheater prewarming, domestic heating, tank farm heating and auxiliary boiler oil heating.

Fig 1.48 shows the system installed for a typical 2000MW oil-fired station. The fuel oil heaters, burner atomisation and bled steam airheaters provide a continuous demand for auxiliary steam along with the domestic heating, tank farm heating and auxiliary boiler services.

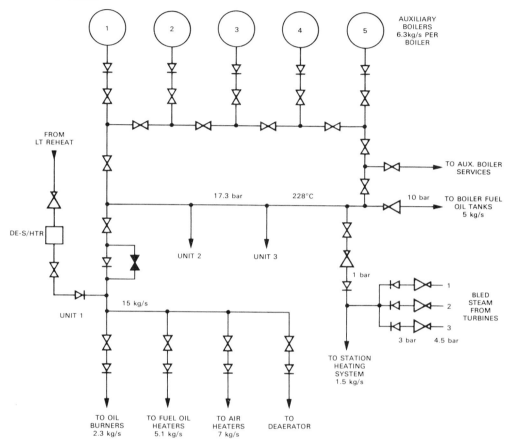

FIG. 1.48   AUXILIARY STEAM SUPPLY SYSTEM FOR 2000MW OIL-FIRED STATION

On a sequential 3-unit hot start the steam demand would be approximately 29kg/s. In the event of one unit being delayed, the demand could exceed 31kg/s. Therefore an auxiliary boiler installation of 5 × 6.3kg/s rating is required.

Once a unit has reached 50 – 60% CMR, its auxiliary steam requirements can be supplied from its own steam lines, thus making the unit completely independent of the auxiliary boilers.

The auxiliary boilers which are typically shell boilers with twin burners and combustion chamber deliver the steam to a ring main. Each unit has its own auxiliary steam range fed from either the auxiliary steam ring main or from the unit cold reheat line via a pressure reducing valve and desuperheater. Under certain operating conditions, the unit cold reheat line can be used to support the auxiliary boiler steam ring main.

14.3 General Service Water

This is a general purpose water supply for station plant where high quality water is not necessary. Examples of demand are, the storage tank and pumps for fire fighting purposes, air compressor cooling, chemical monitoring plant coolers, large motor and transformer coolers and miscellaneous water supplies to the coal and ash plant. A system installed for a 2000MW fossil-fired station is shown in Fig 1.49. The system is arranged in two sections, with the source of water taken from a suction dock of the cooling water pumphouse via two 100% duty self cleaning strainers. The latter provide water to a manifold, which distributes the supply to:
(1) Suction pipework of fixed station internal and external fire fighting pumps and fire hydrant mains.
(2) Four 33% duty general service water pumps which supply water to high level tanks. These tanks provide a near constant hydraulic head on a number of services and, at the same time, give a storage quantity of approximately 50 minutes in the event of failure to restore a tripped pump promptly. Alarm conditions of low water level are signalled to the control room and at this stage there is 30 minutes supply available.
The tank storage system supplies water to coal and ash plant services, cooling water for air compressor heat exchangers, cooling water for a number of large electric motors, air coolers and chemical monitoring coolers. The maximum system demand occurs during base load operation of the station, when the needs are $0.13m^3/s$.

The strainers, which supply water for both the general service water pumps and the fire fighting plant, are rated at $1.3m^3/s$.

Schemes continue to be developed to meet specific station designs. For example, it would not be satisfactory to adopt the scheme described for raw water as it would lead to fouling of heat exchangers. In this case, a pump recirculating system would be used with domestic water and the tanks would serve for storage and the introduction of make-up water to replenish system losses.

14.4 Air Supplies

Systems are required for:
(1) General maintenance purposes, mainly for pneumatic tools.
(2) Pneumatically operated regulating valves and dampers.
(3) In the case of nuclear stations, for breathing air in association with special suits needed for operators to gain access to contaminated and high temperature areas.
To cater for these categories of air supply, separate systems are employed. The general purpose duty is not a continuous demand and it must also supply a widely varying use of tools during normal operation and statutory overhaul periods. Instrument air is vital to boiler and turbine operations and must be very reliable. Similarly, breathing air supplies must also have a high reliability.

A typical schematic arrangement of the general service air supply distribution is shown in Fig 1.50. This has a twin compressor and pressure vessel storage design which, experience has shown, is quite satisfactory. The duty of the compressors is normally about $28m^3/min$ at a delivery pressure of 7 bar. Much investigation has been made to decide on the most reliable type of compressor, and there is a preference for reciprocating machines, although other types are not excluded.

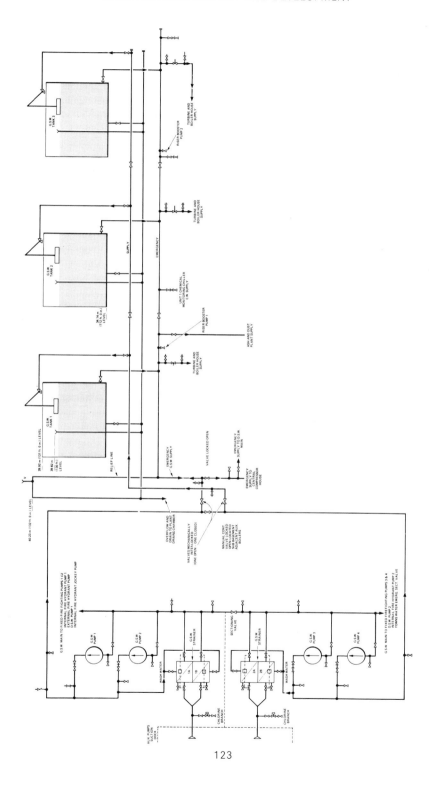

FIG. 1.49   GENERAL SERVICE WATER SYSTEM FOR 2000MW FOSSIL-FIRED STATION

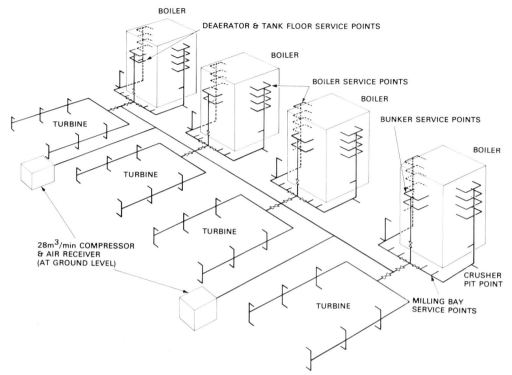

FIG. 1.50   GENERAL SERVICE AIR SUPPLY SCHEMATIC

A schematic arrangement of an instrument air system for a fossil-fired and nuclear station is shown in Fig 1.51. In order to obtain satisfactory availability of supplies, the equipment comprises three 100% duty compressors and pressure vessels. The demand for air would be approximately $42m^3/min$ to $57m^3/min$ at 7 bar. For these conditions, since air is provided for sensitive and vital controls, it must be free of any oil deposits from the compressor and have means of extracting moisture. The specifications for this type of air compressing plant, call for compressors to be equipped with oil-free piston rings, dryers and moisture traps to be inserted at key points in the system pipework.

Breathing air systems are applicable to nuclear stations to cover two conditions, a low pressure system for contaminated zones with ambient temperatures and a high pressure system for reactor entry working at elevated temperatures. The high pressure system is made up of three compressors and associated pressure vessels; each compressor is rated at approximately $4m^3/min$ with a 10 bar available pressure. In some designs, the low pressure system is either an extension, appropriately regulated, or a separate system with three compressors of $2m^3/min$ each and system pressure approximately 7 bar.

In all cases, a breathing air back-up is provided in the form of a cylinder facility. The compressors are fitted with carbon rings and the system is provided with filters, dryers and, in the case of the low pressure system, pressure reducers for ensuring system high quality air.

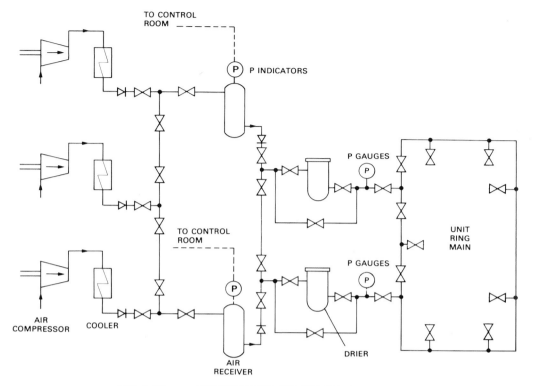

FIG. 1.51   INSTRUMENT AIR SUPPLY SCHEMATIC

Air supplied to fossil-fired boiler sootblowers is provided from a separate unique source.

14.5   Fire Protection

The design process for fire protection is an extensive process for both fossil-fired and nuclear stations. The main difference in approach however, is that in addition to considering the preservation of life and economic loss, the procedure for a nuclear station also considers the effects of fire on the shutdown capability of the reactor and the release of activity to the atmosphere.

In both cases, however, the analysis considers the following points:

(1) The need to prevent a fire occurring by controlling the use and disposition of combustible materials and sources of ignition.
(2) The need to reduce the severity of a fire that does occur by segregating combustible materials and plant into fire compartments to prevent spreading, and installing fire detection and extinguishing systems.
(3) The need to limit the consequences of a fire by protecting essential plant and providing access for personnel.

Discussions take place with a number of bodies, which includes the Health and Safety Executive, Local Authority representatives, the Fire Research Station, the Fire Insurers Research and Testing Organisation, the Fire Officers Committee and local fire brigade representatives. Also, in the case of nuclear stations, consultation is necessary with the Nuclear Installations Inspectorate. The following sub-sections describe some of the salient systems installed.

14.5.1 *Turbine House*

The steam conditions employed in modern turbine generators, coupled with the type, quantity and pressure of the lubricating and control hydraulic fluids used throughout the unit, identify the plant as a major fire risk zone. The risks arise from leakage and escape of the combustible lubricants and hydraulic fluids contained in the separate control fluid, lubricating and seal oil systems.

The risks also extend to main plant ancillary equipment and floor areas, which house lubricating systems likely to lie within the radius of an oil spray pattern escaping from the failure of an oil bearing or pumping supply header. Hydrogen used for generator cooling constitutes a separate risk.

The initiating factors for an oil fire could be an escape of fluid onto hot steam surfaces, impregnation and absorption of oil into lagging and leakage into electrical equipment.

Fixed high velocity waterspray systems cover all oil systems, oil piping, pumps, coolers and all similar associated equipment including adjacent floor areas. The waterspray systems are divided into convenient zones and each zone has sufficient numbers of projectors to cover the risk adequately. The protection principle is that an adequately designed high velocity waterspray system with correctly positioned nozzles, effectively extinguishes a pressurised jet oil fire within the protection zone.

Fire detectors for the zones are of the frangible glass bulb type. Additional selective rate-of-temperature rise detectors are provided at strategic points and arranged to give early warning of any unusual high temperature conditions in the area. Zone control deluge valves are sited so as to be operable, without fire, smoke or fume hazard to the operator, during a fire in any of the associated protection zones. This requirement is achieved by locating valve assemblies away from the protection zones or by enclosure of the valves in protective cubicles.

Despite its large volume, a turbine house can become smoke-filled, restricting access to plant and in some cases making such access impossible. To enable fire fighting to be carried out, facilities are provided whereby smoke can be dispersed. Controllable top and bottom ventilation is provided by main access doors, turbine loading bay doors and windows.

14.5.2 *Boiler House*

The principal risks at the boiler front are spillage or leaks of pre-heated fuel oils onto hot boiler casings, the rupture of flexible connections on burner fuel oil systems and the leakage of propane from boiler lighting-up systems. A minor fire involving oil leakage can quickly escalate into a major fire by enveloping and destroying flexible fuel oil leads and compressed air piping serving the boiler front. Power and control cables may also be affected and their resultant loss could incur a prolonged and costly boiler outage.

Automatic high velocity waterspray projectors are provided to groups of burners, each group or zone having separate control valves for water supply. The area of detection for each group includes the immediate floor and gallery areas, and initiation of the protection system automatically shuts off air, gas and fuel oil supplies to the boiler front.

14.5.3 *Oil Storage and Transfer Equipment*

The main hazard involves the ignition and explosion of fuels contained in oil storage tanks, oil pump houses, pipelines and associated equipment. Heavy 'black' or 'residual' oil before heating is a low fire risk. Lighter oils are a comparatively higher risk and propane presents a severe explosion risk.

Tank segregation, spacing and layout follow the recommendations outlined in the CEGB Standards and the requirements of the Health and Safety Executive and Local Fire Authorities. Major oil tank fires are beyond the resources of the station fire fighting equipment and reliance must be placed on County Fire Brigade services to tackle such fires. However, fixed protection equipment is provided in the form of low velocity tank shell cooling sprays and foam protection coverage to tank internals, and tank bund areas are provided to contain a tank fire and prevent damage to other plant and adjacent tank structures. Foam application to tank internals is by base injection or tank top pourer systems.

Oil transfer pumps, heaters, pipework and valves are protected by automatic waterspray, with dry powder and mechanical foam type portable fire extinguishers sited at strategic positions.

14.5.4 *Coal Plant*

Generally speaking fire protection to coal plant is usually catered for by the installation of hydrant points together with detection devices located over conveyors and in coal mills. Heavy reliance is placed on good housekeeping and operating procedures. The only other effective way of extinguishing a fire is to rake out the coal and damp down the hot fuel.

14.5.5 *Electrical Equipment*

The following protection systems are provided:
(1) Oil-filled transformers above 100kVA — high velocity water sprays.
(2) High voltage switchgear — extinguishing gas (portable appliances).
(3) Cable tunnels — combination of fast acting (semi-deluge) sprinklers and pre-action sprinklers.
(4) Computers, computer suites and data processing equipment — combinations of automatic Halon extinguishing gas and portable equipment.

Cable tunnels are equipped with a very comprehensive fire detection system, consisting of heat detecting cables which traverse the entire tunnel. A further precaution in cable tunnels to limit the effect of any fire, is the use of cable segregation between units, control and power cables, together with a system of fire barriers.

The automatic Halon extinguishing gas system utilises the extinguishing agent bromotrifluoromethane (Halon 1301) which is a halogenated methane compound. It is colourless, odourless and electrically non-conducting at ambient temperatures, and has a boiling point of $-58°C$ and a freezing point of $-168°C$. Under pressure, the Halon is liquified and stored in steel cylinders. Nitrogen is used as a propellant to improve the discharge of the extinguishing gas into the protected areas.

The extinguishing gas is dry and does not damage machinery or materials with which it comes into contact. Its extinguishing property is based on its ability to break the combustion chain reaction, rather than the effects of physical cooling or dilution of oxygen vapour concentration.

14.5.6 *Nuclear Stations*

Of special concern are circulators on gas-cooled reactors and active waste storage areas. Circulator protection is achieved by a combination of equipment segregation, bund wall enclosures and high velocity water sprays. Active waste areas which have a low risk be on detection and manual fire fighting, while high risk areas are equipped with high velocity waterspray protection. The emphasis however must rely on the segregation of combustible materials and essential plant such that a fire in any particular area cannot spread to another area causing a release of radioactivity to the atmosphere or a fault which compromises the safe shutdown capability of the reactor.

14.5.7 *Fire Pumps*

The water storage, and pumping system for a high and medium velocity waterspray, sprinkler and internal hydrant system requires a trunk main to be maintained at a water pressure of 10.5 bar at all times to supply waterspray and sprinklers in the event of a fire.

Fire pumps are hydraulically sized to meet the required demand with configurations of four 33.3%, three 50% and four 50% duty pumps in combinations of diesel and diesel electric driven pumps.

A separate strategic water reserve is provided to maintain the largest waterspray or sprinkler demand over the largest single risk. This reserve is from a clean, potable water supply which will satisfy the demand of the largest single risk for minimum a period of 40 minutes.

14.6 Heating, Ventilating and Air Conditioning

14.6.1 *Introduction*

The increasing importance of environmental standards has considerably raised the status of this equipment. To match these changing circumstances, site investigations are undertaken to analyse heat loss from equipment and air flow movement in small and very large buildings including turbine and boiler houses, and calculation codes for these high heat losses and massive buildings have been derived.

Another aspect concerns control and instrumentation systems. Modern equipment requires a scrupulous limit on the temperature and humidity range. This has placed emphasis on a better understanding of air movement and the control and sensing systems for regulating plant performance.

Nuclear stations need the system design to contain the spread of radioactive contaminated air. At nuclear stations it is necessary to have plant which is serving safety equipment capable of operating under seismic conditions.

The following sections briefly outline some of the more important features of heating, ventilating and air conditioning design and performance.

14.6.2 *Turbine and Boiler House*

Surveys of heat loss have been made at a number of power stations to gain data for 500MW to 660MW boilers and turbines; eg at Drax, the losses for a turbine and boiler unit are 3MW and 22MW respectively.

In the main building, ventilation air flow, during normal operation, enters through doors and ventilators at low level in the peripheral walls of the buildings. It is warmed by heat emitted from the plant and rises by convection to the higher levels of the turbine and boiler house when it reaches the intake of the forced draught fans. The general air circulation pattern is shown in Fig 1.52. Surveillance has shown that as the air warms, it rises and the pressure varies from negative to positive, thus forcing air out of the top of the building. This gives rise to a neutral zone within the building that is at atmospheric pressure. The height at which this zone occurs is altered by adjusting the ventilators.

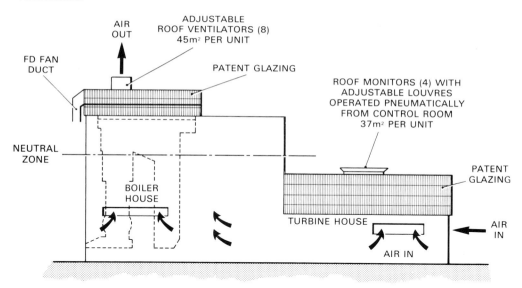

FIG. 1.52   GENERAL AIR CIRCULATION PATTERN IN TURBINE BOILER HOUSE

A survey has been carried out on comfort and heat stress levels for station personnel and this data leads to the adoption of a 15°C to 20°C rise in air temperature across the buildings, with an ambient temperature of 20°C. At these conditions, the air flow for one Drax type unit is approximately 1000m³/s.

It is also necessary to consider air flows and temperatures when fossil-fired boilers are shutdown. In this condition, the forced draught is reduced in proportion to the units on load. During the shutdown period, the boiler heat losses remain at a high value for many hours. This could result in the normal operational air temperatures being exceeded. For new stations, either special vents are included above the boiler or provision made for their installation at a later date, when operational experience is available.

Severe condensation in the turbine house could result if the high level vents are incorrectly positioned or adjusted, giving rise to an incorrectly positioned neutral zone. With winter temperatures, outside air will move into the turbine house above the neutral zone and condense on glazed roof sections and cold surfaces. This is eliminated by positioning the turbine house ventilators below the neutral zone as shown in Fig 1.52.

Ventilation of the building during a fire has been examined and discussed with specialist authorities, including the Fire Research Station. For small fires, smoke is exhausted via the boiler house. Air movement in these conditions is not significantly different from the normal operating condition and any ventilators positioned in the turbine house roof are ineffectual, because they are lower than the neutral zone. During a major fire, one solution is to accept the collapse of roof cladding, which is designed for this condition and does not exacerbate the original fire.

14.6.3 *Control, Computer and Relays Rooms*

A system is provided to remove heat generated by equipment and personnel and limit the spread of fires.

The external conditions for a station vary, depending on its location, but at an inland site the design conditions are as follows:

Summer maximum	28°C 45% relative humidity
Winter maximum	−2°C 100% relative humidity

Internal design conditions are:

Space temperature	20°C ± 2°C
Humidity	50% ± 10%
Air velocity	0.5m/s maximum
Pressure	positive with respect to outside air

These conditions are only obtained by a combination of refrigeration plant, air heating exchangers using auxiliary steam as the heating source and mechanical draught cooling towers. A schematic arrangement of the system to serve a central control room and computer suite is shown in Fig 1.53.

In the central control room, the air fans and duct work are designed so that the system operates in a full fresh air mode, following detection of smoke in adjoining rooms or interconnected parts of the ducting system.

14.6.4 *Nuclear Stations*

Additional features are needed for nuclear stations such as:
(1) Plant having a safety role must have a reliability factor higher than that which is normally acceptable; it is now designed to meet numerical standards of reliability.
(2) Areas with potential radioactive contamination, must have a ventilation system which does not release the contamination to other parts of the building or the outside air.

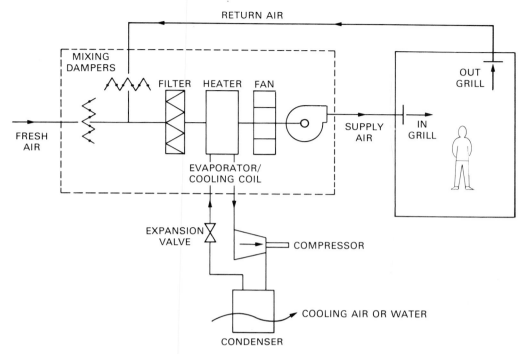

FIG. 1.53   HEATING, VENTILATING AND AIR CONDITIONING SYSTEM FOR CONTROL AND COMPUTER ROOMS

External environmental conditions for heating, ventilating and air conditioning (HVAC) plant in rooms having safety equipment, must be related to the reactor safety criteria. This leads to the adoption of more extreme ambient temperatures for this section of the HVAC plant. At Heysham 2, this was selected at −17°C to +36°C.

For radiological contamination control purposes the buildings are sub-divided into zones, which are classified according to UK Standards of potential contamination level. In order to assist in controlling the classification, the flow rate and direction of air passing through the interfaces are be designed to the criteria given in Table 1.2.

These criteria apply to the designed condition, ie with doors, hatches, etc closed. When one of two sequential doors protecting an orange to red interface is open, the average air velocity through the door space must not be less than 0.1m/s. The same criterion applies as a target for a green to orange interface.

In the design of main systems, the following principles are applied to ensure a high degree of availability:

(1)   Failure of any one fan, filter or heater does not cause total failure of a system.

(2)   Spurious or intended closure of a damper, including that for fire protection, is alarmed and is clearly detectable if it prevents design operation of a system.

(3)   The design, wherever practicable, minimises the time taken for repair or maintenance by choice of components, by specifying standard fittings and items of equipment, by permitting ready access to components (including dampers) and by grouping of similar items.

TABLE 1.2. Air Velocities at Interfaces Between Zones of Differing Contamination Level

Zone Interface	Air Velocity Through Interface Openings
(1) White to green	Generally from the white to the green zone. No air velocity criteria are applicable, subject to any local containment requirements arising from hazard analysis.
(2) Green to amber White to amber	Towards the amber zone, with an average velocity, not less than 0.5m/s.
(3) Amber to red	From the amber to the red zone, with an average velocity not less than 1m/s.
(4) Between identified areas of differing contamination within an amber zone.	Air flow towards the area of greatest contamination. No velocity criteria are applicable, but as far as the design position allows, the velocities should tend towards (and not exceed) interface (2) above.

The zones, as defined in the UK Atomic Energy Code of Practice, are as follows:

White — means a clean area free from radioactive contamination whether surface or airborne.

Green — means an area substantially clean by reason of regular monitoring and cleaning.

Amber —means an area in which some surface contamination must be expected and airborne contamination will exceed the maximum permissible concentration occasionally, such that provision must be made for its control.

Red — means an area to which there is normally no access.

(4) The plant is arranged such that it can be fully-isolated for repair or maintenance without the necessity to shutdown the system.

(5) It is possible to change filters without having to shutdown the ventilation system. During this operation, the integrity of the extract system is not below that of the rest of the system.

(6) Where a room pressure differential exists by design for ensuring clean conditions and maintaining the contaminated areas at different pressure levels, this is not removed or reversed upon failure or during maintenance of any single part of the HVAC equipment.

(7) Operation of the systems is fully automatic and they are capable of continuous running while unattended.

(8) For clean areas, one standby is provided, except where the consequences of shortfall are not serious (charge hall and turbine hall). Systems may be three 50% duty or two 100% duty according to plant economics and available space.

(9) Contaminated HVAC is generally based on one standby. For a three 50% duty system, operation at one 50% duty provides a reduced capability in many cases.

(10) Where loss of contamination control potentially leads to rapid contamination of normally occupied areas, eg pile cap to chargehall, three 100% duty extract fans are installed.

Much importance is placed on the design of ductwork and in general all materials are fire resistant. Extract systems serving high humidity areas, or those containing materials or equipment which may emit corrosive fumes, are made of, or lined with, corrosion resistant materials. Areas for which this is required include the fuel storage pond and flask loading and handling bay.

Systems are allocated such that areas segregated by major fire barriers are ventilated by independent systems. This also applies, with minor exceptions, to areas associated specifically with reactor units and those which are required to be segregated against hot gas release. HVAC penetrations of fire barriers over and above these segregations, are protected by automatic dampers.

Other than the pile cap extract, no safety claim is made for forced ventilation cooling as a means of removing excess heat, generated by hot gas release or fire, in order to maintain plant operation. Any such cooling resulting from continued ventilation provides additional margins.

14.7   Water Treatment Plant

Very pure water is required for boiler make-up and a typical specification is as follows:

Conductivity	Not greater than 0.10 micro Siemen/cm
Sodium	Not greater than 0.015mg/litre as Na
Silica	Not greater than 0.02mg/litre as SiO_2

For a fossil-fired station, the quantity of water needed varies between 3% and 4% of the total boiler evaporative capacity. This demand is influenced by the type of station operation, whether it is on base load or shift duty and whether the boilers use steam operated sootblowers. At nuclear stations, with their predominantly base load operation and omission of soot blowers, make-up varies between 1.5% and 2.5%. This leads to a requirement for storing water on site to safeguard the operation of the station. In general it is normal practice to store approximately 24 hours supply of raw water for domestic and boiler make-up requirements and a reserve of treated boiler feedwater amounting to the equivalent of 24 hours normal make-up supply to the boilers.

The main treatment of water to give make-up of the quality described is by the demineralisation process. Evaporators are installed at some fossil-fired stations for treating river water and they have been used at one nuclear station site for treating sea water. The performance of sea water evaporators has been very disappointing and it is not intended to install this type of equipment in future without putting in hand some development work. For nuclear stations with once-through boilers, 100% flow mixed bed polishing plants are installed.

15. SUMMARY

This chapter has, within the constraints of the available space, attempted to give the reader an insight into the factors, considerations and processes involved in progressing the initial concept of a power station to an engineering reality. The evolution of different station types, the development of main plant and systems and the selection of components and materials have developed against an increasing background of environmental, safety and cost considerations. Together they represent an increasingly complex picture requiring balanced judgements founded on sound engineering and proven performance, to enable the CEGB to fulfil its statutory obligation in the supply of electricity.

2.

Littlebrook D Oil-fired Power Station

1. INTRODUCTION

The decision to design and build oil-fired power stations is influenced by many factors, one of great importance being the fuel cost. The CEGB programme for the building of large oil-fired power stations was conceived in the 1960s, when fuel oil was comparatively cheap. The initial programme consisted of 500MW unit stations at Fawley, Pembroke and Kingsnorth (dual-firing oil or coal), the latter being completed in 1973. These were followed by a further three stations, Ince B (500MW units) and Grain and Littlebrook D (660MW units) which were built and completed between 1974 and 1984.

The lead time from initial conception to commissioning spanned the period when world oil prices began to rise substantially. The last of the oil-fired stations, Littlebrook D, may not have been completed, had the future shift in fuel availability and costs been known earlier. However, it is the policy of the CEGB to maintain some flexibility of fuel usage so that electricity supplies are less vulnerable to problems in the fuel markets. In 1984 oil-fired stations accounted for about 20% of the total generating capacity.

This Chapter is devoted mainly to the technical and constructional description of the Littlebrook D Power Station plant.

Approval to build a 2000MW oil-fired station at Littlebrook was given in July 1972 and authority to proceed with design and construction in November 1973.

The Littlebrook D power station is situated immediately west of the existing B and C power stations, on land owned by the CEGB. The D power station, consisting of three 660MW units, was built to help meet the increasing demand for electricity in south east England and particularly Greater London.

Although in its original concept Littlebrook D was designed as a base load station, ie it would have been supplying power to the grid continuously, the use of oil makes the station more suitable for 2-shift operation. Assuming that coal maintains a cost advantage over oil, it is likely that the station will be used increasingly on a 2-shift or single-shift routine.

2. LITTLEBROOK D SITE SELECTION (FIG 2.1)

The Littlebrook D station stands on the south bank of the River Thames, immediately to the west of the Dartford tunnel approach and adjacent to the existing C station site. Littlebrook A, B and C stations were coal-fired and are now closed. To the north of the site lies the Thames and industrial areas, eg Dagenham, Tilbury and Rainham, and to the south lies a residential area. The south east of England has a shortage of suitable sites for a modern power station and the CEGB decided that it could only meet the anticipated demand in the time available by the development of Littlebrook D.

The CEGB owns some 77 hectares of land at this site and the area of land for which Section 2 Consent was received for Littlebrook D, was 21 hectares. The area of the A, B and C Stations for which the CEGB had already obtained consent was 56 hectares.

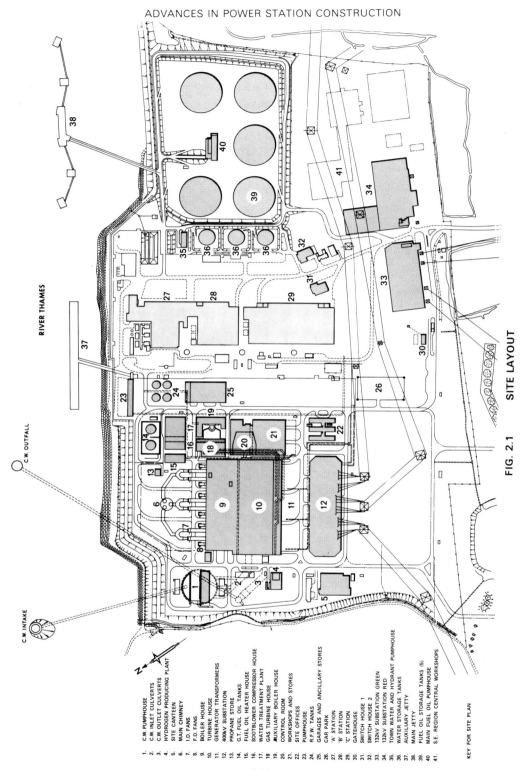

RIVER THAMES

C.W. OUTFALL

C.W. INTAKE

1. C.W. PUMPHOUSE
2. C.W. INLET CULVERTS
3. C.W. OUTLET CULVERTS
4. HYDROGEN PRODUCING PLANT
5. SITE CANTEEN
6. MAIN CHIMNEY
7. I.D. FANS
8. F.D. FANS
9. BOILER HOUSE
10. TURBINE HOUSE
11. GENERATOR TRANSFORMERS
12. 400kV SUBSTATION
13. PROPANE STORE
14. G.T. FUEL OIL TANKS
15. FUEL OIL HEATER HOUSE
16. SOOTBLOWER COMPRESSOR HOUSE
17. WATER TREATMENT PLANT
18. GAS TURBINE HOUSE
19. AUXILIARY BOILER HOUSE
20. CONTROL ROOM
21. WORKSHOPS AND STORES
22. SITE OFFICES
23. PUMPHOUSE
24. R.F.W. TANKS
25. GARAGES AND ANCILLARY STORES
26. CAR PARK
27. 'A' STATION
28. 'B' STATION
29. 'C' STATION
30. GATEHOUSE
31. SWITCH HOUSE 1
32. SWITCH HOUSE 2
33. 132kV SUBSTATION GREEN
34. 132kV SUBSTATION RED
35. TOWN WATER AND HYDRANT PUMPHOUSE
36. WATER STORAGE TANKS
37. AUXILIARY JETTY
38. MAIN JETTY
39. FUEL OIL STORAGE TANKS (5)
40. MAIN FUEL OIL PUMPHOUSE
41. S.E. REGION CENTRAL WORKSHOPS

KEY FOR SITE PLAN

FIG. 2.1   SITE LAYOUT

Negotiations with the Dartford Municipal Borough Council were completed for both the lease of some 59.5 hectares adjoining the southern boundary of the site and for the sale of sand and gravel within this area.

2.1 Access

Road access to the general vicinity of Dartford was via the A2 London Road, A206 Dartford to Erith Road, A207 Blackheath to Dartford Road or Dartford Tunnel. To enable the Dartford Tunnel approach road to be used by construction traffic, it was agreed to construct a bridge over the approach road and the necessary roads to connect the bridge to the main road.

No rail connection existed to the power station site and the coaling jetty on the River Thames at the existing Littlebrook Power Station, was only suitable for berthing two 4,600t colliers. A new jetty was constructed so that a tanker of up to 30,000t could be berthed.

2.2 Noise and Vibration

The Secretary of State included the following condition in the Section 2 Consent, "The station shall be so constructed as to avoid as far as may be reasonably practical, noise and vibration in the operation there".

Acoustic and initial noise surveys were carried out before construction commenced and further surveys were carried out to ascertain seasonal variations. During construction and commissioning noise levels were periodically reviewed. Careful attention was given to sound proofing the auxiliary gas turbine installation, particularly the air intakes and exhaust to avoid nuisance. The west end of the turbine hall had an internal acoustic insulation fitted because of the close proximity of a hospital and a housing estate.

In addition, plant was designed so that noise levels within the station comply with CEGB standards and the Department of Employment's code of practice for reducing noise to people employed on the site. Attention was also given to minimising noise levels during construction of the station and to providing adequate silencing during the pre-service steam cleaning of the boilers and associated pipework.

2.3 Environmental Aspects

The problem of atmospheric pollution on the lower reaches of the River Thames was such that the greatest care was taken in the design and subsequent operation of the station to avoid, as far as practicable, noxious chimney emissions. An atmospheric pollution survey for the routine measurement of sulphur dioxide at ground level was carried out, which commenced two years before the programmed commissioning date of the first unit and continued until after the station achieved full operation.

The Civil Aviation Authority accepted the CEGB proposal to construct a station with a single chimney 210m high, provided obstruction lighting was installed.

There was sufficient cooling capacity in the River Thames for direct cooling of the station, but in order to minimise recirculation at the then operating B and C stations, the design and location of the cooling water intake and discharge headworks was critical. The Port of London Authority, whilst confirming the acceptability of a direct cooled station, reserved the right to ask for compensatory aeration equivalent to the heat load imposed upon the river by the station.

3. SYSTEM CONSIDERATIONS

Littlebrook D was aimed at filling a mid-merit role between the low fuel cost base load nuclear and coal stations and the peak lopping gas turbines. The principal objective was to provide reliable proven plant having high availability coupled with flexible operating characteristics at low capital cost. In this role a slight loss in thermal efficiency due to the selection of a 2-cylinder exhaust turbine design proved more cost effective than the 3-cylinder machines used elsewhere.

The plant was expected to operate continuously at varying loads between 50% and 100% continuous maximum rating (CMR) and would only be shutdown for routine outages in which inspection, repair and maintenance work would be carried out. The plant was also designed to be deloaded rapidly, stored at temperature and reloaded rapidly for 2-shift operation. Thermal cycling and fatigue stresses were taken into account over its life. An automatic control system enables the plant to respond to incremental load changes due to normal grid system demands and also to emergency pickup and frequency regulation during system excursions.

4. OPERATIONAL ROLES

The plant is capable of:
(1) Continuous operation at any load between 50% and 100%.
(2) Output regulation in response to system frequency variations at any steady operating load.
(3) Providing system immediate reserve capacity by responding rapidly to large system frequency changes resulting from incidents involving loss of generation.
(4) Providing frequency control in the event of the emergency isolation of the plant with a small local load.

4.1 Control Range

When operating at part load, the plant is capable of achieving load changing rates up to 5% CMR/minute in either direction in the load range 50% to 100% CMR. The periods of part-load operation at 50% CMR could be of up to 10 hours duration. The control equipment ensures output regulation in response to system frequency variations at any load in the control range and ensures rapid corrective response to sudden frequency changes due to imbalance between system load and generation.

4.2 2-Shift Operation

The plant is capable of being unloaded in a controlled manner from full load conditions over a period of not more than 20 minutes. After unloading in this manner, followed by a shutdown period of up to 8 hours, full load on any unit can be achievable within 30 minutes of synchronising.

The auxiliary plant arrangements are designed such that for 2-shift operation all sets in the station can achieve full output within 90 minutes from first set synchronisation following an overnight shutdown. It is also a requirement that a failure of one set during the starting sequence or a failure in the control arrangements does not prevent other sets in the station achieving full output within the programmed time.

4.3 Load Rejection

Each main unit and associated auxiliary plant and control system is capable of withstanding a sudden loss of demand from 100% CMR to 25% CMR without shutting down the unit. This facility meets the situation where an exporting group of units becomes disconnected from the main supergrid system. The duration of this abnormal operating condition could vary widely.

4.4 Overload Capacity

The boiler plant and turbine generator are capable of providing overload capacity by means of bypassing the final stage of feedwater heating.

4.5 Design Life

The life span of the generating plant is anticipated to be in the order of 35 years and the typical number of cycles of operation during this lifetime on which component design is based is as follows:

Regulating Function	Lifetime Operations
Total number of starts	5000
Number of warm starts (36–48 hours)	1000
Remainder (mainly hot starts)	4000
Number of periods of operation at 50% CMR	3500

4.6 Fuel Oil Supplies

Table 2.1 gives some of the design fuel oil specification details.

4.7 Conditions of Operation

Table 2.2 gives the specified design parameters for steam flows, temperatures and pressures at selected loads.

TABLE 2.1. DESIGN FUEL OIL SPECIFICATION DETIALS

Property	Units	Typical Basic Fuel		Range of 95% of Annual Supplies Intended Limits	Specification Limits	
		Heavy	Light		Minimum	Maximum
Gross calorific value	kJ/kg	42,330	43,030	41,870–43,500	41,870	—
Net calorific value	kJ/kg	39,780	40,470	39,310–40,940	39,310	—
Kinematic viscosity at 82.2°C	cSt	46	10.5	6.6 – 65	6.6	115
Nominal viscosity-Redwood 1 at 38°C	sec	2000	200	100 – 3500	100	6500
Temperature to give 80 secs Redwood 1	°C	110	63.5	45 – 120	—	130
Specific gravity at 16/16°C		0.940	0.920	0.900–0.995	0.900	0.995
Closed flash point (Pensky Martens)	°C	greater than 70	greater than 70	—	65	—
Pour point	°C	20–50	38	50 Max.	—	50
Water content	% vol	0.2	Trace	0.5 Max.	—	0.5
Ash	% wt	0.05	Trace	Trace – 0.15	—	0.15
Sediment	% wt	Negligible	Negligible	Negligible – 0.1	—	0.25
Ultimate Analysis	%					
Carbon		85	85	84–86	—	—
Hydrogen		11.5	12	11–12.5	—	—
Sulphur		3.0	3.0	Max 3.0	—	4.0
Chlorine		Trace	Trace	Trace	—	—
Vanadium (as metal)		0.015	Trace	Trace–0.03	—	0.05
Sodium (as metal)		0.01	Trace	Trace–0.02	—	0.02
Ash		0.05	Trace	Trace–0.15	—	0.15
Oxygen, Nitrogen		Trace	Trace	—	—	—

5. STATION LAYOUT (FIG 2.2)

The site area of 21 hectares which was available for the construction of the new station is a comparatively small area on which to build a 2000MW power station.

The factors which influenced most of the station layout and plant orientation were:
(1) The limiting boundaries for river and road access.
(2) The suitable locations of construction storage and contractors' areas.
(3) The need to commission gas turbine plant early in the overall construction programme.
(4) The effect of the extensive cooling water civil works location and access.
(5) The need to complete the construction by working generally from north west to the access in the south east of this restricted site.

TABLE 2.2. SPECIFIED STEAM CONDITIONS

		Short Time Overload	Boiler CMR	94.8% Boiler CMR	75% Boiler CMR	50% Boiler CMR
Evaporation	kg/s	600	600	569	450	300
Steam pressure at outlet connection of stop valve at superheater outlet	bar	166.5	166.5	166.5	166.5	166.5
Superheater outlet temp	°C	541	541	541	541	528
Reheater steam flow	kg/s	510	441	418	330	220
Reheater outlet pressure	bar	46.2	41.3	39.4	31.4	20.9
Reheater inlet temp	°C	352	339	337	331	321
Reheater outlet temp	°C	541	541	541	541	515
Economiser inlet feed temp	°C	195	252	252	239	218

5.1 Main Plant

The existence of the transmission routes, together with the knowledge that fuel would be delivered by sea, determined that the boiler house and therefore the chimney should be located near to the river. Consequently, the location and orientation of the boiler drum, turbine house and generator compounds together with their access routes were established.

The width of the boiler house and the boiler centres were determined by boiler dimensions and the sootblower lance withdrawal requirements.

The choice between transverse, longitudinal and diagonal layouts for turbines was optimised primarily on turbine house crane span, high pressure and reheat pipework runs, and the type of low pressure and high pressure feed heaters to be used. The choice of tubular horizontally mounted heaters and the laydown requirements for the main turbine casings led to a transverse turbine arrangement.

To reduce the civil engineering costs of a large boiler house the forced draught fans were located outside the boiler house with the suction ducting protruding into the boiler house roof.

5.2 Auxiliary Plant

The location of construction storage areas and contractors' and CEGB site offices influenced the location of reserve feed water tanks and the water treatment plant which were located at the north east corner of the site. The same considerations influenced the location of gas turbines and their associated fuel tanks. The three gas turbine exhaust flues were directed into a single chimney which also included the flue from the auxiliary boiler, thereby influencing the location of the auxiliary boiler house.

The fuel oil heater house was located between the fuel oil storage tanks and the boiler house, with the sootblower air compressor house also in close proximity.

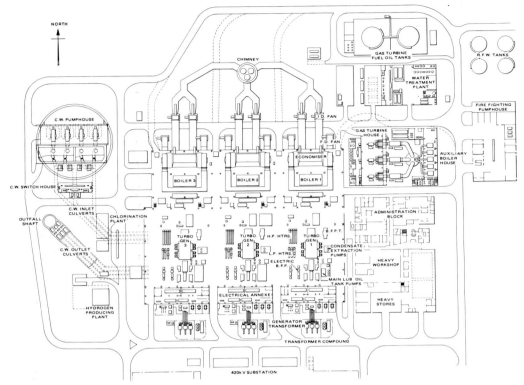

FIG. 2.2   STATION LAYOUT

5.3 Cooling Water Pumphouse

The location of the cooling water intake works in the River Thames dictated the location of the pumphouse on the west side of the site. The outfall shaft was placed at the same end as the pumphouse so that the culvert excavation did not seriously affect access to the boiler house. The chlorination plant was consequently located adjacent to the pumphouse.

5.4 Workshop and Administration Block

The administration block, which also contained the central control room, and the workshop were required to be as close as possible to the turbine house and therefore were located in the area to the south east of the main plant buildings.

6. FOUNDATIONS

The geology of the site generally consists of soft alluvium, including layers of peat, overlying a stratum of river gravel and sand, which varies in level and thickness over the whole site; beneath this lies chalk, which is initially soft but progressively improves with depth. The thickness of the gravel varies between 1m and 5m and the alluvium between 8m and 15m.

To raise the level of the site by some 3m and to provide a firm working surface, gravel fill from an adjacent pit was placed over the power station area and construction storage areas before piling commenced.

Because of these ground conditions, almost all building foundations are piled. A combination of step-taper driven piles with in-situ cores and large diameter (1200mm and 1050mm) steel-cased in-situ piles were used, depending on the concentration of load to be carried. All piles had to be permanently cased due to the instability of the alluvium and the chalk stratum, into which the piles were driven. Down-drag forces from the settling of alluvium reduce the useful load capacities of the piles. The only structures not piled are those of small size and light weight, in areas of pre-consolidation which could accept settlement relative to the piled buildings.

7. STRUCTURAL STEELWORK

The total weight of structural steelwork at Littlebrook D Power Station is about 33,000t. The boiler house structure, 64m high and 180m long is subdivided into three boiler bays adjacent to one deaerator bay within its 70m width and is designed to support three top hung boilers each of 5500t as well as a mass of ancillary plant, cables and pipework. The suspended floors around the boilers are designed for a superimposed load of $12kN/m^2$ at the lower level and $7.5kN/m^2$ at the upper levels. All major plant loads are specifically accommodated within the design. In addition to dead and superimposed loads the structure was checked for the effect of a boiler failure causing an implosion.

Superimposed floor loads in the deaerater bay vary from $7.2kN/m^2$ up to $24kN/m^2$ according to floor duty and equipment loading.

Wind loads on the structure are resisted by stiff moment connections between beams and columns, thus transferring the horizontal shear forces to foundation level. Provision for expansion in the length of the building is by sliding joints at two cross-sectional locations.

The turbine house, 35m high, 58m wide and 188m long is a single bay structure, structurally connected to the deaerator bay by its roof beams which are tubular lattice girders spanning the full 58m width. The steelwork supports two 175 tonne capacity overhead electric cranes which run on welded plate girders supported from the main box columns. The surge loading from these cranes is resisted by horizontal girders located at the crane rail level.

Site connections of steelwork members in the main building are generally made either by black bolts or high strength friction grip bolts, and utilise seating brackets on shear plates welded to the webs of beams on the faces of columns. Moment connections are of traditional form with high strength friction grip bolts transferring flange tension forces and web plates transferring vertical shear. At certain connections of large girders to columns, neutral axis connections are used to prevent bending moments being transferred to the columns.

The ancillary building complex, 48m long, 46m wide and 27m high is linked to the main building by a 3-storey bridge and has a conventional multi-storey steel frame constructed from standard sections using moment-connections. Two lines of vertical diagonal bracing are incorporated immediately below the roof level. Conjoint with the ancillary building complex is the workshops and stores building, 74m long by 60m wide and 9m high, which is a multi-bay portal frame structure with limited diagonal bracing.

The auxiliary buildings are, in general, portal frame structures constructed from standard sections. The larger ones incorporate diagonal bracing in an end-on intermediate bay and have rail girders supported from the columns carrying electric overhead cranes.

The structural steelwork is fabricated from steel to BS 4360 grades 43A, 43C, 43D and 50B. Standard rolled sections used range from 50mm × 50mm angles up to 914mm × 305mm universal beams and welded plate girders are from 1.5m up to 4m in depth with flanges of up to 75mm thick. Welded box columns vary from 1m × 1m to 2m × 1m, using plates from 25mm to 80mm thick and incorporating plate diaphragms.

Fabricated units are manufactured in normal manageable lengths and the weights arranged to suit both erection and availability of materials; the larger units average about 30t.

8. CIVIL DESIGN AND ARCHITECTURAL FEATURES

8.1 Buildings

The main building (boiler house, turbine house, deaerator bay and electrical annexe), the control block, workshops and stores and the majority of the auxiliary buildings, are steel framed structures with pre-cast concrete cladding at ground level and aluminium cladding above. The main fuel oil pumphouse and towns water and hydrant pumphouse in the tank farm area, and the gas turbine fuel oil transfer pumphouse in the main station area, are entirely of reinforced concrete to provide 4 hours fire resistance.

In the main building, the cooling water culverts, cable tunnels and service tunnels are incorporated in the structural foundation slab, whilst service trenches are within a mass concrete infill slab on top, which forms the ground floor level. The turbine generators are on steel blocks founded on heavy concrete bases, separated from the surrounding foundations by a 25mm joint of resin-bonded cork to isolate vibration. To allow for differential movement the cooling water culverts passing through these bases are articulated at each side.

Following usual practice, suspended floors in the boiler house are mainly open grid type and in other areas are of reinforced concrete. The concrete floors were cast on permanent steel formwork, because of the large storey heights and to allow early access for construction to the areas beneath.

The turbine house roof, apart from a strip adjacent to the deaerator bay which is of concrete, is of built-up felt with insulation on aluminium decking. The deaerator bay, boiler house and electrical annexe roofs are constructed from reinforced concrete with asphalt as waterproofing, precast concrete units being used where possible for speedy erection and economy. Over each boiler is a pent-house structure incorporating ventilating louvres and with a single skin aluminium decking roof to provide rapid venting in the event of a serious fire.

The auxiliary buildings have felt insulation and aluminium decking roofs, except where reinforced concrete is necessary to meet loading or fire resisting requirements. The aluminium cladding of the hydrogen plant house is designed to become partially detached in the event of an internal explosion, to release the pressure and prevent serious damage to the main structure.

The gas turbine house, whilst similar to the majority of the auxiliary buildings in its form of construction, incorporates three reinforced concrete air intake shafts and three reinforced concrete acoustic cells to house the gas turbines. The cell walls and roof are lined with steel plate to prevent blade penetration and precast units forming the roof can lift a limited amount to release sudden excessive pressure, but are restrained to ensure they return to their original positions.

The water treatment plant house incorporates two large effluent sumps, extending to 8m below floor level. The sumps are of reinforced concrete double-skin construction with bituthene between, lined with asphalt protected by acid and alkali-resisting brickwork, which is in turn covered by a film of epoxy resin.

8.2 Chimneys

The station has two chimneys, the main chimney being 210m high and the auxiliary chimney 70m high. The main chimney comprises a reinforced concrete wind shield enclosing three brickwork flues and founded on a reinforced concrete base, supported by 1200mm diameter bored piles. The brick flues are built in 9m high sections supported on floors spanning across the wind shield. The top sections of each flue are constructed from stainless steel and are heated to minimise the deposit of acid smuts. A rack and pinion hoist is provided within the wind shield to facilitate maintenance of the flue heating equipment.

The auxiliary chimney comprises a reinforced concrete wind shield enclosing three steel gas turbine flues and five steel auxiliary boiler flues. The reinforced concrete base is supported on step-taper piles.

8.3 Tank Farm

There are five fuel oil storage tanks in the tank farm, each having a capacity of 110,000t. Each tank has a reinforced concrete base 86m in diameter, supported on step-taper piles and covered with a 50mm thickness of bitumen sand mix on which the tank base is laid.

There are two auxiliary buildings situated in the tank farm, the main fuel oil pumphouse and towns water and hydrant pumphouse. Both are of piled reinforced concrete construction throughout and are designed for 4 hours fire resistance. The whole tank farm is surrounded by a gravel bund incorporating a pulverised fuel ash core.

8.4 Cooling Water System

Cooling water for the station is extracted from the River Thames via an intake situated some 200m from the river bank and flows through a tunnel driven in the chalk stratum to an onshore pumphouse. (See Fig 2.1). After being pumped through box section reinforced concrete culverts to the condensers, the water returns through similar culverts to the seal pit, whence an outlet tunnel leads to an outfall structure in the river some 300m downstream of the intake. The intake and outfall structures were designed as free air reinforced concrete caissons, which were sunk into temporary sand islands to the required level of −23m OD and became well embedded in the chalk stratum. Shafts were sunk from the bottom of each caisson to a level of approximately −48m OD and connect with the tunnels driven from onshore shafts. Both tunnels are situated in sound chalk which permits free air driving and are lined with precast concrete bolted segments with precast concrete infill 'pellets' to give a smooth interior bore. All shafts are also lined with bolted segments with a mass concrete internal lining to give the required hydraulic profiles.

The cooling water pumphouse, together with the drum screens and forebay were designed to be founded directly on the chalk stratum and built inside a 70m diameter circular cofferdam. Part of the cofferdam, which was constructed by the diaphragm walling technique, forms the front wall of the forebay. This design enabled the pump and screen chambers to be constructed within a clean, dry working area, in spite of their depth below ground level.

The cooling water system design incorporates a number of gates to allow the isolation of individual sections for maintenance purposes. The ports of both offshore structures can be closed by stop gates to enable the tunnels and shafts to be dewatered. The screen chambers can be isolated from the pumphouse forebay and gates, adjacent to the outlet seal pit, isolate this and the outlet tunnel from the culverts under the main building.

8.5 Main Jetty

Fuel oil is delivered to the tank farm via a jetty designed to receive fully laden tankers up to 30,000t dead weight or part laden up to 40,000t dead weight. The berth is dredged to accommodate tankers at all times, but there are sand bars between the berth and the River Thames navigable channel, which restrict the period for berthing.

The jetty head is of in-situ concrete construction supported on steel piles. At the berthing points the deck incorporates a deep beam to which steel fender piles are fixed via rubber energy absorbers. The jetty head is flanked by four mooring dolphins with solid concrete decks on piles.

The jetty approach is designed to a Ministry of Transport heavy loading standard and the main structural elements are precast concrete. Cross head beams span between piles and support longitudinal beams carrying precast slabs, while an in-situ reinforced slab on top ties the structure together.

8.6 Cable Routes

The majority of cable routes external to buildings and other piled areas, take the form of precast concrete cable troughs laid in the gravel fill, with 'snake pits' provided at the transitions from troughs to piled structures to allow for ground settlement. 132kV oil-filled cables are carried in reinforced concrete troughs spanning between piled ground beams. Inside buildings, cables are housed in tunnels at foundation level and in cable risers and races at higher levels. In the control block the races have blockwork walls and reinforced concrete floors and roofs. In the main building, races in the turbine house are steel-framed with reinforced concrete floors and roofs and steel cladded walls, those in the boiler house are steel-framed with steel cladding all round.

8.7 Services

The external fire hydrant system is designed as several interconnecting ring mains, encircling the following:
(1) Main building.
(2) 400kV substation.
(3) Control block, workshops and stores.
(4) Gas turbine house and auxiliary boiler house.
(5) Water treatment complex and gas turbine fuel tanks.
Each ring can be isolated. The system is fed by electrically driven pumps in the cooling water pumphouse. Two emergency pumping-in points are provided where mobile pumps can pressurise the system, either from the raw feed water tanks or from the lake, south of the station. There are also international tanker connections on the auxiliary jetty, where vessels can draw water from the system to fight onboard fires, or, if required, can pump water into the system. Pipework is ductile iron where supported by piled foundations and UPVC Class E for some runs in the ground subject to settlement. Flexible connections are provided at transitions from piled to unpiled areas.

The fuel oil tank farm is served by a ring main encircling the tanks with one internal spur to the main fuel oil pumphouse and an external spur onto the main jetty. The system is fed by electric pumps in the towns water and hydrant pumphouse and an emergency pumping-in point is provided, adjacent to No 3 water storage tank. There are international tanker connections on the main jetty head and two further pumping-in points are provided on the jetty approach for use by river craft. Pipework is cast iron Class 3, apart from the jetty spur which is UPVC Class E.

The sewage system is based on a series of sewage stations which collect sewage from each building by gravity flow and eject it into a series of common mains by compressed air. This arrangement gives flexibility in mode of operation and allows the installation required for the construction period to evolve into the system required for station operational conditions.

Surface water drainage in the main station area discharges to a creek on the west of the site. All flows which might be polluted with oil pass through separators either of the American Petroleum Institute design or, for larger flows, a tilting plate design. The creek empties to the River Thames through a flap valve, and acts as an intertidal storage area. To avoid the risk of flooding under storm conditions at high tide, pumps are provided which automatically pump excess water into the cooling water outfall system. The fuel oil tank farm area is free draining as the top fill layer is 300mm of rubble. Under storm conditions, excess water runs via a central spine drain to an oil separator and then to the river. Contaminated water which has been used for washing the boiler prior to access is chemically treated before discharge.

8.8 Architectural Concept

The siting of the power station within an area of grey and visually confusing industrial landscape on the south bank of the River Thames, was an opportunity to introduce a colourful and visually explicit focal element, particularly when viewed from the Temple Hill Housing Estate at Dartford.

Previous experience of the problems associated with power station construction led to the adoption of a simple building cladding system, utilising two principal elements. The predominant material is profiled alkyd amino coated aluminium sheeting in large areas of simple shapes. At low level, precast concrete plinth units are used to withstand rough treatment and provide for the fixing of small plant items. The units, cast on site using local aggregate, are manufactured in two widths, 1.2m and 2.4m, with heights of 1.2m, 3.5m and 5.5m. They are designed to be stable immediately they are placed into position by provision of a splayed base, which has the virtue of keeping persons and vehicles away from the exposed aggregate finish, thus minimising the risk of damage. See Chapter 1 Fig 1.8.

These two principal elements together with a range of co-ordinated components are used in a variety of combinations to unify visually buildings of diverse functions, expressing a comprehensive design philosophy throughout the development. Colour is used to identify the various buildings. The boiler house and the deaerator are dark yellow ocre with dark brown detail, the turbine hall and ancillary building complex is white and the auxiliary plant buildings are dark brown with bright red and green detail. The problem of scale posed by the juxtaposition on the site of buildings of widely differing sizes, is resolved by varying the scale of such items as the pitch and form of corrugations in cladding, louvres, ventilators and doors.

8.9 Littlebrook Nature Park

To facilitate the building of Littlebrook D Power Station it was necessary to have adequate space for the storage of materials and plant during construction. This was achieved by leasing land adjacent to the station site from Dartford Borough Council. This land was gravel bearing and permission was obtained to extract the gravel for use as fill and aggregate for concrete during the early construction period. When the gravel extraction was completed the workings were to be allowed to flood and form a lake, some 500m long by 300m wide. The terms of the lease included an obligation on the CEGB to landscape the area to Dartford Borough Council requirements before handing it back.

Dartford's initial intention was to use the land, some 48 hectares, as a public park with the new lake as a central feature. However, it was subsequently suggested that it would make an excellent nature park and could be used principally for educational rather than recreational purposes. For many years, the area had been of interest to ornithologists and it was noted that the edges of the newly flooded gravel workings were already supporting a flourishing wildlife community, in addition to those already existing on some other parts of the leased land. Members of Dartford Borough Council were in full agreement with the new proposal and a scheme for landscaping was prepared. This included modifying the contours of the originally flat site, the formation of dipping pools and a shallow water pond, enlargement of a marsh area and creation of wooded enclosures, open grassed areas, and islands and a lagoon at the east end of the main lake. The park also embraces a second lake, about 280m long by 150m wide, which existed prior to the CEGB leasing the land.

The scheme, in general, preserved the existing vegetation and lake to cause minimal disturbance to existing wildlife. The new features were designed to provide a habitat to encourage the wildlife to flourish, expand and to attract new species, if possible. Also, the scheme was designed to conceal or camouflage the visually unattractive elements on the site and generally to provide an attractive foreground to the station when viewed from the higher ground to the south.

9. BOILER AND ASSOCIATED SYSTEMS

A section through the station showing the general location of the boiler and turbine generator is given in Fig 2.3.

9.1 Boiler Construction (Fig 2.4)

The boiler plant consists of three 660MW oil-fired reheat boilers designed to meet the requirements given in Section 4 - Operational Roles. The boilers are single furnace, front wall fired, assisted circulation reheat type.

Each boiler is fitted with 32 steam atomising oil burners arranged in four rows of eight in the burner windbox. Fuel oil for the burners is drawn from the storage tanks by six pressure pumps located in the transfer pumphouse and is then pumped to the fuel oil heater house where it is heated by three groups of heaters. Each group, comprising 15 individual steam heated units, is arranged in sets of three, from which the heated oil discharges into a common main which then divides into three to serve each boiler.

The heat recovery system incorporates a plain tube economiser and twin rotary gas airheaters. Steam airheaters ensure that the gas airheater metal temperature is maintained above the acid dew point of the boiler gas.

The draught plant incorporates two forced draught (FD) fans and two induced draught (ID) fans.

Low pressure (LP) and high pressure (HP) chemical dosing equipment is incorporated in the LP feedwater and boiler circulating systems respectively for the adjustment of alkalinity and reduction of scale-forming impurities.

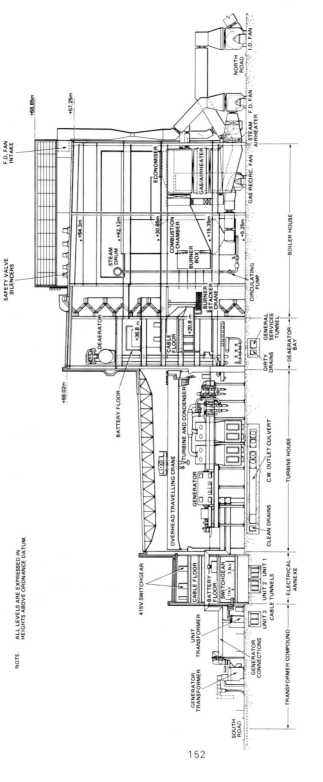

NOTE: ALL LEVELS ARE EXPRESSED IN
HEIGHTS ABOVE ORDNANCE DATUM.

FIG. 2.3   SECTION THROUGH STATION

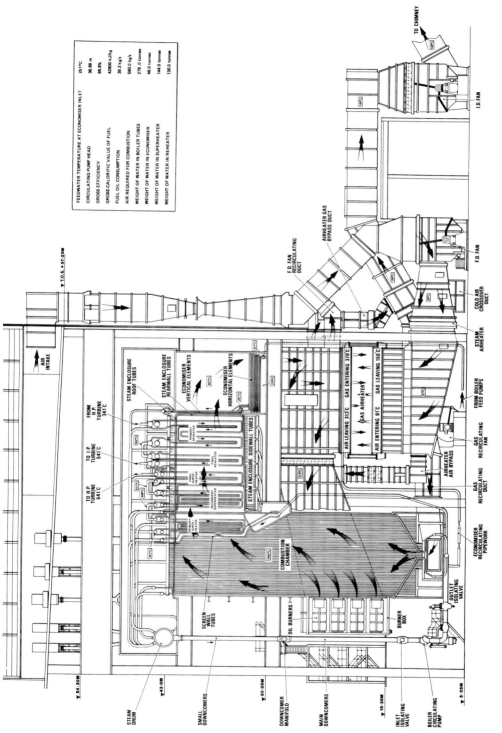

FEEDWATER TEMPERATURE AT ECONOMISER INLET	251°C
CIRCULATING PUMP HEAD	36.89 m
GROSS EFFICIENCY	88.8%
GROSS CALORIFIC VALUE OF FUEL	42800 kJ/kg
FUEL OIL CONSUMPTION	39.3 kg/s
AIR REQUIRED FOR COMBUSTION	560.0 kg/s
WEIGHT OF WATER IN BOILER TUBES	278.0 tonnes
WEIGHT OF WATER IN ECONOMISER	46.0 tonnes
WEIGHT OF WATER IN SUPERHEATER	144.0 tonnes
WEIGHT OF WATER IN REHEATER	130.0 tonnes

FIG. 2.4   SECTION THROUGH BOILER

Eight electrically-operated sootblowers are installed, four on each side of the boiler, for high pressure air cleaning of the economiser elements.

Water washing, compressed air sootblowing and fire protection water facilities are provided for each of the gas airheaters.

An integrated system of controls and instrumentation is provided to allow safe and efficient operation of the plant. All the main plant items are operated from the central control room. Local control is provided for certain items not directly affecting station output.

A sequence control system allows single switch startup and shutdown initiation of functional groups of plant. This includes starting and tripping interlocks to prevent unsafe plant conditions occurring at any time during normal startup and shutdown or due to any plant fault condition resulting in a trip.

A modulating control system comprising a number of independent control loops ensures that the performance of the plant automatically matches the megawatt output demand. Auto/manual stations are incorporated in the control loops to allow the operator to control sections of plant manually when necessary.

9.2 Feedwater and Circulating System (Fig 2.5)

The feedwater and circulating system for each boiler incorporates an economiser unit, steam drum, four circulating pumps, downcomers and water walls, HP and LP dosing equipment, water sampling equipment and instrumentation for indicating and control duty.

Each boiler is supplied with feedwater drawn from the deaerator outlet by a feed pump. The feedwater passes through the economiser to the steam drum and then is drawn by the boiler circulating pumps and distributed through the discharge pipework to the boiler water wall circuits and returns to the drum as steam/water mixture. During startup and periods of low load, pipework connecting the circulating pump discharge manifold to the economiser inlet pipes allows the economiser to be recirculated with hot water.

In the steam drum, the steam and water are separated by cyclone separators and steam scrubbers. The water, the chemical content of which is adjusted by the HP chemical dosing equipment, returns to the boiler circulating system and the steam passes to the superheater system. Drains from the various points in the system are fed to the dirty drains vessel.

9.3 Boiler Circulating Pumps

Each boiler is provided with four glandless motor/pump units. Each unit consists of a single stage centrifugal pump, driven by a 671kW induction motor of the wet stator type operating at 1450 rev/min from a 3.3kV 3-phase 50Hz supply. Each pump has a double discharge and delivers 725kg/s at 363°C.

The motor/pump unit is mounted vertically with the pump casing at the top welded into the circulating system pipework. The motor casing, complete with the pump internals, is flange bolted to the pump casing and can be easily removed.

The pump bearings are lubricated by boiler water at system pressure which fills the motor casing. To keep the motor cool, its water content is circulated through the bearings and windings to a side-mounted heat exchanger by an auxiliary impeller

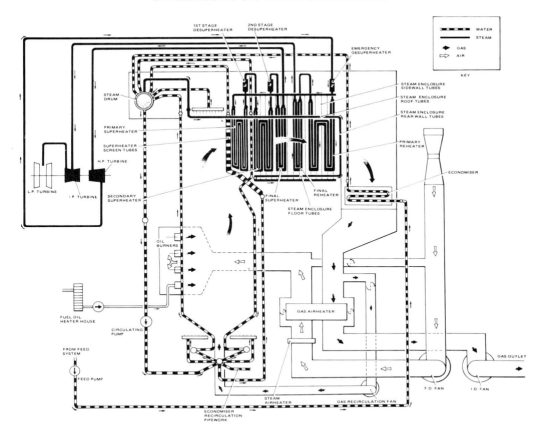

FIG. 2.5   **BOILER FLOW DIAGRAM**

integral with the thrust disc at the base of the motor. The flow of heat and boiler sediment from the pump to the motor is limited by the close clearance between the shaft and the motor casing neck. Should any foreign matter pass into the motor enclosure it is filtered out by a strainer located at the base of the motor.

9.4   Steam Drum (Fig 2.6)

The steam drum is situated at the front of the boiler at the 43.8m level and is suspended from the boiler steelwork by two cradle type slings. The drum which is 31.4m long is of welded construction and consists of 11 rolled steel sections 2.2m internal diameter by 123mm thick and two 75mm thick dished end plates. The drum end plates each incorporate an elliptical manhole which is sealed by a hinged cover. The shell and end plates are manufactured from material to BS 1501-271B.

Welded to the outer surface of the drum are a total of 93 nozzles to which are welded the steam pipes, water pipes, safety valves, etc.

The steam/water cyclone separators, steam scrubbers, downcomer anti-vortex baffles, feed pipe, chemical injection pipe and pipework for the water level indication equipment are fitted inside the drum. A plenum chamber is formed along the front and rear walls of the drum to contain the incoming steam/water mixture before it passes through the cyclone separators.

Steam/water mixture enters the drum plenum chambers through the riser pipes from the front, rear, side and screen wall outlet headers. From the plenum chambers it passes through the cyclone separators, where water particles are separated from the steam; the water falls to the bottom of the drum and the steam passes up through the cyclone scrubber boxes and then to the take-off pipes through the outlet scrubbers, where further separation takes place. Water particles separated from the steam by the outlet scrubbers are collected in a series of troughs and returned by a drain pipe to the bottom of the drum.

9.5 Steam System

The steam system for each boiler comprises superheater and reheater circuits through which the steam passes to the high pressure and intermediate pressure (IP) turbines respectively. There are four stages of superheat and two stages of reheat employing roof tubes and pendant banks of tubes arranged in a single horizontal pass. (See Fig 2.4). Superheat temperature control is by first and second stage desuperheaters situated respectively in the outlets from the primary and secondary superheaters. Reheat temperature control is by gas recirculation to the bottom of the combustion chamber, with desuperheaters at the inlet to the primary reheater for emergency use.

Saturated steam from the steam drum passes through the combustion chamber/steam enclosure roof tubes, which form the radiant superheater surface, and continues through the steam enclosure rear wall and floor tubes to the floor outlet header. The steam then passes upwards through the steam enclosure side walls to the primary superheater inlet and through the secondary and final superheaters before being fed as superheated steam to the HP turbine, through four steam mains.

After passing through the HP turbine the steam is returned to the boiler to be reheated by the primary and final reheaters, before passing to the intermediate pressure (IP) and LP turbines.

9.5.1 *Superheater Circuit*

The superheater circuit consists of primary, secondary and final sections suspended in the boiler horizontal pass. The primary section is suspended in the combustion chamber at the entrance to the horizontal pass, with the secondary and final sections behind.

Before entering the primary superheater the saturated steam from the steam drum passes through the steam enclosure tubes. These comprise roof, rear wall, floor and side wall tubes which enclose the secondary and final superheaters together with the primary and final reheaters. The roof tubes provide the radiant superheater surface.

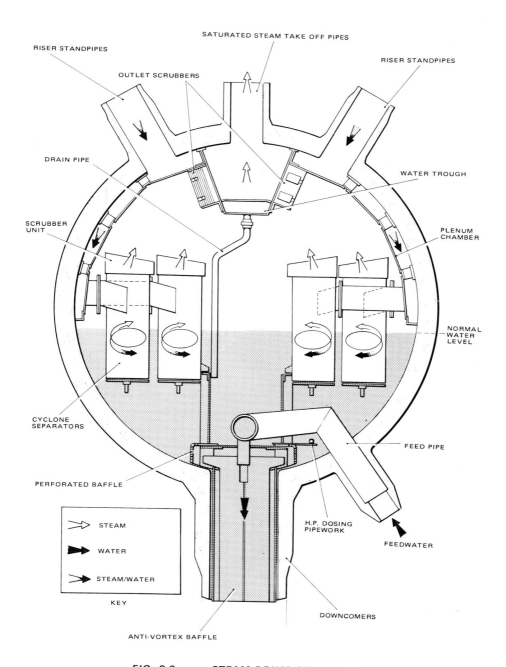

FIG. 2.6   STEAM DRUM OPERATION

9.5.2 *Reheater Circuit*

The two reheater sections are situated in the horizontal pass and are suspended within the steam enclosure. They are arranged against the gas flow, ie the final reheater is located immediately to the rear of the final superheater and is followed by the primary reheater.

After passing through the HP turbine the steam is returned to the boiler through four inlet pipes. The steam passes through the primary and final reheaters before being conveyed by the outlet pipes to the IP turbines.

9.5.3 *Auxiliary Steam Back-up System*

The use of steam-atomised fuel oil burners in the main boilers risks the loss of a main boiler, and hence the unit, due to the loss of auxiliary steam. To safeguard against this, the main boilers have a connection from the primary superheater outlet headers, such that main boiler steam can instantaneously feed the auxiliary steam system. In order to match the main boiler steam conditions to the auxiliary steam conditions, a pressure reducing/desuperheater valve is installed which incorporates a spraywater connection from the HP feed system. The nominal pressure of the auxiliary steam system is 17.5 bar and the back-up steam system is set to operate automatically when this pressure drops to 16 bar.

9.5.4 *Desuperheaters and Spray water*

Temperature control of steam entering the secondary and final superheaters is by first and second stage spray desuperheaters. The first stage desuperheaters are incorporated in the pipework connecting the primary superheater outlet and secondary superheater inlet headers. The second stage desuperheaters are incorporated in the crossover pipework connecting the secondary superheater outlet and final superheater inlet headers. Desuperheaters for emergency use are also incorporated in the inlet pipes to the primary reheater. The desuperheaters are welded-in to form a section of the respective pipes.

Spraywater is taken from the boiler main feed line, downstream of the feed regulator valves, and delivered by a pump to the desuperheater nozzles through the spraywater pipework. Isolating and non-return valves are provided in the spraywater mains to the first stage, second stage and reheat emergency desuperheaters with isolating and control valves in the branch pipes to individual desuperheaters.

9.6 Gas System (Fig 2.7)

Two interconnected systems, each of approximately 70% CMR capacity, capable of independent operation, are provided for each boiler to convey boiler flue gas to the chimney. The systems are identical, each incorporating an induced draught (ID) fan and gas airheater connected by ducting and discharging into a flue leading to the chimney. An interconnecting crossover duct is fitted between the two ID fan inlet ducts.

A proportion of the flue gas is utilised for reheat temperature control purposes (8% at CMR) and is recirculated from the economiser outlet duct to the bottom of the combustion chamber by two gas recirculating (GR) fans, one at each side of the boiler.

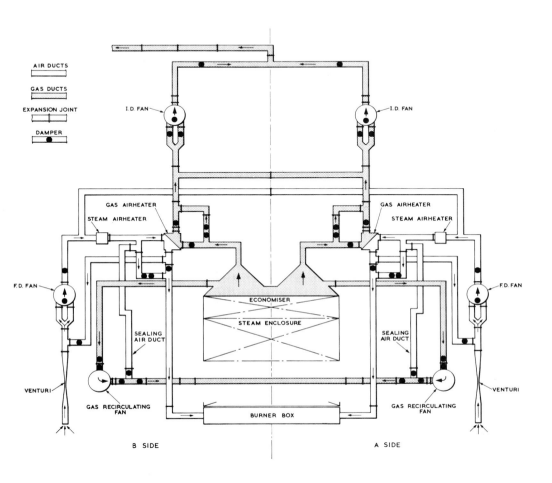

FIG. 2.7   GAS AND AIR FLOW

Isolating and control dampers are installed in the ducting to control the gas flow. They are arranged to permit startup and part load operation of the boiler with both gas airheaters and one ID fan isolated and shutdown.

9.7   Combustion Air System (Fig 2.7)

Two interconnected systems, each of approximately 70% CMR capacity, capable of independent operation, are provided for each boiler to supply combustion air to the burner windbox. The systems are identical, each incorporating an FD fan, steam airheater and gas airheater connected by ducting and supplying one side of the windbox. An interconnecting cross-over duct is fitted between the two FD fan discharge ducts.

Isolating and control dampers are installed in the ducting to control the air flow. They are arranged to permit startup and also part load operation of the boiler with both gas airheaters and one FD fan isolated and shutdown.

Air from the atmosphere is drawn into the system by the FD fans and discharged through the steam airheaters and the air side of the gas airheaters to the burner windbox. The gas airheaters are bypassed until specified operating temperatures are reached.

10.   HP PIPEWORK

10.1   Main and Reheat Steam Systems

The pipework is designed for a station operating life of 160,000 hours. The high temperature pipework, ie main steam and hot reheat, are manufactured in Cr/Mo/V steel and the welds are made with 2.25% Cr 1% Mo electrodes. No allowance is made for the lower strength of the weld metal since, in both cases, the weld strength resulting from the use of minimum bend thickness as the basis of design throughout was not less than 90% of that for the parent Cr/Mo/V pipe material.

10.1.1   *Main Steam Pipework*

There are four main steam pipes connecting the boiler to the HP turbine steam chests, each being 279.4mm bore by 50.85mm minimum thickness. The design conditions are $17.5 \times 10^6 \text{N/m}^2$, 541°C and a maximum design stress of 62.85N/m^2.

10.1.2   *Reheat Pipework*

The cold reheat steam from the HP cylinder of the turbine exhausts into two 660mm bore pipes each of which discharges into twin 508mm bore pipes forming the reheat inlet of the boiler. The reheat inlet piping is manufactured in 27t carbon steel suitable for design conditions of $5.1 \times 10^6 \text{N/m}^2$, and 339°C.

The hot reheat piping from the boiler reheater to the IP turbine steam chests is in Cr/Mo/V steel suitable for design conditions of $4.86 \times 10^6 \text{N/m}^2$ and 540°C. There are four 482.6mm bore by 21.25mm minimum bend thickness pipes having a maximum stress of 64N/m^2 running parallel under the boiler.

There is an overall requirement that the pressure drop through the reheat system shall not exceed 7.25% of the pressure at the HP turbine exhaust. This requirement is based on an optimisation of capital cost and pressure drop as this affects the thermal performance of the unit as a whole. The reheat inlet and outlet pipework has been sized to match this concept allowing for minimum capital cost and production requirements for the pipework.

10.1.3 *Measurement of Creep at High Temperature*

Provision has been made on the main steam pipework and hot reheat pipework of creep measurement 'pips' to be provided so that the behaviour of the pipes over the life of the plant can be monitored.

The measurement 'pips' are placed on the major axis, near the boiler stop valve, near the turbine and about mid-point of the pipe run on one leg of each service either side of the boiler.

11. TURBINE AND CONDENSER

11.1 General and Design Data (Fig 2.8)

The power plant comprises three multi-cylinder, impulse-type, 660MW turbine generators, each mounted on a steel foundation.

Steam for each turbine is supplied from an adjacent boiler. The steam is fed initially into the HP cylinder and then passed through the reheat system before entering the IP cylinder; this steam is then passed to the LP cylinders before exhausting into a twin, underslung condenser.

Condensate extracted from the condenser is pumped through the condensate system and feed heating system where it is progressively heated before entering the boiler.

The turbine design is as follows:

Type	impulse, disc and diaphrams, single reheat
Speed	3000rev/min
Heat rate	8212kJ/kWh
Power output	660MW (at generator terminals)

Steam conditions at turbine inlets:

HP cylinder	
pressure	159.6 bar
temp	538°C
IP cylinder	
pressure	39.00 bar
temp	538°C
LP cylinder	
pressure	4.63 bar
temp	259°C

Vacuum (back pressure):

LP1 cylinder	61mbar
LP2 cylinder	61mbar
Electrical turning gear speed	37rev/min

11.2   High, Intermediate and Low Pressure Cylinders

Each turbine set consists of a single-flow HP cylinder, a double-flow IP cylinder and two double-flow LP cylinders. The rotors in each cylinder are coupled together and form one line of shafting to drive the generator. There are two bearings supporting each rotor and these are housed in pedestals, separate from the turbine cylinders.

Steam from the LP cylinders exhausts into a spring-mounted, twin-shell condenser. Fig 2.9 shows the general arrangement of the LP cylinder.

11.3   Bearing Pedestals

The main purpose of the bearing pedestals is to support the rotor in a fixed relationship to the cylinders. By this means, all gland clearances are kept constant during all phases of operation. All pedestals and covers are rigidly constructed from fabricated steel, stiffened with ribs and gusset plates.

11.4   Bearings

The HP, IP and both LP rotors are rigidly coupled to form, in effect, a single shaft. The complete assembly is supported on eight journal bearings and is located axially by a single thrust bearing situated between the HP and IP cylinders.

11.5   Turbine Expansion

The temperature differentials occurring between different parts of the turbine in service cause a differential expansion or contraction between the turbine components. Construction features are provided whereby the various components of the turbine can move, relative to each other, whilst still maintaining correct alignment of the rotors and the correct clearances between stationary and rotating parts.

11.6   Electrical Turning Gear and Hand Barring Arrangement

The HP, IP and LP rotors must be turned continuously throughout the process of warming through prior to startup and also during cooling after shutdown. This must be done to avoid distortion of the rotors.

The electrical turning gear drives the rotors at a constant speed of 36rev/min through a reduction gear and a synchronous, self-shifting clutch. Power is provided by an electric motor. In the event of electrical turning gear failure and for maintenance purposes, hand barring facilities are provided to turn the rotors manually.

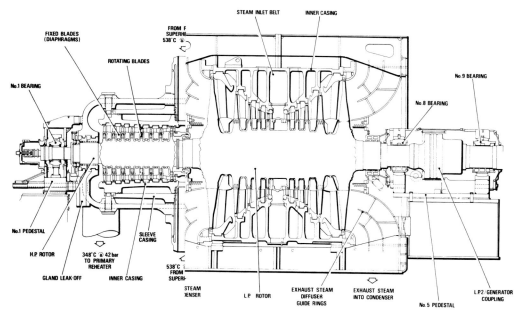

FIXED BLADES (DIAPHRAGMS)

ROTATING BLADES

STEAM INLET BELT

INNER CASING

FROM F SUPERHI 538°C @

No.9 BEARING

No.1 BEARING

No.8 BEARING

No.1 PEDESTAL

SLEEVE CASING

H.P ROTOR

348°C @ 42 bar TO PRIMARY REHEATER

GLAND LEAK·OFF

INNER CASING

538°C FROM SUPERI

STEAM ·DENSER

L.P ROTOR

EXHAUST STEAM DIFFUSER GUIDE RINGS

EXHAUST STEAM INTO CONDENSER

No.5 PEDESTAL

L.P.2/GENERATOR COUPLING

H.P. TURBINE

L.P.2 TURBINE

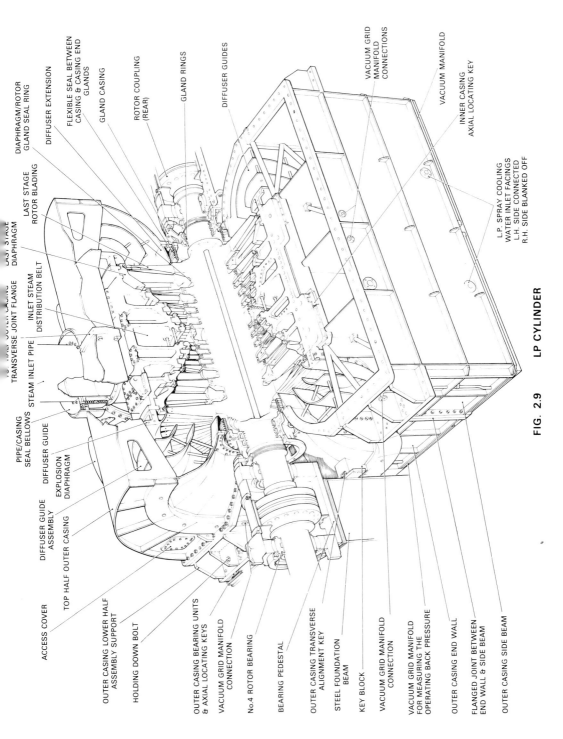

FIG. 2.9   LP CYLINDER

DIAPHRAGM/ROTOR GLAND SEAL RING

DIFFUSER EXTENSION

FLEXIBLE SEAL BETWEEN CASING & CASING END GLANDS

GLAND CASING

ROTOR COUPLING (REAR)

GLAND RINGS

DIFFUSER GUIDES

VACUUM GRID MANIFOLD CONNECTIONS

VACUUM MANIFOLD

INNER CASING AXIAL LOCATING KEY

L.P. SPRAY COOLING WATER INLET FACINGS L.H. SIDE CONNECTED R.H. SIDE BLANKED OFF

LAST STAGE ROTOR BLADING

LAST STAGE DIAPHRAGM

INLET STEAM DISTRIBUTION BELT

TRANSVERSE JOINT FLANGE

PIPE/CASING SEAL BELLOWS

STEAM INLET PIPE

DIFFUSER GUIDE

DIFFUSER GUIDE ASSEMBLY

EXPLOSION DIAPHRAGM

ACCESS COVER

TOP HALF OUTER CASING

OUTER CASING LOWER HALF ASSEMBLY SUPPORT

HOLDING DOWN BOLT

OUTER CASING BEARING UNITS & AXIAL LOCATING KEYS

VACUUM GRID MANIFOLD CONNECTION

No.4 ROTOR BEARING

BEARING PEDESTAL

OUTER CASING TRANSVERSE ALIGNMENT KEY

STEEL FOUNDATION BEAM

KEY BLOCK

VACUUM GRID MANIFOLD CONNECTION

VACUUM GRID MANIFOLD FOR MEASURING THE OPERATING BACK PRESSURE

OUTER CASING END WALL

FLANGED JOINT BETWEEN END WALL & SIDE BEAM

OUTER CASING SIDE BEAM

11.7 Forced Cooling

A significant reduction in the time of running the electrical turning gear and hence a quicker access for maintenance purposes, can be achieved by increasing the cooling rate of the set after shutdown, while maintaining the limitations imposed by the differential contraction of rotors and casings.

The increased rate of cooling is achieved by passing a controlled flow of air through the space between the inner and outer casings and between the inner casing and the rotor. This applies to the HP and IP cylinders only.

The air supply is obtained from a self-contained cooling unit mounted on the turbine floor. This unit supplies air at a pressure of 1.5 bar g to any set as required.

11.8 Governing and Control System

Turbine operation is governed by a computer-controlled system whose signals regulate the operation of the steam valves. This system governs the fluid supply to the hydraulic actuators which position the valves.

The system provides monitoring and protection equipment, controlled run-up from start and maintains the turbine speed within specified tolerances under the demands of changing generator load. In addition, it provides emergency stop facilities for turbine and generator protection.

11.9 Turbine Protection

Protection devices are installed to prevent damage to the turbine generator in the event of a malfunction. These consist of mechanical devices fitted into the emergency trip gear on the turbine HP end bearing pedestal, trip mechanisms initiated from the electric governor cubicle and explosion diaphragms in the LP cylinders.

In addition to turbine protection, the set is also tripped if a malfunction develops in the stator winding cooling water system, thereby preventing damage to the generator.

11.10 Turbovisory Equipment

Turbovisory equipment comprises a group of sub-units which, in conjunction with turbine mounted detectors, provide continuous monitoring of the turbine generator set under all operating conditions. Some measurement outputs are fed to recorders, thus producing a permanent record of turbine data. Others are passed to indicators and/or data processing equipment. Signal monitors are included in some functions to provide audible and/or visual alarms if certain pre-set values are exceeded.

11.11 Flange Heating

During a start and run-up, particularly from cold, the walls of the HP and IP cylinders could heat up more rapidly than the thicker metal of the flanges. Flange heating is provided therefore to reduce the temperature differentials of the cylinder walls and flanges by passing high temperature steam through passages in the horizontal joint flanges of the HP and IP cylinders.

11.12 IP Rotor Cooling

The use of IP rotor cooling is provided to reduce rotor temperature, thereby extending the material creep life. Cold reheat steam from the HP cylinder is passed via an orifice plate and rotor cooling steam pipe, through the inner casing nozzle box to the centre of the IP rotor, then dispersed axially outwards through the first stage discs.

11.13 Gland Sealing

Gland sealing prevents the escape of steam and the leakage of air into the shaft ends of all the turbine cylinders including the boiler feed pump turbine. This arrangement also performs a similar function on the spindles of the HP and IP steam valves and on the inlet steam valves of the boiler feed pump turbine.

Shaft glands consist of labyrinth type seals formed by spring mounted, finned segments operating with a fine clearance against the turbine rotor. The last pocket on each gland is connected to the gland steam condenser to ensure that no steam leaks into the turbine hall.

When steam pressure in a turbine cylinder is above atmospheric pressure, the sealing clearance spaces in the seals are packed with steam passing from inside the cylinder into the gland pocket. When the steam pressure within the cylinder is below atmospheric pressure, sealing steam is supplied to seal the glands and prevent the inward leakage of air.

11.14 Lubricating Oil System

Each turbine generator set has a separate lubricating oil system supplying oil at a pressure of 1 bar g to the turbine and generator journal bearings and thrust bearing. This system also supplies oil to the generator shaft seals and the boiler feed pump turbine bearings.

The system comprises a main oil tank, main and auxiliary oil pumps, duplex filters, oil coolers, purifier, valves and interconnecting pipework.

The main oil pump is located on a pedestal at the rear of the turbine generator set, adjacent to the exciter, where it is driven directly from the end of the exciter shaft.

11.15 Pedestal and Cylinder Alignment

Each rotor is accurately aligned to its adjacent rotor. Measurements are taken to ensure correct peripheral alignment and parallelism. The accuracy of this critical alignment is dependent upon a constant relative position between individual bearing supports, the supporting columns and the foundations. It is vital, therefore, to obtain and record any foundation movement which may occur. Variations in horizontal alignment are measured by equipment which assesses the difference in fluid levels at strategic points on the turbine generator sets. Equipment for measuring these parameters is permanently installed. Two portable sets are provided for back-up purposes. Vertical expansion or contraction is measured by means of an invar rod assembly installed at the rear of each set.

11.16 Condenser (Fig 2.10)

Two single-pass condensers are positioned under each turbine generator set, one beneath each LP exhaust casing. The steam flow from each LP cylinder exhausts into the condenser and is directed across two single-pass tubenests.

Inlet and outlet waterboxes, identical in shape, are located at each end of the condenser.

11.17 LP Exhaust Spray Cooling

When the turbine generator is running at no-load or light-load, the steam flow is not sufficient to work efficiently in the LP cylinder blading. In the no-load condition, a very low pressure drop over the blades causes them to transfer energy to the steam, thus reversing the normal process when under full load running. This transfer of energy raises the temperature of the steam and excessively high temperatures would be reached if the turbine was allowed to run for prolonged periods under these conditions, to prevent this, exhaust spray cooling is provided.

11.18 Air Extraction

The function of the air extraction plant is to raise and maintain vacuum conditions in the turbine main condenser by removing air and other incondensable gases vented into the condenser from various parts of the turbine and feed heating system.

12. GENERATOR (FIG 2.11)

The generator stator is constructed, for transport purposes, from two main assemblies, a multi-ribbed pressure tight outer frame and a skeleton inner frame which embodies the laminated stator core and the stator winding. The generator stator is assembled on site by sliding the stator inner frame into the outer frame.

The core is supported on spring mountings to minimise the transmission of vibrations to the foundation steelwork, and the generator rotor is supported in plain journal bearings which are housed in the rear of the LP turbine and the exciter endshield.

The generator shaft seals are supported in carriers mounted off the endshields and provide sealing to prevent the cooling hydrogen escaping along the shaft at the ends of the gas tight outer frame. The seals are supplied with oil at a pressure of approximately 1.38 bar more than the frame hydrogen pressure. The oil is both the sealing and the lubricating medium.

The generator endshields are of one piece construction and are designed so that the generator rotor can be removed from the stator without removing the endshields.

The generator rotor and stator are direct hydrogen-cooled. The stator winding, phase end connections, terminal bushings and the neutral shorting bar are direct water cooled. The neutral terminals are mounted on the top of the machine and the main terminals are located under the machine.

The generator is solidly coupled to the turbine and, both mechanically and electrically, to the air-cooled exciter.

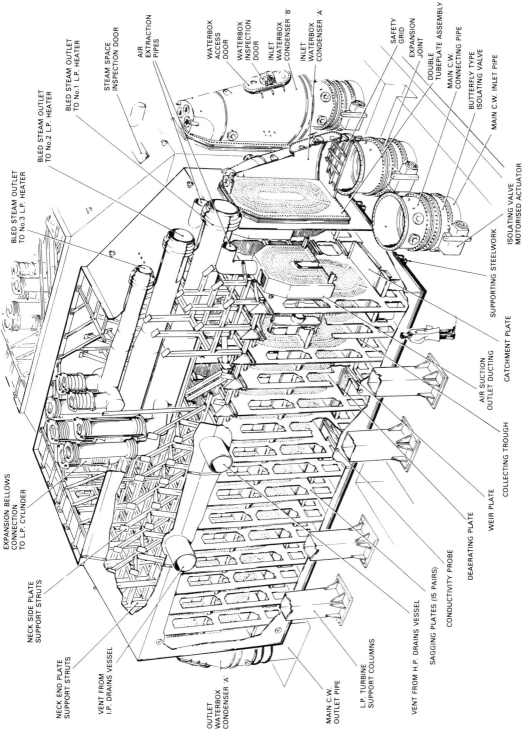

BLED STEAM OUTLET
TO No.2 L.P. HEATER

BLED STEAM OUTLET
TO No.1 L.P. HEATER

STEAM SPACE
INSPECTION DOOR

AIR
EXTRACTION
PIPES

WATERBOX
ACCESS
DOOR

WATERBOX
INSPECTION
DOOR

INLET
WATERBOX
CONDENSER 'B'

INLET
WATERBOX
CONDENSER 'A'

SAFETY
GRID

EXPANSION
JOINT

DOUBLE
TUBEPLATE ASSEMBLY

MAIN C.W.
CONNECTING PIPE

BUTTERFLY TYPE
ISOLATING VALVE

MAIN C.W. INLET PIPE

ISOLATING VALVE
MOTORISED ACTUATOR

SUPPORTING STEELWORK

CATCHMENT PLATE

AIR SUCTION
OUTLET DUCTING

COLLECTING TROUGH

WEIR PLATE

DEAERATING PLATE

CONDUCTIVITY PROBE

SAGGING PLATES (I5 PAIRS)

VENT FROM H.P. DRAINS VESSEL

L.P. TURBINE
SUPPORT COLUMNS

MAIN C.W.
OUTLET PIPE

OUTLET
WATERBOX
CONDENSER 'A'

VENT FROM
I.P. DRAINS VESSEL

NECK END PLATE
SUPPORT STRUTS

NECK SIDE PLATE
SUPPORT STRUTS

EXPANSION BELLOWS
CONNECTION
TO L.P. CYLINDER

BLED STEAM OUTLET
TO No.3 L.P. HEATER

FIG. 2.10   CONDENSER GENERAL ARRANGEMENT

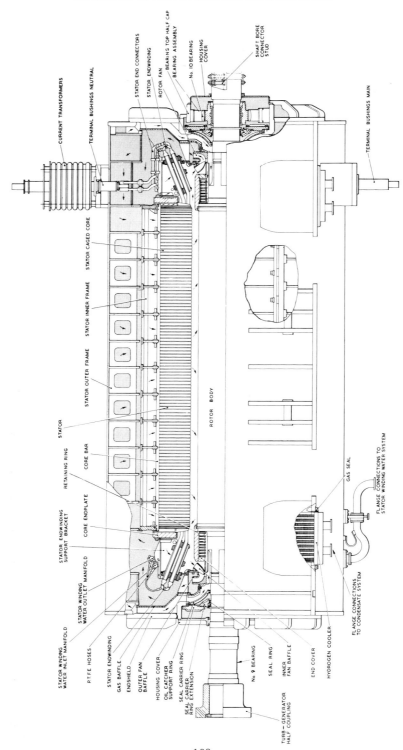

FIG. 2.11   GENERATOR GENERAL ARRANGEMENT

12.1 Design Data

The generator design data is as follows:

Rated output	660MW
	776MVA at 0.85pf lagging
Terminal voltage	23.5kV
Speed	3000rev/min
Stator current	19,076A
Hydrogen pressure	4.14 bar
Volume of hydrogen space	140m³

Weights:

Inner stator frame	236t
Outer stator frame	118t
Rotor	76t
Excitation current	4805A
Excitation voltage	569V

12.2 Hydrogen Producing Plant

The hydrogen producing plant is a completely self-contained system requiring only the input of electrical power and demineralised water to produce the hydrogen necessary for cooling the main generators.

The plant is located in a separate building as shown in Fig 2.1. It will generate a maximum of 34m³/h of hydrogen, compress it to a pressure of 31 bar, purify, dry and store it in storage vessels ready for use. 17m³/h of oxygen are formed as a by-product of the process. This is not stored but is allowed to vent to atmosphere as a low pressure gas.

The gas-producing process consists of a bank of 17 electrolytic cells each containing two nickel plated iron anode plates and three iron cathode plates in a 28% solution of potassium hydroxide. A 56.5V 3000A dc supply from a transformer rectifier is applied across the 17 cells in series and the flow of current through the electrolyte causes the water to 'dissociate' and form hydrogen and oxygen gases. From the cell bank the hydrogen passes through a water seal and is either vented to atmosphere or flows into a wet seal gas holder when the generated gas purity is satisfactory.

13. FEED HEATING AND BOILER FEED PUMPS

The feedheating plant layout is shown in Fig 2.12 and a simplified diagram of the feedheating system in Fig 2.13.

The feedwater heating system raises the temperature and pressure of feedwater returning from the turbine to the boiler and, if necessary, it accepts make-up supplies from the reserve feedwater system.

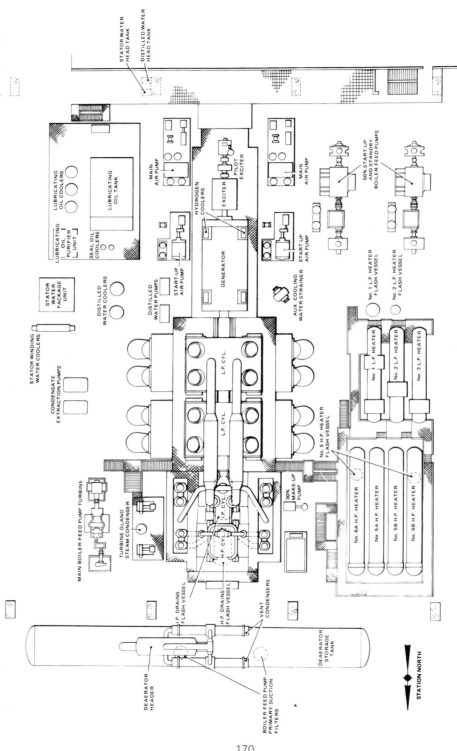

FIG. 2.12   FEED HEATING PLANT LAYOUT

STATION NORTH

Briefly, steam exhausts from the two LP turbine cylinders into two single pass condensers where it is condensed and falls to the bottom of the shell and into a hotwell. Condensate is drawn from the hotwell of each condenser by extraction pumps. After the pump discharge, part of the condensate passes through a turbine gland steam condenser. This condensate then rejoins the main supply and is passed directly to the LP heaters.

Each of the three LP heaters is of the 2-pass surface type in which feedwater flows through the tubes and bled steam is passed into the shell. Steam for feedwater heating is bled from various stages of the two LP turbine cylinders. On leaving the LP heaters, feedwater passes up to a deaerator storage tank via two vent condensers and a deaerator header. The deaerator is incorporated in the system to remove dissolved gases and to reduce the oxygen content of the feedwater. The storage tank caters for any fluctuation in system flow and provides water at a suitable suction head for the boiler feed pumps.

After deaeration, feedwater enters a downcomer and flows through one of two 100% duty strainers. It is then taken to either the starting and standby pumps or the main boiler feed pump (BFP), where the pressure is raised to feed the water into the boiler. Feedwater discharged from the boiler feed pump passes through the boiler feed regulating valves into a common main feeding four HP heaters. The heaters are arranged in two parallel banks of two and during normal operation, each bank receives half the total feedwater flow. Steam for HP feedwater heating is bled from tappings off the boiler feed pump turbine and from the HP turbine cold reheat lines. After leaving the HP heaters, feedwater is delivered to the boiler.

Valves in the condensate and feed heating plant which are subject to vacuum in service, with the exception of valves with special packing, are water sealed. The water supply to the valves is from two manifolds, which receive their water from the condenser double tubeplate sealing head tank. Tapping points from each manifold are fitted with manually operated isolating valves.

13.1 Design Data

The feedheating system design data is as follows:

Final feedwater temperature (at CMR)	251.1°C
Feed pressure from last HP heaters	199 bar
Feed flow from last HP heaters (at CMR)	564.5kg/s
Deaerator oxygen content not exceeding	0.007mg/litre
Time taken to heat deaerator storage tank water from 15°C to 110°C using steam heating coils	2 hours

13.2 Condensate Extraction

Each condenser has its own condensate hotwell, into which the falling condensate from each tubenest is collected. The condensate is extracted from the hotwells through a pipe and suction strainers to one of the two 100% duty caisson type vertical extraction pumps, which discharge the condensate through the LP feedwater heating system up to the deaerator.

13.3 Condensate Auxiliary Service Supplies

Tappings are taken off the condensate extraction pump discharge main and the recirculation-to-condenser pipe to supply a proportion of the condensate for the following auxiliaries:

(1) Desuperheating spraywater to the deaerator low load bled steam reducing valve and to the BFP turbine exhaust steam pipe to the deaerator.

(2) Surplus condensate to reserve feedwater (RFW) tank.

(3) Supply to LP spray cooling. A tapping off the RFW tank outlet pipe provides emergency LP spray cooling via a pump.

(4) Gland seal supply to main and starting and standby boiler feed pumps. An emergency supply is provided by one of two 100% duty pumps from a tapping off the RFW tank outlet pipe.

(5) Spraywater to HP and IP drains flash vessels.

13.4 Condensate Recirculation

The recirculation of condensate back into one of the condensers is necessary under no-load and low-load conditions, in order to provide sufficient condensate to maintain a continuous flow through the extraction pump and turbine gland steam condenser. As load is applied, recirculation is reduced and finally stopped. The control is achieved by control valves working in conjunction with a level controller sensing condenser water level. The recirculation line is tapped off the condensate extraction pump discharge line downstream of the turbine gland steam condenser.

13.5 Condensate Make-up and Surplus

During the normal running of the turbine, a continuous supply of make-up (treated water) amounting to approximately 1% CMR is flowing into the condenser through the condensate control valve bypass. The requirement for further make-up is indicated by the deaerator water level falling. The control of further make-up supply is achieved by operation of the condensate control valve working in conjunction with the deaerator storage tank level controller, which can provide an additional 4% make-up. In the event of a high water level in the condenser, a pneumatic float switch closes a make-up control isolator to prevent flooding of the condenser.

During dumping of contaminated condensate, a supply of make-up is pumped into the system through a connection in the extraction pump discharge pipe situated downstream of the sectionalising valve. The maximum amount of make-up that can be delivered with the make-up pump is 30% CMR flow when the deaerator is at CMR pressure.

When the level in the deaerator storage tank rises and becomes excessive, the deaerator level controller starts to open the condensate surplus control valve. This allows a proportion (up to a maximum of 6.3% CMR flow) of the condensate being delivered by the extraction pump to pass to the reserve feedwater tank. During this condition, the make-up control valve will have closed completely, preventing any make-up entering the condenser other than the fixed 1% through the bypass valve.

13.6   Low Pressure Feedwater Heating

The LP feedwater heating plant consists of three horizontal heaters, mounted in three tiers on supporting steelwork at the 7.7m, 8.9m and 10.1m levels. During normal operation, approximately 40% of the feedwater first passes through a turbine gland steam condenser. The feedwater then passes in succession through the tubes of each LP heater (see Fig 2.14) up to, and through the vent condensers to the distribution header of the deaerator.

Feedwater flow through the LP heaters is controlled by motor operated valves. Each control valve is fitted with a manually operated bypass. Each heater can be bypassed on the feedwater side, the drains are automatically diverted. Bypass operation is brought in automatically by high water level switches or manually by local control.

The pipework up to the deaerator incorporates a feedwater protection device known as the 'absence of prime' switch. This ensures that on loss of prime due to failure of both extraction pumps, it will not be possible to restart either extraction pump automatically or remotely, without first establishing that the system is fully primed and that no possible damage could occur through water hammer.

Steam supplied to the LP heaters is bled from various stages of the two LP turbine cylinders. Isolation of bled steam is automatically carried out by closure of the appropriate solenoid controlled, pneumatically operated butterfly valves, together with isolation of feedwater flow should the level of water in the heater shell rise to a high level. Initiation is from either of two duplicate level switches mounted on individual standpipes on each LP heater.

Various hot drains from the steam system are routed either to an HP drains manifold or to the turbine HP and IP drains flash vessels. The LP bled steam pipe drains are collected and taken into the condenser via the condenser flash vessel.

Residual condensate from the HP drains manifold and from the turbine HP and IP drains flash vessels is taken into the bottom of the condenser. Vented steam from the turbine drains flash vessels and the condenser flash vessel is taken into the steam space of condensers.

13.7   Deaerating Plant

The primary purpose of the deaerating plant is to reduce the residual oxygen content of the feedwater to no more than 0.007mg/litre. The plant incorporates two 50% duty vent condensers, a deaerator header and a storage tank fitted with four steam heaters to provide a source of 'off load' heating. The storage tank also acts as a surge vessel which caters for immediate fluctuations in demand by the boiler feed pumps.

The deaerator is essentially a direct contact LP heater capable of dealing with the total feedwater flow. Water is pre-heated in the vent condensers by exhaust steam and gases from the deaerator header. This pre-heated water is then piped from each vent condenser to an inlet on each side of the deaerator header, where it impinges on a flow plate and falls down through a double tier of perforated trays and separates into small droplets which are swept by the circulating steam to release the oxygen. Vented steam, together with the incondensable gases from the vent condensers, is used to drive the air ejectors on the main air extraction pumps, or during startup conditions, is taken to the suction of the condenser air extraction plant.

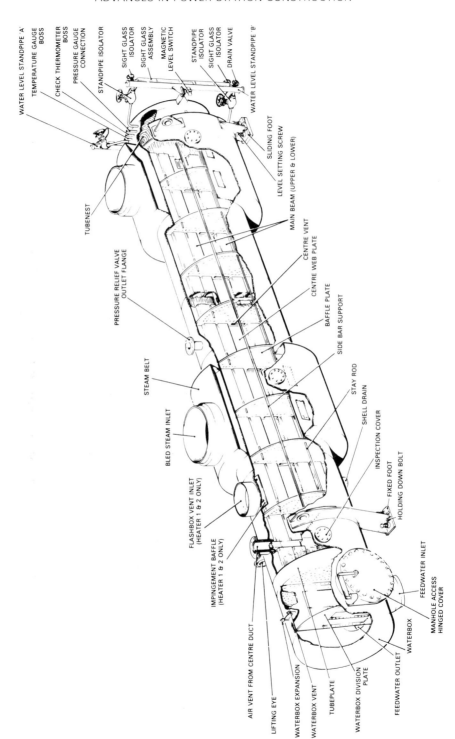

WATER LEVEL STANDPIPE 'A'
TEMPERATURE GAUGE BOSS
CHECK THERMOMETER BOSS
PRESSURE GAUGE CONNECTION
STANDPIPE ISOLATOR
SIGHT GLASS ISOLATOR
SIGHT GLASS ASSEMBLY
MAGNETIC LEVEL SWITCH
STANDPIPE ISOLATOR
SIGHT GLASS ISOLATOR
DRAIN VALVE
WATER LEVEL STANDPIPE 'B'
SLIDING FOOT
LEVEL SETTING SCREW
MAIN BEAM (UPPER & LOWER)
CENTRE VENT
CENTRE WEB PLATE
BAFFLE PLATE
SIDE BAR SUPPORT
STAY ROD
SHELL DRAIN
INSPECTION COVER
FIXED FOOT
HOLDING DOWN BOLT
FEEDWATER INLET
MANHOLE ACCESS HINGED COVER
WATERBOX
WATERBOX DIVISION PLATE
FEEDWATER OUTLET
TUBEPLATE
WATERBOX VENT
WATERBOX EXPANSION
LIFTING EYE
AIR VENT FROM CENTRE DUCT
IMPINGEMENT BAFFLE (HEATER 1 & 2 ONLY)
FLASHBOX VENT INLET (HEATER 1 & 2 ONLY)
BLED STEAM INLET
STEAM BELT
PRESSURE RELIEF VALVE OUTLET FLANGE
TUBENEST

FIG. 2.14   LP HEATER

Feedwater from the deaerator outlet passes through a downcomer and two test loop pipes connected in parallel to a manifold. Under normal conditions, feedwater is taken through the downcomer to the BFP filtering system. When required, water is passed through test loops to measure the flow from the deaerator.

Steam for deaeration is supplied from the IP turbine or boiler feed pump turbine exhaust. Low load steam from the cold reheat line is supplied via a pressure reducing and desuperheating valve supplied with spraywater. Spraywater connections are also located in the vertical leg of the BFP turbine exhaust steam supply pipework to the deaerator. Isolation of the deaerator bled steam, feedwater and spray water is automatically carried out should the deaerator water level rise to the emergency high level.

13.8   Low Pressure Feedwater Filtering

An LP feedwater filtering system is incorporated in the pipework between the deaerator storage tank discharge and the boiler feed pump suctions. The system comprises two 100% duty primary suction strainers and a secondary strainer.

Element cleaning of both strainers is accomplished by back flushing a small percentage of the main flow. The resultant backwash effluent is reclaimed by discharging into the secondary strainer before delivering it to the boiler clean drains tank via its flash vessel. Sludge drains from the secondary strainer are taken to waste.

The feedwater outlet pipe from the filtering system to the boiler feed pump suction, incorporates a branch pipe fitted with a control valve and carries recirculating water to the condenser. During shutdown periods water may become contaminated. Until conductivity is back to normal the recirculating valve is used to return water back to the condenser for dumping.

13.9   Boiler Feed Pumps

From the filtering system in the deaerator downcomer, feedwater is supplied to the boiler via the regulating valves and HP heaters by a turbine driven 100% duty main pump and two motor driven 50% duty starting and standby pumps.

Under normal conditions, the main boiler feed pump (BFP) delivers the complete feedwater load to the boiler, while the starting and standby pumps are used during the startup procedure and low-load conditions to supply the boiler until sufficient steam is available to operate the main pump turbine.

The main BFP is of tandem design consisting of a single-stage suction pump and a 2-stage high pressure pump. The former is driven through a speed reducing gearbox from the opposite end of the BFP turbine shaft and the latter is directly driven by the BFP turbine. Feedwater is taken by the suction stage pump from the deaerator via the filtering system and discharged through an interconnnecting pipe to the suction of the pressure stage pump, which after raising the pressure of the feedwater, discharges through the boiler feed regulating station and HP heaters to the boiler.

Each starting and standby BFP is also of tandem design consisting of a 2-stage suction pump and a 2-stage high pressure pump. The former is directly driven through a flexible coupling from one end of the electric motor shaft, while the latter is driven from the other end through a combined fluid coupling and speed increasing gear unit.

13.10 Boiler Feed Pump Turbine

The single cylinder BFP turbine is of the horizontal variable speed impulse design with five expansion stages. It is arranged to drive directly the pressure stage of the 100% duty main feed pumpset through the front pedestal and the suction stage of the pump through the rear pedestal via a speed reducing gearbox. Flexible couplings are used to transmit the drive from each end of the turbine.

The BFP turbine operates on either bled steam from the cold reheat line of the main turbine or live steam taken after the main boiler stop valve. Live steam from the boiler enables the BFP turbine to be operated until the load on the main turbine increases to 33% CMR, when a changeover from live steam to cold reheat steam takes place.

Steam to the HP heaters is bled from tappings on the BFP turbine cylinder. With the loss of the BFP turbine, the bled steam supply to the heaters is lost.

Steam exhausts from the BFP turbine at a pressure corresponding to the transfer pressure between the IP and LP cylinders of the main turbine. The bulk of the exhaust steam is fed to the deaerator storage tank, but a tapping taken off this line supplies steam to the boiler bled steam air preheater located at the FD fan inlet. The remainder of the exhaust steam is discharged to the exhaust section of the IP cylinder. When the steam demand of the deaerator is greater than can be supplied by the BFP turbine exhaust, the deficit is made up with steam from the IP cylinder exhaust.

When the main set is not in operation and during startup of the BFP turbine, the steam route to the IP cylinder is isolated and the surplus exhaust flow above the demands of the deaerator is diverted to the main condenser via the IP flash vessel. The steam route to the IP cylinder is also isolated on turbine trip or acceleration.

13.11 High Pressure Feedwater Heating (Fig 2.13)

Four horizontal heat exchangers, arranged in two parallel banks of two are provided to heat the feedwater discharged through the boiler feed regulating valves by the BFPs. If necessary, a bypass in parallel with the two heater banks, allows one or both banks to be bypassed.

The steam for one pair of HP heaters is bled from tappings on the BFP turbine cylinder, and for the other pair of HP heaters a supply is bled from the main HP turbine cold reheat line.

Under normal operation, each bank of heaters receives half the total feedwater flow from the main BFP. Either bank can be isolated manually or automatically should the drain water level in one or more heaters reach a high level.

In each bank, feedwater enters No.5 HP heaters (Fig 2.15) through a motor operated valve and flows through the heater tubes commencing with the drain cooling section at the inlet into the condensing section of the heater. The feedwater absorbs the sensible heat from the bled steam drains water as it flows through the tubes, absorbing the latent heat from the bled steam condensing on the outside of the tubes.

The feedwater flows from the outlet of No.5 HP heaters upwards to No.6 HP heaters inlet. Due to the higher temperature of bled steam to No.6 HP heaters, they are fitted with an extended desuperheating section in which the superheat is progressively absorbed by the partially heated feedwater in its passage through the last section of the heater. The high temperature bled steam to each heater enters the desuperheating

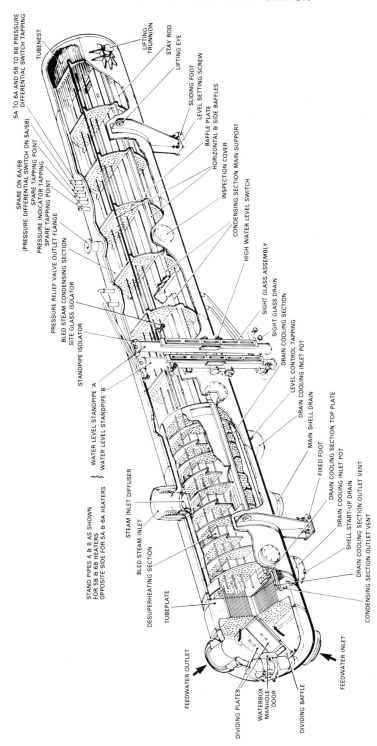

FIG. 2.15   HP HEATER

section through an expansion piece and after giving up the greater part of its superheat, passes into the annulus of the condensing section. The condensed steam drains water flows along the bottom of the heater shell to the drain cooling section.

The arrangement of pipework throughout the system is such as to permit adequate flexibility of pipes without undue loading on any plant item. Pipework before the feed regulating valves is designed for a pressure of 31.6×10^6 N/m^2 and 19.6×10^6 N/m^2 after the valves.

13.12 High Pressure Heater Drains (Fig 2.13)

During on-load operation the flow of drain water from both banks of HP heaters is disposed of either to a condenser or to the deaerator storage tank. Distribution of drains is automatic and is controlled by pneumatically actuated valves, which operate in response to signals from upper and lower level controllers on the heater shells.

Under steady conditions with the load above 298MW (45% CMR) the drain flow from No.6 HP heaters is cascaded to No.5 HP heaters via a flash vessel in which the flashed steam is taken into the top of No.5 HP heaters and the condensed steam drains into the bottom of the same heater. The drain from No.5 HP heaters in each bank is carried up to the deaerator storage tank.

Two standing drains, each taken from a drain cooling pot at the bottom of each HP heater, are commoned to join a single drain back to condenser A. Each commoned standing drain is controlled by a motor operated valve which opens when the bled steam isolating valve on the heater is fully shut and closes when the bled steam valve is open.

14. COOLING WATER SYSTEM

The cooling water (CW) system supplies water from the River Thames for cooling the condensers of the three 660MW turbine generators and also supplies various auxiliary coolers via the auxiliary cooling water systems. Fig 2.16 shows the system diagram.

River water enters the system through the intake ports of a reinforced concrete caisson constructed offshore and flows under gravity through a concrete tunnel, of 4.27m internal diameter and approximate length of 280m, to the forebay of the CW pumphouse. Fig 2.17 shows a section through the CW intake and pumphouse. The latter is constructed within a circular reinforced concrete coffer, approximately 68m in diameter, and comprises three broad divisions containing the forebay, drum screen chamber, pump chamber and delivery manifold. The water is drawn from the tidal forebay by four vertical-spindle, concrete volute pumps, through their respective rotating drum screens and discharged into the rubber-lined steel delivery manifold. From the manifold the water leaves the CW pumphouse via three reinforced concrete culverts to supply the two condensers associated with each of the three turbine generators and also to supply the separate auxiliary cooling water (ACW) systems. The discharge from the condensers and the ACW systems is returned to the river via three

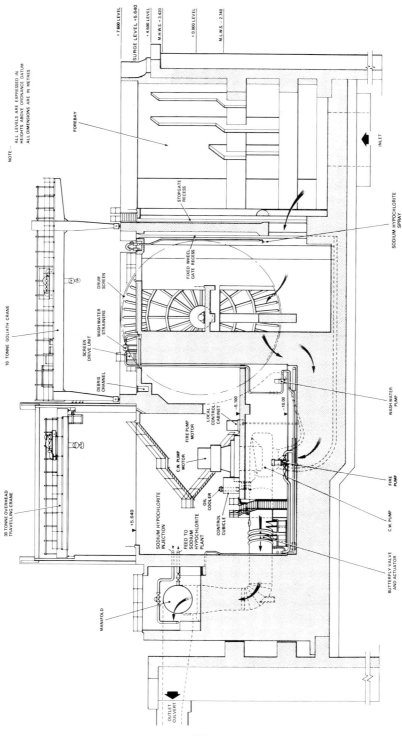

FIG. 2.17   CW INTAKE AND PUMPHOUSE

concrete outlet culverts to a seal weir and seal pit, then through a concrete tunnel, of 4.7m internal diameter and approximate length of 478m, to an offshore outfall caisson of reinforced concrete construction.

Debris is prevented from entering the CW system by screens positioned across the river intake ports and rotating drum screens at the pump suctions from the forebay. Trash enmeshed on the drum screens is dislodged and removed from the system with the use of jet wash equipment installed above each drum and served by two water pumps.

Water can be prevented from entering the system for the purpose of inspection and maintenance by the positioning of stop logs (barrier) across the river intake and outfall ports. Fixed roller gates and stopgates are also provided for closing the drum screen chamber from the forebay.

Automatically controlled air admission valves are fitted to the condenser downstream water boxes. Their purpose is to reduce pressure surges in the culvert system when the water box pressure falls to near vapour pressure following any CW pump trip.

An auxiliary pump is provided to supply water to the ACW systems when the main CW pumps are not operating. The pump is located in the No.3 main CW pump pit and draws water from upstream of the main CW pump butterfly valve and discharges it downstream of the valve.

14.1 Design Data

The CW system design data is as follows:

Main CW Pumps:

Type	Vertical spindle, concrete volute
Number off	4
Flow	14.6m^3/s
Head at full flow	16.5m

Auxiliary CW Pump:

Type	Single entry mixed flow volute
Number off	1
Flow	1.5m^3/s
Head at full flow	12m

CW Screens:

Type	Double entry drums
Number off	4
Design flow	14.25m^3/s at lowest water level

CW Valves:

Type	Butterfly
Bore	Discharged 2.4m dia hydraulically operated
	Manifold 3.2m dia electrically operated

14.2 Hydraulic Design Principles

14.2.1 *System design*

For the three 660MW units, consideration was given to the provision of three 100% duty or four 75% duty CW pumps.

The advantages of the 4-pump system as installed are as follows:

(1) Greater flexibility of operation, by enabling the station to take one pump out of service and maintain full load under suitable low river water temperatures and tidal conditions that occur during the winter months.

(2) Greater availability of the station arising from pump failure or maintenance and hence lower losses of generation.

Based upon the anticipated load factors at the time of the design, it was estimated that the savings due to higher generating availability using the 4-pump scheme would equate to the higher capital costs after only two years of operation.

The CW screens are integral with the pumps and are isolated and drained down as units. This simplifies the isolation arrangements without sacrificing much reliability, since drum screens have a high reliability.

The pump/screen units can be isolated with double valve isolation. Each screen inlet is furnished with double stop gates and the pump outlet valve can be closed in conjunction with manifold valves to achieve double isolation of the discharge side.

In order to achieve double isolation for the two outer pump/ screen units, it is necessary to have one turbine generator unit out of commission. The inner pumps can be double isolated with all units on load but with some loss of pump distribution.

To allow maintenance of culverts and condenser inlet and discharge valves without a complete station shutdown, each unit is provided with individual inlet and outlet culverts.

Single inlet and outlet tunnels are provided for the whole station. Valve maintenance is not associated with tunnel isolation. If it proves necessary to clean out a tunnel, the complete station is shutdown and the tunnel is isolated by inserting stop logs at the intake and outfall caissons.

14.2.2 *Hydraulic Design (Fig 2.18)*

Tunnels and culverts were optimised by comparing construction and pumping costs for the lifetime of the station. This led to the installation of 4m diameter tunnels and 3m × 2m rectangular culverts. The latter were made rectangular to reduce excavation in the difficult ground conditions at Littlebrook.

The system is designed to operate syphonically and it is necessary to limit the syphon to approximately 9m by a weir. The need to seal the culverts above the design weir level led to the use of a syphonic seal weir. This reduces the back pressure on the system by approximately 2.5m when running as a syphon, but keeps the culverts full on shutdown. A pipe is provided connecting the top of the syphon with the top of the outlet valve chamber to break the syphon smoothly on shutdown.

The syphon is secure at all flows down to 2-pump operation. At lower flows the syphon will break and the system will operate with an increased head of approximately 2.5m, thus reducing the pump flow. For this reason, the CW system is not run under operating conditions with less than two pumps in service.

The inlet to the forebay was modelled to establish an even flow distribution to the four screen chambers. The modelling showed the need for vertical wall ribs and baffles and entrance fairings to the screen inlet gates. Subsequent screen modelling also showed that cross walls were needed in the screen inlet channels to improve flow to the rear of the screen. This in turn improved distribution and rotation in the draft tube.

Severe oscillations were encountered in the outfall of another station system, which involved deep tunnels similar to Littlebrook. It was established that this oscillation was caused by a large hang-up of entrained air in the landshaft, which periodically vented either back up the shaft or along the tunnel. The air entrainment was caused mainly by the large fall of water over the weir into the standing water level within the shaft. At Littlebrook, the introduction of a second weir at the outlet to the river limited the fall over the first weir, at extreme low water conditions, to about 2.5m at low flow and about 1m at full flow.

Following modelling of the downtake shaft, it was necessary to introduce cross beams to reduce rotation-induced vortices, which occurred particularly when only one unit out of three was in operation.

14.2.3 *Pressure Surge Analysis and Surge Suppression*

Computer simulation analysis shows that it is not possible to keep the system pressures within the design limits, for all pump trip and tide level combinations, by altering only the pump discharge valves' closing characteristics. It is also necessary to introduce air into the condenser outlet water boxes under specific conditions.

The pump discharge valves are hydraulically operated, with two rates of closure. The initial phase is at a fast rate, then at a preset angle the rate of closure is slowed and maintained until the valve is completely closed. For the Littlebrook system, the best closure characteristics are for a changeover of closing rates at 45° from fully open and the times for the fast and slow closing phases are 18s and 27s to 42s respectively. The changeover is achieved by a cam attached to the valve spindle, controlling a solenoid valve which alters the discharge route of the hydraulic oil from the cylinder. The rates of closure can be adjusted by throttle valves in the oil discharge lines.

Air admission into all outlet water boxes is required at a maximum rate of $3m^3/s$ at normal temperature and pressure. On the top of each box, two 200mm nominal bore flanged connections are provided for power operated butterfly valves. The addition of air into the water box acts as a cushion when cavity collapse occurs and reduces the peak pressures that would otherwise be generated. It is desirable to omit air only when excessive pressure occurs, so the signals for the air valves to open are related to the size of cavity in the water box. Pressure tappings are installed at the bottom of the box and connected to a pressure switch at approximately 4m below the top of the box. The switch is set to initiate opening of the valve when the pressure reaches −280mbar falling. The valves on each water box are operated only from the switch associated with that box. A fast response is required from the air admission valves, so a pneumatic system is used.

A simulation of tripping one pump out of four, with the pump valve failing to close, shows a maximum steady state reverse speed of approximately 100% (175rev/min) at low tide. CW pumps are specified to accept a reverse speed of up to 130% without damage occurring.

The water level fluctuations in the forebay and outfall downshaft following a 4-pump trip were analysed for extreme high tide, ie with the Thames Barrage closed. In the forebay the maximum level reached was +7.28m OD, which is approximately 0.3m below the rim for the forebay.

The analysis of the seal pit revealed an unlimited level of +12.2m OD, whereas the seal pit wall reached only +10m OD. However, at mean high water spring tide, the maximum water level in the downshaft reaches only +9.4m OD, so it was decided to accept the small overspill which would occur in the unlikely combination of extreme high tide and full pump trip, since this would occur rarely, if ever, in the life of the station.

14.3   Structural Design Principles

A deep tunnel in good chalk was chosen in preference to a deep water culvert or a shallow tunnel. A culvert would require temporary works in the river, causing navigation hazards. A shallow tunnel would require the use of compressed air, which would create difficult working conditions.

The tunnel is formed from precast concrete bolted segments, with precast concrete lining. This method requires the minimum excavated diameter and is quick and easy to install.

The original site level was approximately +1.0m OD with a top stratum of alluvium to a depth of 8m to 15m. The site level has been raised to approximately +4.16m OD, using local gravel. This removed the necessity to excavate in the alluvium for the culverts and determined the invert level at approximately +2.00m OD. The sealing weir level necessary to maintain the outlet culverts full of water for no-flow conditions is +4.25m OD. The head over this is recovered for flow conditions by constructing it as a syphonic weir, thus making the floor in the gate chamber at datum the actual system weir. There is also a weir in the river outfall structure to limit the drop from the gate chamber into the CW downshaft at low tides, minimising the amount of air taken down into the tunnel.

A circular diaphragm wall, used to provide the area necessary for construction of the pumphouse, has been lined with an internal attached reinforced concrete wall to form the permanent wall of the forebay area.

14.4   Dewatering System

A dewatering facility for the CW pumps and delivery culverts is provided, using the two screen wash water pumps and an arrangement of pipework and valves to discharge the water into the pumphouse forebay. When dewatering is required, the valves which are normally open for the wash water system are closed and dewatering valves opened. Alternatively the inlet culverts can be drained back through a 'unitised' CW pump into the forebay. Dewatering of the intake tunnel, forebay, outlet culverts and outfall tunnel is carried out by portable pumps.

14.5 Air Extraction System

An air extracting plant is provided to assist with priming the CW system. It comprises two main liquid ring air extraction pumps serving a bus main to assist in priming the unit condensers. Six smaller capacity liquid ring pumps, two for each of the three turbine generator units, provide continuous extraction of air from the condenser outlet water boxes.

Isolating valves, separator and interceptor tanks with associated level and pressure switches automatically control the operation of the plant.

To achieve sufficient air extraction from one empty condenser and the ACW system within a period of 20 to 25 minutes requires the running of the two main air extraction pumps. For normal run-up situations, eg after overnight shutdown, one main air extraction pump is sufficient to evacuate one turbine generator unit.

The continuous air extraction pumps are provided for abnormal CW pump/unit operations combined with low tide level conditions.

14.6 Operation and Control

A full load flow of 58.4m^3/s is provided when all four CW pumps are running at their design duty point. Normally, a minimum of two pumps are in service at all times to provide security of CW supply when the station is on load.

The controls for CW pump startup and shutdown are normally governed by sequence control, initiated from the central control room. The sequence control equipment monitors the plant condition and if satisfactory, the startup sequence is carried out.

To protect the CW pumps, interlock and intertripping facilities are provided independent of the sequence control equipment. These prevent CW pump operation during conditions of low lubricating oil pressure, the CW pump discharge valve remaining closed, excessive differential pressure across the drum screens, or any plant faults resulting in the operation of the pump motor protection.

Operation of the CW pumps, drum screens and main valves are controlled locally from cubicles and panels in the pumphouse.

14.7 Treatment

To combat the formation of organic slimes and prevent the growth of shellfish within the system, the design incorporates the use of a treatment plant for the introduction of sodium hypochlorite into the water at a number of points.

14.8 Flow Measurement

The preferred method of determining the flow through the CW system is the procedure using the radioactive isotope constant dilution technique. The principle of the constant dilution method permits the injection of a radioactive isotope at a known rate into the system at the suction inlet to CW pump and the measurement of isotope concentration at the turbine generator condenser. Tappings into the CW system are provided for the flow measurement of each of the four pumps and each of the three condenser units.

14.9 Maintenance Facilities

Two cranes are provided for general handling and maintenance of pumphouse equipment. A travelling overhead crane of 30t capacity is mounted above the pump chamber and loading bay and is operated by a pendant control. A Goliath crane positioned above the drum screen chamber has a capacity of 10t and is also operated by a pendant control.

To provide a means of support for a CW pump and facilities for maintenance work after removal from its installed position, a cradle and servicing pit with adjacent space sufficiently large for a pump lay-down are located at the west side of the pump chamber.

15. FUEL OIL SYSTEM (FIG 2.19)

The installations concerned with the supply and storage of the fuel oil are located to the east of the station site. They comprise the main jetty, a tank farm of five storage tanks, a pumphouse and two flowmeter stations. There is also an auxiliary jetty where fuel oil can be loaded into barges for export to other power stations.

The main jetty comprises a jetty head and four mooring dolphins. Three off-loading arms mounted on the jetty head can be connected to the manifolds of tankers for importing fuel oil. The main jetty installations include a hydraulic system for power operation of the off-loading arms and a system for draining the off-loading arms of fuel oil.

Access to the main jetty from the shore is by an approach road approximately 200m long, which joins the tank farm perimeter road.

The imported oil can be fed to any selected tank by opening and closing appropriate valves. Main fuel oil pumps in the main fuel oil pumphouse can draw suction from any tank for supply to heaters and boiler house; hot oil from the boilers is recirculated and can be fed to any tank.

Export pumps in the main fuel oil pumphouse can also draw suction from the tanks and discharge the oil through the export flowmeter station to the auxiliary jetty. The system is designed to handle fuel ranging in viscosity from 24 centistokes to 1500 centistokes at minimum storage temperature. Separate pumping facilities are provided for off-load circulation and for exporting oil.

The oil is stored at a tank farm consisting of five storage tanks each of 110,000t capacity. These tanks are connected to a common outlet manifold. Isolating valves are provided in the manifold and interconnecting pipework to permit separate tank isolation and connection to the pressure, circulating and export pumps. Each tank is also connected to a return line which permits recirculating oil flow from the boilers and heater house.

From the tank farm the oil passes to the pumphouse which contains six pressure pumps, two recirculating pumps and two export pumps. During normal boiler operation, the oil is pumped by the pressure pumps through a single main to the heater house. A flywheel is incorporated in the drive to each of the pressure pumps so that, in the event of interruption of the electrical supply to the pump drive motors, the pumps will 'run on' to maintain the oil discharge pressure, thus ensuring that the burner flame remains stable for a minimum period of 3 seconds.

In the heater house the fuel oil is heated by three banks of steam heaters, each bank comprising 15 individual heaters arranged in sets of three. The heating sets are connected in parallel and are brought into service as required, depending on the oil flow rate and grade of fuel used. The temperature of heavy fuel oil and distillate fuel is controlled at 140°C and 43°C respectively. The fuel oil heaters are heated by saturated steam at 17.24 bar and 210°C from the auxiliary steam supply.

Electric trace heating is provided to maintain an adequate oil temperature in the supply pipework and remains in operation at all times during the use of heavy fuel oil. The trace heating is not used when distillate fuel is being used.

Heated fuel oil passes from the heater house to the boiler house where the fuel oil main divides into three, one to each of the three boilers. In each case, the oil is distributed by the boiler front pipework to 32 steam atomised fuel oil burners arranged in four horizontal rows of eight in the burner windbox. Each burner is capable of firing an oil flow of 1.3kg/s at CMR and 1.44kg/s at overload. Fuel in excess of burner requirements is diverted and returned to the storage tanks through separate pipework.

Propane gas is used for the initial light-up of the burners. Each burner, fitted with a gas/electric igniter, is controlled independently from either the control room or the local control panel.

A burner removal hoist is provided for each of the three boilers to facilitate the quick replacement of burner sprayer units. The hoist is underslung in front of the burner galleries and is able to traverse the boiler front to service each of the 32 burners.

16. GAS TURBINE GENERATORS (FIG 2.20)

There are three single-ended gas turbine generating sets installed on site. Each set consists basically of a pair of gas generators discharging into a power turbine which is directly coupled to the rotor of an ac generator, the output of which is 35MW at 11kV.

The sets are normally operated by signals initiated in the remote station control room and fed to the automatic sequencing and control circuits in the cubicles of the local control panel. The circuits automatically start, synchronise, load and shutdown the set. When required, the automatic sequences can be initiated from the local control panel where a manual starting facility is also provided.

During starting and throughout normal operation, the sets are continuously monitored by automatic protection circuits. The circuits provide audible and visual alarms for serious faults and initiate shutdown of the set in the event of a serious malfunction; vital alarms and indications are repeated in the remote station control room.

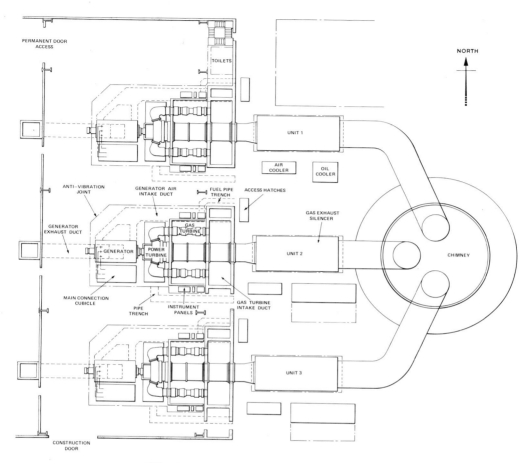

FIG. 2.20   GAS TURBINE INSTALLATION

17.   WATER TREATMENT PLANT

The fully automatic water treatment plant provides treated water and polished recycled condensate for boiler feed make-up and reserve feedwater for the three boilers.

The plant located in the water treatment and condensate polishing plant house is capable of producing make-up water at a continuous rate of $273m^3/h$ when supplied with towns water from the raw water storage tanks. In addition, de-ionised water from the reserve feedwater tanks, which receive water from the clean drains tank and turbine condensate, may be passed through the polishing mixed bed section at $427m^3/h$. Fig 2.21 shows in simplified form the process flow.

187

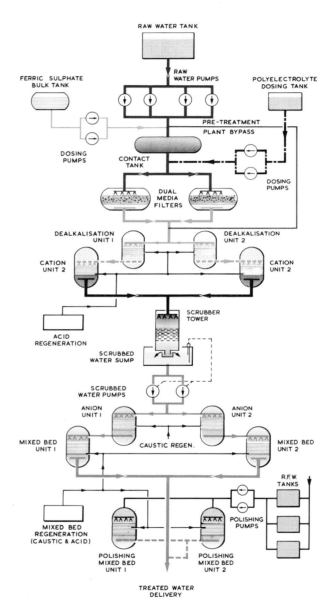

FIG. 2.21   WATER TREATMENT FLOW DIAGRAM

The make-up treatment removes, as far as possible, all the dissolved solids in the raw water supply. To minimise silica deposits in the turbine equipment, the water treatment plant is capable of producing make-up water containing less than 0.02mg/litre of silica.

Following treatment within the make-up section or condensate polishing section, the water passes into the treated water bus-main which provides the following services:

(1) Normal unit make-up water into the turbine condenser.

(2) Make-up water to the general service water system.

(3) Recirculation of treated water to the reserve feed tanks for storage.

The bulk chemical storage tanks are provided with sufficient capacity for at least six weeks continuous full operation of the plant.

18. AUXILIARY BOILER PLANT

The auxiliary boiler plant is required to provide auxiliary steam for the main generating plant and station auxiliaries. It is located to the east of the main generating station and is housed in a steel and concrete building. Fig 2.22 shows the layout of the auxiliary boiler house.

The plant comprises five identical boilers supplying superheated steam at 17.5 bar and 228°C to a steam ring main in the auxiliary boiler house. This steam is used for atomising fuel oil and purging the main fuel oil burners on a total loss system, and steam heating, with condensate recovery, for the following:

(1) Fuel oil for the main boiler.

(2) Main boiler air at low load levels.

(3) The main deaerator at low-load and off-load.

(4) Station buildings.

(5) Fuel oil tank farm.

Within the auxiliary boiler house, pressure reduced desuperheated steam from the main boilers is used for heating the fuel oil day tanks and fuel oil heaters. Four boilers are required for peak demand with one on standby.

Each of the boilers is of the multi-tubular, wet back, three pass type, having twin furnaces each fitted with a single rotary cup burner. The burners are fitted with a 'voluvalve' system which controls the volume of fuel oil passed either to the rotary cup atomiser or re-circulated back to the pump suction. Air is supplied to the burner from a forced draught fan mounted adjacent to the burner. Some of this air is supplied to a primary fan mounted on the burner shaft. A fuel/air ratio motor controls the quantity of air and fuel required for combustion. This motor is operated by electrical signals from a position controller, which receives a desired value from the steam range pressure controller or from manually operated pushbuttons. Initial ignition of the fuel oil is from a flame of propane gas ignited by an electric spark.

Each boiler is fitted with shell and superheater safety valves which discharge through silencers to atmosphere and all necessary valves and fittings for safe and efficient operation.

189

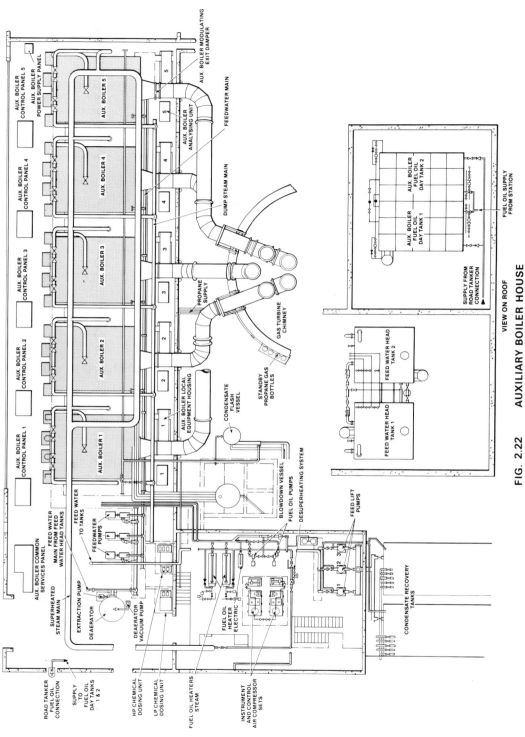

FIG. 2.22   AUXILIARY BOILER HOUSE

19. FIRE PROTECTION SYSTEMS

The fire protection systems comprise a number of separate systems each designed to offer the maximum protection for the types of fire risks. Because of the diverse nature of the plant, separate self contained fire fighting systems have been provided, including a comprehensive coverage of selected hand portable and mobile equipment. Fire protection philosophy and system application were subject to approval by the Chief Fire Officer for Kent County Council. The systems provide the following protection coverage.

19.1 Water Systems

The main fire protection agent is water-based, comprising high velocity waterspray, medium velocity cooling sprays, sprinkler and selected internal/external hydrants. These cover the turbine generator plant and structures, unit and auxiliary transformer groups, boiler feed pumps, boiler burner front, gas turbines and gas turbine bulk fuel oil storage systems and transfer pumps, main fuel oil heaters and pumps, propane stores and hydrogen generating plant, cable floors, risers and tunnels. The systems also supply high velocity water to serve low expansion foam base injection equipment protecting the gas turbine fuel oil tanks and medium expansion foam protection to boiler windbox internals.

These fire protection water systems are supplied by pressurised trunk mains from four 33.3% duty fire pumps. The pumps draw towns water from raw water storage tanks which are allocated for fire fighting duties only. An emergency pumping-in point is provided close to these tanks to allow County Fire Brigade tenders to pump-in and reinforce the system in emergency conditions, such as failure of the fire pumps.

A separate high velocity waterspray system protects the auxiliary and services transformers serving the main fuel oil transfer pumphouse. This system is supplied via two 100% duty fire pumps with water taken from separate towns water storage tanks.

A primary external water hydrant system made up of underground interconnected ring mains and spur branches supply numbers of strategically placed pit hydrants. These hydrants are designed to provide adequate protection coverage to the main and auxiliary plant and building fabrics including the 400kV supergrid transformer compound. Two 100% duty electrically-driven hydrant pumps serve this system and are located in the CW pumphouse forebay taking water from the River Thames.

A secondary external hydrant system protects the fuel oil storage tanks and tank farm bunded area with a spur connection serving the main jetty. These hydrants are designed to support the application of water/foam cannons. Two 100% duty electrically-driven hydrant pumps supply the system from three towns water storage tanks.

An internal hydrant system serves to reinforce the fixed water protection systems in the turbine and boiler houses; a total of 24 strategically-placed hydrants are provided. This system is supplied by two 100% duty electrically-driven fire pumps located in the main fire pumphouse and which take their supply from the towns water storage tanks.

A secondary internal hydrant system within the boiler house supplies six 'dry riser' mains and these extend above boiler drum level, having hydrant outlets at each floor including the turbine house roof. These dry risers can be charged at four pumping-in points.

19.2 Gas Systems

Auto/manual Halon 1301 (BTM) extinguishing gas systems protect the common services and unit computer suites and the control and transmission simulator workshops. These Halon systems are designed to protect the room and roof voids in the 'total flood' concept. Similar extinguishing gas flooding systems protect the cable marshalling cubicles. In addition, Halon 1211 (BCF) extinguishing gas systems provide protection to the gas turbine unit enclosures. The systems are also designed to operate to the 'total flood' concept.

A wide range of portable extinguishers and mobile fire fighting equipment positioned at strategic fire points supplement the fixed systems.

20. HEATING AND VENTILATION EQUIPMENT

The requirement for heating and ventilation relates to two separate areas of the station buildings, the main buildings and the ancillary buildings complex.

The main buildings do not require heating but do require ventilation. The ancillary building complex requires heating, cooling and carefully controlled ventilation.

20.1 Main Buildings

Ventilation of the boiler house is primarily to provide air for cooling the boiler house generally, and also to supply combustion air to the forced draught fans which draw air from above the boiler. This is achieved by fixed air inlet louvres along the length of the north face of the boiler house at a low level. The total area of the louvres allows air flow to meet the combustion requirements. The flow path commences below the +15.35m OD level which is a chequer plated floor at the front of the boiler. Flow is upwards through the open grid floors to the top of the boiler house and the forced draught fan suction. Variable louvres located at higher levels on the north face of the boiler house allow additional general cooling air to enter.

Ventilation of the turbine house is primarily to reduce condensation. The turbine house roof does not contain voids where moisture could collect and louvres ensure a good cross flow of air just beneath the roof level.

In general, ventilation of the main buildings ensures a reasonable environment for plant operators and certain items of plant susceptible to the effects of heat. The ventilation equipment also ensures the egress of smoke and hot gases in the event of fire.

20.2 Ancillary Buildings Complex

The buildings consist of five storeys containing the main control room, offices, laboratories, cable floor, plant rooms, workshop and stores. The depth of the building is such that continuous cooling in the central and upper floor areas is required. The heat gains emanate from equipment, lighting and personnel housed in the buildings.

Windows, where fitted, are sealed and double glazed. There are no windows in the cable floor.

The large workshop area is naturally-ventilated but the stores area, which is beneath the main plant room, has a ventilation system. These areas are serviced separately from other parts of the main building which have an air conditioning system.

The differing conditions in the various areas of the building require specific heating and ventilating conditions to ensure the correct working environment for personnel and equipment.

The heating and ventilation system provides:

(1) A conditioned environment in all areas of the control room floor for personnel and equipment, and also a means of exhausting smoke from the computer rooms in the event of a fire.

(2) In the cable floor, a temperature of not less than 10°C and not more than 8°C above the outside shade temperature.

(3) A conditioned environment for personnel on the lower floor levels, workshop and stores, consistent with the nature of their duties.

20.3 Design Conditions

The design parameters for achieving the objectives were taken as follows:
Outside weather considerations:

   Winter   −1°C dry bulb at 100% relative humidity
   Summer   29.4°C dry bulb at 50% relative humidity

Room air movements:

   Within the range 0.1m/s to 0.2m/s measured at not less than 1.5m from any external wall and at a height of 1.5m above floor level. Relative air pressures are arranged so that the air flows from clean areas, eg computers, to less demanding areas, eg corridors.

Internal temperature gradient:

   Shall not exceed 3°C from the floor level to 2m above the floor level.

The environmental requirements of the various rooms and areas within the ancillary buildings complex are given in Table 2.3.

20.4 Control Philosophy

Control of air conditioning was chosen as a combination of variable volume units dealing with the large areas, constant heat gains, etc, and terminal heater units dealing with perimeter heat losses and smaller rooms.

TABLE 2.3. HEATING AND VENTILATION DESIGN CONDITIONS

Room Area	Design Temp °C Max	Design Temp °C Min	Relative Humidity %	Air Changes Per Hour
Control room	22	18	45–55	3
Computer rooms	22	18	45–55	27
Equipment rooms	22	18	45–55	12
Cable marshalling	—	10	—	2
Conference room	22	18	40–60	17
Toilets	—	15.5	—	6
Library	22	18	40–60	19
Laboratories	22	18	40–60	7
Lecture room	22	18	40–60	26
Lockers	—	21	—	6
Ablutions area	—	21	—	6
Canteen & dining rooms	22	18	40–60	16
Kitchen	—	15.5	—	10
Medical centre	23	19	40–60	16
Heavy workshop	—	15.5	—	2
Light workshop	—	18	—	2
Welding shop	—	15.5	—	2
Rigging shop	—	13	—	1
Main stores area	—	13	—	1

Workshops and stores are heated by means of floor-mounted warm air blower units arranged around the perimeter of the building and some hot water radiators in smaller rooms. The hot water is supplied from a heat exchanger in the plant room. Natural ventilation occurs by means of wall-mounted inlet louvres and roof-mounted outlet ventilators.

Terminal reheat units in the canteen, reception and medical centre areas are fed with air at a constant 12°C by two 50% duty air handling units located in the plant room. Air is introduced by ceiling-mounted circular and strip diffusers and extraction is generally by ceiling-mounted grilles. The supply air handling units each consist of a filter section, cooling coil, heating coil and fan. Pneumatically-operated dampers in the fresh air inlet are arranged to adjust the volume of air handled in accordance with external temperatures.

Similar arrangements exist of the kitchen, ablutions, lockers and associated areas except that the extract systems terminate with roof-mounted fans discharging to atmosphere.

Perimeter rooms on the offices and laboratories floors at the 15.25m and 10m levels are served by terminal reheat units supplying air at a constant 12°C. All other areas on these floors are served by variable volume units which also provide air at 12°C. Extraction is generally from ceiling-mounted grilles but in some cases discharge is direct to atmosphere. There are two main air handling units each of which consists of a filter section, cooling coil, heating coil and fan. Pneumatically-operated dampers adjust the volume in accordance with supply duct and external temperatures.

Cable flats at the 20.88m level are served by a single air handling unit which feeds inlet grilles on each of the three unit cable areas and the two station cable areas. Fire dampers are strategically located together with extract fans which can be speeded up, for rapid removal of smoke and fumes in the event of fire.

At the control floor level (25.5m) east facing perimeter rooms and the control room are supplied with air at a constant 12°C. The computer and equipment rooms are served by variable volume units supplying air at 12°C. Extraction is via ceiling-mounted grilles and duct mounted grilles. Silencers are fitted at duct entry and exit points. Dampers in the ductwork can be set to 100% recirculation in the event of smoke entering the system from an outside source, and a 'firemans switch' is provided to give 100% fresh air input if required to purge smoke or extinguishing gas.

21. ELECTRICAL SYSTEMS AND EQUIPMENT

21.1 Introduction (Fig 2.23)

The electrical supply system comprises three main generating units each connected via a generator transformer and high voltage circuit breaker to the 400kV grid system.

The existing 132kV substation at Littlebrook provides the necessary station supplies for normal startup and standby running of the three generating units, normal and abnormal station supplies and is also used, as required, to export gas turbine output.

In addition to supplying the 400kV grid system, the generating units are connected to an auxiliary electrical system based on three voltage levels, 11kV, 3.3kV and 415V, and comprise unit and station boards with associated transformers. Three gas turbine generating units connected to the 11kV unit boards are provided as a standby power source for dead station start conditions (no 132kV supply available), peak loading conditions or supplementing the 132kV grid system. The choice of connection and duplication of supplies permits standby running at the three voltage levels giving good system flexibility.

System interlocks for operational and maintenance purposes are provided at all voltage levels.

Instrumentation, controls and other essential services such as unit and station computers are supplied at 110V single phase ac on a unit and station basis.

Power supplies to turbine and boiler valve and damper actuators are distributed by actuator power centres strategically located throughout the station.

A dc system comprising batteries, battery charging units, distribution boards and associated switchgear provides supplies ranging from 50V dc to 240V dc for guaranteed instrument supplies, emergency lighting, switchgear control circuits and gas turbine starting equipment.

Computerised and conventional alarm systems are incorporated in the control room.

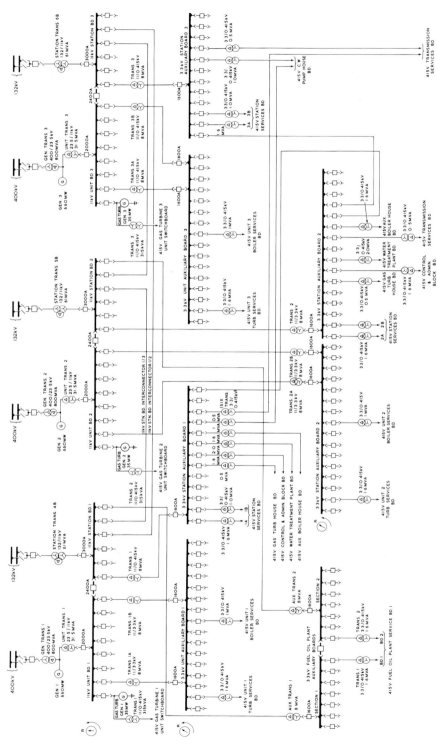

FIG. 2.23   UNIT AND STATION SUPPLIES

21.2 High Voltage Connections

Each of the 660MW generating units is connected to the 400kV grid system via a three single-phase unit type, delta/star connected generator transformer. There are three 400kV double circuit lines from Littlebrook, two via 400kV/275kV transformers and the third at 400kV. The existing 132kV substation supply is connected to the three station transformers via 132kV circuit breakers.

An outdoor 400kV substation accommodates type SF6 switchgear and provides a switching arrangement.

21.3 Auxiliary Electrical System

The auxiliary electrical system is based on three voltage levels, 11kV, 3.3kV and 415V and comprises unit and station boards and associated transformers.

Each unit has a 31.5MVA 23.5kV/11kV unit transformer and associated 61MVA 132kV/11kV station transformer. Both the unit and station transformers supply an 11kV switchboard, comprising unit and station sections interconnected by a 2400A circuit breaker. Each unit transformer is connected to the 11kV unit board which supplies the necessary auxiliaries for normal on-load running of that unit.

The capacity allows a margin for varying fuels and will allow the export of 28MVA of gas turbine output via the unit transformer, when the gas turbine is operating on peak load and is supplying the normal unit load.

The station transformer is connected to the 11kV station board and its capacity has been chosen to fulfil one of two roles, either simultaneously supplying:

(1) The normal CMR auxiliary load of one unit.

(2) The start load of a second unit from cold.

(3) The normal load of two thirds of the general station services.

   or:

(4) The normal CMR auxiliary load of one unit.

(5) The running load of one 50% duty starting and standby feed pump.

(6) The direct on line starting of one 50% duty starting and standby feed pump.

Two auxiliary boards are provided; a unit board at the 3.3kV level is supplied from two 8MVA 11kV/3.3kV transformers connected to the associated 11kV unit board, and a station board supplied from a single 8MVA 11kV/3.3kV transformer connected to the associated 11kV station board.

In addition, a 3.3kV fuel oil plant auxiliary board is provided, having two sections each supplied from its own 8MVA 11kV/3.3kV transformer and interconnected by a bus section switch. The two transformers are supplied from 11kV station boards.

The 415V unit and station services boards are supplied from their respective 3.3kV unit and station auxiliary boards via 3.3kV/415V transformers of various ratings. Additional 415V services boards are supplied via fused switches and circuit breakers, from the 415V unit and station services boards.

Following a successful start, each 415V gas turbine unit board is supplied directly from its respective gas turbine generator via an associated 315kVA 11kV/415V transformer.

21.3.1 *Interconnections*

Certain interconnecting facilities exist at all voltage levels of the auxiliary electrical system. At the 11kV voltage level each 11kV station board can be interconnected with either of the other 11kV station boards, allowing any 132kV/11kV station transformer to be made available to any station service or unit service. At the 3.3kV voltage level the 3.3kV station auxiliary boards can be interconnected. At 415V each unit turbine services board can be interconnected to its associated 415V station services board. Similarly each 415V unit boiler services board can be interconnected to its associated 415V station services board.

21.3.2 *System Interlocks*

System interlocks are provided at all voltage levels for operational and maintenance purposes. At the 11kV and 3.3kV voltage levels, the operational interlocks are electrical, utilising auxiliary contacts associated with the respective switchgear. At the 415V level, the operational interlocks take the form of coded keys associated with the switchgear. The operational interlocks perform the following functions:
(1) Prevent incorrect sequential closing of switchgear for synchronising purposes.
(2) Prevent paralleling of more than two 3.3kV station auxiliary transformers.
(3) Permit closure of two from three circuit breakers at the 415V level, thus ensuring that the fault capabilities of switchgear are not exceeded.

The maintenance interlocks at all voltage levels, including those for earthing purposes, take the form of coded keys associated with the respective switchgear.

21.3.3 *Fault Level Indication Equipment*

The design of the electrical auxiliary system allows, with certain combinations of power infeed and fault power contribution from motor circuits, the fault capability of 11kV switchgear to be exceeded.

The fault level monitoring equipment gives visual indication in numerical form of the prospective fault level on the 11kV switchboards at any particular time. It may be arranged to prevent the closing of any circuit which would cause the fault capability of the switchgear to be exceeded.

In addition, the equipment provides signals to initiate alarms in the station/unit alarm systems. Signals also operate light emitting diodes (LED) mounted adjacent to the switchgear control switches. The LEDs are illuminated on all switchgear which, if closed, would cause the prospective fault level limit of the switchgear to be exceeded. This signal is repeated to inhibit the closure of the appropriate switchgear, if required.

21.3.4 *Gas Turbine Generating Units*

The three gas turbine generating units each develop 35MW at 11kV, each generator being connected to its associated 11kV unit board and, via a 315kVA 11kV/415V transformer, to an associated gas turbine unit board. Because of their quick starting capability (approximately 3 minutes) the gas turbine generating units are used in a standby role, primarily to maintain station services at all times.

The rating and design are such that the gas turbines will fulfil all their required operating criteria.

21.3.5 *Switchgear*

The 11kV switchgear is of the air break type, having rated normal currents of 2000A and 3000A respectively and a fault capability of 750MVA break and 900MVA make (symmetrical).

The 3.3kV switchgear is of the air insulated, air break pattern. Circuit breaker normal current ratings are 1600A, 1200A and 800A, having a fault capability of 150MVA symmetrical. Switching devices are used for motor circuits up to 1MW and for transformer feeders up to 1MVA.

The 415V ac and dc switchgear comprises circuit breakers, contactor control gear, isolator, selector and fuse gear. The ac switchgear has a short circuit rating of 31MVA.

21.4   Remote Control and Indication

Remote controls, indication and instrumentation associated with the three generating units are located in the control room. The electrical supplies to these services are provided at 110V single phase ac and are arranged on station and unit basis; one station supply system and three unit supply systems.

21.4.1   *Station Guaranteed Instrument Supplies System (Fig 2.24)*

The station guaranteed instrument supplies (GIS) system supplies the following:
(1) Control and instrumentation.
(2) Station computer.
(3) Spare computer.
(4) Transmission main data processing system.
(5) Transmission back-up data processing system.

The station GIS system comprises two battery/battery charging/ inverter units 1 and 2, supplying sections 1 and 2 respectively of a station 415V GIS board, the two sections interconnected by a bus section switch. A mechanical key interlock prevents the incoming supplies to sections 1 and 2 being paralleled.

Under normal operating conditions, inverter units 1 and 2 are supplied from their respective battery charging units 1 and 2, which are in turn supplied at 415V ac from sections 1 and 2 of the control and administration board. If charging unit 1 fails, battery 1 maintains the supply to section 1 of the 415V GIS board and similarly, if charging unit 2 fails, battery 2 maintains the supply to section 2.

In the event of an inverter failure, the 415V station GIS board may be fed directly from the 415V station services board via a static switch and maintenance bypass switch.

Duplicate supplies are taken from both sections of the station 415V GIS board to 415V/115V transformers which in turn supply the 110V GIS distribution boards. Automatic changeover arrangements to the alternative supply exist at the distribution boards. The 415V/115V transformers and 110V distribution boards form composite GIS load distribution centres, located throughout the station.

For each computer, duplicate supplies from either section of the station 415V GIS board are taken to 415V/240V isolating transformers with final termination at the computer cubicle.

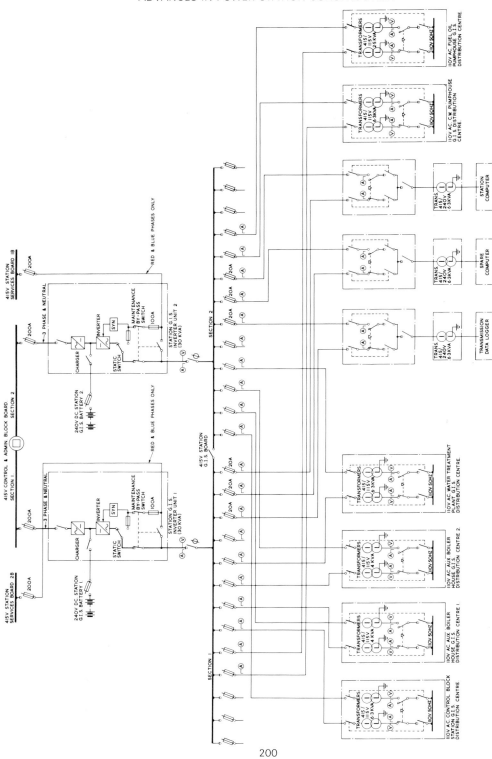

FIG. 3.34 GUARANTEED INSTRUMENT SUPPLY SYSTEM

21.4.2 *Unit Instrument Supply System (Fig 2.25)*

The unit instrument supply system comprises both GIS and non-GIS supplying the control and instrumentation, and the unit computer.

The unit instrument supplies system is arranged in a similar manner to that of the station GIS system, with the following exceptions:

(1) For each unit, the 415V GIS board forms one section of the instrument supply board, the other section being a non-GIS board.
(2) The GIS section is supplied from a single battery/battery charging/inverter unit, the latter deriving its supply from a unit boiler services board.
(3) The non-GIS section is supplied direct from a station services board.

21.4.3 *Actuator Power Centres*

The actuator power centres provide a convenient means of distributing supplies for turbine and boiler valve and damper actuators. They are strategically located throughout the station in close proximity to the centre of load for a group of actuators. Each power centre is based on a modular arrangement, each module catering for 10 actuators. A single actuator cable provides power for both control and indication connections between its associated power centre and actuator junction box. In addition to distributing power, each centre marshals the control cables into a pair matrix system for all the actuators served by that actuator power centre.

21.5 DC Systems

Three dc battery systems provide distribution at 240V, 110V and 50V. In each case, the batteries are installed on a unit basis with interconnections between each unit dc switchboard.

The 240V battery and its distribution system provide emergency supplies to the various unit services, emergency lighting, dc standby auxiliary drive motors, etc in the event of loss of the normal ac supplies.

The 110V battery provides the dc supplies for circuit breaker closing and tripping and for essential plant interlocks.

The 50V battery system provides supplies for remote control and indication, sequencing, alarms and telecommunications.

21.6 Operating the Auxiliary Electrical Supply System

The auxiliary electrical system is designed on the principle that only two prime power sources at the 11kV voltage level are paralleled at one time.

21.6.1 *Normal Starting of a Unit*

The power supply for normal unit starting is derived from the 132kV substation via the associated 61MVA 132kV/11.5kV station transformer.

When a unit is required, the interconnector between its associated 11kV unit and 11kV station boards is closed, thereby making power available to all unit auxiliaries. The boiler is flashed and when steam is available the generator is run up and synchronised to the 400kV grid system via its 400kV circuit breaker. The unit transformer and associated station transformer are then synchronised and unit

201

auxiliaries load transferred to the unit transformer. Finally, the interconnector between the unit and station boards is opened, leaving the station transformer to supply station auxiliaries only.

21.6.2 *Normal Running*

Normal running conditions are dictated by the load demand on the 400kV grid; if one unit is running and a second unit is required, the procedure for starting is as previously described. In the event of a station transformer being out of service the associated station board can be interconnected to either of the two available station transformers via their respective station boards.

21.6.3 *Standby Running*

The choice of connection and duplication of supplies permits standby running of all three voltage levels.

The rating of the station transformers allows any one station transformer to act as standby to a unit transformer, thus allowing the latter to be taken out of service. The gas turbine generators can be utilised to supply unit auxiliaries, however under this condition they run lightly loaded and below their optimum capability.

Each 3.3kV unit auxiliary board is supplied from two 8MVA transformers, capable of supplying the normal load. Normally both are in service but they can act as standby to each other. At the 415V level, each unit board has a standby supply from a station services board.

21.6.4 *Dead Station Start*

The dead station startup condition arises when the station output is required but all units are shutdown and the 132kV substation supply is not available. Under this condition, the gas turbine generating units are started and used to supply unit auxiliaries. Once a main generating unit is available, the sequence of events is as for normal starting of a unit, with the exception that the interconnector between the 11kV unit board and 11kV station board remains closed until an alternative supply is available to the station board.

21.6.5 *Station Supplies*

The arrangement of station supplies is adequate for normal station running.

Each of the 11kV station boards is supplied from a 61MVA transformer, which normally runs lightly loaded. The interconnector arrangements between 11kV station boards allows two station transformers to be safely paralleled, thereby providing a high degree of security of supplies.

At the 3.3kV level, the three station auxiliary boards are supplied from 8MVA auxiliary transformers which can be safely paralleled through interconnectors. In order to prevent the possibility of excessive circulating currents, no provision is made to parallel 3.3kV station auxiliary boards with 3.3kV unit auxiliary boards.

At the 415V level, unit and station services boards are supplied from their associated 3.3kV unit auxiliary boards.

21.6.6 *Shutdown of a Unit*

Normal shutdown of a unit is achieved by reducing to an acceptable level the proportion of the load being supplied to the grid by the associated unit, and then opening the 400kV circuit breaker. The unit is then synchronised to the 132kV substation supply via the interconnector between the 11kV unit board and the associated 11kV station board. The unit auxiliary load is next transferred to the station transformer and the generator disconnected from the unit board. The generating unit can then be shutdown.

Emergency shutdown of a main unit following, for example, operation of an automatic trip, initially utilises the dc system to maintain supplies to essential services necessary during generator run down, eg lubricating oil pumps. The supplies to these essential services and other auxiliaries are then maintained by the associated station transformer, by the operator closing the interconnector between 11kV unit board and 11kV station board.

21.6.7 *2-Shift Operation of Main Units*

During 2-shift operation of main units, when one unit is already on load and a second unit is required to be put on load, the auxiliary electrical system is capable of bringing a second unit to full load within 30 minutes, following synchronisation with the 400kV grid. In addition, all three main units can be on full output within 90 minutes of the first unit synchronising with the 400kV grid.

21.6.8 *Transformer Outage Conditions*

The rating of the station transformers allows each to act as standby to its associated unit transformer or alternatively act as standby to each other. Therefore, a unit transformer can be taken out of service without loss of output from the unit.

The rating of the 8MVA unit and station auxiliary transformers allows one to be taken out of service without disrupting unit or station 3.3kV and 415V supplies.

All units can start with a transformer out of service at all voltage levels. With any one station transformer out of service, the remaining two station transformers start the three units simultaneously by distributing the load, one station transformer starting two units, the second station transformer starting the third unit and supplying station loads.

With all units shutdown, the gas turbine generator units can be synchronised to the 132kV substation via the three station transformers. However, fault level conditions do not permit the output of two gas turbine generators to be taken via one station transformer. Therefore, with any one station transformer out of service only two gas turbine units can be utilised to supplement the 132kV substation.

21.6.9 *Main Unit Operation with Gas Turbines Operating on Peak Load*

With all units running and supplying full load, additional load demands can be met by utilising the gas turbine generator units. In addition, with the main units shutdown, the gas turbine generator units can be used to supplement the 132kV grid system. If during this condition the main units are required, the gas turbine generator outputs in parallel with the station transformer supply unit auxiliaries. When the main unit output is available, it is synchronised to the 400kV grid system and the unit auxiliary load transferred to the unit transformer by first opening the interconnector between the 11kV unit and station boards and then synchronising the gas turbine generator with the unit transformer.

21.6.10 *Low or Falling Frequency Operation of the Main Unit*

The electrical supply system is designed to maintain full output from all units within a frequency band of 49.5Hz to 51Hz and pro-rata decrease in station output within the range 49.5Hz to 47Hz.

The motors for auxiliaries are capable of continuous operation down to a frequency of 48Hz and for 15 minutes down to a frequency of 47Hz.

In the event of the frequency falling to 49.7Hz the gas turbine generator units start automatically and synchronise to their respective unit boards. If the frequency continues to fall, operator action is required.

In the event of rapid falling frequency, each gas turbine is automatically started and synchronised, synchronisation being possible down to 40Hz. However, due to the specified startup time of 3 minutes, it is recognised that a condition of rapidly falling frequency cannot be arrested solely by the gas turbines and operator action is therefore necessary.

21.6.11 *Cabling System*

A concept of cabling for the control and instrumentation system, known as a 'pair network' system, was first introduced at Grain Power Station and later employed at Littlebrook D.

This system utilises large main marshalling cubicles fitted with jumpering fields, situated in strategic locations and interconnected by large trunk cables of up to 500 pairs with small marshalling boxes in the fields connected to the main marshalling cubicles by smaller trunk cables. This has the advantage of enabling the long trunk cables to be designed and installed as early as the construction programme will allow, without having to wait for the detailed design of plant circuitry which can follow later. Detailed interconnectors are completed in the jumpering fields as and when the information becomes available.

A similar marshalling and jumpering arrangement is used for cored $2.5m^2$ connections from switchgear and other circuits, for which paired $0.5m^2$ cable is inadequate.

22.   CONTROL AND INSTRUMENTATION

22.1 Introduction

The plant is operated from a central control room, having a control desk from which a single operator controls the unit hot startup automatically, from burner ignition to full load operation. There is sufficient indication and control to run the plant in a steady state, deal with emergency conditions in safety and shutdown under normal and emergency conditions. The control room staffing concept is one operator per unit, with one supervisor and one trainee operator per control room.

The design is aimed at the maximum conservation of manpower and the highest utilisation of plant by the avoidance of cumulative plant damage. The operator provides the unifying link and is responsible for the control of the process at all times.

To achieve these requirements, the plant is equipped with sequence systems, automatic run-up and loading and modulating control. To prevent plant damage, it is essential to monitor the state of the plant constantly to ensure that operating parameters are within limits. The use of a computer system for monitoring these was an obvious choice.

The computer system is designed on a unit basis with no redundancy of central processors or main peripherals. In the event of failure, a standby computer system may be brought into service via switches and permanently installed cables. It is necessary to provide back-up equipment to continue operation during a computer outage and therefore, the instrumentation systems are designed to permit startup and operation notwithstanding a computer fault. However, in view of the infrequency of total computer failure, the back-up instrumentation and alarms are limited for economic reasons and provide only a degraded service during the computer system outage.

22.2 Extent of Automation

22.2.1 Sequence Control

Sequence control is applied to discrete plant groups, to reduce the number of actions required from the operator, while leaving him in control of the initiation of startup or shutdown of major plant items.

22.2.2 Automatic Run-up and Loading

The turbine generator has the normal facilities for manual run-up to synchronising speed. In addition, provision is made for this to be carried out automatically. Synchronisation can be carried out by automatic synchronising equipment, after manual initiation by the operator. After synchronisation, the turbine generator can be loaded automatically by the computer to a target load selected by the operator. This facility is provided with a manual alternative.

22.2.3 Variable Load/Frequency Characteristics

The effect of system frequency variations on the unit load can be controlled automatically by the computer to provide selected load/frequency characteristics.

22.2.4 Modulating Control

Where necessary, modulating control is provided to enable the required unit response to be obtained and to reduce the need for the operator to monitor and adjust the plant continuously. The modulating control system also enables the plant to be operated at maximum efficiency, within the specified plant constraints.

22.2.5 Display and Recording of Measured Information

Measured information on the state of the main plant and auxiliaries, is provided on visual display units which form part of the unit computer system. Logging of selected measurements is carried out by the computer. Information from the plant, which is essential for short-term operation, or which is used in close association with specific desks or panel controls, is displayed independently of the computer, by indicators and recorders. These back-up instruments provide sufficient information to enable the plant to be run up, loaded, normally operated and shutdown, without the computer being available.

22.2.6 Display and Recording of Alarms

All alarms are annunciated audibly and on visual display units, forming part of the unit computer system. The all-alarm display is designed to provide detailed information on a plant area basis. Computer logging of alarms and selected plant states is carried out automatically or on demand. In addition, there are sufficient alarms displayed on facias at the desk and panel to allow short-term operation of the unit, independently of the computer.

22.3 Central Control Room Instrumentation (Fig 2.26)

22.3.1 General

The central control room (CCR) is located at one end of the turbine house. Adjacent to the CCR there is accommodation for unit operators and supervisory staff. On the same level, there are separate rooms for the computer equipment, telecommunications, metering and records.

Additional accommodation is provided for equipment cubicles used for modulating control, unit alarms and station services alarms.

Equipment rooms are segregated on a unit basis, so that the risk of a fire affecting more than one unit is minimised. Each room is air-conditioned, to remove the heat generated by the equipment and to reduce the presence and mobility of dust.

22.3.2 Panels and Desks

The CCR contains the following panels and desks:
(1) A unit control desk and associated unit panel, for each generating unit.
(2) A station services panel/desk, for control of the cooling water system and monitoring of services, together with fire alarms and general alarms.
(3) A supervisor's desk.
(4) Transmission control panels (400kV, 132kV, 66kV).
(5) Auxiliary electrical supply panel, for control of the station electrical system and gas turbines.

The facilities provided enable both hot and warm starts to be carried out from the CCR. Provision is also made for normal on load operation, normal shutdown and emergency action.

22.3.3 Control of Each Unit

Each unit control desk and associated panel, have controls and indications for the following:
(1) Startup.
(2) Shutdown.
(3) Normal operation.

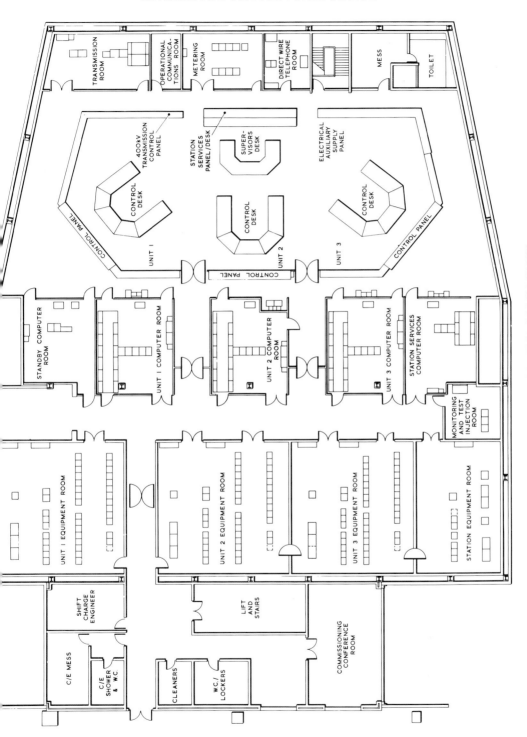

FIG. 2.26   CENTRAL CONTROL ROOM LAYOUT

(4) Emergency operation.

When a unit is shutdown for an overnight or weekend period, all operations to shut the unit down take place from the CCR.

After an overnight or weekend shutdown, the unit is started up entirely from the CCR. If maintenance or other work has taken place during the shutdown, the permit-for-work system ensures that plant is restored to a normal state for operation, prior to the startup. .

For a completely cold start, after suitable checks, manual operations are undertaken local to the plant as well as in the CCR.

22.3.4 *Unit Control Desk (Fig 2.27)*

The controls and instruments on the unit control desk are arranged on a modular basis, to facilitate maintenance and any alterations necessary. The modular system used is based on DIN Standard 43700 using a 72mm square module.

All the controls for a hot or warm startup, and for shutdown, are on the unit control desk, in addition to controls requiring frequent adjustment during steady operation of the unit. Controls and instruments are arranged in functional control areas, related to plant groups.

22.3.5 *Unit Panel*

A unit panel, associated with each unit control desk, provides controls and instruments which require less frequent operation or observation, eg recorders and sootblower controls. Also incorporated in the unit panel is a master section, displaying turbine stop valve steam pressure, main and reheat steam temperatures, drum level and MW load. Flame viewing television monitors provide surveillance of the flames produced by individual oil burners.

22.3.6 *Station Services Panel/Desk*

The station services panel/desk uses modular construction where possible and accommodates controls and instruments which are not associated with a particular generating unit. The plant served includes:

(1) CW system.
(2) Fuel storage tanks.
(3) Fire protection.
(4) Auxiliary boilers   } Extension indications
(5) Water treatment plant } and alarms only

In the case of the CW system, the panel/desk is the normal position for the control of the plant. For the water treatment plant and auxiliary boilers, the panel/desk accommodates extension indicators and alarms from the plant, which are controlled from local panels. The panel/desk also accommodates the station services and fire alarms.

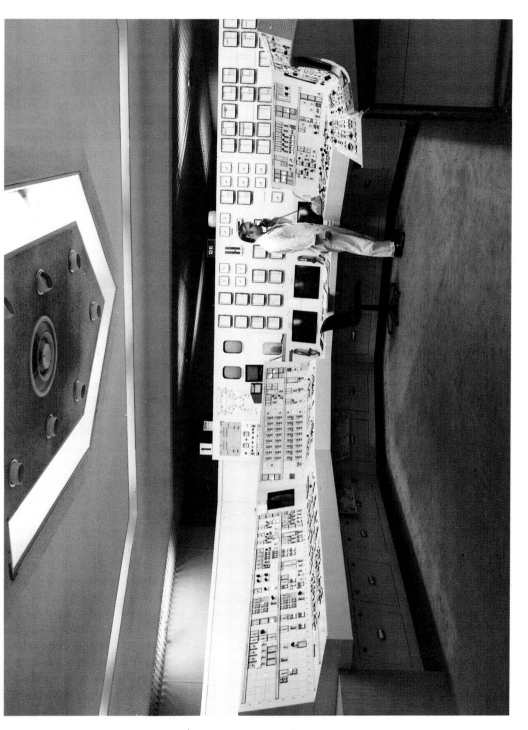

FIG. 2.27   UNIT 1 CONTROL DESK

22.3.7 *Supervisor's Desk*

The supervisor's desk provides facilities for use by the control room supervisor. These include communications to grid control, unit desks and plant areas. The desk is also fitted with visual display units, which display selected information from the unit computer systems.

22.4 Sequence Control

Sequence control is applied to discrete plant groups to reduce the number of control actions required by the operator, particularly during a warm or hot start, leaving him in control of the initiation of startup or shutdown of major plant items.

22.4.1 *Sequence Scheme*

The sequence scheme is based on a logical progression of command followed by a completion check. The equipment for the startup and shutdown sequences for a particular plant group is separate from the sequence equipment for other plant groups. Sharing of equipment between startup and shutdown sequences of the same plant groups is limited.

The equipment uses reed relays to carry out the logic switching. Standardised circuits, mounted on printed circuit boards, carry out functions such as stepping and checking of time delays.

During normal operation, each sequence, once initiated, proceeds to completion, each command instruction normally followed by a completion check. Each sequence is designed to complete in a predetermined time interval, as monitored by a timing unit. A 'sequence in progress' indication is provided while the sequence proceeds, and this is automatically extinguished at the completion of the sequence. Separate indication is provided of sequence completion and this must be manually cancelled by the operator. Remote indication on the sequence plaque is also provided which shows the step-by-step progression of each sequence, and which, under fault conditions, identifies the current state of the sequence equipment and the plant.

Under abnormal conditions, such as if the sequence fails to complete in the specified time, the timing unit initiates an alarm and inhibits further progress of the sequence. When the fault has been corrected, the sequence is restarted by the operator, thereby resetting the timing unit and the sequence is arranged to recommence from the beginning. If the sequence fault is not corrected an alternative plant group may be selected and the faulty sequence shutdown. If there is a persistent plant fault or sequence equipment fault then the plant within the particular sequence is operated by manual controls at the switchgear, or local to the plant.

22.4.2 *Sequence Operation*

Each sequence, in the shutdown state, monitors a number of pre-start checks, which must be satisfied prior to the startup sequence being initiated. The condition of the pre-start checks is indicated on a digital indicator fitted to the sequence control plaque on the unit control desk or, for CW pump control, on the station services control panel.

If a pre-start check is not satisfied, it is identified on the digital indicator by illuminated numerals. If required, the operator then demands a list of sequence checks to be displayed on a visual display unit of the unit computer, in order to identify the failed pre-start check in plain English. Action is then required to clear the fault before the sequence is initiated.

Each sequence is initiated by the operation of a single switch, at the unit control desk. The equipment then carries out a check of the plant and, if conditions are satisfactory, automatically initiates the next operation required by the sequence. This cycle is repeated, without additional operator intervention, until the sequence is completed, provided that all check signals are correctly received from the plant.

If a check signal is not correctly received, indicating that the plant is not in a correct state for the next step, the progress of the sequence is inhibited and a flashing lamp, combined with an audible alarm, operates at the unit control desk.

The faulty sequence step is identified by the digital indicator on the sequence control plaque. On the clearance of a fault, the operator initiates the sequence from the beginning.

Under certain fault conditions, it may be necessary to complete a sequence by the operation of local manual controls at the switchgear feeding the plant. Individual plant items are capable of sequence-free operation at the switchgear positions, which also facilitates maintenance work.

To assist in testing and commissioning, a test panel is installed on the sequence equipment cubicles.

22.5 Conventional Alarms

22.5.1 *Alarm Systems*

The alarms discussed here are provided by relay equipment and illuminated facias in the CCR. They are referred to as 'conventional' to distinguish them from the larger number of computer alarms, which are displayed on visual display units forming part of the computer system. Conventional alarms are provided for:
(1) Each of the three generating units, one set for each unit.
(2) Station services, including fire alarms.
(3) Electrical auxiliary systems.
(4) Transmission plant.

In addition to these systems which annunciate at the CCR, there are several local alarm systems, eg auxiliary boilers, water treatment plant and main turbine lubricating oil system. The local systems annunciate local to the plant but are 'supervised' by the main CCR alarm systems, such that a few local alarms are repeated to the CCR alarm systems. These inform the operator of the existence of at least one local alarm and also selected alarms of a particularly important kind.

The arrangement of the equipment ensures that when one 'repeat' alarm is annunciated at the CCR and remains displayed, it does not obscure the annunciation of subsequent repeat alarms from the same local system.

22.5.2 *Duty of Alarm Equipment*

The conventional alarms are provided primarily to back-up the computer alarm system and provide the necessary alarms for continued plant operation during periods of computer outage.

The equipment provides visual and audible alarms in the CCR, in response to signals received from the plant. The latter are provided by the closure of contacts in plant operated switches or other devices which sense the alarm conditions.

As the alarm equipment operates in conjunction with the computer equipment, they are arranged to share the same alarm initiating contacts, by means of interposing relays. In general, conventional alarms are also duplicated by the computer, the main exceptions to this being station services fire alarms and 'grouped' alarms.

Alarm annunciation is provided by facias mounted in the CCR supplemented by separate audible devices and illuminated beacons. The facias contain windows, arranged to provide flashing or steady illumination as necessary and are complete with pushbutton switches for accepting and resetting alarms and for the testing of window lamps.

Selected alarms are designated as 'urgent' and are provided with windows of distinctive colour. They are arranged to operate a distinctive audible device in each set of alarm equipment. All other alarms are designated as 'non-urgent' and arranged to operate a separate audible device.

22.5.3 *Grouping of Alarms*

In appropriate cases, a single common alarm window is allocated to a group of alarms and the equipment arranged so that the occurrence of any one or more alarms within the group causes a group alarm to be annunciated by the common window. When a group alarm is annunciated and has been accepted, the occurrence of additional alarms within the group causes the group alarm to be re-annunciated. The single alarms forming the group are annunciated individually by the separate, computer alarm system.

23. COMPUTER EQUIPMENT

23.1 Introduction

Each generating unit has a separate unit computer, arranged to drive visual display units (VDU) mounted in the unit control desk. In addition, a VDU is mounted in the supervisor's desk to provide extension displays from any unit system. A single spare computer, with permanently installed cables which enable it to be selected in place of a failed computer, is also provided. The spare computer is kept energised, ready for immediate use when required.

For the monitoring of station services plant, a further separate computer is provided, with a VDU on the station services panel.

In addition, a separate computer associated with the transmission equipment provides alarm and data display, logging and post incident recording, with a VDU on the supervisor's desk.

There is no built-in redundancy of processors or input/output equipment; the equipment is repaired by replacement of the plug-in modules. There are two printers and five VDUs for each system and it is possible to transfer duties between the printers and VDUs to cover for limited failures. The computer system is designed to enable short repair times, typically 30 min to 60 min.

The control and instrumentation system design allows the continuous safe operation of the plant, including controlled startup and shutdown, with the computer partially or totally out of commission.

The computer systems are powered from the guaranteed instrument supplies and are designed to re-start automatically after a power supply interruption, or excursion outside limits.

Each computer system is located in a separate fire proof room, adjacent to the CCR in the ancillary building complex. The environment of each computer room is controlled. The keyboards, which are mounted to form desk modules, and the VDUs are located on the supervisor's desk , unit control desks and the station services panel in the CCR. Log printers, with stands, are positioned in the unit control room. The environment of these locations is the same as the computer rooms.

23.2 Computer Facilities

The following is an outline of the basic tasks carried out by the unit computer system:
(1) Alarm displays.
(2) Alarm and event recording.
(3) Monitoring.
(4) Data display.
(5) Post incident recording.
(6) Efficiency data collection.
(7) Routine data logging.
(8) Turbine automatic run-up.
(9) Steam feed pump automatic run-up.
(10) Automatic unit loading.
(11) Load/frequency control.
(12) Unit load control.

23.2.1 *Alarm Displays*

Alarm messages are displayed on VDUs in the order of detection. The appearance of the alarm message is accompanied by a flashing identifier as part of the message. When accepted by the operator, the identifier stops flashing. A marker indicates when an individual alarm condition clears and can be re-set. An audible alarm is also provided for each unit and sounds until it is accepted. The audible alarm is part of the conventional alarm equipment.

The alarms provided by the computer are classified as 'urgent' and 'non-urgent', as appropriate, and arranged to operate the corresponding audible alarm units.

The annunciation of a computer alarm also causes the alarm beacon, located on the unit control panel, to be illuminated. The beacon is also shared with the conventional alarm equipment.

23.2.2 *Alarm and Event Recording*

All alarms are recorded in order of detection and printed out in plain English statements identifying plant item, location, state of alarm and time of detection.

23.2.3 *Monitoring (Computer Derived Alarms)*

Measured information from the plant is scanned and compared with pre-determined alarm limits, which are set by the computer program and may be high or low, or a combination of high and low. When the measured information passes one alarm limit, indicating that plant operation is becoming abnormal, an alarm of the condition is annunciated. This facility is provided on all analogue inputs to the computer and is activated, for an individual input, by providing the program with the alarm limits which are to be worked to, on that input.

All analogue inputs operating in the range 4mA to 20mA are tested in each scan. An input of less than 3mA is taken as faulty and the alarm associated with the analogue input is initiated automatically as an indication that the input is invalid.

23.2.4 *Data Display*

Plant measurements, calculated values and control program information is available for display on data formats. Examples are as follows:
(1) Analogue inputs from the plant.
(2) Plant digital states.
(3) Control program calculated values and logically derived states.
(4) Header life program calculated values of life factor.
(5) A reference list of sequence step titles, for use in conjunction with the digital step numbers displayed on the sequence control plaques.
    The following main types of format are provided:
    (a) Alpha-numeric formats with and without line drawing.
    (b) Mimic diagrams.
    (c) Historical data traces.
    (d) Boiler metal temperature distribution formats.
    (e) Variable format.

23.2.5 *Post Incident Recording*

A single, continuously updated post incident record (PIR) is maintained of the measured values of 150 selected analogue inputs; 60 values are stored for each of the 150 inputs.

Following the detection of a major plant incident, the stored values immediately begin to be printed, commencing with the oldest values first. The record continues to be maintained during printing of the record, over-writing values that have been printed. The record ceases to be printed after 15 minutes from the time of the incident or after the 'Stop PIR' control is operated.

An incident consists of either the operation of the 'Start PIR Printout' control, or one of 16 pre-defined combinations of plant inputs occurring together. It is possible to gate together logically up to four analogue and digital inputs to define each of the 16 incidents. An identification of which incident caused the commencement of the PIR printout is printed with the record.

23.2.6 *Efficiency Data Collection*

A separate log is made, containing all main unit plant data required for completion of operating data returns. This accumulates hourly, daily, weekly and monthly totals for the required parameters.

For this purpose, data is input via the analogue and digital scanners, with the exception of the electrical metering values which are input via a serial communications link from the metering equipment.

23.2.7 *Routine Data Logging*

Selected measurements are printed at regular intervals. The selection of measurements is for specific purposes and provision is made for the selection to be varied, as required by changing circumstances of plant operation. In addition, the operator is able to compose a log of selected measurements for printout on request.

Automatically every hour on the hour, the program stores the current values and at eight hour intervals a printout occurs of the previous seven stored logs plus the current log.

The log is also available at any time for printout on operator demand in which case it contains the previous seven stored logs plus the current values of the inputs at the time of demand.

23.2.8 *Turbine Automatic Run-Up*

Provision is made for automatic run-up to speed of the turbine generator, as part of the 2-shifting control concept, to enable a hot start to be carried out in the specified time, at the optimum rate for the prevailing plant conditions.

The unit computer is used for the automatic run-up requirements. The computer carries out pre-start checks and calculates the optimum run-up rates, from the initial thermal conditions of the plant. The operator selects the rate and initiates the run-up, which then proceeds automatically, with the steam supply to the set being controlled to achieve the required rate, until control is taken over by the turbine speed governor.

During the run-up the computer monitors the plant and automatically modifies the run-up programme, when necessary, to prevent dangerous conditions arising. Provision is made on the unit control desk for the manual control of run-up, when the automatic equipment is not in use. Separate equipment is provided for automatically synchronising and connecting the turbine generator to the external power system. This equipment is entirely independent of the unit computer and its operation is manually initiated by the operator after the completion of run-up.

23.2.9 *Steam Feed Pump Automatic Run-up*

The steam feed pump is automatically run-up to speed in a similar way to the turbine generator. The equipment for controlling the run-up is also similar but, in this case, the run-up terminates at a speed slightly below the value at which the boiler automatic feed water control equipment can assume control of the pump speed.

23.2.10 *Automatic Unit Loading*

Automatic loading of the unit is provided as part of the 2-shift operation control concept, aimed at achieving hot startup in the specified time, and also achieving smoother and more consistent loading of the turbine, with the intention of reducing the cumulative effects of thermal stressing.

The modes of loading from the unit computer are as follows:

(1) Fuel Ramp (startup) — the raising of the 'boiler firing level' in accordance with a ramp function generated in the unit computer.

(2) Pressure Loading (startup) — the automatic raising and lowering of the speed set-point of the narrow range governor, by the unit computer, to control the pressure of the steam at the final superheater outlet. It contains functions for the application of a 'block load', for pressure raising in line with startup requirements and for monitoring plant conditions, particularly turbine supervisory monitoring, with consequent modification of control actions.

(3) Fixed Ramp Loading — the raising of the speed set-point of the narrow range governor in accordance with a ramp function generated in the unit computer. This includes plant monitoring as for 'pressure loading'.

(4) Manual Loading — full provision is made for manual loading, from the unit control desk, when the automatic loading is not in operation; this includes full facilities for manually starting and stopping burners.

23.2.11 *Load/Frequency Control (System Regulation)*

To enable the load/frequency characteristics of the unit to be varied to suit changing system operation requirements, the generated electrical output can be continuously controlled by the unit computer.

A signal, representing the deviation of grid system frequency from 50Hz, is used in two ways. Firstly, the 'boiler firing level' is raised or lowered in response to variations in grid frequency. This facility is provided within the unit computer and includes a means of adjusting the gain. The boiler firing level is the rate at which heat is supplied to the boiler for control purposes it is measured in terms of total fuel flow and total air flow to the burners. Secondly, the 'pressure governing' desired value is modified by the frequency signal. This facility is a component part of the boiler control equipment which automatically raises and lowers the speed set point of the turbine governor to control the pressure of steam at the final superheater outlet.

23.2.12 *Unit Load Control*

Unit load control is achieved by the automatic raising and lowering of the governor set-point, by the unit computer, to control the electrical power delivered by the turbine generator under nominally steady state conditions. The effective load/frequency droop associated with this loop is selected at the unit desk, together with load and frequency limits. It has the effect of introducing a frequency dependent offset in the demanded set-point.

24. CONSTRUCTION METHODS

There were no major problems in getting the large and heavy items of plant to Littlebrook D site. Various routes were adopted by contractors of which the principal ones are as follows:

(1) Structural steelwork (approx. 33,000t) — by road.

(2) Boiler drums (32m long, 256t) — by road and sea to Kingsnorth, then by road to site.

(3) Generator stators (outer 121t, inner 266t) — by road and sea to Kingsnorth, then by road to site.

(4) Generator transformers (nine at 185t) — by road and sea to Kingsnorth, then by road to site.

(5) Boiler packs and panels (180 at average 35t) — by road.

Because of the limited size of the site at Littlebrook D it was essential that contractor's storage areas were used by more than one contractor, in sequence. However, as mechanical and electrical plant started to arrive on site, basically to programme, it became evident that additional storage would be required. The additional space was provided both on and off-site. On-site storage used the A station turbine and boiler houses from which the plant had been removed, a large, heated temporary shed built for the storage of switchgear, a large air-balloon storage for insulation materials, and reclaimed land for outdoor storage.

Additional off-site indoor storage, both heated and unheated, was provided at warehousing contractors for the use of all contractors' plant which became available ahead of the programme date for site access.

Construction work was assisted by the introduction of a few on-site labour saving erection techniques. For example, in order to reduce considerably the amount of site welds, the boiler contractor constructed large sections of the boiler in his factory into 'packs' and 'panels'. These were then transported to site where they were lifted into position and then welded to adjacent packs or panels. Fig 2.28 shows the sequence of erection of these boiler parts. There were 60 packs and panels in each boiler with an average weight of 35t per pack.

The turbine contractor shipped the inner and outer cylinders as a combined 'modular' erection unit for the HP and IP turbine. Very stringent checks were carried out in the factory before the 'modular' unit was ready for delivery to site, so that it could be lifted into position as a completed unit.

The construction of the CW pumphouse walls was carried out by forming a diaphram wall excavated and concreted under a bentonite solution. Extensive use of profiled galvanised steel permanent shuttering considerably reduced the requirements of temporary shuttering and scaffolding.

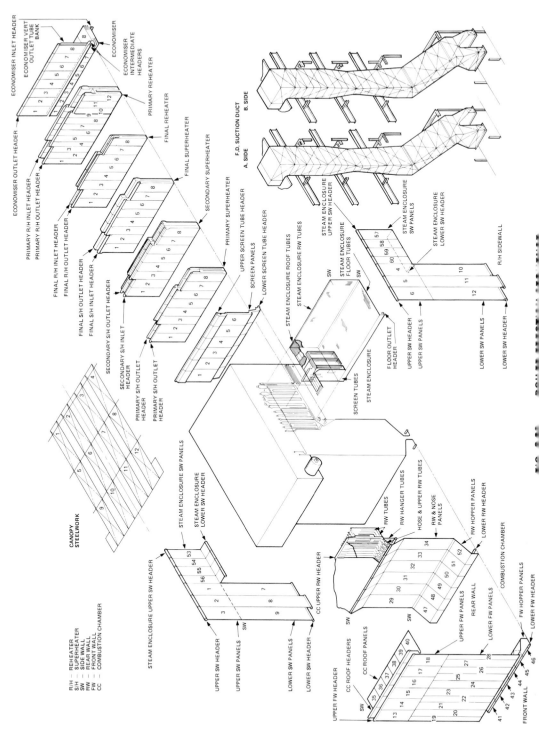

25. COMMISSIONING/OPERATIONAL ACHIEVEMENTS AND AVAILABILITY

The units at Littlebrook D were designed in the late 1960s and early 1970s when the price of heavy fuel oil was attractive against the other fuels. However, since their construction the price of oil has risen dramatically and consequently the units are not required to perform the base load role originally envisaged. Nevertheless the units have been commissioned very rapidly to a rating in excess of their design output and have exhibited unusually high flexibility, reliability and efficiency. They have made a major contribution to the South Eastern Region's generation in the 2-shift and 4-shift operating modes and have confirmed an ability to operate on base loads returning a very high supplied thermal efficiency. In spite of the relatively high fuel cost of fuel oil, Littlebrook D units still have an important role to play in providing power to the National Grid.

Because of the relatively high price of fuel oil at the time of commissioning Littlebrook unit 1, a proposed 'minimum time' programme was set up (Fig 2.29). This programme was based on experiences and tests at Grain Power Station and reflected the best and most economical use of generation time in commissioning the units. Based on the experience obtained at Grain a significant feature of these units has been a very short period of time between synchronising and the unit achieving design rated output. The time scale shown on the programme of 5 to 6 days has been achieved without any significant out-of-merit running costs. All these units have been able to make an immediate contribution by 2-shift operation to system requirements.

The units have been operated primarily to meet daytime peak demands. In order to meet this requirement and to minimise costs it has been necessary for the station to operate immediately following commissioning to a 2-shift regime, moving very rapidly on to a 4-shift regime. A typical load curve responding to variations in the national load profile is shown in Fig 2.30. Times for bringing these very large units on load have been reduced by nearly 50% from the design concepts which, apart from cost benefit, has enabled the station to respond more rapidly to system requirements.

The availability and generated efficiencies of the units since commissioning are as follows:

Unit Number	Availability Since Commissioning %	Load Factor Since Commissioning %	Generated Efficiency %
1	94.94	33.2	38.44
2	96.56	42.8	39.13
3	94.44	83.9	39.09

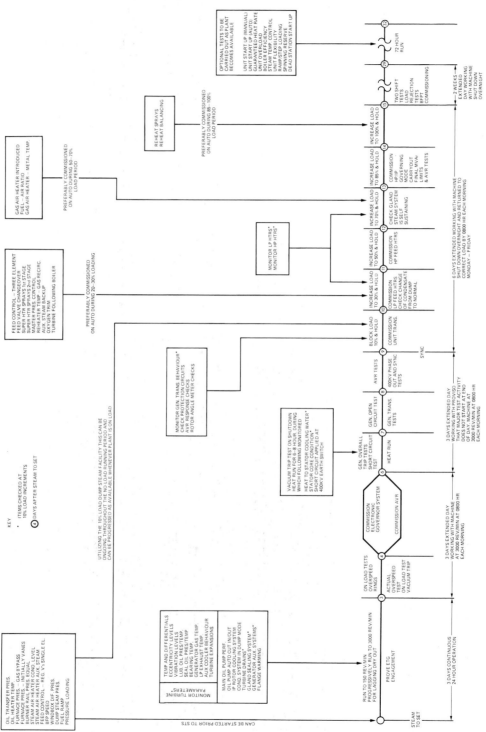

FIG. 2.29 MINIMUM TIME PROGRAMME FOR COMMISSIONING A UNIT

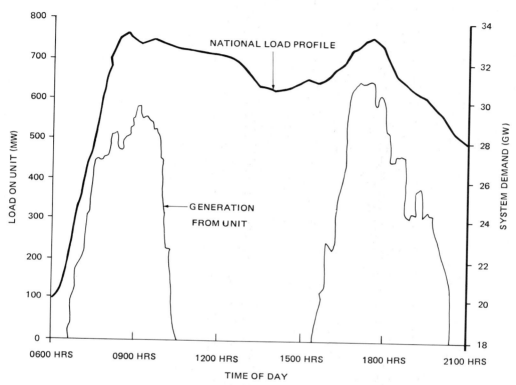

FIG. 2.30   A TYPICAL LOAD CURVE SHOWING UNIT'S RESPONSE TO
VARIATIONS IN NATIONAL LOAD PROFILE

During most of 1984 it was necessary to operate this station at maximum capacity for prolonged periods. During this period the units regularly achieved outputs far in excess of the design rating of 645MW. In fact they achieved, a continuous generated output of 690MW without reduction in thermal efficiency and a sustained overload capacity of 745MW against a design of 705MW.

26. MAIN CONTRACTORS, MANUFACTURERS AND PLANT SUPPLIERS

Mechanical

Turbine generators, condensers and feed heating	GEC Turbo Generators Ltd
Feed pumps	Weir Pumps Ltd
Condensate extraction pumps	Mather & Platt
Main boilers and fuel system	NEI Nuclear Systems Ltd
Circulating pumps	Hayward Tyler Ltd
Fans and gas airheaters	J. Howden & Co Ltd
Steam airheaters	Accles & Pollock
Sootblowers	Diamond Power Speciality
Sootblower compressors	Howden Godfrey Ltd
CW pumps	APE Allen Pumps Ltd
CW valves	Boving Ltd
CW screens	H.S. Brackett
Fire fighting equipment	Mather & Platt
Fuel oil pumps	Stothert & Pitt
	Hamworthy Ltd
Fuel oil tanks	Motherwell Bridge Eng. Ltd
Gas turbines	GEC Gas Turbines Ltd
Water treatment Plant	NEI Nuclear Systems Ltd
Auxiliary boilers	George Clark & NEM Ltd
	Energy Equipment Ltd
Air compressors	APE Bellis
Insulation	Cape Contracts
Cranes	Wharton Cranes Ltd
HP pipework and valves	Pipework Engineering Developments
Emergency pumps and mechanical services	Aiton & Co Ltd

Electrical

Transformers	Hawker Siddeley Power Transformers
	NEI Parsons Peebles Ltd
	GEC Transformers Ltd
11kV switchgear	GEC Switchgear Ltd
3.3kV switchgear	Whipp & Bourne Ltd
415V & DC switchboards	Lawrence Scott & Electromotors Ltd
Batteries and chargers	Chloride Batteries Ltd
Main connections	Watson Norie Ltd
Main cabling	Pirelli Cables Ltd
	James Scott Electrical Contractors Ltd

Control and Instrumentation

Computer systems   Honeywell Information Systems
Control and instrumentation equipment   Babcock Bristol Ltd
Sequence control equipment   STC Ltd
Conventional alarm equipment   D. Robinson & Co Ltd

Civil

Main civil works   John Laing Construction Ltd
Structural steelwork   Cleveland Bridge & Engineering
                                                Co Ltd
Chimneys   Bierrum & Partners
Main piling   Raymond Piling International
                                                (UK) Ltd
Superstructure   M.J. Gleeson (Contractors) Ltd
Landscaping   Blakedown Landscapes

Consultants

Consulting architects   Architects Design Group
Consulting civil engineers   Sir Alexander Gibb & Partners
Consulting landscape architects   Derek Lovejoy & Partners

3.

Drax Coal-fired Power Station

1. INTRODUCTION

The initial concept of Drax power station first appeared in the CEGB's 1962 Development Plan. Various factors enabled a power station of large capacity to be considered and Statutory Consent was initially sought and obtained for a 3000MW station, which was the largest capacity ever planned for a single power station in this country. It was subsequently decided, however, that advantage should be taken of the development of major plant and that the CEGB's first 660MW units should be sited at Drax. The planned capacity of the station could, therefore be increased and approval was granted to the necessary changes in the consent to allow for an ultimate capacity of 4000MW.

1.1 Background to Drax First Half

Authority to carry out preparatory work on Drax was first granted in December 1964. In March 1966, authority to proceed with the project was issued. This stated that the station should be designed to accommodate six 660MW units but initially, three units (numbered 1, 2 & 3) would be constructed and referred to as 'Drax First Half'. Work started on these in 1967; two were synchronised in 1973 and the third in 1974. Because of a shift in the planning of new generating capacity from coal-fired to oil-fired and nuclear-based generation, authority to proceed with the second half of the station (referred to as Drax Completion) was deferred for some years.

1.2 Background to Drax Completion

Based on assumptions of load growth in the 1977 Corporate Plan, the CEGB planned to commence work on the three units of the Drax Completion project (numbered 4, 5 & 6) in 1979 with the intention of commissioning in 1985/86. This formed part of a balanced programme of coal-fired, oil-fired and nuclear stations.

In July 1977 however, the Government of the day requested that Drax Completion should proceed some two years ahead of the CEGB's requirements. The primary object of the decision was to secure jobs in the manufacturing industry in the north east of England. Subject to the Government providing compensation, the CEGB agreed to complete Drax power station in advance of its requirement to meet the growth in demand for electricity.

1.3 Development of the Two Construction Stages

It was a planning requirement by the Alkali Inspectorate that flue gases from the six boilers should be exhausted to atmosphere through a single chimney containing three flues. This meant that construction of the Completion had to be phased such that minimum disruption was caused to the operation of the First Half units.

The provision and layout of ancillary services, eg coal handling plant, ash and dust handling plant, cooling water make-up and purge systems, etc, had to take into account the requirement of early operation for three units with the later addition of a further three units. The physical size of the whole station however, led to the adoption of a split recirculating cooling water system, each half having its own self-contained system.

As far as was practically possible, the principle of replication of the First Half was followed in the construction of Drax Completion. Unit output and operating conditions are the same and the same basic pattern of layout was followed. In order to ensure a maximum level of replication, contracts for boilers and turbine generators were placed with the same First Half suppliers.

Since a period of 10 years had elapsed between design of the First Half and commencement of the Completion, it was not possible to replicate all aspects of the station. Suppliers of some equipment had merged with other companies or had ceased trading; in some instances items were no longer in manufacturers' product ranges. Changes to legislation in the intervening years meant that many items being produced were to a different standard.

More importantly, operating experience over a number of years led to design improvements in equipment and plant which have been incorporated into the Completion project. Important areas of non-replication and design differences are highlighted in the plant descriptions.

2. SITE SELECTION

2.1 Site Location (Fig 3.1)

Drax power station is located on the south bank of the River Ouse, mid-way between Selby and Goole in the County of North Yorkshire. The site was selected following detailed investigations of three large areas in the vicinity — two of these areas were on the south bank of the Ouse near Barlow and Newland, the third area was north of the river at Barmby. The investigations showed that there was little to choose between the areas on foundation conditions. The Drax site, which forms a part of the Barlow area of investigation, was selected in preference to the others primarily because of the shorter rail link required for coal haulage to the station.

A number of factors enabled a power station of large capacity to be considered, the main factors being:
(1) The proximity of the site to the North Yorkshire coalfields means that transportation costs of fuel are kept to a minimum.
(2) Good rail connections for fuel transportation.
(3) Comparatively short transmission connections were needed to the existing 400kV grid network.
(4) Drax is a cooling tower station and the River Ouse is capable of supplying the make-up water requirements.
(5) Adjacent land at Barlow provides an adequate ash disposal facility.
(6) Freedom from restrictions on site area and structure height.

Fig 3.1 shows the location of the site and Fig 3.2 shows an aerial view of the station.

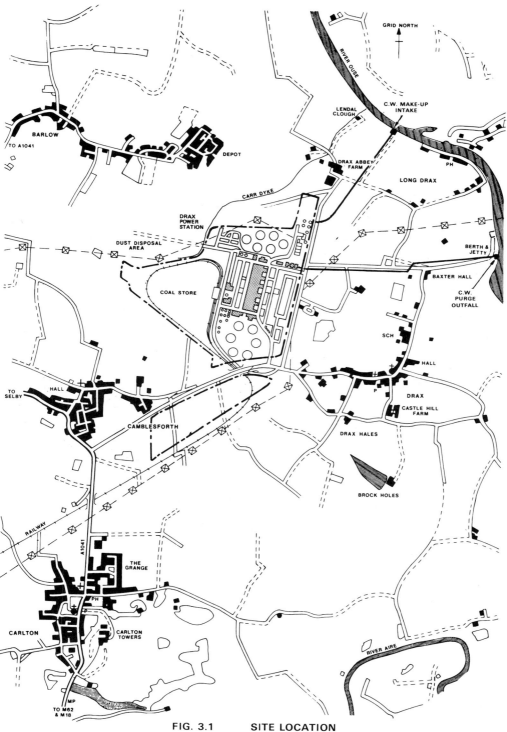

FIG. 3.1   SITE LOCATION

FIG. 3.2   AERIAL VIEW OF STATION

2.2 Ground Conditions

Geologically, the site is within the Vale of York and the strata includes lacustrine deposits overlying triassic sandstone. The bed rock lying at a depth of 25m is of bunter sandstone the surface of which was planed by the passage of ice during the Ice Age and overlain by sands and laminated clays deposited in a glacial lake which extended over this part of Yorkshire. Above these glacial deposits an alluvial mud known as warp was laid down by regular flooding, and this forms the flat low lying countryside. The nature of the strata made it necessary to carry all heavy buildings and installations on piles taken down to the rock and sand strata, but light or flexible structures could be carried on foundations placed on top of specially compacted or stabilised fill.

3. SYSTEM CONSIDERATIONS

Drax power station, by nature of its advanced design, is a high merit station with an overall plant thermal efficiency in the order of 38%. Operationally, therefore, it can be expected to have a high utilisation factor in the early years of operation. The changing patterns of load demand and the inherent characteristics of plant, whether hydro, nuclear, coal or oil-fuelled, require a periodic review of operational roles. Consequently the operational flexibility of the plant must be adequate to meet the changes in requirements over the station life. These factors have been taken into account in the development of the Drax units and adequate design margins have been incorporated which, it is considered, relate to real service conditions. For instance, in order that thermal cycling of the turbine is minimised, so as to avoid cracking of casings and shaft distortions, the boiler was designed to be capable of maintaining high steam temperatures at part load. In addition the shutdown procedures avoid excessive cooling of the turbine to ensure that the target loading rates are achieved following an overnight shutdown. Quality assurance standards have been maintained at a high level to ensure reliability of the plant in service. Also, protective devices and systems are incorporated which minimise the risk of plant failure.

The security standards for generating stations laid down by the CEGB require the transmission system to be designed so that the generators should remain stable following credible faults and credible outage conditions at all times when the units are running in accordance with the planned requirement of the station. The transmission connections have accordingly been designed so that with maintenance and fault outages it is possible to operate the station at maximum output with a prolonged outage of any two of the three connecting circuits.

4. OPERATIONAL ROLES

The plant at Drax is required to be capable of continuous operation over a load range as wide as possible, the design intent being that it is not necessary to shutdown the plant for overhaul, repairs or cleaning of gas passes except at statutory intervals of inspection.

4.1 Control Range

As a minimum requirement, the plant is specified to be capable of operating continuously at full load on weekdays for a period of three months during which the availability exceeds 85%. The load range at other periods of continuous operation may be 50% to 100% continuous maximum rating (CMR). When operating at part load, the plant must be capable of responding to system frequency changes to raise or lower load at rates up to 5% CMR per minute in either direction in the load range 50% to

100% CMR. The periods of part load operation down to 50% CMR for system reasons could be of up to 10 hours duration. The control equipment is required to ensure output regulation in response to system frequency variations at any load in the control range and to ensure rapid corrective response to sudden frequency changes due to imbalance between system load and generation.

4.2 2-Shift Operation

The plant is required to be unloaded in a period of approximately 30 minutes so as to minimise cooling of the boiler and turbine and to ensure a rapid re-load after an overnight shutdown of 6 hours to 8 hours. The target rate of loading is from synchronisation to full load in 30 minutes. No restraint is placed on the duration of the startup prior to synchronisation. This should, however, be as short as possible to assist meeting unforeseen operational demands and to reduce the heat loss during startup. For periods of shutdown in excess of 8 hours, a lower rate of turbine generator loading can be accepted.

In a multi-unit station such as Drax, the main and auxiliary plant is designed so that following an overnight or weekend shutdown all units can achieve full output within 2 hours from the first unit synchronisation. Failure of one unit during the startup and loading should not prevent the other units in the station achieving full output in the programmed time.

4.3 Load Rejection

Each unit has the ability to reject load to the extent of 20% CMR on sudden partial loss of demand from any load level in the range of 50% to 100% CMR without initiating a plant trip. Additionally the plant is capable of tripping without damage on loss of grid supplies or as initiated by a plant fault.

4.4 Design Life

The life span of the generating plant is anticipated to be in the order of 40 years when operating at a load factor required by the system, and the typical number of cycles of operation during this lifetime on which component design is based is as follows:

Regulating Function	Lifetime Operations
Total number of starts	6000
Number of warm starts (36 hours – 48 hours)	1000
Remainder (mainly hot starts)	5000
Number of periods of operation in the range 50%–70% CMR	4000

TABLE 3.1 – DESIGN COAL SPECIFICATION DETAILS

Coal Composition	Basic Design Fuel	Range of 95% Supplies	Range of 100% Supplies
Proximate Analysis %			
Moisture (inherent and free)	11	7–14	5–16
Ash	17	7–27	7–30
Volatile matter	28	26–30	26–33
Fixed carbon	44	42–66	35–48
Hydrogen ⎫	3.8	–	–
Sulphur ⎪	2.0	1.0 – 3.5	0.5 – 4.0
Chlorine ⎬ as fired	0.4	up to 0.55	up to 0.65
Nitrogen ⎪	1.1	1.0 – 1.4	–
Phosphorous ⎭	0.01	–	–
Ultimate Coal Analysis %			
Carbon	59.2		
Moisture	11.0		
Hydrogen	3.8		
Sulphur	2.0		
Chlorine	0.4		
Nitrogen	1.1		
Oxygen	5.5		
Ash	17.0		
Gross CV kJ/kg	24,300	22,000–27,000	
Net CV kJ/kg	23,300	21,000–26,000	
Ash in Reducing Atmosphere °C			
Initial deformation	1150	1100–1200	1050–1400
Hemispherical	1320	1270–1400	1210–1400
Low temperature	1350	–	1290–1400
Ash in Oxidising Atmosphere °C			
Initial deformation	1300	1260–1400	1240–1400
Hemispherical	1400	1400	1400
Flow temperature	1400	1400	1400
Hardgrove index	50	45–63	–

4.5 Coal Supplies

A design coal specification for Drax power station is given in Table 3.1. The figures given in this table show the wide range of coal characteristics which are met by CEGB plants burning UK coals. The plant is required to give full load output when burning coals within the range covering 95% of supplies and to be capable of continuous operation, if necessary at reduced output, for coals within the remaining 5% range of supplies. Plant performance guarantees are given on the basic design fuel.

4.6 Conditions of Operation

Table 3.2 gives the specified design parameters for steam flows, temperatures and pressures at selected loads.

TABLE 3.2 SPECIFIED STEAM CONDITIONS

	Short Time Overload	CMR	94.8% CMR	70% CMR	50% CMR
Evaporation (kg/s)	563	563	534	394	281
Superheater outlet pressure (bar)	166.5	166.5	166.5	166.5	166.5
Superheater outlet temp (°C)	568	568	568	568	525
Reheater steam flow (kg/s)	505	436	414	305	218
Reheater outlet pressure (bar)	46.2	41.3	39.4	29.5	20.9
Reheater inlet temp (°C)	373	365	363	355	295
Reheater outlet temp (°C)	568	568	568	552	490
Economiser feed temp (°C)	207	252	252	239	220

5. AVAILABILITY

The availability for service of generating plant is normally quoted in percentage terms. For a given period of time, it is a measure of the output which the plant is capable of producing over that period compared with the output which it could have produced if working continuously at full rating. Two factors why the plant may not be available to generate at maximum capacity are taken into account. These are random breakdown of components which require immediate shutdown or reduction of load, and planned outages for statutory inspection and maintenance work. These factors are assessed by analysis of the actual operation of plant in the CEGB's stations.

In addition, availability is divided into:

(1) Winter peak availability, which normally takes into account the random breakdown factor.

(2) Settled-down average annual availability which takes into account both (1) and planned outages.

On this basis the CEGB has assumed for investment appraisal purposes that for Drax the availability figures are 86% for (1) and 72% for (2). In practice these figures have been bettered in operation of the First Half units.

6. STATION LAYOUT

The layout for the complete six unit station is shown in Fig 3.3. The main buildings run in a generally north-south direction with the boiler house to the west and turbine house to the east. As stated earlier, construction of the station was carried out in two stages. The First Half, consisting of three units and associated ancillaries occupies the southern end of the site. The Completion stage of a further three units extends the main buildings in a northerly direction. The coal store with a capacity of some 25% of maximum consumption per year occupies the westerly side of the main site. The transmission compound and outgoing distribution lines are to the east. The main site area is flat and has been raised to a general level of +6.1m OD to provide protection against flooding.

The requirements to house six units led to buildings of considerable length. During the development stage, it was found that the minimum length of the boiler house would be 433m and the turbine house varied in length depending on the arrangement of the turbine generators. The arrangement which produced a turbine house of the same length as the boiler house had the axis of the turbines transverse to the longitudinal building line. A disadvantage of this arrangement was that it required a turbine house crane of greater span and capacity than those for which designs were available.

Turning the turbines through 90° into the longitudinal position alleviated the crane problem but considerably increased the length of the turbine house. A somewhat novel arrangement was finally adopted with the axis of the turbines at an angle of 45° to the building line. This significantly reduced the width of the turbine house without making it longer than the boiler house, thus giving the most economic building and providing a compact HP pipework configuration.

The unusually long main buildings made personnel movements of particular importance and in order to minimise the distances between plant and operational centres, the administrative, maintenance and control complex was placed as close as possible to the centre of the station. It was also considered necessary for the workshops to be available to unit 1 without access being impeded by units under construction. This led to the locating of the workshops and stores between units 2 and 3.

6.1 Cooling Water System

For the first stage development of three 660MW units, six cooling towers each 114.3m in height with a base diameter of 91.44m were located to the south of the main station buildings. Four cooling water pumps housed in a pumphouse located between the towers and the main buildings serve the cooling water requirements of the First Half units on a closed circuit system.

For the Completion stage, a further six cooling towers, each 115m in height and base diameter of 92.7m were located to the north of the station. By nature of their size, the towers are a prominent feature of the site and careful grouping was necessary to ensure efficient functioning together with acceptable visual impact.

Make-up water for the cooling systems and for boiler water is extracted from the River Ouse at a make-up pumphouse located on the river bank. Purge water is discharged to the river via an on-site purge pump chamber.

6.2 Main Chimney

The tallest structure on the site is the main chimney at 259m. It is located midway along the length of the main plant buildings and it provides the waste gas discharge from all six main boilers.

6.3 Ash and Dust Disposal

It was originally envisaged that ash and dust produced at Drax would have to be transported away to a suitable remote disposal area and a scheme was proposed utilising the disused airfield at Burn. This would be similar to that already being developed at Gale Common where an artificial hill is being formed with pulverised fuel ash produced by the Ferrybridge C and Eggborough power stations. However, a Public Enquiry held in 1970 resulted in the decision to dispose of ash and dust onto the Barlow site immediately adjacent to the station.

7. FOUNDATIONS

All major foundations at Drax are piled and constructed of reinforced concrete. Unpiled foundations are used only for structures with net ground pressures less than $100kN/m^2$ and where settlement is not critical.

Piled foundations are supported on pretensioned prestressed concrete piles with working load vertical capacities of 1000kN, 800kN and 600kN at the top of the pile. In parts of the site, fill is placed on existing ground. This causes settlement of the existing ground and hence applies dragdown forces (negative shaft friction) on any pile which passes through it. For this reason and because driving a large mass of solid displacement piles into cohesive material can of itself generate negative shaft friction, there is a uniform allowance of dragdown load on all capacities of pile. This additional load allowance is taken to be 300kN with a required factor of safety of two. Hence the maximum loads acting on the piles were specified as 1300kN, 1100kN and 900kN, these loads being considered to act at the top of the fill.

Pile lengths were determined on the basis of investigative probe boreholes and also on the results of preliminary pile loading tests to 2.5 times maximum specified pile capacity.

The reinforced concrete foundations for the main turbines have been designed to meet the tight limits of deflection limitations, as specified by the plant contractor, by finite element analysis. The structure is supported by piles carried to bedrock.

Special attention has been paid to the design of the pulverised fuel mill foundations to minimise vibration and load stresses transmitted to surrounding floor areas and plant. Each mill is mounted on an individual bedplate and foundation inertia block which is supported at the four corners by bearing blocks. The First Half mills were supported by flat rubber bearing blocks, but operational experience indicated a need for improvement in the clamping arrangements. The Completion mills are supported on a mounting system consisting of four rubber bearings arranged in a pyramid formation, one pyramid being positioned at each corner of the block (Fig 3.4). The design criteria for vibration levels were defined by the plant contractor. A mathematical model was formulated to design the mounting system which was then tested on a full size rig. Additionally the mass of the concrete inertia block was increased and the bearing blocks located in a high position. This brought the centre of gravity of the block and mill closer to the supporting system, thus improving the damping effect.

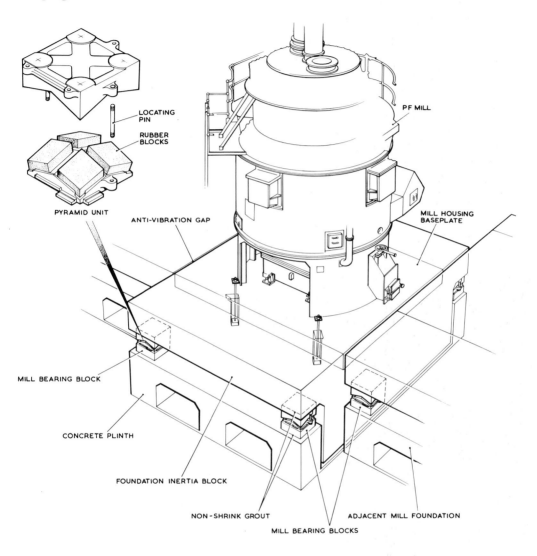

LOCATING PIN

RUBBER BLOCKS

PYRAMID UNIT

ANTI-VIBRATION GAP

PF MILL

MILL HOUSING BASEPLATE

MILL BEARING BLOCK

CONCRETE PLINTH

FOUNDATION INERTIA BLOCK

NON-SHRINK GROUT

MILL BEARING BLOCKS

ADJACENT MILL FOUNDATION

FIG. 3.4   MILL FOUNDATIONS

All concrete used in the site works contains selected pulverised fuel ash of between 20% and 30% of the cement content. The only exceptions to this are the six cooling tower shells with their support legs, and also the structures in the 400kV compound.

For the whole station a total of approximately 48,000 piles were driven and a volume of 520,000m^3 of concrete poured in the construction.

8. STRUCTURAL STEELWORK

With the exception only of the administration building and control block which are reinforced concrete structures, the main and ancillary buildings are built in constructional steelwork clad with preformed precoated steel sheeting.

The main building is one of the largest single structures in the CEGB. It measures 433m × 132m × 76m high. The layout of the building follows a pattern developed over the last 20 years for modern coal-fired stations. Its function is to house and support six 660MW coal-fired boilers together with their respective turbine generators and auxiliary plant.

The design of the main building structure is heavily influenced by the need to limit sway of the structure, at the roof level, to within 50mm of datum location. This is achieved by using parts of the structure as stiff frames. The mechanical annexe and the bunker bay act as the stiff part of the structure.

The boiler house which is the largest part of the main building is sandwiched between the mechanical annexe and the bunker bay. The turbine house 'leans' from the mechanical annexe and the electrical annexe from the bunker bay.

Thermal expansion of the main building structure, which is almost half a kilometre long, is catered for by expansion joints, one at each third distance, along the building.

Vertical bracing of the structure is avoided except in specific areas as it restricts plant layout and maintenance, and it complicates erection of the building.

8.1 Cladding and Decking

The main and ancillary buildings for the Completion are clad in plastic-coated steel sheeting. The roofs are built up from aluminium decking with vapour barrier, insulation board, felt and chippings laid in bitumen.

8.2 Fabrication

All steelwork was fabricated off-site with the exception of the bunker sides which were assembled on site and welded in-situ. The majority of large pieces of steelwork were transported to site in single pieces. The turbine house roof trusses which were over 55m long were despatched to site in three pieces, assembled below their location and lifted by tandem lift into the structure.

8.3 Erection

Erection of the plant at site was undertaken employing three 60t capacity tower cranes especially built for power station steel erection. In addition, 40t, 20t and 16t tower cranes were used in support, these operating from the mechanical annexe roof. The cranes were rail mounted from support beams located on top of the 6.1m level suspended concrete floor. Trestles situated in the basement supported the erection crane tracks from the main pile caps. When very heavy lifts were undertaken, two cranes operating in tandem were employed; this increased the capacity of lift of a single item to 90t.

In total, Drax has been built from 80,000t of constructional steelwork in the form of 64,000 pieces held together with 800,000 bolts. The heaviest single piece lifted weighed 90t. The longest piece erected was over 55m in length. The biggest column section weighed 59t and measured 20m in length. Altogether the main building is clad with 86,000m^2 of coated steel cladding and is covered with 74,000m^2 of built-up decking.

9. BUILDINGS AND CIVIL WORKS

9.1 Turbine House

The turbine house provides cover to the six turbine generators and their associated plant with a roof structure which spans more than 55m. The roof structure is constructed from tubular roof trusses which are supported at the ends by steel box columns.

The turbines and feed heating plant are serviced by three overhead travelling cranes varying in capacity from 210t to 110t. The cranes span the width of the turbine house and run on steel crane girders. The girders are situated on each side of the turbine house approximately 20m above the laydown floor.

The turbines are angled in plan at 45° to the longitudinal direction of the building. Between the turbines are laydown areas where component parts of the turbine are stored or serviced.

Access to the turbine house is from the east through six large doors which each give access to a loading bay.

9.2 Mechanical Annexe

The mechanical annexe forms the main lateral support part of the building. The annexe, which is situated centrally within the building, has concrete floors which support feed heating plant, general services, water tanks, deaerator, stairways, lifts and general services.

The concrete floors were cast in-situ on permanent steel shuttering. The floors are supported from deep plate girders which also form part of the main lateral support frame. The plate girders in turn, are supported from heavy box columns which were made in sections up to 20m in length, brought to site and erected to form the 76m high structure.

9.3 Boiler House

The steam raising sections of each boiler hang by steel rods from the roof of the building, and impose a load on the structure of approximately 13,000t when operating. Hanging the steam raising sections allows for a thermal expansion in the furnace of approximately 285mm in the downwards direction. Supporting each boiler at roof level are twelve 33m long plate girders 3.5m deep. Each plate girder is supported at its ends by box girders which are located on top of the main building columns. In turn, the columns are founded on steel grillages and concrete pile caps.

There are three main floors within the boiler house which support the boiler ancillary plant. The largest of these items is the air heaters. There are two air heaters per boiler located at the back of the boiler on the lower floor. The upper two floors of the boiler house are of open mesh construction which assists air flow throughout the building. The lower floor is concrete cast in-situ.

9.4 Coal Bunker Bay

The coal bunker bay forms another lateral support region of the main station structure. Its roof is lower than the boiler house and mechanical annexe.

Each of the boilers is served by five 5000t capacity coal bunkers. Fabricated from mild steel plate and trapezoidal in section, the bunkers run full height between the 13.72m and 29.90m floors. The bunkers are fed by shuttle conveyors which run on rail beams over the top of the bunkers. Coal is brought into the bunker bay by the two conveyors housed in a single conveyor gantry. Beneath each bunker is a coal feeder which supplies the coal to the milling plant which crushes it before it is blown into the boiler furnace.

9.5 Electrical Annexe

The electrical annexe is situated at the back of the main building and is the smallest building constructed from rolled sections which are encased for added fire protection. It has two elevated floors which house and support switchgear and batteries.

9.6 Auxiliary Boiler Houses

A boiler house located on the south gable wall of the main plant building houses three oil-fired auxiliary boilers which initially served the First Half station requirements. A second boiler house located on the north gable wall of the main plant building houses two oil-fired auxiliary boilers which, together with the other three, meet the total station requirements.

9.7 Ancillary Buildings

There are numerous ancillary buildings that stand independent of the main building. They all conform to standard features, which include steel framework, metal cladding as the main building, built-up roof and brickwork walls to the lower parts of the main elevations.

Some buildings such as the cooling water pumphouse, gas turbine house, coal and ash workshop and compressor buildings have electrical overhead travelling cranes for maintenance purposes.

9.8 Chimney (Fig 3.5)

The basic parameters of height and flue size, for the Drax chimney, were established initially with the Alkali Inspector to suit a gas emission of $5100m^3/s$ at an efflux velocity of 26m/s.

Since the chimney serves six boilers it was necessary to provide three eliptical flues 13.7m × 9.15m. The most important criterion for economy was the need to keep the overall diameter and thickness of the shell to a minimum. The chimney is 259m high and 26m in overall diameter. The outer shell, called the wind shield, varies in thickness from 229mm at the chimney top to 610mm at the top of the base plinth which itself is up to 3.96m thick.

The flues are of reinforced concrete divided into 22m sections supported on open reinforced concrete platforms which are up to 3m thick. To prevent corrosion of the concrete from the sulphurous condensate, the concrete is coated with a fluoroelastomer material.

An inside passenger lift is provided, which is capable of rising to the top in 10 minutes, to facilitate inspection and service of the aircraft warning lights.

A unique system of laminated rubber bearings has been installed under the reiforced concrete platforms, to act as dampers and reduce the lateral movement at the top of the chimney produced by wind induced oscillation.

9.9 Architecture

Although no particular amenity value could be claimed for the locality in which the power station was to be sited, the existence of Eggborough and Ferrybridge power stations situated within a few miles of the site on the same flat plain made the appearance and setting of the new station of fundamental importance. Furthermore, since Drax was to be much larger than any other power station so far designed, it would be visible from considerable distances and it was essential therefore that great care be taken to ensure the best possible appearance of the complete works from the medium and long range view points.

The dimensions resulting from the 4000MW development were unprecedented in power station design. Some form of modulation of such an enormous building envelope was clearly required. The architectural expression depended on a break in height and sculptural modelling. This was achieved by giant louvres at high level in the boiler house and along the east front of the turbine house. A similar stylistic approach governed the elevational treatment of the administration block where windows were sandwiched between splayed walls simulating louvres.

10. BOILER

10.1 General

The boilers are of the radiant heat type with natural circulation and consist of a single steam drum and a water walled furnace unit, complete with an economiser, three superheater stages and two reheater stages. They are automatically controlled, fired by

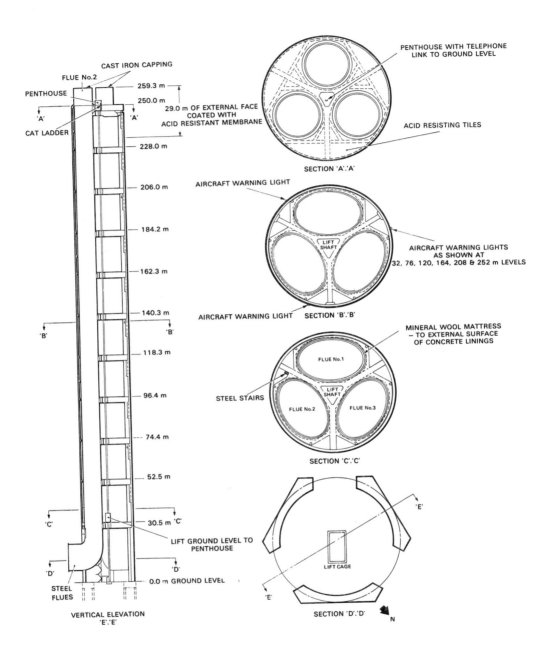

FIG. 3.5   CHIMNEY

pulverised coal and designed to meet the requirements given in Section 4 — Operational Roles. The general arrangement of the boiler and its auxiliary plant is shown in Fig 3.6.

10.2 Boiler Development

Since the general introduction of pulverised coal-fired boilers in the 1950s there has been a considerable increase in unit size. The major steps have been through 120MW, 200MW, 350MW, 500MW to 660MW units. The rapid increase in size at one stage resulted in some cases of excessive extrapolations from earlier designs with a consequent loss in the standard of reliability. To counter this a programme of extensive testing was embarked upon on units in stations such as Ferrybridge, West Burton and Eggborough. This programme provided information which resulted in the attainment of higher standards of plant reliability. The resulting information was used in the design and development of the Drax boilers for units 1, 2 and 3 to give a greater degree of confidence in their satisfactory performance.

On-going development and improved manufacturing techniques led to the introduction of further design changes for unit 4, 5 and 6 boilers. Advantage was also taken of operating experience to improve plant performance. Important changes can be summarised as follows:

(1) Operational experience showed that excessive stresses were imposed on the economiser inlet header during the boiler startup mode. This was caused by stratification of feed water in long horizontal legs as comparatively cold water entered the system. To obviate this problem, short inlet headers were incorporated with modified supports to allow flexing. The feed inlet system was redesigned to minimise horizontal legs where stratification could occur. Extensive stress analysis calculations were carried out to confirm the suitability of the revised design.

(2) As a result of design and manufacturing development, steam drum internals have been modified to improve steam/water separation and to allow interchange of components between drums.

(3) Changes were made to the superheater and reheater tube thicknesses and materials to give increased design life and improve corrosion allowances. Platen designs were altered due to operational experience and constructional changes recognised improved manufacturing methods and practices. Because of the difficulty of fabrication, a change was made from reheater outlet drums to a system of headers with a steam outlet at each end of the headers. This also gave better steam distribution to the hot reheat lines. Fig 3.7 and 3.8 show the detailed changes which have been made to the superheater and reheater arrangement.

(4) Changes to furnace internals were made to give better heat transfer characteristics and to simplify construction procedures.

(5) Boiler tubes have been subjected to 100% ultrasonic examination to comply with more stringent specification requirements and case histories have been prepared relating to all pressure parts. A complete record of the design, manufacture and quality aspects of each component is thus provided which gives confidence in the long term operation of pressure parts.

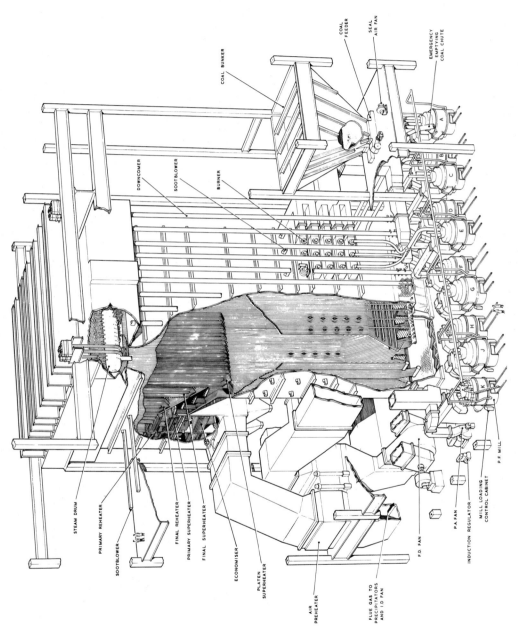

COAL BUNKER

COAL FEEDER

SEAL AIR FAN

EMERGENCY EMPTYING COAL CHUTE

DOWNCOMER

SOOTBLOWER

BURNER

STEAM DRUM

PRIMARY REHEATER

SOOTBLOWER

FINAL REHEATER

PRIMARY SUPERHEATER

FINAL SUPERHEATER

ECONOMISER

PLATEN SUPERHEATER

AIR PREHEATER

FLUE GAS TO PRECIPITATORS AND I.D FAN

F.D. FAN

P.A FAN

INDUCTION REGULATOR

MILL LOADING CONTROL CABINET

P.F. MILL

FIG. 3.6 BOILER AND AUXILIARY PLANT

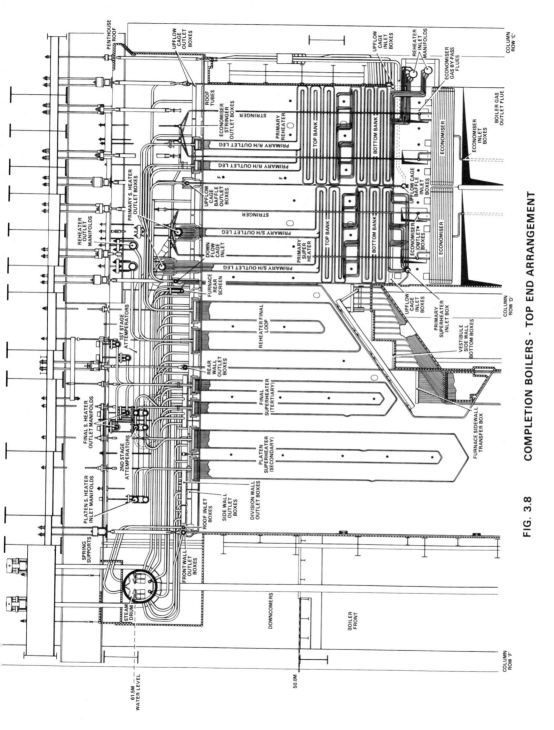

FIG. 3.8   COMPLETION BOILERS - TOP END ARRANGEMENT

(6) A number of detailed modifications were made to the First Half mills after commissioning and these were incorporated into the design of the Completion mills. Changes were made in the layout of pulverised fuel (PF) pipework to minimise risk of blockages, and to comply with higher safety standards the pipework was manufactured from ductile materials to a higher design pressure.

(7) Operational difficulties with induction regulators and brushgear on the First Half primary air (PA) fan drives led to a change to hydraulic couplings for fan control on the Completion units. Impellers and casings were of heavy duty construction to resist erosion from the dust picked up from the air heaters.

(8) Operational experience on a number of power stations demonstrated that a 2-speed motor on forced draught (FD) fans and induced draught (ID) fans was of limited advantage and, because of a potential trip on changeover, tended always to be operated in the high speed mode. Consequently it was decided that single speed motors should be fitted.

(9) In order to meet the Alkali Inspector's requirements the efficiency of the precipitators was increased from 99.3% as installed on the First Half to 99.5%. This resulted in a change of manufacturer. A change from a concrete casing design to a steel casing design was also made.

10.3 Feedwater and Saturated Steam Systems (Fig 3.9)

The feedwater for steam generation is normally supplied to the boiler by a steam driven main feed pump which is rated at 100% CMR. Additionally, two electrically driven feed pumps, rated at 50% CMR, are incorporated in the system to feed the boiler during startup and shutdown, and also to provide a standby means of feeding the boiler in the event of a failure of the main feed pump.

The discharge from the feed pumps is fed to a feed water valve station, where the supply of water to the boiler, from startup to full load, is regulated by a series of manual and automatic control system operations.

The feedwater discharge from the valve station is fed, via two parallel banks of high pressure (HP) feed heaters to the economiser check valves which are located at the 19.8m level on each side of the boiler. The feedwater economiser inlet pressure is approximately 184 bar and the temperature is 254°C.

The feedwater flows from each check valve, via T-connected branch pipes, to one pair of the four economiser inlet manifolds. The check valves are non-return type and are fitted to prevent high temperature steam and water from the boiler passing back through the feedwater supply system in the event of the loss of feedwater pressure. The feedwater then passes from each inlet manifold to five adjacent economiser inlet boxes, each inlet box being fed by a single pipe. From the 20 inlet boxes, the flow is upwards through the front and rear economiser bank tubes, in 20 segregated parallel paths, where the water temperature is raised by the transfer of heat from the flue gases passing over the outside of the tubes. The transfer of heat is enhanced by steel gills welded to the outside of the tubes.

The feedwater passes from the economiser banks to 40 economiser outlet boxes, two outlet boxes being connected to each economiser element block. Stringer tubes, connected to the economiser outlet boxes, convey the water up through the main reheater and primary superheater sections of the cage enclosure to 20 stringer outlet

246

DRAX COAL-FIRED POWER STATION

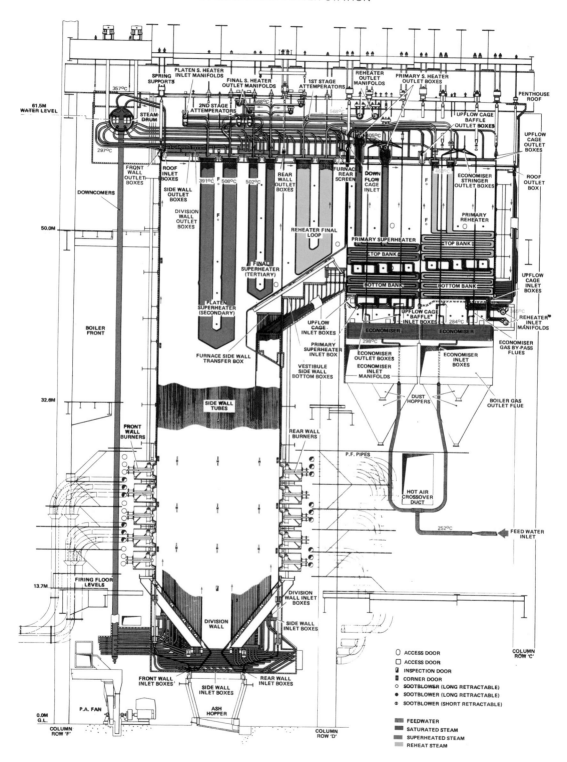

FIG. 3.9   BOILER FEED FLOW

boxes which are sited above the furnace roof tubes, in the penthouse. From the stringer outlet boxes, the feedwater flows, via 20 economiser riser pipes, to the feedwater pipework located inside the steam drum. The two rows of feedwater pipes, each row extending the length of the drum and consisting of 10 short pipes, ensures an even distribution of feedwater throughout the steam drum.

Inside the steam drum, the feedwater replaces the steam generated in the furnace which has been discharged to the superheater circuit.

The water level in the steam drum is monitored by a 'hydra-step' level gauge mounted at each end of the drum. The 'hydra-step' principle relies on the differing electrical conductivities of steam and water. One vessel initiates the boiler trip circuits when the drum level drops below the safer lower limit and the other vessel provides remote level indication in the central control room.

From the steam drum the water flows down 11 large bore downcomers which are welded to connections on the underside of the drum. A combined vortex inhibitor and debris screen fitted over the inlet of each downcomer reduces any whirlpool effect caused by the high rate of flow into the downcomers and prevents ingress of foreign matter. The downcomers extend vertically downwards from the steam drum to the bottom of the furnace enclosure, when they are connected to supply pipes to the furnace wall bottom inlet boxes.

The circulating water passes from the 11 downcomers via a total of 176 supply pipes to the 34 furnace wall bottom inlet boxes: 50 supply pipes feed the 10 front wall bottom boxes; 60 supply pipes feed the 10 rear wall bottom boxes; 20 supply pipes feed each set of five side wall bottom boxes and 24 supply pipes feed the four division wall bottom boxes. A further two supply pipes feed the water direct to the division wall wing tubes.

From the furnace wall bottom inlet boxes, the flow is upwards through the furnace enclosure and vestibule walls to top outlet boxes located above the furnace roof; the water, at this stage, is changing state to a combined steam and water mixture. Water rising in the rear furnace wall is directed two ways; part of it passes through the rear wall support tubes and the remainder is diverted through the tubes which form the nose, vestibule floor and rear screen. Water passing through each of the furnace side wall rear panels is fed via a transfer box which enables a proportion of the water to be directed to the three vestibule side wall inlet boxes, and hence through the vestibule side wall on its side of the boiler. Finally, the combined steam and water mixture passes from the 48 furnace and vestibule top outlet boxes and the division wall wing tubes, via 192 riser pipes, into the annular compartment formed by the girth baffle in the bottom half of the steam drum.

10.4 Steam Drum (Fig. 3.10)

The steam drum is an all welded cylindrical construction supported by two U-slings from the steelwork. The drum was manufactured in one piece at the factory and was transported to the site by sea. It has a length of 30.40m and an internal diameter of 2.29m. Its weight with the internal fittings is 335t.

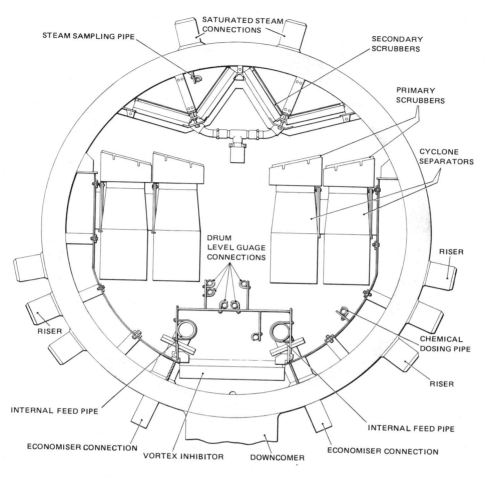

FIG. 3.10   SECTION THROUGH STEAM DRUM

Four rows of 65 conical cyclone separators are fitted in the drum and above each separator is an inclined primary scrubber. Secondary scrubbers are fitted at a higher level close to the saturated steam takeoffs. The drum is made from high tensile steel and no welding is permitted on the drum after it has been stress relieved.

Apertures in a girth baffle allow the combined steam and water mixture to enter cyclone separators which are arranged in pairs, in two rows along the front and the rear of the steam drum. The steam and water mixture enters each cylone body tangentially and whirls around the inside of the cylinder producing a powerful vortex, within which a positive separation of steam and water occurs. The steam collects in the centre and passes out at the top, whilst the water moves downward in a helical path and passes back into the drum water space for recirculation through the furnace walls. The saturated steam flows upwards through the primary inclined scrubbers which remove water residue that may be carried over by the steam after leaving the separator. The steam then passes through secondary scrubbers, fitted at the top of the drum, which provide a final separation of water from the steam before it leaves the drum and flows into the superheating circuits.

10.5 Superheater Steam System (Fig 3.11)

The live steam flow from the steam drum passes in succession through the roof, cage rear and side wall tubes to the inlet manifold of the primary superheater. From there, the steam passes through the platen superheater to the final superheater. The steam temperature is finely controlled by attemperator sprays at the outlet manifold of the primary and platen superheater.

A separate control system is provided for each of four parallel superheater streams. The control system for each stream shares a common inlet drum pressure signal and final steam temperature desired-value setting.

The primary superheater is horizontal and is fitted just above the economiser occupying half the gas outlet area. The platen and final superheaters are the pendant type suspended through the furnace roof and extending the width of the boiler in the top zone of the furnace. Both the platen and the final superheater are manufactured from austenitic steel type Esshete 1250; the significant factor being the higher creep rupture stress which permits most of the calculations to be based on yield stress criteria.

After passing through the superheater, the superheated steam is supplied via motorised controlled stop valves through four steam mains to the HP turbine.

10.6 Reheater Steam System (Fig 3.12)

After passing through the HP turbine, the steam is returned to the boiler in four steam mains where it is passed through the reheating circuits which raise its temperature before delivery to the intermediate pressure (IP) turbine.

Two stages of reheater (primary and final) are fitted. The primary is fitted in the cool gas, immediately before the economiser, thereby avoiding the need for turbine bypass systems to cool the reheat surface during startup. The final reheater is suspended in the vestibule section. The four reheater outlet streams share a combined steam temperature control system which is effected by sequential operation of the primary superheater and primary reheater gas pass dampers that are located at the economiser outlets.

On leaving the reheater, the steam is supplied through four hot reheat steam mains to the IP turbine.

10.7 Draught System (Fig 3.13)

The draught system comprises two forced draught (FD) fans, two induced draught (ID) fans, two rotary airheaters and three electrostatic precipitators. Air for combustion is supplied to the burner registers by the two FD fans by way of air preheaters where the boiler flue gases heat the incoming air. The effluent gases are drawn through three electrostatic dust precipitators by the ID fans which then discharge the gases to the chimney. The furnace is balanced by the FD and ID fans to maintain a pressure just below atmospheric. Provision is made for expansion of the

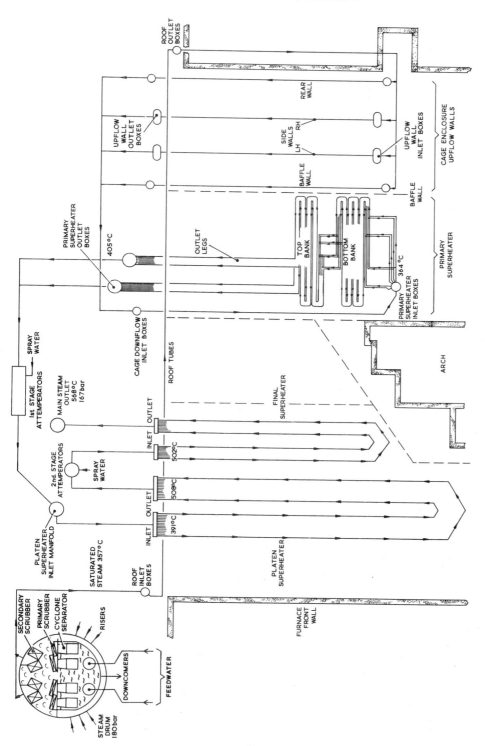

FIG. 3.11   SUPERHEATER STEAM FLOW THROUGH BOILER

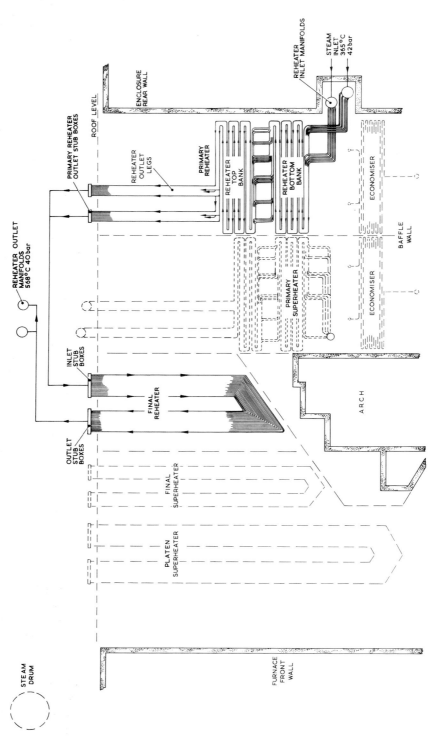

FIG. 3.12 REHEATER STEAM FLOW THROUGH BOILER

ducts by fitting leaf type joints and mounting the trunking on sliding-feet type bearings. Dampers are fitted in both air and gas trunking to isolate parts of the system or to provide alternative modes of operation.

The FD and ID fans are mechanically similar, the ID fans having a higher maximum duty specification. The fans are electrically driven at constant speed and the air or gas volumes are controlled by radial inlet guide vanes.

Gas air heaters are used to provide a means of heating the air supplied to the furnace in order to assist combustion and provide for a more economical use of fuel. They are supported by the main boiler structural steelwork at the firing floor level, and located at the rear of the furnace enclosure, one fitted on each side of the boiler beneath the gas outlet flues.

The air heaters are of contra-flow design, the combustion air from the FD fans passing upwards through the units to the burners, via the hot air ducts, whilst the furnace gases from the boiler outlet flues pass downwards through the units to the precipitators. The two streams, which flow through diametrically opposite segments of the rotor, are separated from each other by a small blanking section fitted with sealing plates which forms a division between them.

Each of the air heaters is fitted with its own combined sootblowing and water washing gear, fire detection and fire fighting equipment and a pressure fed oil system for lubricating the rotor bearings.

The rotor, which is 14.4m in diameter, is the central part of the air heater and contains the heat transfer matrix. Each rotor has a nominal heating surface of 51,826m^2. Radial plates extending from the hub divide the rotor into 24 sectors which in turn are sub-divided at the hot and intermediate ends by sector division plates which strengthen the rotor and carry the hot and intermediate end element containers. At the cold end of the rotor, grids welded between the radial division plates perform the same duty as the sector division plates, but enable the cold end element containers to be radially withdrawn from the air heater. The weight of the rotor is carried on the underside by a spherical roller thrust bearing whilst at the top a spherical roller guide bearing is provided to resist radial loads.

The rotor is driven by a small electric induction motor coupled to the rotor hub through a double worm reducer and spur gearing. These, together with the top steady bearing, comprise the drive unit which is mounted centrally on the top frame member.

10.8 Precipitators

Three electrostatic precipitators per boiler are fitted in each gas discharge line to collect dust from the boiler flue gases. The precipitators achieve a collecting efficiency of 99.5% with a gas inlet temperature of 120°C and volume per boiler of 803.23m^3/s. Dust-laden flue gases are directed through the zones of each precipitator in which discharge and collecting electrodes are situated. A variable high voltage dc current is supplied to the electrodes.

The precipitators, which are mounted on steel support structures above ground level, comprise six rows in series of dust-collecting plates forming multi-parallel gas paths and with discharge electrode wires suspended vertically within these paths. Boiler flue gases are directed by baffles in the inlet flare to flow evenly through the precipitator via gas passages formed by the spaces between the collecting plates. The cleaned gases are extracted by the ID fans and vented to atmosphere through the station chimney. The dust which accumulates mainly on the collecting plates but also on the discharge wires is removed at intervals by mechanically-rapping the plates and wires. The dislodged dust falls into hoppers suspended below the precipitators and supported by the steel structure. Accumulated dust in the hoppers is removed by the dust disposal plant to which the hopper outlets are connected.

The generation and control of the high voltage (HV) systems, rapping control systems and heating systems are located at ground level on the gas inlet side of the precipitators.

Electrically-operated pairs of isolating dampers are provided in the inlet flues from the three airheaters and in the outlet flues that lead to the ID fans. These isolating dampers may be operated locally or from the appropriate unit control panel in the main control room. When closed, the space between pairs of dampers is vented to atmosphere.

10.9 Coal-Firing System

Each boiler is fired with a mixture of pulverised fuel (PF) and air by 60 circular type PF burners which are arranged in groups of six sited at five levels on the boiler front and rear. Each horizontal group of six burners is served by one mill. (Fig 3.14).

The burners for each boiler are served by 10 independent mill groups each of which comprises a pulverising mill, a coal feeder and a primary air (PA) fan. The 10 pulverising mills are located in front of the boiler they serve, at ground level, and are arranged in staggered formation. Each mill is driven by an electric motor via a reduction gear unit and is equipped with its own lubrication system. The associated PA fans are also sited at the same level, to the rear of the mills they supply. Each mill is supplied with coal from its own coal feeder.

The coal feeders are arranged in side-by-side pairs on the 13.72m level above their associated mills, each pair of feeders being supplied from a common supply bunker. Each coal feeder is driven by a variable speed, compound gear drive unit powered by an electric motor and functions separately to supply its respective mill.

Coal for the system is gravity fed from the bunkers to the coal feeders. The bunker outlets are fitted with close-pitched rod type outlet gates. (Fig 3.15). In the coal feeders the coal is conveyed in a controlled flow to the inlet chutes of the mills they serve, where it is pulverised into a very fine powder; this fine powder is conveyed from the mills to the burners by the high velocity air supplied by the associated PA fans.

The main structure of the pulverising mill (Fig 3.16) is fabricated from mild steel in three cylindrical sections, which house the rotary grinding elements, classifier assembly and the outlet turret. The grinding elements comprises 10 hollow cast steel balls which run between two troughed grinding rings. The lower grinding ring is keyed to a yoke which is driven by the mill drive motor via a gear unit. The upper grinding ring is keyed to a cast iron spider which is located by guides in the mill centre section. The spider and upper griding ring are free to move vertically, but are prevented from rotating. The grinding rings and balls are loaded by means of eight hydro-pneumatic units which apply downward thrust on to the spider.

DRAX COAL-FIRED POWER STATION

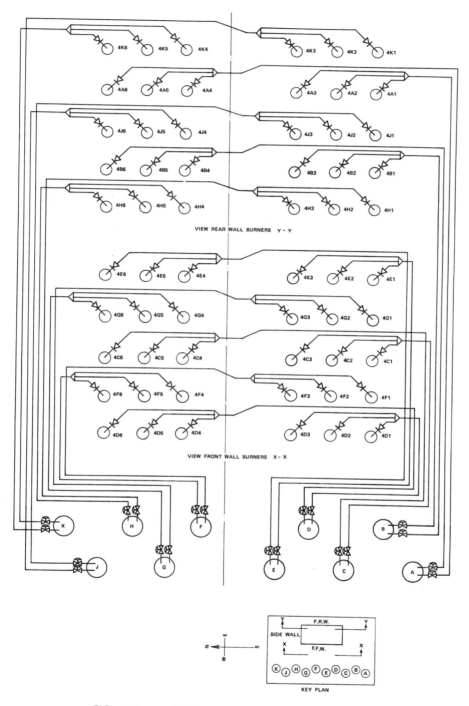

FIG. 3.14   PF BURNER FIRING ARRANGEMENT

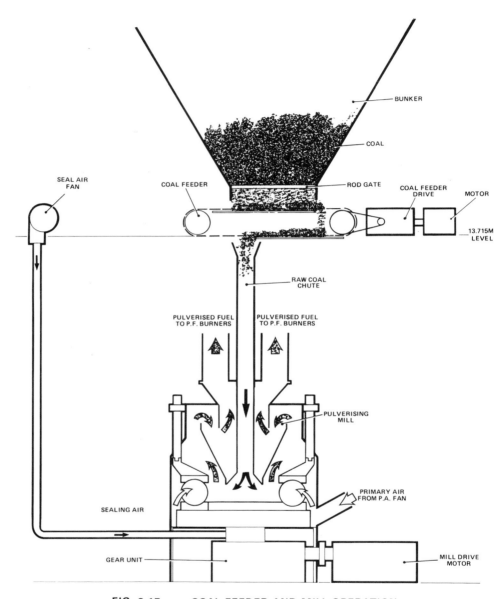

FIG. 3.15   COAL FEEDER AND MILL OPERATION

The classifier assembly is of the multi-vaned cyclone type comprising an inverted open cone and incorporating 16 pivoted vanes around the top. The lower part of the cone is closed by a returns skirt formed by a number of pivoted flaps which hang vertically downwards against the belled end of the coal inlet chute which passes through the classifier cone.

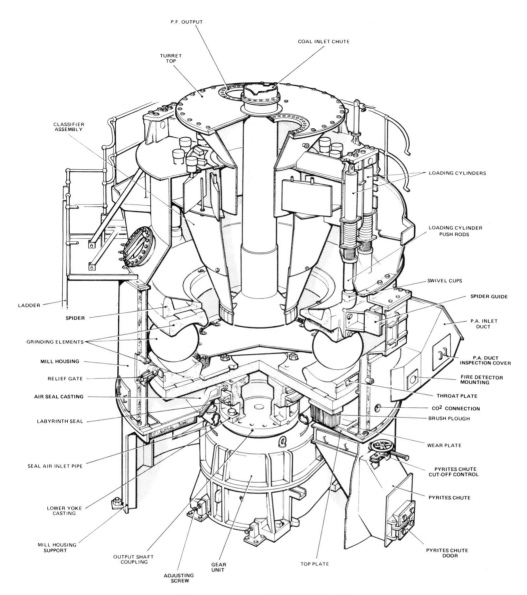

P.F. OUTPUT

COAL INLET CHUTE

TURRET TOP

CLASSIFIER ASSEMBLY

LOADING CYLINDERS

LOADING CYLINDER PUSH RODS

SWIVEL CUPS

SPIDER GUIDE

LADDER

SPIDER

P.A. INLET DUCT

GRINDING ELEMENTS

P.A. DUCT INSPECTION COVER

MILL HOUSING

FIRE DETECTOR MOUNTING

RELIEF GATE

THROAT PLATE

AIR SEAL CASTING

CO_2 CONNECTION

BRUSH PLOUGH

LABYRINTH SEAL

WEAR PLATE

SEAL AIR INLET PIPE

PYRITES CHUTE CUT-OFF CONTROL

PYRITES CHUTE

LOWER YOKE CASTING

MILL HOUSING SUPPORT

PYRITES CHUTE DOOR

OUTPUT SHAFT COUPLING

GEAR UNIT

TOP PLATE

ADJUSTING SCREW

FIG. 3.16   PULVERISING MILL

Raw coal supplied by the feeder enters the mill through the inlet chute and falls into the grinding zone. Primary air is fed to the mill through the inlet ducts in the centre section of the mill, then passes into the throat formed by the annular space around the lower grinding ring and the mill housing, and flows upwards through slots machined in the throat plate.

257

Ground fuel particles are picked up by the primary air stream and carried upwards towards the classifier. The larger particles are initially carried upwards by the air stream and circulate over the upper grinding ring before falling back into the grinding zone by virtue of their weight. The coal/air mixture then passes into the classifier, where any remaining oversize particles are separated out and fall down to the returns skirt until their cumulative weight is sufficient to deflect the flaps and return them into the grinding zone. The setting of the classifier vane angle controls the fineness of the ground product.

Heavy material such as pyrites and unwanted iron which has passed through the grinding zone without being pulverised is carried around the throat plate and discharged through a counterbalanced relief gate into the space below the yoke.

The PA fans draw air from two sources; one from the FD fan discharge before the airheaters (tempering air) and the second after the airheaters (hot air). The air from both sources is mixed automatically and delivered to the mills at a controlled temperature sufficient to dry the fuel and prevent cohesion of the coal particles. The primary air that carries the pulverised fuel to the burners is approximately 25% of the total air supply required for complete combustion of the fuel, the balance of air supply being made up by secondary air taken from the airheater outlets to the secondary air crossover duct and thence to windboxes on the boiler.

10.10 Pulverised Fuel Burners (Fig 3.17)

Three pulverised fuel (PF) burners are installed side-by-side in each of the 20 windboxes on the front and rear of the boiler. Each windbox acts as a manifold for the secondary air supply to the three burners and also serves as a supporting framework for them.

The burners for any one mill are arranged in a horizontal row across the width of the furnace so that whichever combination of mills is in operation, the heat input to the furnace is distributed evenly across the width, thus reducing as far as possible temperature variations across the width of gases entering the superheater.

Each PF burner consists of a burner tube, through which the mixture of primary air and pulverised fuel passes, installed concentrically in a cylindrical secondary air register. The flanged rear end of the burner tube is bolted to the burner front plate. On the wing burners the same flange is used to connect to a PF inlet elbow but the inner burners are provided with an additional flange for connecting to the front plate and the inlet elbow. The inlet elbow is itself flange-connected to the PF supply piping.

The PF inlet elbow and outer end of the burner tube are both lined to reduce erosion. A detachable cover plate is fitted to the inlet elbow, through which passes an impeller/cone support tube. Three adjustable supports hold the burner tube steady and centrally aligned in the throat of the air register.

The impeller/cone support tube serves also to house the oil burner which is located concentrically within the support tube.

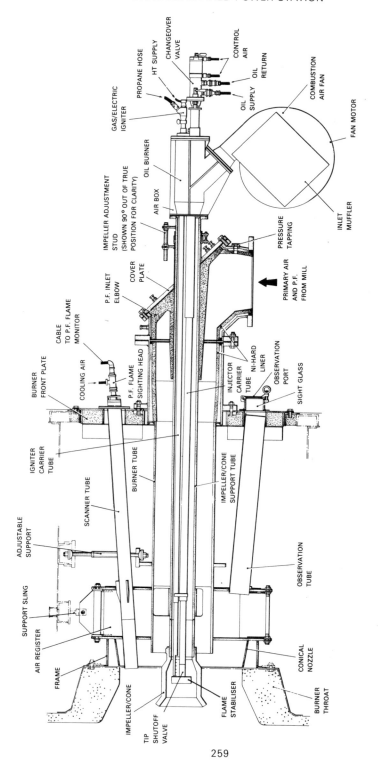

PROPANE HOSE

HT SUPPLY

CHANGEOVER VALVE

CONTROL AIR

OIL RETURN

OIL SUPPLY

OIL BURNER

COMBUSTION AIR FAN

FAN MOTOR

GAS/ELECTRIC IGNITER

IMPELLER ADJUSTMENT STUD (SHOWN 90° OUT OF TRUE POSITION FOR CLARITY)

AIR BOX

INLET MUFFLER

PRESSURE TAPPING

COVER PLATE

P.F. INLET ELBOW

PRIMARY AIR AND P.F. FROM MILL

CABLE TO P.F. FLAME MONITOR

BURNER FRONT PLATE

COOLING AIR

P.F. FLAME SIGHTING HEAD

INJECTOR CARRIER TUBE

NI-HARD LINER

OBSERVATION PORT

SIGHT GLASS

IGNITER CARRIER TUBE

BURNER TUBE

SCANNER TUBE

IMPELLER/CONE SUPPORT TUBE

ADJUSTABLE SUPPORT

OBSERVATION TUBE

SUPPORT SLING

AIR REGISTER

FRAME

IMPELLER/CONE

TIP SHUTOFF VALVE

FLAME STABILISER

CONICAL NOZZLE

BURNER THROAT

FIG. 3.17   PULVERISED FUEL BURNER

10.11 Oil-Firing System

The fuel oil firing system on Drax Completion is based on tip recirculating burners and is designed to meet the requirements of these burners under all operating conditions. The First Half oil burners were steam atomised, but pressure atomised burners were fitted on the Completion units.

Transfer pumps draw oil from the fuel oil storage tanks and discharge it through oil heaters to the pressure pumps. The pressure pumps are of the centrifugal type and take their hot oil suction through duplex hot oil filters, and discharge it to the boiler delivery line. The delivery line to each boiler includes two unit hot oil filters arranged in parallel and a flowmeter. At each boiler the delivery line divides to provide separate feeds to the front and rear burners.

The oil burners are coaxially mounted with their associated PF burner. They are normally used for boiler light-up and provide sufficient capacity for PF flame stabilisation and for carrying a nominal load. Each burner houses an oil injector. See Fig 3.17.

The oil injector is a tip shutoff recirculating type. The operation of the injector is dependent upon a continuous supply of recirculating fuel oil. The fuel oil is used to cool the atomiser when the burner is not in use. This prevents distortion of the atomiser through overheating and also prevents the formation of carbon deposits on it. The injector comprises a tip valve assembly, changeover valve assembly and interconnecting tubes. The changeover valve is operated by pneumatic control from a local actuator unit, it is calibrated for flow rates in the firing mode of operation of the injector. Service connections to the injector are made with flexible hoses which have their coupling fittings sized differently to facilitate correct connection; the coupling fittings are of the self-sealing quick-release type.

Under normal operating conditions the injector is set to one of two modes, either firing or recirculating, by the changeover valve. Within the changeover valve, oil ports in the inlet chamber and outlet sleeve enable the oil flow to the tip valve to be reversed.

Overall control of burner operations is exercised from the unit control desk, where starting and stopping of burner groups can be initiated by means of the appropriate push buttons. In addition local control of individual burners can be initiated by push button operation at the burner control cubicles. The burners of a group are timed to commence their individual starting sequences at 2s intervals from the time of initiation. On shutting down a group all burner services are shutdown simultaneously.

10.12 Sootblowers

Some of the ash formed by the combustion of the coal adheres to the external surfaces of the boiler tubes; this ash must be removed in order to maintain boiler efficiency. To achieve this end, deposits on the boiler tubes are prevented from building up by automatic, retractable sootblowers which are located strategically around the boiler. The sootblowers are electrically operated by a control system which provides sequential and selective control of the 106 sootblowers fitted and has provision for the control of a further 22 sootblowers if experience dictates a requirement for more.

Compressed air is used as the blowing medium and this is supplied from a common main which serves all boiler units. The air for the sootblowers of each boiler unit is supplied via remotely controlled, electrically operated isolating valves.

11. HIGH PRESSURE PIPEWORK (FIG 3.18)

The function of the high pressure (HP) pipework system is to carry the high pressure steam from the boiler main steam outlet valves to the turbine, exhaust to and return from the boiler reheater, and then to return the feed water back to the boiler.

A summary of the pipework sizes and materials of construction together with the design criteria are given for the Completion stage in Table 3.3.

TABLE 3.3 HIGH PRESSURE PIPEWORK DATA

Item	Design Conditions Press Bar	Temp °C	Type of Steel	Thickness mm	Bore mm
Main steam	175.2	568	½%Cr/½%Mo/½%V	77.72 min	266.7
HP steam to BFP turbine	175.2	568	½%Cr/½%Mo/½%V	31.0 min	103.2
Cold reheat	50.3	385	Carbon Steel	14.2 mean	510
Hot reheat	47.0	568	½%Cr/½%Mo/½%V	33.27 min	482.6
HP feed	206.8	255	Carbon Steel	24.7 min and 53 min	304.8 and 432.0

The HP pipework system is a substantial duplicate of that on the First Half. A significant change has been due to the boiler reheater outlet which incorporates a header design rather than a drum. This has resulted in a more symmetrical pipework layout between the reheater and the turbine, since the headers are on each side of the boiler. With the original reheater drum design all the reheat pipes were on one side of the boiler giving rise to problems relating to expansion design.

The final steam and reheat temperatures are 565°C at the turbine. The material chosen for the high temperature duty is a Cr/Mo/V steel. This is based on cost, ease of fabrication, availability and flexibility.

Thermal expansion and contraction of the HP pipework is accommodated by the inherent flexibility of the piping configuration, together with cold pull-up, and constraints applied to the pipework where necessary. Cold pull-up is the term applied to prestressing a pipework system in the cold condition, such that it is in a neutral condition after expanding into its hot position. To permit the measurement and recording of the pipework expansion, a system using a pointer whose movement, relative to fixed marks on the steelwork, is provided. In addition to this, facilities for measuring the amount of creep which has occurred are provided. This takes the form of measuring radial changes from reference points on the outside of the pipework. This is carried out initially and at regular intervals during the life of the station.

12. TURBINE AND CONDENSER

12.1 General (Fig 3.19)

The turbine generators are single line 660MW units fitted with twin pannier condensers and running at a speed of 3000rev/min. Each machine consists of one single flow HP turbine of triple casing construction, one double-flow double casing IP turbine and three double-flow single casing LP turbines. The turbine rotors which are rigidly coupled together are each supported by two bearings housed in their corresponding bearing blocks whilst the whole rotor assembly is located axially by a thrust bearing mounted at the HP end of the IP turbine rotor. The machine is supported on a steel foundation block to which the bearing block pedestals are secured. The turbines are arranged such that the bearing blocks are free to slide in their pedestals thus accommodating the movement of the cylinders and rotors during thermal changes. A system of guides, keys and anchor points is employed to ensure that longitudinal and transverse alignment is maintained at all times, whilst accommodating freedom of movement due to thermal expansion.

The machine is driven by HP superheated steam, which is reheated between the HP and IP turbines to give greater thermal efficiency. Steam from the boilers is admitted to the HP turbine by two HP steam chests which house the steam admission and control valves. The steam chests, which are mounted one on each side of the turbine, are connected to the HP turbine by interconnecting pipes designed to absorb expansion as it occurs, thus relieving the chests of additional stress. After passing through the HP turbine the steam is exhausted to the boiler reheater and is returned to the IP turbine through the twin reheat steam chests which house the reheat steam control valves. The reheat chests, which are mounted one on each side of the turbine, are connected to the IP turbine by interconnecting pipes which absorb expansion as it occurs, thus relieving the chest of additional stresses. Relief valves, provided in the steam lines between the reheater and the reheat steam chests, discharge the steam in the reheater to atmosphere, when the governor and intercept valves close on overspeed. This eliminates the risk of the HP blading overheating in the event of rejection of full load. Exhaust steam from the IP turbine is led to the three LP turbines by four interconnecting pipes two of which connect to no.1 LP turbine and one connected to each of no.2 and no.3 LP turbines respectively. The steam flows in both directions through each of the LP turbines and is exhausted into the condensers which are under a constant vacuum.

Two surface type condensers, which for maximum thermal efficiency operate at vacuum conditions, are provided to condense the steam exhausted from the LP turbines and also to provide the means for air and vapour release from the heaters and other equipment associated with the condensate system.

Five 25% duty air extractor units are provided to maintain the vacuum in the condensers. A quick-start air pump is also provided for use during the initial starting up period.

12.2 Turbine Design Data

Type	5 cylinder tandem-compound single reheat
Speed	3000rev/min
Steam consumption at CMR	548.2kg/s
Steam pressure	159.6 bar
Heat rate	8117kJ/kWh
Power output	660MW (at generator terminals)
Steam conditions at turbine inlets:	
HP Cylinder	159.6 bar and 565°C
IP Cylinder	40.1 bar and 565°C
LP Cylinder	6.32 bar and 308°C
Vacuum (average)	54.2mbar
Barring gear speed	2rev/min

12.3 Turbine and Auxiliary Plant Development

As with boiler development, information obtained from experience of 500MW units was used in the design and development of the Drax First Half turbines. Casings of simpler but stronger shape evolved and rotors of solid single-piece construction were adopted which were substantially stiffer than on earlier units and provided greater stability.

On-going development and changes in manufacturing techniques led to the introduction of further design changes for the Completion turbines. Advantage was again taken of operating experience to improve plant performance. Important changes can be summarised as follows:

(1) IP and LP turbine diaphragms were manufactured to improved techniques and standards but rotor interchangeability for all six units was maintained.

(2) The turbine gland sealing system was redesigned to incorporate detailed modifications and improvements to eliminate problems with automatic control and so minimise operator attention and maintenance.

(3) A flange warming system was incorporated on the HP and IP turbine casings to overcome plant limitations and facilitate flexible operation of the unit whilst avoiding excessive differential expansion on cold startup. Detailed changes were also made to HP rotor and casing geometry to improve their thermal fatigue characteristics.

(4) Modified and improved designs of first stage HP and last stage LP turbine blades were fitted.

(5) The lubricating oil system was modified to reduce bearing leakage. The design of oil coolers was also modified to improve the cooling performance.

(6) The main and boiler feed pump turbine governor and trip system were manufactured to an improved design.

(7) The high silt content of river water causes erosion problems in the condenser tubes. To avoid the need for re-tubing during unit life, the tube material was changed from aluminium brass to titanium.

(8) The change from direct contact to tubular LP heaters in the feed heating system required a condenser extraction pump with a higher generated head to overcome additional system losses. Advantage was taken of the latest design practices to reduce cavitation and air ingress and the type of pump was changed from horizontal to vertical caisson type.

(9) The air extraction plant was uprated from four to five vacuum maintaining units to ensure continuous satisfactory performance. The quick-start pumps were fitted with air ejector augmentors.

Many of the changes and improvements were backfitted on the earlier units following satisfactory development for the new units.

12.4 Steam Chests and Control Valves (Fig 3.20)

An HP steam chest is located at operating floor level on each side of the HP turbine. Similarly, a reheat steam chest is located on each side of the IP turbine. These contain the control valves.

The flow of steam into the HP turbine is controlled by four emergency stop valves (ESV) and four governor valves. Control of the re-admission of steam into the IP turbine is accomplished by four reheat emergency stop valves and four intercept valves. In addition, two hot reheat relief valves are also provided to release the steam in the reheater system to atmosphere when the governor and intercept valves close in an emergency.

Each steam admission valve is operated by its own electro-hydraulically operated relay using fire resistant fluid supplied by two motor-driven positive displacement pumps (one standby and one service) from separate tanks. To enable any large demands for control fluid to be met, hydraulic accumulators are incorporated in the system adjacent to the governor and intercept valve relays. These accumulators serve to prevent any excessive transient flows through the supply pipes and also to maintain system pressure during automatic startup of the standby pump.

To operate the steam admission valves against the opposition of steam pressure, each valve relay is provided with a power piston. The power piston of each HP and reheat emergency stop valve relay is directly coupled to its respective valve spindle, whereas the governor and intercept valve relay power pistons are connected by means of links and variable ratio levers to their respective valve spindles. Opening of the valves is achieved by power fluid acting against the power piston of each relay. Closure of the valve is achieved by releasing power fluid to a drain. A powerful return spring compressed on opening the valve assists closure of the valve against steam pressure.

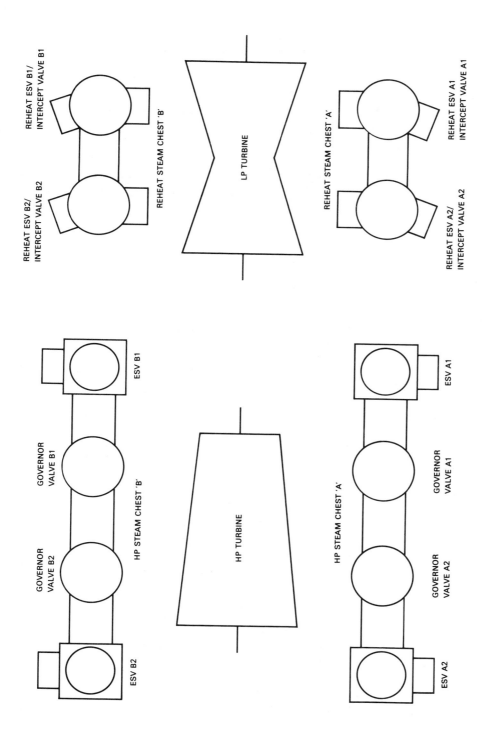

FIG. 3.20   STEAM CHESTS AND CONTROL VALVES LOCATIONS

The flow of power fluid to the relay power pistons is regulated by control units bolted to the relay bodies. The fire resistant properties of the fluid enable high pressures to be utilised which results in each valve being controlled by a relatively small and compact relay unit. Tripping fluid supplied from a comprehensive tripping system controls the position of a relay plunger assembly located in each control unit and hence regulates the flow of power fluid to the relay power pistons. In the event of the tripping system being activated in an emergency, all the steam admission will close automatically as tripping fluid pressure falls and power fluid is released to drain. The emergency stop valve relays and the reheat emergency stop valve relays open their respective valves immediately the tripping system is pressurised. Runup of the machine from rest to synchronisation is achieved by applying servo control to the governor and intercept valve relay control units using signals from the turbine electronic governing system.

12.5 High Pressure, Intermediate Pressure and Low Pressure Cylinders

The HP turbine is of the single flow type and comprises an outer casing which includes inlet and exhaust branches and housings for labyrinth type glands, an inner casing with a fully bladed bore, an intermediate sleeve and a bladed rotor. Steam enters the HP turbine from the HP steam chests by means of four pipes, two connected to the top steam inlet assembly and two connected to the bottom assembly. The steam expands through the casing and rotor blades, thus causing the rotor to rotate, before being exhausted to the reheater.

The intermediate pressure (IP) turbine is of the double flow type and comprises an outer casing which includes inlet and exhaust branches and housings for labyrinth type glands, an inner casing and blade rings with bladed bores and a bladed rotor. Steam from the reheat steam chests enters the IP turbine by means of four pipes, two connected to the top steam inlet assembly and two connected to the bottom assembly. The steam expands through the casing and rotor blading, thus causing the rotor to rotate, before being exhausted to the low pressure (LP) turbine.

The three double-flow LP turbines each comprise a bladed rotor and bladed cast steel cylinder centre section, to each end of which is bolted a fabricated steel exhaust, which incorporates a housing for vernier type glands. Steam enters the LP turbines from the IP turbine and flows towards the exhaust at both ends of each turbine, expanding through the blades and thus causing the rotor to rotate. During the passage of steam through the LP turbines a proportion is bled-off to the feed heating plant. Steam is leaked-off from the following points on the LP turbine:

To no.1 heater after stage 4 in each LP turbine
To no.2 heater after stage 3 in LP1 turbine
To no.3 heater after stage 2 in LP2 turbine
To no.4 heater after stage 1 in LP3 turbine
To the water extraction condenser after stage 5 in each LP turbine.

After passing the last stage of blading in each LP turbine the steam exhausts into the main condensers mounted one on each side of the LP turbine.

Flange heating has been incorporated on the HP and IP cylinders with the objective of controlling the rate of heating at the horizontal joint of the outer casings during startup.

To minimise the effects of uneven heating of the LP casing and possible rotor misalignment due to the churning effects of the long blades on the LP turbine shaft in the steam space at low load, a spray cooling system is installed in each exhaust space.

12.6 Rotors and Bearings

The turbine rotors and the generator rotor are connected together by five flanged couplings to form a complete shaft which is supported in 12 bearings.

Equipment is installed to monitor and record the conditions under which the rotors and bearings are operating at any time. The use of such equipment allows the machine to be operated remotely by enabling the operator to detect any abnormalities and to keep the relationship between rotors and casings within definite design limits, particularly during the runup and loading period. As a result of this, running up and loading times are kept to a minimum and loading rates under 'hot start' conditions may be adjusted according to the behaviour of the machine. The equipment provided is responsible for measuring rotor eccentricity, differential expansion (eg thermal expansion of a cylinder relative to a rotor) and bearing vibration.

All bearings are individually supplied with lubricating oil from the bearing oil manifold and drainage oil is directed via a drain manifold to the main oil tank.

12.7 Lubricating System

Efficient lubrication of the turbine generators is necessary under all operating conditions and this is maintained by three pumps which draw oil from the lubricating oil storage tank and discharge it through coolers and filters to the machine bearings. The three pumps are: the main oil pump driven off the generator shaft, which supplies the total lubricating oil requirements when the machine is at operational speed; the ac motor driven auxiliary lubricating oil pump which supplies the total lubricating oil requirements when the machine is running up or shutting down or upon failure of the main oil pump; and the dc motor driven lubricating oil pump which fulfils the duties of the ac motor driven auxiliary lubricating oil pump if failure of the ac supply occurs. The lubrication system is a closed circuit in which the oil supplied to the machine is collected in a reservoir, purified and returned to the oil storage tank for re-use.

12.8 Turbine Gland Sealing System

The gland sealing system is designed to supply steam to seal the turbine shaft glands at all operating conditions and to extract leak-off steam from the glands. The system is divided into two similar parts to accommodate the range of temperatures experienced throughout the turbine; one part supplies steam to the HP and IP turbine glands and the other part supplies steam to the LP and boiler feed pump turbine glands.

Two steam supplies are connected to each system: one supplies steam at superheater outlet conditions (tapped off the HP steam chest steam inlet crossover pipes) which is referred to as live steam and is used during startup, shutdown and low load operation; the other supply is leaked off from the HP turbine glands and is employed on the LP and IP glands when the turbine is operating at more than approximately half load. The use of HP glands leak-off steam produces a slight thermal gain over the permanent use of live steam.

12.9 Governing, Control and Protective Systems (Fig 3.21)

The efficient running of the machine is ensured by the comprehensive control, supervisory, protective and instrumentation systems which are employed to guard against malfunction of any part or item of equipment in the system. The control of the turbine steam admission and release valves is by a hydraulically operated system which employs a fire resistant fluid as its operating medium. An electronic governing system feeds signals to servo units which control the pressure and flow of the fire resistant fluid to the hydraulic relays of the steam valves. Fluctuations in these signals precipitate action of the appropriate steam valve in controlling the machine.

The protective system provides an automatic safeguard against damage as a result of failure or emergency conditions occurring during all operating conditions. It provides audible and visual alarms for the warning of failure or emergencies as they occur, thus ensuring that the appropriate action may be taken either in the control room or locally on the machine. This system automatically trips the machine under certain emergency conditions, eg low vacuum, high condenser water level or loss of steam at the stop valve, whilst at the same time affording full protection to all other ancillary equipment.

The auto runup and loading equipment is supplied to assist the operator in the running up, loading and unloading of the machine in such a manner as to reduce substantially the likelihood of over-stressing or maloperation of the plant. It provides standard pre-determined patterns of acceleration and loading, leading to greater reliability, availability and reduction in maintenance costs.

The equipment will accelerate the machine from barring gear speed within the range of the automatic synchronising equipment and will load or unload the machine to a value selected by the operator. It will also carry out a series of pre-start checks on the state of the plant and will warn the operator of any malfunction which would prevent the initiation or completion of runup. At the same time the equipment will also select the optimum runup rate and loading rate and display these to the operator.

The comprehensive instrumentation system fitted to the machine provides the operator, either in the control room or at the machine, with visual indication of the condition of the machine during all operating conditions.

12.10 Turbine Forced Air Cooling

A turbine forced air cooling system is provided so that, when required, the time taken to cool the turbine at shutdown can be reduced. A reduction in cooling time allows maintenance work to be effected earlier than would normally be possible. The system cools the HP and IP turbines by injecting air into the interspace between the inner and outer casings of each turbine, thereby cooling the casing more rapidly than the rotor while maintaining control of differential expansion.

12.11 Turbine Generator Foundations

The turbine generator foundations consist of an arrangement of longitudinal beams, transverse beams and vertical columns. A central structure supports the turbine generator and steam chests and the condenser is mounted on support columns on each side of the turbine generator.

The central structure forms an open framework which provides excellent ventilation round the turbine generator thus minimising thermal mis-alignment between the individual parts of the assembly. The structure is designed to ensure that its natural frequencies are very low in relation to the machine rotation of 3000 cycles per minute in vertical, transverse and torsional modes. The arrangement of transverse beams, with support columns at the ends and in the same plane, which support the bearing housings, ensures that there is no coincidence between normal running speed and the natural vibration frequency of the bearing support structure.

12.12 Condenser (Fig 3.22)

Twin pannier condensers are installed to condense the steam exhausted from the LP turbines. For maximum thermal efficiency, they operate at vacuum conditions. They also provide the means for air and vapour release from the heaters and other equipment associated with the condensate system. As the condenser is of single pass design on the cooling water side, each LP cylinder has a varying vacuum due to the water temperature rise along the length of the condenser tube. This varies from 45.07mbar at the cold end to 53.67mbar in the centre and 63.90mbar at the hot end.

Each condenser contains cooling tubes fixed at each end in double tubeplates which are vacuum sealed. Inlet and outlet waterheads at each end of the condenser are arranged such that cooling water flow through the condenser is single flow. Secured to the underside of each condenser are drain troughs, a hotwell and two flashpots all of which are interconnected by pipework.

Steam exhausted from the LP turbines enters the condenser through the trunking linking the condensers to the LP turbines and is guided over and around the cooling tubes by baffles arranged within the condenser. As the steam is cooled and turned to condensate it is collected in the condenser floor and drain troughs and passed to the hotwell from where it is removed by extraction pumps.

The condenser tubes, which are made of titanium are arranged in the condenser shell to provide a single-flow system through the condenser. Each end of the tubes is expanded into the inner tubeplate then secured to the outer tubeplate by an arrangement of metallic and fibre rings. Sagging plates support the tubes within the condenser shell, and are arranged to give the tubes a curvature along their length which allows natural drainage when emptying the system.

The cooling water surfaces of the tubeplate carriers, the tubeplates and the waterheads are protected against corrosion by a neoprene coating.

To eliminate the possibility of cooling water entering the condenser steam space, each tubeplate interspace is maintained at condenser vacuum during normal condenser operation. The vacuum is established by interconnecting the tubeplate interspace by pipework, through a drain vessel, to the condenser air extraction system.

12.13 Condenser Extraction Pump

Condensate is removed from the condenser by two 100% duty vertical spindle caisson type extraction pumps. Each pump is designed to deliver 550kg/s of condensate from suction conditions of −10m to a generated head of 250m when running at 985rev/min. The caisson which incorporates the suction branch, is situated below floor level. The pumping stages are suspended from the top flange and

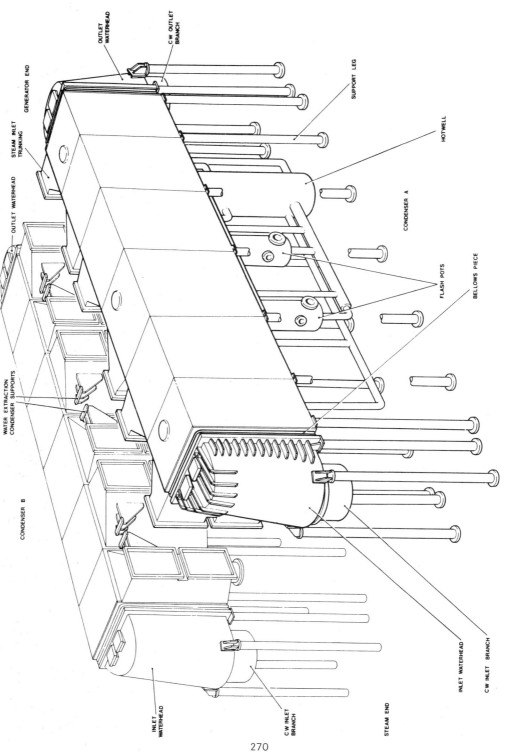

FIG. 3.22   ARRANGEMENT OF CONDENSERS

submerged within the contents of the caisson. To minimise cavitation problems the suction stage impeller is double entry. The remaining four stages are hydraulically balanced single entry impellers working in diffuser/return vane bodies. The pump is driven by a stool mounted, direct coupled, motor. The pumps pass the condensate through the LP feed heating system to the deaerator.

13. GENERATOR (FIG 3.23)

The generator is of the water and hydrogen cooled type, coupled to direct-driven main and pilot exciters.

The generator stator is constructed in two parts, comprising an inner frame which supports the core and windings and an outer casing containing the hydrogen coolers. The stator winding is of the double layer, diamond coil type and formed from hollow section copper conductors contained in slots in the stator core.

The stator windings are cooled internally by demineralised water circulated around a closed circuit system. The coolant is circulated through the winding conductors from the current carrying terminals at the exciter end to the non-current carrying terminals at the turbine end of the generator. The coolant is circulated by one of two 100% duty ac stator coolant pumps. A dc stator coolant pump is employed to circulate the coolant in the event of failure of the ac supply. Coolers are incorporated in the system to extract heat from the stator coolant after it has passed through the winding. A purifier plant located in the system maintains the purity of the coolant.

In addition to the stator coolant flowing through the winding conductors, the stator core and the generator rotor are cooled by hydrogen which is circulated within the stator casing by two centrifugal fans mounted one at each end of the rotor.

Hydrogen seals of the thrust collar type prevent gas leakage at the points where the rotor emerges from the stator casing. Oil is supplied to the seals by one of two 100% duty ac seal oil pumps, at a pressure which exceeds that of the hydrogen within the stator casing. In the event of failure of both oil pumps, a back-up supply is available from the lubricating oil system. Should there be failure of both the oil pumps and the back-up supply, a dc seal oil pump will supply oil to the seals.

The rotor is machined from an alloy forging with forged steel sliprings located at the exciter end. The rotor windings are formed from hard drawn silver bearing copper strip and directly cooled by hydrogen. Brushgear is provided to feed current into and out of the rotor windings through the sliprings. A forced air ventilation system is employed to dissipate the heat generated by friction between the brushes and the sliprings.

Each end of the generator rotor is supported in a spherically seated journal bearing, housed in a pedestal and lubricated from the turbine lubricating oil system. The pedestal at the turbine end also houses the LP3 turbine shaft bearing. The circulation of currents through the bearings and the oil film is prevented by insulation between the exciter end pedestals and other equipment, including lubricating oil pipework.

Excitation for the generator rotor field is provided by direct driven main and pilot exciters.

The pilot exciter provides a 3-phase power supply to control equipment (thyrister converter) which regulates the excitation of the main exciter field. The 3-phase power output from the main exciter is fed to a static rectifier which supplies dc excitation power for the generator rotor field via a main field breaker.

Cooling for the pilot exciter is provided by a fan on the rotor shaft which circulates air through an open circuit ventilation system that also supplies cooling for the generator brushgear and the main exciter brushgear.

The main exciter ventilation consists of a closed circuit system. The air is circulated through the system by two centrifugal fans, one being located at each end of the exciter rotor shaft.

Housed in a pedestal positioned between the generator sliprings and the main exciter is the barring gear drive which is provided to rotate the shafts of the machine slowly before runup and after shutdown to prevent distortion or uneven heating or cooling of the shafts. The drive comprises a clutch secured to a barring gear shaft and triple reduction gearing driven by an ac motor. Also housed in the pedestal are the generator rotor steady bearing and the main exciter inboard bearing.

13.1 Design Data

Generator output	660MW
MVA	776MVA at 0.85PF
Voltage	23.5kV
Current	19,076A
Efficiency	8.59%
Hydrogen pressure	4 bar
Hydrogen to charge system	765m^3
Rotor volts and amps	560V at 4500A
Hydrogen coolers gas flow	37.76m^3/s
Hydrogen coolers water flow	70.8 litres/s
Main exciter	3870kVA and 5170A
Pilot exciter output	90kVA and 260A
Stator coolant pressure	3.8 bar
Stator coolant pump capacity	36 litres/s

14. CONDENSATE AND FEED HEATING SYSTEM (FIG 3.24)

14.1 General

From the condenser extraction pumps discharge, the condensate passes through the gland steam condenser, three water extraction condensers and the LP heater drains cooler. A condensate recirculating loop is located immediately downstream of the water extraction condensers. Having picked up temperature, the fluid is then termed 'feedwater' and it passes forward to the LP heaters.

Each of the five LP heaters on units 4,5 and 6 is of the inverted U-tube type in which feedwater flows through the tubes and bled steam is passed into the shell. Steam for feedwater heating is bled from various stages of the three LP turbine cylinders and from the exhaust of the main feed pump turbine. On leaving the LP heaters, feedwater passes up to a deaerator storage tank via a vent condenser and a deaerator. The deaerator is incorporated in the system to remove dissolved gases and to reduce the oxygen content of the feedwater. The storage tank caters for any fluctuation in system flow and provides water at a suitable suction head for the boiler feed pumps. The bled steam supply to the deaerator is from the cold reheat system.

After deaeration, feedwater enters a downcomer and flows through one of two 100% duty strainers and is then taken to either the starting and standby, or the main feed pump where the pressure is raised to that necessary to feed the water into the boiler. Feedwater discharged from the boiler feed pump passes through the feed regulating valves into a common main feeding four HP heaters. The heaters are arranged in two parallel banks of two and during normal operation, each bank receives half the total feedwater flow. Steam for HP feedwater heating is bled from tappings off the main feed pump turbine and from the HP turbine cold reheat lines. After leaving the HP heaters, feedwater is delivered to the boiler.

14.2 Plant Development

The LP feedwater heaters installed in the First Half are direct contact type which required a number of pumping stages in the system. During operation, problems occurred associated with water level instability and control. As a consequence, for the Completion units a decision was made to depart from replication with the inclusion of five surface type LP heaters instead of direct contact heaters. A unit evaporator was not included as operational experience showed that it did not provide make-up water as required for startup and 2-shifting modes. In addition, changes were made in the configuration of feed pump suction lines to reduce vibration and the design of feed suction pumps and strainers was changed to give improved performance and greater integrity.

14.3 Condensate System

The condensate system extracts the condensed steam from the main condenser hotwells and discharges it to the feed heating system. The condensate is drawn from the main condenser hotwells by one of two 100% duty extraction pumps. In the event of failure of the pump in service, the standby pump is automatically started by contacts operated by a falling discharge pressure. From the pumps discharge main the condensate passes through the gland steam condenser and the moisture extraction condensers, where it is employed as a cooling medium, followed by the condensate control valves, section valve and forward feed isolating valve to the feed heating system. The condensate control valves function to throttle the flow so as to maintain a constant pressure at the extraction pump, while the section valve is used during system flushing and dumping. A recirculating line is also provided which connects the condensate discharge line to the main condenser to maintain a flow through the extraction pump during low and off load conditions.

Provision is included in the system for reserve feed water (RFW) tanks and a clean drains tank. The reserve feed tanks serve to hold a supply of demineralised water for system makeup, etc and also to receive the surplus water when there is an excess in the system. The clean drains tank is a common receiving vessel for miscellaneous drainage.

An outsurge line enables surplus condensate in the system to be discharged via a non-return valve and the outsurge valve group to the reserve feed water tanks. Any make-up required by the system is provided by the water treatment plant and is supplied via the insurge valve group to the main condenser.

14.4 Condenser Make-up and Surplus

Under normal operating conditions the insurge and outsurge control valves operate on a split range from a level controller in the deaerator storage tank to control the level of water in the storage tank within predetermined limits.

The insurge control admits up to 5% CMR flow from the polishing plant as the deaerator storage tank water level falls below the normal level. A motor operated isolating valve is installed in parallel with the insurge control valve to cater for a startup demand of up to 15% CMR. In parallel with these two valves is a small bore line with an isolating valve and orifice plate designed to provide a constant system insurge of 1% CMR flow.

The outsurge control valve expels up to 20% CMR flow from the system to the RFW tanks as the deaerator storage tank water level rises above the normal. A motor operated isolating valve is installed in parallel with the outsurge control valve to provide a coarse form of emergency control should the outsurge control valve be inoperative.

If it is required to dump or polish condensate due to the water quality being unsuitable for admission to the boiler, the section valve is closed and condensate is discharged either to waste or to the polishing plant depending upon the degree of condensate contamination. Condensate from the polishing plant is pumped back into the system downstream of the section valve by the polished condensate pump. The water level in the condenser is at all times controlled by the condenser level control system.

14.5 Low Pressure Feed Heaters

The system includes five low pressure heaters and a separate flashing drains cooler arranged in a single line. There are three bypass systems covering the drains cooler and LP heater 1, LP heaters 2 and 3, LP heaters 4 and 5. Heat is recovered from the LP heater drains by means of a gravity cascade through flashboxes into the LP drains cooler and finally a gravity drain into the main condenser.

The five LP heaters are all the same physical size and contain similar surfaces. The elevation within the station is critical to allow heater handling by the station crane and still permit a drain system gravity cascade direct to the main condensers. A limitation has had to be placed on the overall heater length due to the height limitation and this has resulted in a marginal increase in optimum terminal temperature differences. The varying terminal temperature differences have been imposed by keeping the tubenests standard.

The LP heaters are all condensing heaters with no drain cooling or desuperheating sections. The heaters are arranged vertically and head down, apart from the flashing drain cooler which is head up. Duplicate high water level alarm and trip switches are fitted to the heaters, and also a facility to enable on-load testing of switches via a standpipe.

The LP feed heating system is designed to have an inherent security against backflow of steam or water to the turbine. This includes designing the system for predictable variations in operating conditions and the individual items of plant for long term integrity.

14.6 Deaerator

A high level deaerator is mounted 33.7m above the basement floor level and comprises a deaerating chamber and a vent condenser mounted on a storage tank which contains 246,300kg of water providing a 7.5 minute supply at CMR flow.

This assembly is incorporated in the system to reduce the oxygen content of the feedwater to a figure not exceeding 0.007mg/litre and to provide water at a suitable suction head for the feed pumps. The bled steam supply is taken from the cold reheat system if the main feed pump turbine is out of service.

Feedwater after passing through the LP heaters enters the deaerator through the vent condenser and deaerator header and into the storage tank. The vent condenser is mounted horizontally on two support legs welded to the storage tank shell and has a heat transfer surface area of stainless steel tubes enclosed in a cylindrical shell by which a 2-pass flow of feedwater is preheated to saturation temperature before it enters the deaerator header chamber. The deaerator header unit is also mounted horizontally on top of the storage tank and consists of a cylindrical shell containing a nest of stainless steel distribution plates and has a feedwater inlet at the side, two drain feedwater outlets to the storage tank in the underside, a steam inlet trunk at each end and a vent steam outlet in the top.

Feedwater entering the deaerator header impinges on to the flow plate which breaks the water into droplets for deaeration and scouring by circulating steam. The broken water cascades over and through the distributing and deaerating plates flowing to the bottom of the header and into the storage tank through the outlets.

The droplets are swept by a current of steam supplied through dispersers in each steam inlet tank. The scrubbing action created by the steam breaks up the droplets as they gravitate thus liberating the incondensible gases. Steam and gases thus liberated from the feedwater rise through the deaerator chamber trays and pass through the vapour inlet trunk to the vent condenser for heating the incoming feedwater. The steam is condensed and drains back to the deaerator storage tank. Steam admitted to the deaerator storage tank ensures that the surface of the stored water is constantly swept by steam, thus preventing the formation of air pockets.

Two steam heaters are incorporated at the bottom of the storage tank, each provided with a steam inlet and outlet connection for heating up stored water prior to startup and maintaining temperature during a short shutdown. Each heater consists of an inlet and outlet manifold which is drilled to accommodate a nest of 42 steel U-tubes. The steam supply for these heaters is taken from the auxiliary boilers.

14.7 Feed Suction Filters and Pumps

Feedwater from the deaerator flows through a downcomer via micro-wire filters to three feed suction pumps. The downcomer has two test loops, with appropriate isolating valves, installed in parallel with the main line. These loops have flow orifice devices and are used for efficiency checks only. The main line bifurcates to two 100%

capacity feed suction filters which are automatic in operation, with changeover and backflushing being initiated on high differential pressure across the filter. The outlet from the filters is passed via isolating valves to the inlet of the feed suction pumps.

There are three 33.3% feed suction pumps which generate the head necessary for the suction requirement of the starting and standby or the main feed pump. Each is a centrifugal, cartridge type pump driven by an electric motor through a flexible coupling. The pumps are 9-stage horizontal machines, each having a self contained oil system with an oil tank, duplicate oil pumps and a head tank to ensure lubrication under run down conditions. Bedplate and gland cooling is provided from the general service water system. No pump leak-off system is installed and this means that a flow path through one or other of the feed pumps must be available at all times.

The system can tolerate the loss of one feed suction pump with the unit running at full load without causing a vapour lock at the main feed pump suction.

14.8   Boiler Feed Pumps

From the feed suction pumps, feedwater is supplied to either the main boiler feed pump or the starting and standby boiler feed pumps. There is one steam driven main boiler feed pump of 100% duty and two 50% duty electrically driven boiler feed pumps for starting and standby. Under normal operating conditions the main boiler feed pump delivers the complete feedwater load to the boiler, while the starting and standby pumps are used during the startup procedure and low load conditions to supply the boiler until sufficient steam is available for the main boiler feed pump turbine.

14.8.1   *Main Boiler Feed Pump*

The main boiler feed pump (BFP) is a 2-stage horizontal pump of barrel casing design and is driven by a turbine via a flexible coupling. The speed at full load is approximately 7000rev/min with gland sealing water supplied from the extraction pump. A back-up gland sealing water supply is from a pump drawing directly from the reserve feed water tanks. There is a twin leak-off system to ensure adequate flow under low load conditions. Bearing lubrication is by force feed taken from the main turbine lubricating oil system.

14.8.2   *Main Boiler Feed Pump Turbine*

The main BFP turbine is a single cylinder machine mounted on a bedplate which is secured to a separate foundation block adjacent to the main turbine. The turbine and support assembly is designed to facilitate quick removal and replacement with a complete spare unit. The cylinder is supported at each end by bearing blocks mounted on the boiler feed pump bedplate. A system of keys and guides accommodates the movement of the cylinder and rotor during thermal changes whilst maintaining alignment.

The main BFP turbine is driven by high pressure steam supplied from the boiler when the machine is operating at low load and by steam bled from the HP turbine cold reheat pipes when the main turbine is operating at its rated load. When the main BFP turbine is being supplied with steam from the boiler, the steam is exhausted to the main turbine condenser via a desuperheating system and when it is being supplied with bled steam the steam is exhausted to the LP heater 3. Changeover from the live to bled steam occurs at loads of 30% to 42% on the main turbine.

Steam sealed labyrinth type glands are provided at each end of the turbine where the steam rotor emerges from the cylinder. The glands are designed to provide a suitable seal which will prevent the leakage of steam from the turbine when its pressure is greater than atmospheric and also prevent the ingress of air when the steam pressure is lower than atmospheric. Steam for sealing purposes is tapped off the system supplying the glands of the main turbine.

To control the flow of steam into the main BFP turbine, three steam chests are employed, one for live steam from the boiler and two for bled steam, each chest containing an emergency stop valve and a governor valve. Control of these valves is by means of an electronic system with hydraulic actuation of the steam admission valve relays using fluid supplied by the main turbine fluid control system. To protect the machine, a comprehensive tripping system is provided, the purpose of which is to close all the steam admission valves in response to a variety of emergency conditions either automatically or through operator intervention.

The main BFP turbine bearings are supplied with lubricating oil from the main turbine system.

14.8.3 *Starting and Standby Boiler Feed Pumps*

The starting and standby BFPs are 2-stage horizontal centrifugal pumps of the barrel casing design. The pump shaft, together with the bearings, the mechanical seals and the pump half coupling, form a cartridge. This cartridge can be removed from the casing for maintenance without disturbing the suction and discharge pipework or the alignment of the gearbox and the driving motor. Each pump is driven by an electric motor through speed increasing gears at approximately 7820rev/min and has a rated capacity of 300kg/s, the discharge pressure being 220 bar after the non-return valve. The pumps are similar in design to the main feed pump but are of reduced capacity.

Water from the general service water supply is used to cool the pump mechanical seal coolers, pump baseplate pedestals, gearbox pedestals, gearbox oil cooler, motor air cooler and electrolyte cooler. Water from the reserve feed tanks is utilised in the pump drive and non-drive end seal cooling annulus.

A separate lubricating oil system is incorporated in each pump. A mechanically driven main oil pump provides oil for the bearings and gearbox under normal running and a motor driven auxiliary oil pump is used for startup and shutdown conditions.

14.9 Feedwater Control

Feedwater control is achieved under various operating conditions using a combination of regulating valves and speed control of the feed pumps. There are five regulating valves provided, three 50% duty main and two 20% duty startup valves which are mounted in parallel in the feed pumps discharge pipework between the pumps and the HP feed heaters. Each regulating valve has motorised inlet and outlet isolating valves, which in turn have a small bore manual priming valve in parallel with each. Provision is made for draining the regulating valve in the event of its isolation.

14.10 High Pressure Feed Heaters

After passing through the feed regulating valves the feed water divides and enters the four HP heaters arranged in two parallel banks of two heaters. These are provided to heat the feedwater before its entry to the boiler. Each heater utilises bled steam as the heating medium with the HP heaters 7A and 7B supplied from the main BFP turbine and the HP heaters 8A and 8B supplied from the cold reheat pipework. To protect both the main turbine and the BFP turbine against back flow of steam or water in the event of a main turbine trip or high water levels in the heaters, each steam line is provided with a motorised isolating valve and a power assisted non-return valve.

Each of the HP heaters is vented to the main condenser with the vent lines from each bank commoned together before being routed to the condenser. Included in each vent is an orifice plate for pressure adjustment and also a manual isolating valve for maintenance purposes.

The feedwater flow through each heater bank is controlled by a motorised isolating valve in the inlet and a similar valve in the outlet line. In addition, a bypass line incorporating a motorised valve and a manual valve is provided across the inlet isolating valve. The manual valve, which can be easily replaced, is provided to accept the erosion which occurs due to flashing when the system is initially primed. A feedwater bypass line containing duplicate spring loaded valves is provided across the heater banks to enable the isolation of either bank in the event of a heater flooding.

The drainage arrangement of each HP heater bank is identical, with the HP heater 8 draining to the HP heater 7 via a motorised isolating valve, control valve and a flashbox. The drainage leaving the HP heater 7 is diverted either via a non-return valve, motorised isolating valve and control valve to the deaerator with loading in excess of 85%, or via a motorised isolating valve and control valve to the condenser flash vessel with loading below 80%. Progressive changeover occurs between these two values. To prevent flooding of the HP heater 7 and the resultant isolation of the heater bank when the turbine driven feed pump is out of service, a line is tapped off the drainage line between the HP heater 8 and flashbox to divert the drains via a motorised isolating valve and control valve to the condenser flash vessel.

15. COOLING WATER SYSTEM (FIG 3.25)

15.1 General

The water quantity flowing in the River Ouse is insufficient for a once through or mixed tower and direct arrangement, therefore a closed recirculating tower system has been adopted for the cooling water (CW) system.

The physical size of the station necessitated a split CW system with a CW pumphouse at the north end of the turbine hall supplying the Completion units and a CW pumphouse at the south end of the turbine hall supplying the First Half units.

The cooling water system consists of three sections:
(1) Makeup system.
(2) CW recirculating system.
(3) Purge system.

The cooling water is taken from the River Ouse via the makeup system to both pumphouse forebays. It is pumped from the forebays through the culverts beneath the sub-basement floor level in the turbine hall, through the condensers and is returned to the cooling towers where, after cooling, it gravitates back to the forebays. Approximately 1% of the CW system throughput is lost by evaporation, which results in an increase in the dissolved solids concentration. In order to keep the dissolved solids concentration constant, approximately 2% of the system's water is purged-off back to the river via the purge system. This means that 3% of the system's capacity must be replaced via the makeup water system.

The original chlorination plant was replaced with hypochlorite plants for the injection of sodium hypochlorite solution into the main and auxiliary CW systems to prevent mussel accumulation.

15.2 Design Data for Total Station

Main CW Pumps

Type	Vertical spindle, concrete volute, mixed flow
Number	8
Flow per pump	$14.77 \text{m}^3/\text{s}$
Total generated head	25.88m
Net generated head	24.90m
Pump speed	198rev/min
Gearbox type	Double reduction, three layshafts
Motor type	Vertical skirt mounted cage induction
Motor rating per pump	4776kW
Voltage	11kV
Motor speed	990rev/min

CW Makeup Pumps

Number	4
Type	838mm mixed flow bowl
Capacity per pump	$1.67 \text{m}^3/\text{s}$ at 30°C
Head	16.46m
Speed	490rev/min
Motor rating per pump	355kW at 3.3kV
Pump efficiency	82.5%

Purge Pumps

Number	4
Type	686mm axial flow, vertical spindle
Capacity per pump	$1.32 \text{m}^3/\text{s}$ at 6.4m head
Motor rating per pump	131kW at 415V
Control	Fixed speed
Pump efficiency	78%

15.3   Cooling Water Make-up System

The CW makeup system is designed to meet the requirements of the whole station. The capacity is based on 3% of the throughput to the cooling towers, assuming 1% evaporation and 2% purge, plus $0.1518m^3/s$ for water treatment plant makeup. This represents a total station flow of approximately $3.87m^3/s$.

An intake and pumping station at the River Ouse pumps the water up to the pre-treatment area through off-site makeup lines. The intake comprises coarse bar screens and two rotating drum screens, each of which is designed to take 100% of the flow requirements of the complete station. Four makeup pumps are provided, three of which are variable speed and one fixed speed. Corrosion problems occurred in the two 1.067m diameter steel makeup pipelines installed for the First Half. Consequently, a 1.4m diameter concrete makeup pipe has been installed for the Completion through which 100% total station flow is normally passed. The steel pipes are retained for standby service.

On site, the makeup water enters four circular concrete sedimentation tanks before passing into a distribution chamber, from whence it flows separately to the two CW recirculating systems.

15.4   Cooling Water Recirculating System

The basic concept of the cooling water recirculating system for Drax Completion is as the First Half, but advantage has been taken of operational experience to incorporate design improvements.

Cooling water from the towers flows through individual open culverts into paired flumes which feed the CW pumphouse forebay. See Fig 3.26. Water from the base of the forebay is drawn through a rectangular suction culvert to the pump suction flare which directs the water vertically into the eye of the CW pump impeller. Any tendency to create a vortex at the pump suction flare is avoided by a flow-splitting wall built into the culvert floor at the suction flare.

The four main CW pumps for the Completion are built into a concrete structure. Each pump is surmounted by a 5:1 reduction epicylic gear unit driven from a vertical skirt-mounted motor. The pumps discharge through quick closing butterfly type valves into a 3m diameter manifold.

The CW system requires very large flows through long culverts; consequently there is a danger of water hammer in the system. Starting and stopping the CW pumps and the resultant sequential discharge valve operation inevitably sets up pressure waves in the system. The discharge valve opening and closing times must be set to avoid creating drastic changes of flow conditions which could cause damaging shocks to the system. The discharge valve closing rate is therefore adjusted to satisfy two conflicting requirements; it must close quickly enough to avoid loss of CW flow through a failed pump, yet it must close slowly enough to avoid excessive pressure surges in the system.

There are three take-offs from the CW pump manifold each of which feeds cooling water to a turbine condenser unit via a 3m diameter inlet culvert. The individual culverts bifurcate into two 2.2m diameter culverts feeding pipework to the A and B condensers. The condensers for each turbine nominally require a CW flow of $18.93m^3/s$ and the auxiliaries a further $0.757m^3/s$.

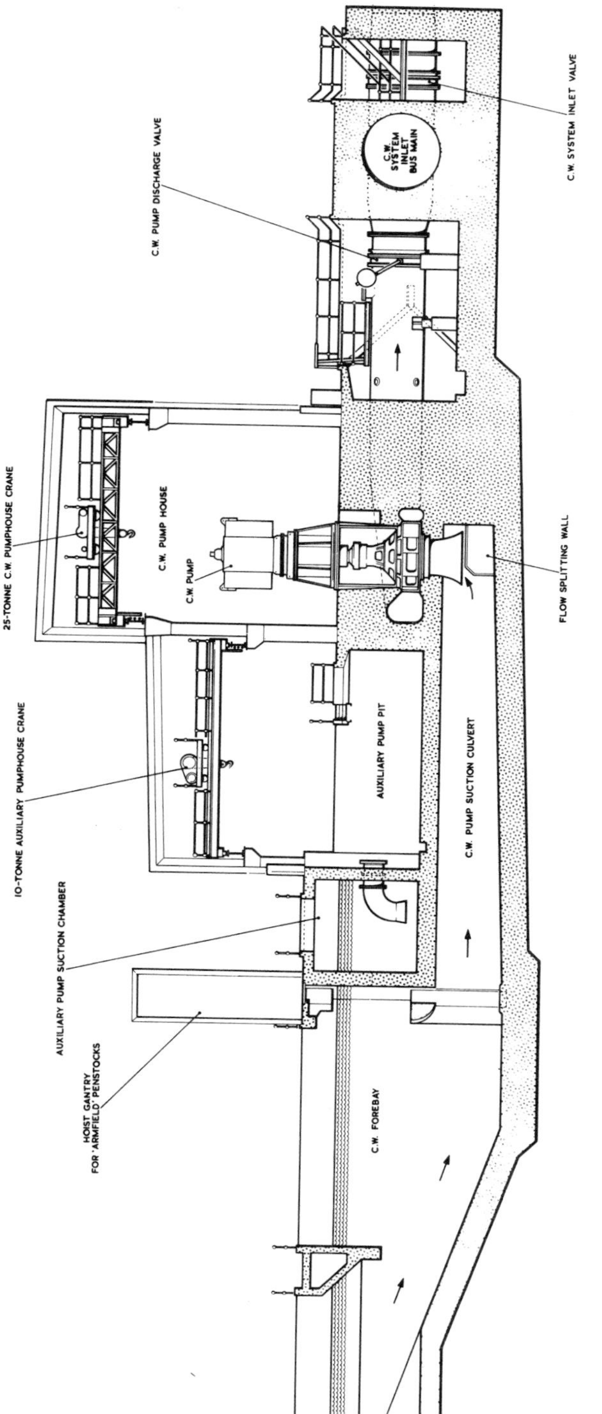

FIG. 3.26   SECTION THROUGH CW FOREBAY AND PUMPHOUSE

It is possible in cold weather to operate three units with three CW pumps; however, for security, two pumps will always be in service when only a single unit is operating.

From each turbine condenser A and B, pipework leads to two 2.2m diameter concrete culverts combining into a single 3m diameter culvert which returns the CW to a pair of cooling towers. An interconecting valve chamber is provided which allows any turbine condenser/tower pair combination to be used.

15.5 Cooling Towers (Fig 3.27)

The six cooling towers for the Completion located at the north end of the site are externally similar to those on the First Half of the station but with modified internal packing arrangements. The cooling tower ponds act as settling areas for any silt carried through the system and provision is made for periodic silt lancing. Cooling water from individual towers flows into paired flumes feeding the CW pumphouse forebay.

Each tower is designed to cool $9.47m^3/s$ of water from $30.21°C$ to $19.6°C$ at ambient conditions of $10°C$ dry bulb temperature and $7.5°C$ wet bulb temperature. The towers are 115m high and 92.7m diameter at pond cill level.

15.6 Purge System

Resulting from operational difficulties experienced with the purge system on the First Half, it underwent a complete redesign. The redesigned system collects water from the purge system, sludge system and water treatment plant effluent in two inlet chambers. One chamber accepts purge water from the First Half, whilst the other accepts purge water from the Completion, sludge from sedimentation tanks and effluent from water treatment plant. A facility is provided for interconnecting the two chambers via a third chamber. Water from the inlet chambers flows through coarse screens into four pump suction chambers. The purge pumps lift the water into two discharge chambers from where it is discharged through an off-site pipeline and outfall flap into the River Ouse.

16. COAL HANDLING PLANT (FIG 3.28)

The coal handling plant, located at the west of the station site is designed to serve the total station capacity of 4000MW. The plant was constructed in two phases. The initial phase was designed to serve the three First Half units but where conveyors and equipment would eventually serve the total station, these were rated accordingly. The plant was augmented at the Completion construction stage by:
(1) An additional bucket wheel machine (north) together with its associated conveyors and junction house.
(2) Providing an additional rail track above an already constructed track hopper to increase the capacity of coal reception.
(3) Extending the shuttles and associated feeder conveyor system above the bunkers to supply the three Completion units.

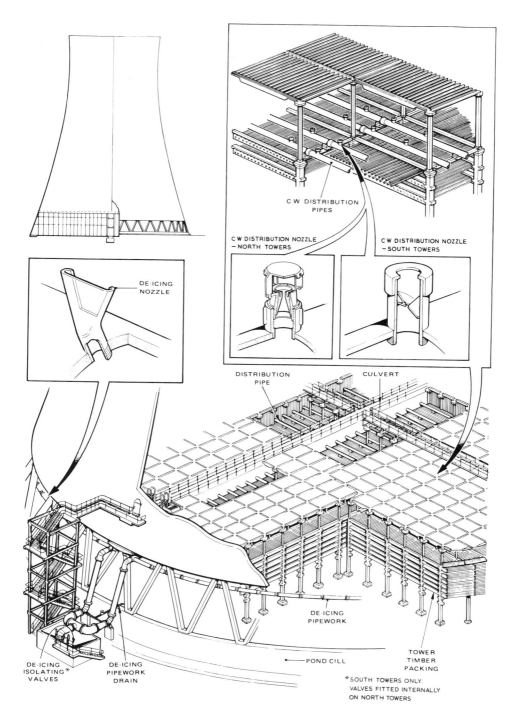

C W DISTRIBUTION
PIPES

CW DISTRIBUTION NOZZLE
– NORTH TOWERS

CW DISTRIBUTION NOZZLE
– SOUTH TOWERS

DE-ICING
NOZZLE

DISTRIBUTION
PIPE

CULVERT

DE-ICING
PIPEWORK

DE-ICING
ISOLATING *
VALVES

DE-ICING
PIPEWORK
DRAIN

POND CILL

TOWER
TIMBER
PACKING

*SOUTH TOWERS ONLY
VALVES FITTED INTERNALLY
ON NORTH TOWERS

FIG. 3.27   COOLING TOWER

16.1 General

Coal is transported to the station in train loads of up to 1016t contained in hopper bottom wagons made up into trains of between 30 and 40 wagons. The trains are automatically weighed at speeds up to 24km/h for gross weight on entering the station sidings and for tare weight on leaving. A computer and teleprinter are installed to record and printout the relevant weights.

The wagons discharge coal into hoppers beneath the rail tracks (track hoppers) whilst moving at a speed of 0.8km/h, the opening and closing of wagon doors being controlled by a system of lineside equipment.

Coal is extracted from the track hoppers by paddle feeders and fed on to associated conveyor belts. From here, the coal is transported on a system of conveyors by which it can be fed to the bunkers or to stocking out areas. Fig 3.28 shows the alternative routes along which the coal can be transported.

Stocking out and normal reclaiming of coal from stock is achieved by the use of two bucket wheel machines located to the north and the south of the main stocking out area. These work in conjunction with mobile plant which also serves to form the long term stockpiles contained on either side of the bucket wheel machines.

It is important that the amount and quality of coal used is known so that station performance can be assessed. To this end, belt weighers are installed on conveyors to monitor the weight of coal en route to both the boiler bunkers and stocking out locations. The weighing units indicate the totalised weight and rate of flow at local and remote positions. A coal sampling plant is installed at the entry to the bunker building to obtain bulk samples which are used to determine the quality of the coal being burned.

16.2 Fuels

The plant is designed to handle mixed fuels with a maximum size of 76mm cube consisting of slack, fines and other grades of washed or unwashed coals. The plant can be operated with the coal in a sticky, frozen or extremely wet condition; the total moisture content of the coal received may be up to 25%. The average coal density is approximately 820kg/m^3.

16.3 Bucket Wheel Machines (Fig 3.29)

Each bucket wheel machine facilitates storage of coal and is used in conjunction with a stockpile conveyor for either stocking out or reclaiming, with easy changeover from one mode to another. Each machine is carried on a powered carriage moving on rails between two storage areas. The stockpile conveyor is located between the rails.

For stocking out, coal is received by the machine from the stockpile conveyor, carried on elevating and boom conveyors and discharged over the end of the boom. The bucket wheel is not used when stocking out. The machine boom can be luffed and slewed so that the stockpile is built up to the required shape and height.

For reclaiming, the boom conveyor and stockpile conveyor are reversed in direction. Coal is picked up from the stockpile by the bucket wheel and carried back to the stockpile conveyor by the boom conveyor. The elevating conveyor is not used when reclaiming.

FIG. 3.29   BUCKET WHEEL MACHINE

All main functions of the First Half (south) machine are electrically powered. On the north machine, boom slewing, luffing and bucket wheel motions are driven by a single hydraulic power pack; other main functions are electrically powered. Automatic and manual control systems are provided on the south machine but in operation the automatic feature was little used during First Half operation. Consequently, a manual control system only is provide on the north machine. The controls on both machines are located in an operating cabin attached to the mast from where a clear view of the stocking and reclaiming operations can be seen.

External power supply cables are connected via a power rotated cable drum mounted on each machine. The drum automatically coils or uncoils the cable when the machine is moved. A similar drum connects control and communication cables to the station main control room.

16.4 Conveyors

The conveyors are all deep troughed with 45° angle sides; the belt width is 1.8m. All belts are driven at a nominal speed of 2.28m/s apart from the shuttle conveyors above the bunkers and the bucket wheel machine boom conveyor which are driven at slightly higher speeds.

In general, the conveyor drives are located at the heads of conveyors. The long reversible conveyors associated with the bucket wheel machines are, however, driven by two separate motors situated at the terminal pulleys.

Shuttle conveyors discharge coal at any desired point along the length of the boiler bunkers. Each shuttle conveyor is integrated with a carriage which travels on rails in a forward or reverse direction, distributing as required over the length of two bunkers. At each end of the desired run, the shuttle is automatically reversed by adjustable reversing switches.

16.5 Rail Unloading Plant

The track hoppers into which the rail wagons discharge are constructed of reinforced concrete and lined internally with glass tiles. Each hopper is fitted with ultrasonic sensor switches to initiate visual indication, at the master control desk and at track hopper consoles, that the hopper is full or empty.

Coal is extracted from each track hopper by a set of four paddle wheels in a combined carriage which is traversed in forward or reverse directions along the length of the hopper. As the carriage moves along, a cam engages a mechanism which raises the hopper discharge doors to allow the horizontal paddle wheels to scoop coal from the hopper apron. The doors are closed on the trailing contour of the cam, under gravity. The travel speed of the paddle wheels and the discharge rate from the hoppers are controlled to keep the level of coal in the hoppers as constant as possible.

Lineside equipment is installed in the track hopper house to unload the coal trains. The equipment is pneumatically controlled and comprises unlatching units, door opening units, door closing units and relatching units, which are spaced at intervals along the track. Wagons are unloaded in batches of four, each batch unloaded alternately in one of two positions half a length apart to give a uniform distribution of coal in the hoppers.

16.6 Emergency Stocking Out and Blending Facilities

Additional plant has been installed to give greater flexibility in the operation of the coal handling plant. Two conveyors provide an emergency stocking-out facility for discharging coal trains in the event of there being no station demand and the normal stocking-out routes being unavailable.

Twin ground hoppers each with an associated conveyor provide a blending facility where lorry-borne inferior coal can be blended with the normal coal stream. This also provides an emergency reclaiming facility to satisfy station coal burn in the event of loss of, or difficulty with, the screening plant, bucket wheel machines, etc.

16.7 Mobile Coal Handling Plant

Heavy duty mobile equipment is employed to transfer coal from the stockpiles to the bucket wheel machines, and also to compact the coal when stocking-out to prevent fire due to spontaneous combustion.

16.8 Controls and Monitoring Equipment

Control of the coal handling plant is centralised at a master control desk located in the coal and ash building. All conveyors (excluding shuttle conveyors) are controlled from a mimic panel as are paddle feeders, trash screens, flap valves, railway signalling and wagon weighing. The bucket wheel machine motions, sampling plants and shuttle conveyor motions are controlled by local control equipment.

Each conveyor drive has a control cabinet incorporating control relays and alarm circuits. These cabinets form an interface between the plant and the remote controls. The electrical controls and interlocks for all conveyors, paddle feeders, screens and shuttle conveyors are designed to ensure that a dangerous pile-up of coal is eliminated and all sections of plant are started and stopped in the correct sequence.

Controls for lineside equipment are located in the track hopper house. The equipment includes alarm initiating devices, interlocks, actuators, position indicators, limit switches and local control panels with instruments. In addition connections are provided for the installation of remote instruments.

The lineside control units are interlocked with the paddle wheels, paddle wheel carriages and the track hopper conveyors so that on failure of any one (or more) of these items the lineside equipment is retracted and the signals are set to stop the discharging train. Eight overfill switches in each track hopper are also interlocked with the lineside equipment to automatically retract the door releasing and opening mechanisms and stop the discharging train. An automatic emergency stop is also initiated if any of the lineside units do not respond to respective control relays.

The conveyor system and bucket wheel machines are equipped with closed circuit television networks to monitor the working conditions of the coal supply. The camera units on the bucket wheel machines are directed to allow the machine operator to monitor the transfer of coal to and from the machine from the main belt system. The entire coal conveyor system of 61 routes can be viewed from 19 camera positions. The operator can select viewing from these camera positions for display on 11 monitor screens located in the coal and ash building.

16.9 Electrical Supplies

Electrical supplies for the coal handling plant are derived from the 11kV station boards. These boards feed the 3.3kV coal plant auxiliary boards via two 11kV/3.3kV, 8MVA transformers. Four 3.3kV/433V, 2MVA transformers, fed from the 3.3kV coal plant board supply low voltage supplies to 415V coal plant services boards and also coal plant lighting and heating boards.

A separate switchhouse accommodates all the boards and transformers supplying the coal plant, with the exception of those boards associated with the bunkers.

17. ASH AND DUST HANDLING PLANT

17.1 Ash Handling Plant

The ash handling plant collects and removes for disposal the furnace ash from the six boilers.

Hot ash falls by gravity from the boiler furnace into an ash hopper located directly beneath. Dipper seal plates attached to the furnace enclosure outlet are immersed in a trough of water around the top of the ash hopper. This maintains a seal between the two and also allows for downward expansion of the boiler.

As the ash falls into the hopper, it passes through a continuous wall of spray water set up by water jets positioned in the front and rear walls of the hopper. This wall of water cools the ash, reducing the formation of sinter and the risk of fire.

Two arrangements of ash hopper are used at Drax. Those in the First Half boilers are divided into two sections, each section being sub-divided into six U-shaped hearths, each sloping down towards the boiler rear. Those in Completion boilers are of continuous section, sub-divided into 15 U-shaped hearths also sloping towards the boiler rear.

The ash in the hoppers is broken up and sluiced away by high pressure water jets through a rotary valve located at the bottom of each hearth into a sluiceway. Jetting nozzles in the sluiceway assist the flow of ash to two 100% duty ash crushers. These reduce the ash to a particle size that is transportable in water and can be discharged by jet pumps through ash conveyance piping to the ash settling pits. Swivelling connections at the end of the conveyance pipes allow the ash to discharge into different sections of the settling pits.

Partial drainage of the ash takes place in the ash settling pits before it is removed by grabbing cranes and transferred to a drainage apron. Here, further drainage takes place before the ash is transferred by the cranes into road vehicles for disposal. As a standby to this facility, the ash can be transferred by the grabbing cranes into moveable feeder hoppers discharging onto a conveyor system which disposes of the ash to the dust disposal area on Barlow land adjacent to the station. See Fig 3.30.

Water draining from the ash settling pits and drainage aprons is discharged by means of pumps into one of two ash water lagoons, where the ash dust is allowed to settle out. The reclaimed water is pumped to the ash plant reservoir which supplies the HP sluicewater pumps with water.

The eight HP sluicewater pumps supply the water for the jetting nozzles and jet pumps. A pressurised water supply from the general services water pump main provides gland sealing water for the pumps.

A control panel located in the coal and ash building incorporates instruments and controls in a mimic flow diagram of the ash handling system. This provides the operator with the means of monitoring and controlling the performance of the plant. The HP sluicewater pumps and sealing water pumps are controlled from this panel. Ash crushers and other pumps are controlled from local panels and each ash grabbing crane from an operator's cab suspended below the crane beam.

Electrical supplies for the ash handling plant are derived from the 11kV level of the station auxiliary system. Two 8MVA transformers are installed and under normal operational conditions each works at approximately half load. Under maintenance outage or fault conditions, the system is arranged to provide full capacity of operation with one transformer out of service. This supply feeds the ash and dust plant auxiliary boards, from which four 3.3kV/415V transformers supply the low voltage supplies to the ash and dust plant services boards.

The boards and transformers are housed in a switchhouse adjacent to the coal and ash building.

17.2 Dust Handling Plant

The dust handling plant collects and removes for disposal the dust in all the six boiler economisers and associated precipitators. The collecting system terminates at the dust storage bunker from where the disposal system conveys conditioned dust to the Barlow land disposal area. The plant is designed to remove to storage in 16 hours the dust produced in 24 hours.

Dust collected in the eight economiser hoppers per boiler is conveyed by motive air from jet blowers into an economiser surge hopper. From here it is conveyed by a pneumatic discharge pump into the precipitator surge hopper which also receives dust from the precipitator outlets (24 per boiler on the First Half and 27 per boiler on Completion) conveyed by motive air from jet blowers. The dust from each of the six precipitator surge hoppers is then pumped into the dust bunker by a pneumatic pump having a discharge capacity of 95t/h. The dust bunker is located above the dust bunker house. See Fig 3.30.

The capacity of the dust bunker is 3556t of dry dust at a specific volume of $1.3m^3/t$. The bunker is divided into two equal compartments by an internal, full depth, division wall. Dust from the precipitator surge hoppers enters the bunker through an inlet unit at the top and is deflected down into the compartments. The floor of each compartment incorporates aeration troughs through which air, supplied by three aeration fans, is passed to fluidise the dust. The fluidisation assists the flow of dust to the bunker outlets. The fluidising air and conveying air is exhausted to atmosphere through a filter situated in the bunker roof.

Four outlets from each bunker compartment connect to drag-link conveyors which discharge into the dust conditioners. Also three retractable discharge chutes are provided for the disposal of dry dust into road vehicles. These chutes receive dust from one compartment of the bunker through individual air slide conveyors.

There are eight conditioners located in the dust conditioning house. Each conditioner consists of an inclined drum which is rotated about a stationary internal assembly of baffles, scraper bars and spray nozzles. Dust enters the conditioner through an inlet duct at the top of the drum and passes through an initial curtain of spray water. As the drum rotates, the wetted dust moves down the drum and is thoroughly mixed. It is then further conditioned by secondary spray nozzles before discharge of the mixture onto an open conveyor system.

The open conveyors transport the conditioned dust to the disposal area on Barlow land. Here the dust is distributed by duplicate ground level conveyors on to a wing conveyor which feeds on to a bridge conveyor, itself feeding a mobile boomstacker mounted on caterpillar tracks. The boomstacker deposits the dust, progressively forming a mound which is graded and compacted by mobile equipment.

The surface of the mound is sprayed with polymer compound to minimise dust dispersal. The area is then progressively landscaped.

Control of the equipment in the dust handling plant is primarily from control panels located in the dust bunker house. Local control panels are provided in some instances to allow individual items of plant to be run out of sequence.

The dust plant receives electrical supplies through the same boards as the ash plant.

18. GAS TURBINE PLANT

18.1   General (Fig 3.31)

The station gas turbine plant consists of two gas turbine houses, each containing three single-ended gas turbine generating sets and located on opposite sides of the gas turbine chimney. The two houses are associated with the two stages of station construction. Each gas turbine set is capable of generating 35MW at 11kV.

The gas turbine sets are normally operated by signals initiated from the station control room and fed to the automatic sequencing and control circuits in the cubicles of the local control panel. The circuits automatically start, synchronise, load and shutdown the set. When required, the automatic sequences can be initiated from the local control panel. A manual starting facility is also provided.

During starting and throughout normal operation the sets are continuously monitored by automatic protection circuits. The circuits provide audible and visual alarms for serious faults and initiate shutdown of the set in the event of a serious malfunction; vital alarms and indications are repeated in the station control room.

A common stack is provided for all six gas turbines but each has a separate internal flue.

18.2   Operational Requirements

A gas turbine set is associated with each main generator unit in order to satisfy the following system objectives:
(1) To provide sufficient auxiliary power to enable station startup without external supplies.

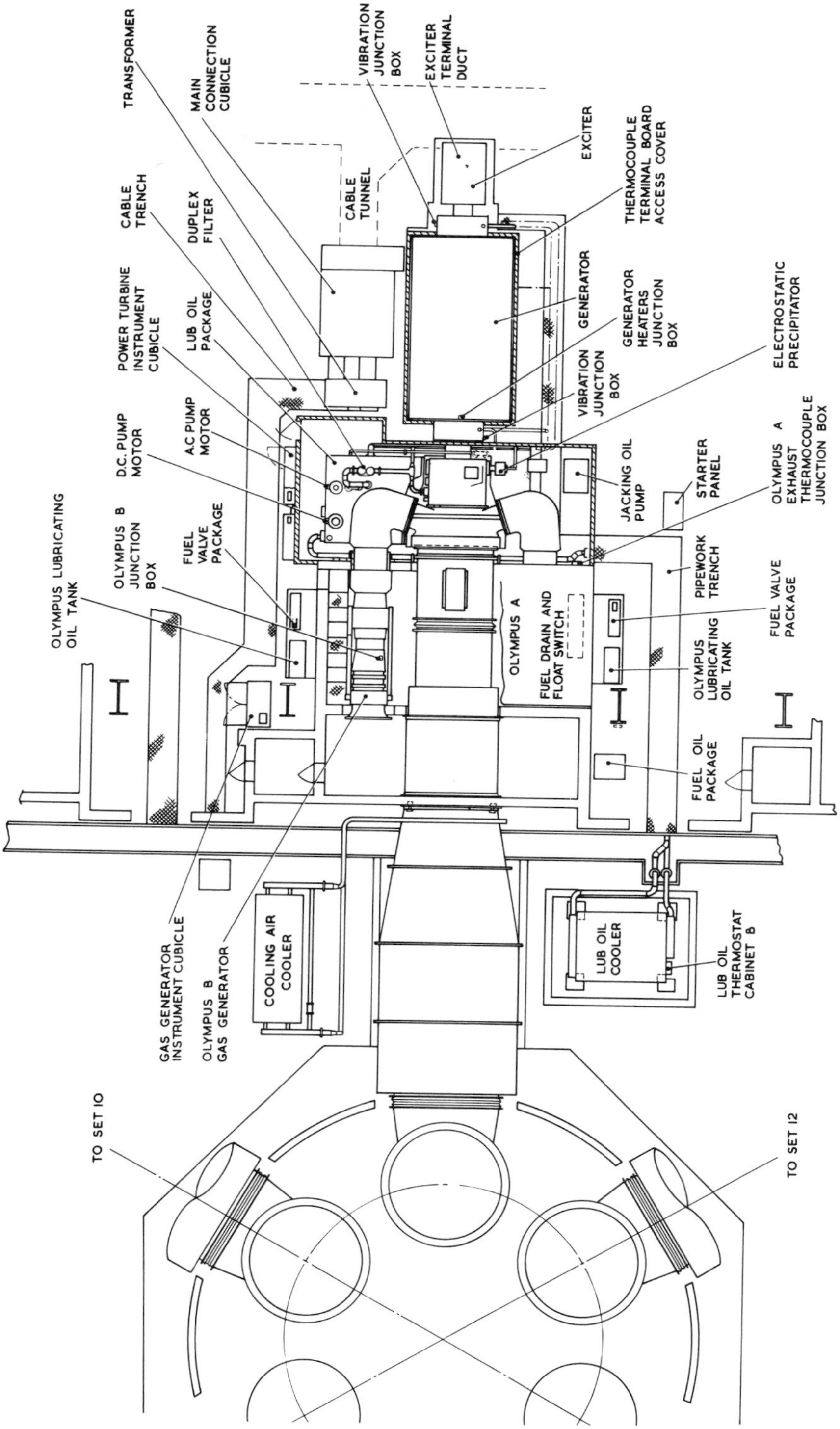

FIG. 3.31   GAS TURBINE SET PLAN VIEW

(2) To provide a stable supply to the electrical auxiliary loads within the frequency range 47-51Hz when the auxiliary supply system is isolated from the grid system, and to be capable of restoring the auxiliary system frequency to 50Hz.

(3) To provide output capacity for economic generation purposes, including contribution to the reserve capacity requirements of the system by means of an emergency start facility.

Operation for economic generation purposes could involve duties ranging from continuous generation for periods up to 16 hours per day to frequent starting followed by short duration runs. In order to satisfy these overall requirements, the following general conditions apply:

(1) Main set operation must not be prejudiced by faults associated with the gas turbine plant.

(2) The output of all gas turbines must be freely available to the grid system regardless of the operating condition of the main generating units, including their being off-load.

18.3 Air Intake Filters

Combustion air for the gas generators and cooling air for the ac generators is passed through a system of filter units housed in the filter house of each set. The system for each set consists of 10 vertically mounted filter units. These units are continuously monitored by a series of differential pressure switches which cause the roll filter material to advance to a clean section when a build-up of dirt occurs. When the roll is exhausted, indications and alarms are provided on a control panel in the filter house and remote alarms and indications initiated on the unit control panel to prompt filter replacement.

18.4 Gas Generators

Two distillate fuelled industrial gas generators are utilised for each set. The gas generators are installed side-by-side in a concrete cell which provides acoustic attenuation and personnel protection.

The gas generator is an axial flow, twin spool, high pressure ratio unit with a 5-stage low pressure compressor and a 7-stage high pressure compressor. Each compressor is driven through coaxial shafts by its own single-stage turbine. The low pressure and high pressure compressors are arranged in tandem. The low pressure compressor compresses the air supplied to the air intakes and delivers it to the high pressure compressor. The high velocity of the intake air causes a decrease in temperature and consequently there is a risk of ice formation during cold weather. This is prevented by an anti-icing system.

Combustion air from the air intake filters enters the front of the combustion chambers where fuel is injected and ignited by an electronic igniter system. The exhaust gases from both gas generators are discharged into the inlet bends of the power turbine.

18.5 Power Turbine

The power turbine is a bedplate mounted, 2-stage axial flow type expansion turbine. It is driven by the exhaust gas efflux from the gas generators, which enters the turbine via two inlet bends and is then directed to the first stage nozzle blades. The gases are then expanded through the first and second stages of the turbine where the linear kinetic energy of the gas is converted to rotational kinetic energy. The spent exhaust gases are directed along ducting to the chimney and discharged to the atmosphere.

The turbine rotor shaft is directly coupled to the ac generator by a flexible coupling. An electric motor driven barring gear assembly is fitted to rotate the rotor shaft, and hence the ac generator rotor shaft, at low speed during the cooling period after shutdown. This procedure eliminates any tendency for the rotor shafts to sag and take on a permanent 'set', which would result in excessive stress and vibration. The barring gear is coupled to the rotor by a flexible coupling.

18.6 AC Generator and Exciter

The 35MW ac generator is of the air cooled, horizontal mounting, brushless type. The cylindrical rotor is supported on two forced lubricated sleeve type bearings, one at each end of the stator. The cooling air from the air intake filters is ducted to a pit below the stator frame.

The main exciter is a revolving armature, brushless machine and is directly coupled to the generator rotor. The exciter output is rectified by silicon diodes which are connected as a 3-phase bridge. The diodes are assembled onto a diode wheel which is shrunk onto the exciter shaft. A permanent magnet pilot exciter is mounted on the same shaft within the bearing span. The exciter cooling air is drawn from the generator air circuit via ducts in the foundation block.

18.7 Fuel Oil Supply System

The fuel oil supply system supplies distillate fuel to the gas generators at a pressure within the limits 0.35 bar to 17.5 bar.

Low pressure fuel pumps transfer filtered liquid fuel from the storage tanks, via a weight-operated fire valve and a hand-operated isolating valve, to a fuel oil package.

The fuel oil package houses a duplex filter, a fuel flowmeter and a low pressure fuel shut-off valve which is activated by auto-sequence control. From the low pressure shut-off valve the fuel passes via pressure regulating valves to the high pressure fuel pumps. The high pressure pumps pass the fuel to a fuel valve package where it is metered in accordance with governor demand signals and conveyed back to the gas generator combustion section through burner assemblies from which fuel emerges as a conical shaped atomised spray.

18.8 Lubricating Oil System

There are three separate lubricating systems for each generating set; one for each of the two gas generators and one for the power turbine and ac generator bearings.

Oil from an auxiliary tank passes through a normally open valve and a Y-type strainer before entering the gas generator at the main oil pump suction filter inlet. Return oil passes from the gas generator to the tank via a duplex filter assembly which filters the oil down to 15 microns. The oil tank breather pipe connects to the gas generator centrifugal separator.

Lubrication for the power turbine and ac generator is provided by a system which comprises a main oil storage tank situated adjacent to the power turbine pedestal, a mechanically-driven pump mounted on the pedestal, an air cooled oil cooler situated outside the turbine hall, an electrically-driven ac standby pump and an electrically-driven dc emergency pump.

18.9 Starting Equipment

The starting equipment for the generating set consists of a starter motor mounted on each of the two gas generators, and a starter cubicle which houses the electrical control equipment associated with the starters. The starter control cubicle is situated adjacent to the appropriate unit. An auto-sequence control system is provided, whereby startup is initiated by switch selections made at the local control panel or the unit control panel. Indications are provided on the control panel to show when the set is in a completely prepared state for startup, whilst other indications and alarms enable the operator to monitor the sequence and take appropriate action if a serious fault occurs.

19. WATER TREATMENT PLANT (FIG 3.32)

The purpose of the water treatment plant, which supplies water requirements for all six units, is to remove all chemical impurities and dissolved gases from the make-up water such that scaling and corrosion of the steam generating plant is prevented.

The need for make-up water results from losses with the turbine/boiler unit due to leakage and purge for quality control purposes.

19.1 Design Criteria

The original design provided for a 2% make-up on all six units with an additional 1% for evaporators on the First Half units when operating at full load.

With the requirement to carry out 2-shift operations immediately after commissioning it was decided not to install evaporators on the Completion. The reason for this is that the peak demand for make-up occurs at startup and evaporators have a very low output under these conditions.

Additional make-up output has been provided by installing an extra stream of deionisation plant, together with increased polishing plant capacity. This now gives a total capacity of 3% make-up even with one stream out of service, plus 2% make-up polishing on two units and 30% recirculation polishing on a third unit. The most adverse make-up requirement is 2.18% when 2-shifting on all six units.

DRAX COAL-FIRED POWER STATION

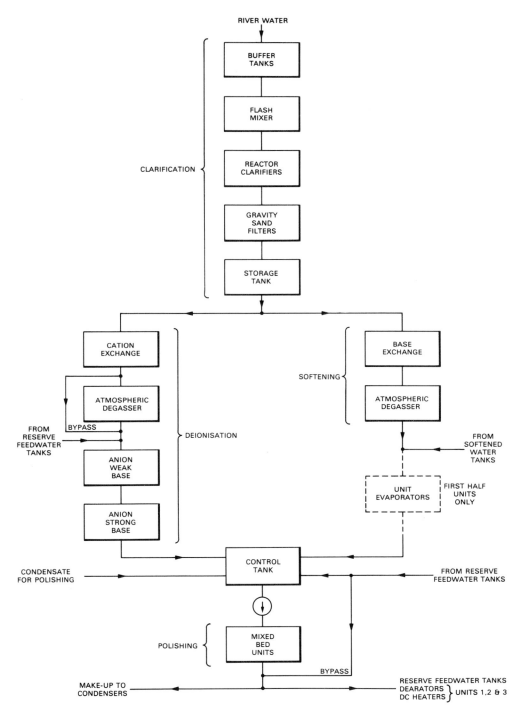

FIG. 3.32   WATER TREATMENT PLANT PROCESS FLOW

19.2 System Description

The supply of raw water is taken from the River Ouse, which can be heavily contaminated at times. So that the water can be handled by the deionisation plant it is initially processed in the clarification plant. This eliminates the suspended solids by means of gravity separation in large buffer tanks, followed by chemical injection for a flocculation process in the flash mixers and reactor clarifiers and finally gravity sand filters for last stage clean up and storage.

The water then passes to the deionisation plant where ion exchange takes place. Output quality is such that it only requires final polishing in the mixed bed units prior to use in the boiler feedwater system, and it is temporarily stored in the control tank depending on demand. Polishing plant feed pumps draw water from the control tank, and pass it through the mixed bed units, where the final exchange takes place such that very high quality demineralised water is produced. This is either used as make-up for the running units or is stored in the reserve feedwater tanks.

Polished condensate pumps circulate contaminated feed water from the units, back to the mixed bed polishing plant via the control tank and return it into the condenser make-up system.

A separate softening plant is also provided which takes water from the clarification plant and converts it to the quality required for feed to the evaporators on the First Half. Excess capacity is stored in the softened water tanks. Output from the evaporators is passed to the control tank.

Facilities for polishing the contents of the reserve feedwater tanks via the control tank are also provided.

The ion exchange resins in the various vessels in the water treatment plant are regenerated when exhausted by the normal acid and caustic rinsing process. This is part of the automatic control of the whole plant which also ensures that the resins are not subject to excessive flows before regeneration.

Treatment of the regeneration effluent is also incorporated such that the discharge into the River Ouse complies with the various statutory regulations. This is achieved by controlling the pH value of the effluent and diluting it via the CW purge system.

20. AUXILIARY BOILERS AND STEAM SYSTEMS

20.1 General

Auxiliary steam is used to provide heating for the deaerator on startup, for fuel oil pumping and heating services and for atomising the fuel oil burners on the First Half main boilers. It is also utilised for hot water and space heating services in the administration and welfare building. The total auxiliary steam demand assuming that the whole station is on full load during severe winter conditions, except for one unit which is starting up, is 10.66kg/s.

The auxiliary steam is derived from three sources:
(1) Three boilers (no. 1, 2 & 3) each of 3.15kg/s capacity on the First Half.
(2) Two boilers (no. 4 & 5) each of 6.3kg/s capacity on the Completion.
(3) A back-up supply from the platen superheater outlet headers on the Completion main boilers. Each main boiler auxiliary steam system is rated at 6.3kg/s.

20.2 Auxiliary Boiler Installation

Auxiliary boilers 1, 2 and 3 are located adjacent to the south gable of the main station buildings. Each boiler is a single combustion chamber oil-fired unit of 3-pass fire tube type. These boilers will ultimately be used purely as reserve capacity.

Auxiliary boilers 4 and 5 (Fig 3.33) are located adjacent to the north gable of the main station buildings. Each boiler is a twin combustion chamber oil-fired unit of 3-pass type incorporating a water cooled transfer chamber at the end of the first pass. Each boiler is fitted with two rotary cup oil burners, one for each combustion chamber. The rotary cup provides the centrifugal force necessary to atomise the oil and will provide a turn-down ratio of 5:1. A single burner management control panel for each boiler ensures that the air-to-fuel ratio is maintained with an oxygen-in-flue-gas content of not greater than 3% and a carbon monoxide level of not greater than 100mg/litre. There are two centrifugal forced draft fans integral with each boiler.

Auxiliary boilers 4 and 5 include a common fuel oil pumping and heating plant, incorporating duplicate pumps and heaters, which supplies oil to the burners of both auxiliary boilers. There are also three multi-stage boiler feed pumps each of which can supply the needs of one boiler. Under normal conditions with two pumps supplying two boilers the third pump is a standby.

20.3 Auxiliary Steam System

The five auxiliary boilers supply steam into a distribution system in the form of three mains ranging along the length of the station. The mains are interconnected by a steam manifold at each end, thus forming a ring loop. One main supplies the hot water and heating services, the second supplies the fuel oil burners on the First Half main boiler and the third supplies the heating coils on all six deaerators. The back-up steam supply from the Completion main boilers is connected into this third line. Steam to the fuel oil pumping and heating plants is supplied from the manifolds.

20.4 Condensate Return

Auxiliary steam condensate recovered from the First Half of the station returns to auxiliary boilers 1, 2 and 3 hotwells. That from the Completion station returns to auxiliary boilers 4 and 5 feedwater tank. These are interconnected by a condensate transfer system. Excess condensate overflows from the feedwater tank into the station blowdown disposal tank.

20.5 System Operation and Control

The back-up auxiliary steam supplies from the Completion main boilers operate in unison with the two auxiliary boilers 4 and 5. All five sources of auxiliary steam are fully automated to trip in or out as demand dictates.

On the main boilers, steam is supplied from the platen superheater outlet header at a pressure of 184.5 bar and 545°C to a mixing and pressure reducing valve, simultaneously with a supply from the feed spraywater system at a pressure of 320 bar and 200°C. This gives an auxiliary steam supply from the pressure reducing valve matching the auxiliary steam system conditions of 6.3kg/s at 17.5 bar and 209°C.

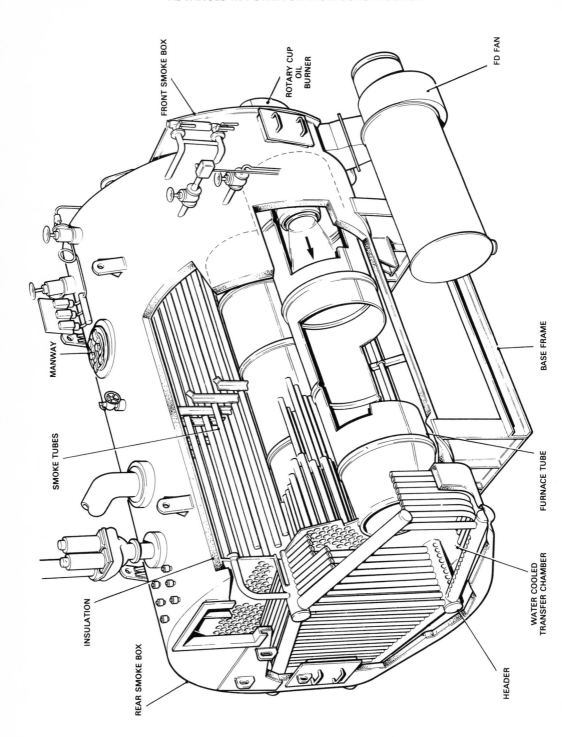

FRONT SMOKE BOX

ROTARY CUP OIL BURNER

FD FAN

MANWAY

SMOKE TUBES

INSULATION

REAR SMOKE BOX

BASE FRAME

FURNACE TUBE

WATER COOLED TRANSFER CHAMBER

HEADER

The control systems for the auxiliary boilers and associated equipment are designed to allow the plant to operate with minimum operator attendance. The overall control concept is aimed at the production of steam at a pressure of 17.5 bar with a flow capability ranging from zero to 31.5kg/s at a temperature of 209°C from a feedwater temperature in the auxiliary boilers of 60°C. The control panels and equipment are housed in the auxiliary boiler control room.

21. FIRE PROTECTION

21.1 General

Fire on the protected plant in the power station may occur from a variety of causes and with differing conditions. The fixed fire protection equipment and systems installed vary in application and method of control according to the nature and size of the plant being protected.

High velocity water spray zoned systems protect major plant areas and items. Halon extinguishing gas is used in high fire risk areas containing the computer suites, electronic service cubicles and marshalling cubicles.

Water supply to the protection systems is maintained under pressure in a graduated diameter pipework trunk main. The charging and maintaining of water pressure in the trunk main is effected by the use of five diesel engine powered pumping units. These units are automatically started in sequence as the demand for water increases. The units, together with ancillary equipment are housed in the south auxiliary CW pumphouse.

21.2 Spraywater Systems

The spraywater systems incorporate equipment which provides for automatic and/or manual discharge of water onto the protected plant. The systems are divided into four main categories. Each category has its own characteristics which relate to the rate of flow, and each has its own range of water distributors, to bring about the required discharge of water on to a particular risk.

The principle of the first category is the application of large quantities of water through projectors which give a high intensity spray of fine water droplets. This completely encloses the risk and also covers its immediate surroundings. The spray is non-conducting and is of sufficient intensity to emulsify the surface of any burning oil, contained in, or released from, oil-filled equipment.

The second category of protection is provided for the major portion of the protected plant and is similar in operation to the first, but will give a less dense coverage. Projectors are able to apply a conical spray of water droplets, delivered at a minimum water pressure of 4 bar.

The third category, in addition to controlling and extinguishing a fire, assists in containing the spread of the fire by cooling the plant exposed to heat in the risk area. By projecting a cooling film of water over the plant adjacent to the fire, absorption of heat is prevented, thus confining the fire to the minimum possible area.

The fourth category of protection, using sprinklers, is provided in various cable races and tunnels, sub-basements, turbine generator sets and boiler feed pump support legs. Each sprinkler is both a detector and water distributor and is mounted on a pipework system which connects back to a control valve. A feature of sprinkler protection is that water distribution is confined directly to the fire, the water being discharged only from sprinklers on which the seals have ruptured.

Two types of installations are used in spraywater systems, a 'wet' installation and a 'dry'installation. The wet installation, as its name implies, has water throughout the pipework, this being retained by quartzoid bulb sprinklers or controls. In the dry installation, the water is retained by a control valve installed in the distribution pipework. The valve opens automatically when quartzoid bulb detectors operating on an independent detector air line, or quartzoid bulb controls installed in the distribution pipework, are fractured allowing compressed air to be released to operate the valve. This releases the water into the distribution pipework to the open distributors located over the protected plant.

Dry installations are employed where the protection is installed outdoors or where low temperature conditions, down to or below freezing point, are likely to occur.

In certain areas of the plant, manually controlled spraywater equipment is considered appropriate in preference to the automatically controlled installations. Manually controlled equipment is installed to enable experienced personnel, qualified to assess conditions on the protected plant, to operate the system as required to allow water to pass only onto selected areas of the plant. The protection system for the air pre-heaters and the water mist curtains for the boilers are manually controlled.

The turbine generator sets and boiler feed pumps are protected against fire by a specially designed manual system. The system consists of a series of control valves protecting selected areas around the machines.

Hydrant valves with hose and branchpipes are installed at strategic positions around the turbine generator sets for use on incipient fires or for backing up the fixed protection.

Sections of pipework and valve manifolds which may be exposed to low temperatures during the winter are electrically trace heated and insulated. The trace heating tape is wound around the pipe/manifold and supplied via a thermostat from a 240V supply.

The instrumentation used with the spraywater equipment falls into three categories:

(1) Alarm indication equipment, which includes pressure switches, acting on system pressures both on hydraulically and pneumatically charged lines and causing an electrical circuit to be made or broken whenever pressures fall or rise from their preset value.

(2) Pressure registering equipment in the form of pressure gauges, for measuring the pressure applied to the system, both on air and water installation pipework.

(3) Control equipment consisting of pressure switches, operating on compressors and fire pump switchgear to cause the motors/engine to run by automatic control and so maintain the installation in a state of readiness at all times.

21.3 Halon Gas Protection Systems

The gas used in the protection systems is Halon 1301 bromotrifluoromethane (BTM). It is a colourless, non-corrosive, liquifiable gas which at a low percentage mixture in air inhibits the spread of fire and will, if maintained, extinguish the fire. It is a stable substance which will not change chemically or physically during long-term storage. When used, it emerges from the discharge nozzles as a mixture of gas and liquid droplets, which rapidly evaporate.

Fixed pipe distribution systems have been installed for the protection of the floor voids and cabinets associated with the various computer and instrumentation rooms. Additionally, the underbox and cable areas for each turbine generator control panel and desk are protected by modular units.

Each distribution system is designed to deliver the BTM gas, automatically or manually through pipework from a storage cylinder. Upon detection of a fire, the cylinder delivers its gas into a manifold assembly and a fixed pipework system via a flexible connector and adaptor. The flexible adaptor facilitates the connection between the discharge valve of the cylinder and the manifold. Initiation of the protection in floor voids is by an automatically operated fire detection system installed in each area. A similar detection system is installed in the cabinets, but is used only to raise alarms to indicate the need for manual operation of the system involved.

Each modular unit is a self-contained assembly comprising a discharge head, supply cylinder and mounting bracket. A separate control and detection system is installed in each underbox and cable area to initiate automatic operation of the modular units. The detection system will actuate all the units within one area simultaneously.

22. HEATING AND VENTILATING SYSTEMS

Heating, ventilating and air conditioning (HVAC) systems are installed in the administration and control room buildings. These provide a controlled environment with regard to temperature humidity, dust and air movement.

Normally occupied rooms and areas containing electronic equipment are served by steam heated air handling units which supply warm air to various distribution points. The units contain fresh and recirculated air mixing boxes, humidifiers and refrigerant sections. Control of the systems is effected — largely automatically — through control centres which respond to temperature and other information from various parts of the system by modulating steam control valves, humidifiers and coolers as necessary.

Other plant and amenity rooms and areas are served by steam heated calorifiers which supply low pressure hot water to natural draught convector heaters, finned radiators and fan air heaters.

Calorifiers are installed to provide hot water to the various staff facilities and amenities.

Fire dampers are installed in the air ducting systems which isolate automatically any section of ducting that may be subjected to excessive heat, thus containing any fire which may occur. Smoke sensors are also located in the computer rooms which when activated, will shut down the associated HVAC system.

Areas containing equipment protected by BTM extinguishing gas fire protection are provided with extraction ductwork and fans discharging directly to atmosphere. The location of the discharge point is fixed to preclude the possibility of the fumes exhausted being drawn into the intakes of any air handling plant.

The main buildings are naturally ventilated and supply the boilers with all the air necessary for combustion. Air is drawn in through louvres to the lower regions of the building, in particular the turbine hall and bunker bay. The air then rises in the boiler house where the intake ducts to the FD fans are positioned close to the boiler house roof. Excess heat is exhausted through roof ventilators situated on top of the boiler house roof.

The area above the bunkers, containing the coal conveyors is segregated and ventilated separately from the remainder of the main buildings.

23. STATION AND UNIT AUXILIARY ELECTRICAL SYSTEMS

The six 660MW generators are connected to the National Grid via the 400kV substation. The generator voltage in each case is 23.5kV, stepped up to 400kV via a 800MVA generator transformer. A 23.5kV/11kV unit transformer takes power directly from the generator terminals to supply the unit auxiliaries via an 11kV unit board. The unit transformers for the First Half are rated at 42MVA; those for the Completion are rated at 48MVA.

Each unit needs a sufficiently large supply to enable it to startup. The unit transformer is obviously not available for this duty, so the unit auxiliaries are taken from station transformers. A 132kV substation supplies four 57MVA, 132kV/11kV station transformers, each transformer then supplying an 11kV station board.

Fig 3.34 shows the station electrical supplies system.

23.1 400kV Substation

The substation is an outdoor main and reserve busbar type, in sections, equipped with pressurised air blast circuit breakers for the First Half, and with SF6 circuit breakers for the Completion. The change from air to SF6 is a result of rapid development of switchgear using SF6, where a much more compact unit is now obtainable because it is a greatly improved insulant compared with air. The main busbar for the First Half is connected to that for the Completion units via a 1320MVA fault limiting reactor, and the main busbar and reserve busbar sections can be connected together with bus couplers. The substation has 240MVA, 400kV/132kV supergrid auto transformers (which supply the 132kV substation) and 400kV outgoing feeders connect to the National Grid.

23.2 11kV System

The system layout came about as a result of the gap of about 12 years between the design of the system for the First Half and its modification to cater for all six units.

Each station transformer feeds onto its own 11kV station board; the boards are interconnected as in Fig 3.35. The two 11kV systems are never connected together through the bus section switches on station boards 3 and 4.

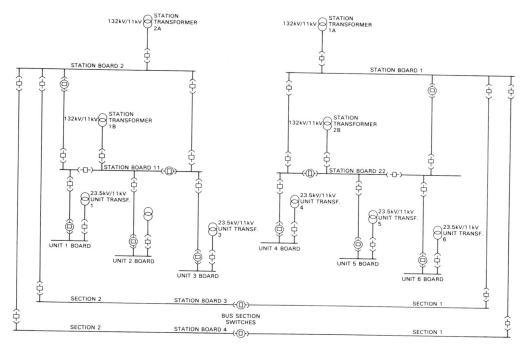

FIG. 3.35   11kV SYSTEM

Each 11kV unit board has three major power sources (and hence high fault level infeeds) connected to it they are the unit transformer, station board interconnector and gas turbine, as well as major drives and auxiliary step-down transformers.

All the 11kV switchgear is rated at 900MVA make, 750MVA break and is of the air break type. The difference in make and break ratings is due to the fault contribution of running induction motors which appears during a make, but not during a break sequence.

23.2.1   *11kV System Microprocessor*

The 11kV system would have been an extremely complex system to operate with an extensive hardwired interlocking system and at the same time it was undesirable to operate a system without interlocking as this could lead to switchgear fault levels being exceeded.

A microprocessor-based unit was designed and installed, which prevents the closure of 11kV circuit breakers, if so doing would exceed the switchgear fault level on either make or break duty. The unit is provided with an override facility which enables the operator to override the interlocking function on a switchboard or a per circuit basis, and a simulation facility which enables him to check what the effect of his action would be without actually carrying out any switching. Two visual display units are provided in the control room, one giving a coloured display of the state of the various 11kV switchboards, which can be selected by the operator, and one giving alarms associated with the unit and system. All 11kV switchgear is channelled through the microprocessor.

There are two exceptions where the microprocessor fault level calculation process is by-passed: an emergency start of the starting and standby boiler feed pumps, where time is of the essence, and gas turbine starts where the unit board is not interconnected to a station board. In this case, the gas turbine has to synchronise in a time window and time spent in the calculation stage can be saved if it is known that fault levels cannot be exceeded.

23.3 3.3kV System

The system comprises an extensive station scheme, derived from the various 11kV station boards via 11kV/3.3kV auxiliary transformers, and a unit scheme derived from each 11kV unit board via two 12.5 MVA auxiliary transformers of similar voltage ratio.

The station system comprises boards for general station services and others for coal plant services, ash and dust plant services, dust disposal services, and other special services, supplied through either 8MVA or 10MVA transformers.

The unit system carries all the equipment and drives that the unit needs for its own purposes, eg pulverised fuel mills, primary air fans, feed suction pumps.

A 3.3kV standby auxiliary board fed from a 11kV station board via a transformer, is provided for the First Half which can be interconnected to each unit board one at a time to provide a supply in the event of non-availability of either or both the unit auxiliary transformers. It also provides a maintenance supply.

23.4 415V System

The 415V system is extensive and complicated by the fact that the switchgear for the First Half and original station auxiliaries was no longer manufactured by the time the switchgear for the Completion was due to be ordered. Some of the original station boards had to be extended to cater for the extra equipment that had to be provided for the three new units; mainly on the coal plant, ash and dust plant and CW system. The new switchgear had to be fitted at the ends of the existing gear, the main area of difficulty being the busbar extensions and the fitting of the panels.

The 415V system for the First Half is derived from 3.3kV/415V transformers, whereas for the Completion, the supplies are derived from 11kV/415V transformers. The reason for the difference in obtaining the unit supplies is because operational experience on the First Half showed that voltage regulation problems existed, and the elimination of an intermediate transformer helps the regulation. In addition, automatic on-load tap changing has been fitted to the 11kV/415V unit auxiliary transformers. The voltage of the 415V unit services board is measured and used to operate the tap changer.

As well as the 415V unit boards, the Completion units also have 'unitised' station boards. This is a new concept whereby some supplies for each unit are taken from the station supplies and is intended to cater for the increase in auxiliary load and gives improved flexibility.

A 415V standby services board is provided for the Completion units and can feed each unit services board (one at a time), providing a maintenance facility and back-up.

23.5 Guaranteed Instrument Supplies System

The guaranteed instrument supplies (GIS) system is a short break system consisting of three station inverters and two unit inverters per unit. All inverters are rated at 30kVA. Harmonic distortion due to computer loads requires considerable derating of the inverters. A 415V ac supply is taken, rectified and then inverted back to a 2-phase 415V supply. A battery is provided at the dc interface which can maintain the inverter supply for 30 minutes should the ac supply fail. Should the inverter fail completely, a raw 415V ac supply can be put directly onto the distribution system. The inverter feeds onto distribution boards fitted with 415V/110V transformers and these provide supplies to control and instrumentation essential equipment which:

(1) has stringent supply requirements,
(2) is required for safe shutdown of plant,
(3) is needed for post incident monitoring and recordings,
(4) is required for dead station startup.

Generally, one unit inverter supplies the computer load and the other supplies the conventional instrumentation load. The three station inverters share the conventional station load and each carries a computer load, one the maintenance and data amendment computers, another the station computer and the third the transmission computer.

23.6 Non-Guaranteed Instrument Supplies System

The non GIS system caters for the remainder of the 110V ac supplies. It consists of a main system and a standby system which supply equipment that can tolerate longer breaks in supply and which is less crucial to the unit.

23.7 DC Systems

Battery maintained systems at 50V, 110V and 240V provide secure supplies to the following:

Essential instruments.
Controls.
Switching closing and tripping functions.
Telecommunications.
Protection.
Interlocks.
Alarms.
Essential standby plant.
Emergency lighting.
Emergency oil pumps.
Gas turbine auxiliaries.

At 240V, each switchboard has two float chargers rated at 300A each and one boost charger. At 110V, there is one float and one boost charger per switchboard. Centre point earthing is used throughout the systems, replacing the earlier biased earthing.

The 50V system on the Completion is provided by 400Ah batteries, one per unit and two for the station supplies. In ancillary buildings, 50V is provided by self-contained battery charger units except in the CW and compressor houses where a transformer rectifier is fed from a 110V ac supply.

23.8 Metering

Generator and transformer outputs are measured by high accuracy meters. Main and check meters are used for the 660MW generators. Transformers have import and export meters. Each meter has a readout and an impulse output to summation equipment. The summation equipment processes these outputs to produce total and sub total power signals for display at the station control room and grid control centre. Summated signals are also transmitted to the unit computers for station thermal efficiency performance calculations. A teleprinter, in conjunction with the meters, prints and punches out half hourly power demands.

23.9 Main Connections

The main generator output is connected to the 23.5kV/400kV 800MVA generator transformer by phase isolated, air insulated, aluminium busbars which are housed in circular aluminium enclosures. Dry air from dehumidifying equipment is fed into the enclosures and busbar trunking to maintain a pressure of 1.25 bar and so prevent the ingress of dust and moisture. Similar arrangements exist for the connections to the unit transformers which are teed-off from the generator main connections.

The delta connections at the generator transformer utilised oil-filled boxes on the First Half units, but these were superseded by air insulated connections on the Completion units.

The main conductors, which are supported within the busbar trunking by insulators, are rated at 20.1kA for the generator output and 1180A for the unit transformer connections. The insulation level is rated at 33kV.

23.10 Station Earthing System

The station earthing system comprises sheet steel piles 16m long which are driven into the ground at selected points over the whole station area. The piles are linked by means of integral connection pits and interconnecting cables. Connections to the earthing piles which were made using the fusion jointing process on the First Half station were replaced by bolted joints at the Completion station construction.

The station electrical supply voltages at 3.3kV and above have earth connections through liquid resistance vessels which limit earth current flow during fault conditions. The 415V system earth is solidly connected to the station earthing system.

An earthing transformer is used to earth each 660MW generator neutral point to the station earthing network.

Instrumentation supply voltages have a separate earthing network which is itself connected to the station earthing system at one location.

Lightning protection is effected on all structures higher than 15m (except the cooling towers) by aluminium air termination strips which are individually bolted to the station earthing network.

23.11 Communication Systems

A UHF radio communication system using hand held transmitter/receivers provides five channels which serve the following applications:

(1) Supervision of maintenance work from the works control desk.
(2) Monitoring of security and common services from the supervisor's desk in the control room and from the security centre.
(3) Coal plant operators to coal plant control desk.
(4) First Half boiler/turbine house operational areas to unit control desks.
(5) Completion boiler/turbine house operational areas to unit control desks.

Telephone communications within the station are served by a stored program control (SPC) private automatic exchange (PAX) with capacity of up to 1000 lines. This is a computer-based system which uses two central processing units to ensure a high level of availability.

The exchange has its own battery charger, air conditioning and Halon gas fire protection system. The batteries provide a minimum of 6 hours normal operation of the exchange system in the event of the loss of ac supplies to the charger.

A teletypewriter terminal in the exchange building provides user access to the exchange system software for entry of system parameters, system status interrogation and system maintenance.

A direct-wire telephone system for strategic station operational purposes terminates at each unit control desk in a 24-way key panel and at a 40-way key panel at the supervisor's position.

High integrity of the system is achieved by the use of short term fire-proof cable. A 2000Ah battery under constant trickle charge provides a secure power supply to the system.

23.12 Cabling

During the period between the construction of the First Half station and the Completion, the requirements of cabling standards became more onerous. In particular, the reduction of fire risks, improvement in supply security and additional safety for personnel led to improved materials and location arrangements.

The development of PVC insulation materials resulted in the reduced spread of fire and the reduction of toxic fumes. Essential tripping and communication circuits use cables which have a short-time fireproof rating. Cable tunnels are equipped with fire barriers at regular intervals along their length and all main cable routes are fitted with a heat detecting zone system which initiates remote alarms and, in some locations, operates fast-response fire protection equipment.

Control and instrumentation circuits use multicore and twisted pair cables. Multicore cables have up to 37 cores and twisted pair cables are in the range 2 pairs to 100 pairs. Control cabling conforms to balanced pair working rather than the 'paired matrix' system used at other stations.

Paired matrix working is where twisted pair cables are routed to large termination racks, terminated by colour code, and the various circuits then connected up by putting jumper wires in the racks. This system allows for cables to be run, glanded and terminated without concern for how the various circuits are to be connected up later. It also makes future modifications easy to incorporate. The system lends itself to large trunk cables but these can be difficult to install.

Balanced pair working is where the cores are run as twisted pair cables to marshalling or junction boxes and connected up directly to form the circuits required. This means that the circuit connections must be finalised at an early stage in order to route cables directly. Smaller cables are involved, terminations are fewer and commissioning is easier. This system was adopted for the Completion units as it had been on the First Half units.

24. CONTROL AND INSTRUMENTATION

The basis of the design of the control and instrumentation (C & I) is such that one unit operator retains overall responsibility for operating the plant via operator-supervised automatic controls rather than the completely automatic system.

Over 700 functions have to be carried out from the control room during a hot startup of a large unit but the number of operator actions are halved by the use of sequences, interlocks and other automatic and semi automatic systems. A balance has been made between manual and automatic control systems.

A full analysis of the startup, shutdown, emergency and normal operational routines was made and this enabled the unit control desk and panel to be designed so that it is capable of being operated by one unit operator with an assistant during periods of peak activity.

The unit operator is assisted by the provision of systems as follows:

(1) Remote manual control with direct acting circuits controlled by the unit operator.

(2) Additional remote control for items normally controlled automatically, if the modulating control system were to fail.

(3) Sequence control with sequence control equipment initiated by the unit operator, where it is required to permit hot startup to reduce the operator work load during startup (or emergency conditions).

(4) Modulating control systems which have a range of control which extends from full load to a very low load (50MW). Additionally these control schemes should be designed to be reconfigurable for future operational strategies.

(5) Automatic turbine and boiler loading systems.

(6) Independent automatic startup for plant items which operate in an emergency, eg starting/standby feed pumps.

(7) Independent protection system to ensure integrity of plant under manual and automatic modes of operation.

(8) Data processing systems to give improved means of display and recording.

All of the above are controlled from a unit control desk and panel, one for each unit on the station.

Each unit control desk and associated panel has controls and indications for startup, normal operation, shutdown and emergency operation.

One operator is able to control a unit hot start, from burner ignition to full-load operation. There are sufficient indications and manual controls to run the plant in a steady state, safely deal with emergency and abnormal conditions, and shutdown under normal and emergency conditions.

The central control room staffing is assumed to be one supervisor with part-time assistant, two assistant operators per station and one unit operator per unit, in addition to roving operators on the plant.

When a unit is started up after an overnight or weekend shutdown, the startup takes place entirely from the central control room. If any maintenance or other work has been carried out during the shutdown period, then the unit operator must ensure that the plant has been returned to its normal operational condition.

For a cold start, the control provided in the central control room may be supplemented by plant local operations.

The C&I systems described here are representative of those provided for the Completion units. The major systems only are described and are covered under the following headings:
(1) Grid system control requirements.
(2) Plant control and regulation.
(3) Unit load control.
(4) Central control room.
(5) Distributed computer control.
(6) Boiler control (DDC centres).
(7) Data processing software.
(8) Auxiliary plant startup and shutdown.
(9) Monitoring of solid and gaseous chimney emission.
(10) Plant mounted modulating control loops.
(11) General measurements

24.1 Grid System Control Requirements

The required plant flexibility and its effect on the design of the control and instrumentation systems arises from consideration of the role the units are expected to play in the grid system.

Section 4.4 sets out the typical cycles of operation during the main plant lifetime that have been taken into account.

The key requirements are as follows:
(1) The need to meet the target MW loading punctually following shutdowns, and the need for the control system, manual systems and staffing levels to minimise the effects of equipment failures.
(2) A single fault under normal operating conditions should not cause the output of more than one main generating unit to be lost.
(3) Each unit is required for 2-shift operation on a regular basis.
(4) The capability of unloading from full load in not more than 20 minutes; and after a shutdown of up to 8 hours, full load will be achieved in 30 minutes after synchronising.
(5) Each unit will be capable of achieving full-load within 160 minutes after light-up.

(6) There should be an absolutely minimal risk of violating metallurgical and other design constraint boundaries under flexible operating conditions.

(7) Gas side corrosion will be minimised by close control of combustion conditions and plant temperature; waterside corrosion will be minimised by monitoring of water chemistry.

(8) The performance criteria of the control system must continue to be met, as far as possible, under any anticipated plant configuration including the outage of plant auxiliaries.

(9) The highest practicable operating efficiency of the plant compatible with the other requirements must be maintained by the control systems.

(10) The plant and its control systems will operate in a safe manner at all times so as to comply with CEGB practices (PF code of practice, safety rules, etc) and Government legislation.

(11) The statutory requirements will be met relating to solid and gaseous emissions as agreed for the Drax site.

The overall system control requirements also have the effect of specifying the unit load control and plant immediate reserve requirements. For example in the case of the immediate reserve, the plant (and its supporting control system) must be capable of controlling frequency within the limits 49.8 to 50.2Hz when a 15% CMR step increase is applied (in the load range 50% to 70% CMR) with limits of transient frequency drop of not more than 1Hz and to restore permanently it to within 0.3Hz of its initial value in 20 seconds thereafter.

The foregoing are typical of the grid system requirements and they have a direct effect on the design of the overall C&I scheme for a coal-fired unit.

24.2 Plant Control and Regulation

In the case of coal-fired units the plant response to grid system needs is dependent upon the quality of auxiliary plant regulation. The method of primary air speed control is a key factor and the fluid coupling response is critical to overall plant response. The means of plant regulation are listed for completeness in Table 3.4. It may be appreciated that the response characteristics of each plant item controlled is an essential factor in meeting the loading requirements. Alternative overall unit load control regimes are provided to meet grid system requirements, namely boiler-follows-turbine or turbine-follows-boiler; these are described in Section 24.6, Boiler Control.

24.3 Unit Load Control

The grid control requirements demand that boiler-following-turbine and turbine-following-boiler control schemes be made available.

The unit load control system which has been provided regulates both the set points of the turbine narrow range governor and the desired values of the boiler total fuel controller in accordance with grid demands and plant limitations. The system employs admission steam pressure governing and master pressure control in an integrated control system that minimises operator intervention. All automatic modes are selectable by the operator; the alternative to automatic control is manual regulation of

TABLE 3.4 PLANT CONTROL AND REGULATION

Item Controlled	Means of Regulation	Method of Actuation	Control Loop	
Pulverised fuel output from mill	Coal feeder speed	Variable speed motor with thyristor switching		Feeder speed loop
	Primary air fan speed	Hydraulic actuator		PA flow loop
Primary air flow through mill	Primary air fan speed	Fluid coupling	Mill loops	PA flow loop
Secondary air flow per mill	Secondary air damper position	Hydraulic actuator		Secondary air loop
Primary air temperature from mill	Linked hot and cold air damper position	Hydraulic actuator		Tempering air loop
Forced draught air flow	FD fan vane position	Hydraulic dampers		Duct pressure loop and combustion air loop
Furnace pressure	ID fan vane position ID fan discharge damper blade position	Hydraulic actuator Hydraulic actuator		Furnace pressure loop
Reheater outlet steam temperature	Superheater & reheater gas bypass damper position (linked)	Hydraulic actuator		Reheater outlet temp loop
Reheater steam flow balancing	Butterfly valves in reheater steam circuit	Hydraulic actuator		Reheater balancing loop
Superheater steam temperature	Spray valves per attemperator	Pneumatic actuator		Superheater steam temp loop
Drum water level	Feed pump speed	Main feed pump governor Standby feed pump liquid resistors		Feed control loop
	Feed regulating valve position 2 × 30% CMR 3 × 50% 30–100% CMR	Hydraulic actuator		Feed control loop
Generator power output	Governor valve position Fuel supply to boiler	Main turbine governor		Admission pressure control loop Total fuel loop
Boiler steam pressure	Governor valve position fuel supply to boiler	Main turbine governor		Admission pressure control loop Total fuel loop
Generator terminal voltage	Exciter field resistance	Motorised rheostats		Dual AVR loop

the set-point of the narrow range governor and/or the desired value of the 'total fuel controller'. The load/frequency characteristic (droop) can be varied at the unit desk to suit the prevailing system operational requirements.

The control system software is installed in control centres in the boiler control system, but will accept demands from the data processing system for regulation of both the turbine narrow range and wide range governor set points as required for turbine run-up and loading.

Provision is also made within the overall unit load control software for automatic loading of the boiler/turbine unit. This is a feature of the 2-shift operation control concept aimed at achieving hot startup in the specified time with only one unit operator. It has the advantage that smoother and more consistent loading of the turbine is achieved with the intention of reducing the cumulative effects of thermal stressing.

Two alternative modes of automatic loading are provided, namely MW ramp loading and pressure startup, these being affected by direct digital control (DDC) programs within the boiler control system, specifically on DDC centre 8. See Table 3.5.

TABLE 3.5 ALLOCATION OF CONTROL LOOPS TO CENTRES

Centre	Function	Number of Control Loops involved
DDC centre 1	Mills A & B control	8
	Feed water control (standby)	4
DDC centre 2	Mills C & D control	8
	Superheat control	8
	Reheat control	4
DDC centre 3	Mills E & F control	8
DDC centre 4	Mills G & H control	8
	Combustion control (standby)	3
DDC centre 5	Mills J & K control	8
	Unit load control (standby)	2
DDC centre 6	Combustion control	3
	Superheater control (standby)	8
	Reheater control (standby)	8
DDC centre 7	Feed control (main) and auto run up of the feed pump turbine	4
DDC centre 8	Unit load control and main turbine run up and loading ramps	2
	Total	86

24.4 Central Control Room

The unit plant is operated from a central control room, having a unit control panel and desk from which it is possible to undertake hot start, warm start, steady load and shutdown procedures, and to supervise other activities such as cold start from burner ignition to full load operation. There is sufficient conventional indication and control for normal operation of the plant and to deal with emergency conditions. Computer-based visual display units are provided to aid the operators but they are not essential to the operation. The room layout together with the equipment rooms and cable flat are shown on Fig 3.36, and Fig 3.37 shows a general view of the control room.

The unit desk is designed on a modular basis using a 72mm grid module and generally with a 'dark panel' presentation. All the controls necessary during a hot or warm startup and for shutdown, are accommodated on the unit control desk. Also accommodated are the controls requiring frequent operation during the steady operation of the unit. Four visual display units (VDU) are provided with a keyboard to enable selection of up to 80 formats.

Supporting instrumentation and controls are housed on the unit panel. This includes recorders and total C&I for the 10 pulverised fuel mills. By the use of a multiplex system the operator is able to select the throughput of any two mills to be controlled from the unit desk.

A control room supervisor's desk is located centrally, on which the main grid telemetry facilities are accommodated together with the chimney monitoring TV and VDUs to allow the supervisor to oversee any problem arising on each unit.

Facilities are provided in the central control room to accommodate desks and panels which house controls and indications which are common to all units. These are:

General services desk (fuel oil, sootblower compressors, towns water, fire pumps, fire alarms, etc).
132kV mimic panel.
400kV mimic.
Electrical auxiliary system desk (11kV, 3.3kV and 415V).
CW desk.

24.5 Distributed Computer Control

The considerable amount of plant data which has to be monitored to achieve efficient control of coal-fired power plant requires a large data processing system so that the information can be structured and presented to the unit operator in a manageable form.

There are approximately 4900 analogue and 10,000 digital signals monitored by a distributed data processing and control system (DPS)for the three units.

The design adopted a distributed computing system as indicated by Fig 3.38 in preference to a single large computer. The advantage of distributed computing with data links between many computing centres, compared with the First Half practice of one large computer, is to lessen the consequences of a single computer failure and enable the plant to continue running. A failure of a single large computer system could have meant shutting down the unit. Distributed computing not only enables the control and display facilities to be remotely located and achieve diversity, but, in the

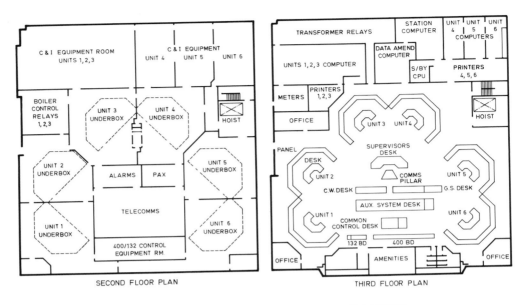

SECOND FLOOR PLAN

THIRD FLOOR PLAN

FIG. 3.36   CENTRAL CONTROL ROOM LAYOUT

event of a single computer failure, it reduces the number of automatic control loops that need to be manually controlled, to a level that can be managed by the operator and permit continued operation.

The major effect of the distributed network has been to allow stage-by-stage commissioning and avoid the problems involved in the large single or dual computer systems of earlier stations. Furthermore if each processor is arranged so that it has spare capacity, this can be utilised when failures occur to carry out the duty of the failed computer. This is particularly important where computers are used for plant automatic control applications. However, distributed computing also necessitates the use of reliable communication links between the various computers. A total of 51 point-to-point links are involved with some additional multi-drop links. The multi-drop connection method is where several centres share a single R422 high integrity high speed data link (HDLC) line which is linked between the same input parts on each centre.

24.5.1 *The Computer Network*

The data processing system and the boiler control system required HDLCs between the main unit computers, data gathering computers (data centres) and automatic control computers (DDC centres). The network is shown in Fig 3.38.

The purpose of the distributed computer system is as follows:

(1) Collect operating data from the boilers, turbines and common services relating to the 660MW power generating units for presentation in the form of displays and printouts in the central control room. A considerable amount of this data, eg boiler and turbine pressures, is derived from electronic transmitters.

(2) Monitor 400kV transmission substation states and event signals from the grid substation where the station output is transferred to the national grid.

FIG. 3.37   GENERAL VIEW OF CENTRAL CONTROL ROOM

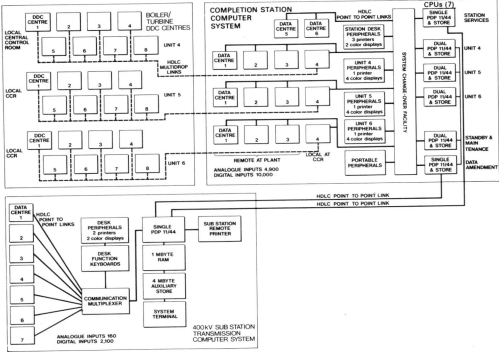

FIG. 3.38   DATA PROCESSING SYSTEM SCHEMATIC

(3) Provide automatic boiler and turbine control schemes for all the boiler and turbine control loops.

As can be seen on Fig 3.38 the systems are required to communicate with each other; the distributed network involves seven central processing units (six station plus one 400kV substation) communicating with 21 microprocessor based data centres (14 station plus seven 400kV substation) and various peripheral printer and display facilities. The network also includes HDLC connections to 24 DDC microprocessors (eight per boiler) which carry out the boiler/turbine control functions.

24.5.2 *General Description of Computing System*

The seven central processing units (CPU) are located in the central centrol room annexe with three of the data centres and also the DDC centres, whereas the remaining data centres are located on the plant close to analogue and digital signal marshalling points. The computing system comprises CPUs and data centres which are described as follows.

Unit Central Processing Units

Each of the three boiler/turbine generating units is supervised by a separate CPU consisting of two PDP 11/44 mini computers operating in a dual processor configuration to provide the necessary computing power and backing store. See Fig 3.39.

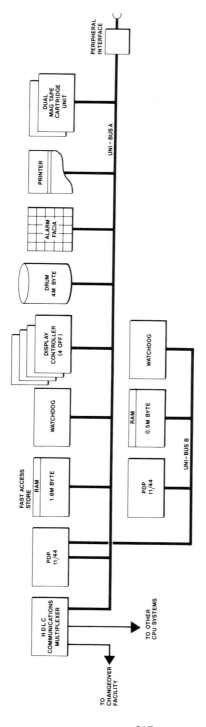

FIG. 3.39   CENTRAL PROCESSING UNIT SCHEMATIC

A station (or common) CPU incorporates a single PDP 11/44 with backing store for monitoring alarms and signals from plant which is common to all three power generating units and the 400kV transmission system. This communicates with two data centres, one in the central control room annexe and the other located near to the signal marshalling points.

A fifth PDP 11/44 CPU, operating in a dual processor configuration, is provided to act as standby processor to the station and unit CPUs in the event of failure and when maintenance is required. The standby CPU has its own data centre and this is designed so that special programmes enable testing of nearly all printed circuit boards used within the network. When failure of a CPU occurs, software enables the standby CPU to determine if it is replacing a station CPU (single PDP 11/44) or a unit CPU (dual PDP 11/44). Changeover is manually selected by switching of relays in the system changeover facility.

A sixth CPU, which incorporates a single PDP 11/44 with drum backing store, is provided for data amendments (host) purposes. It does not have data centres or input/outputs as for the other CPUs because it is used for programme development and storage, modification of applications programmes, amendment of constants and alarm limits. It also 'downline' loads amended data into the other four operational CPUs, the standby CPU and transmission CPU. This CPU, commonly referred to as a 'host' computer, loads compiled programmes down HDLC links to the target computers which actually carry out the tasks directed by the program. The use of the host computer in this particular role is to ensure security when programmes are modified and to permit off-line development without disturbing operational programmes until the modification is proven.

A seventh CPU is provided comprising a single PDP 11/44 with drum backing store. This is required to monitor the status and event signals (largely switchgear status) in the 400kV transmission sub-station. It utilises seven remote data centres and drives VDU and printers. See Fig 3.38.

Other hardware essential to the network is provided as follows:

Rotating drum store	4M byte of memory (32M bits of information).
System terminal	for operator access to the system to facilitate running of diagnostic programmes
Magnetic tape	low cost method of loading software and diagnostics programs.
Battery back-up	20 minutes at least in the event of main power failure.
Colour displays and controllers	a total of 14 colour VDUs are provided.
Communications multiplexer	using HDLC communications
System changeover facility	with selector switch control of CPU changeover to standby and a patch panel for emergencies.
Watchdog	monitoring program for internal faults.

Data Centres

Data centres collect data from the analogue and digital input marshalling points around the plant. This data after processing is fed to the unit CPU via HDLC links at a rate of 125k baud (1 baud is 1 bit per second).

When the unit PDP 11/44 CPU initiates a scanning sequence, the built-in intelligence in the PDP 11/23 data centre carries out a series of monitoring tasks, periodically reporting back to the main CPU. This periodic reporting is a feature of the Drax Completion system, and enables the main computer to do other work without constant interruption by the data centres; it is a system method often adopted in distributed systems.

The data centre typically carries out the following tasks before sending data to the unit CPU:

> Establishes the validity of the data collected.
> Checks thermocouples for open circuit.
> Linearises the data.
> Converts measured data to engineering units.

Digital inputs are monitored by status modules so that the unit CPU is interrupted only when a digital input change-of-state occurs. These particular changes are 'time-tagged' for analysis later when plant faults are investigated.

24.5.3 *HDLC Links between Unit CPU, Data Centres and DDC Centres*

An essential feature of the system is the use of HDLC links between the centres of intelligence which form the complete system. Great emphasis has been put on the need for reliable data links because of the considerable amount of data to be transmitted over long distances between the central control room and the data centres located near the analogue and digital marshalling points out on the plant.

A particular feature of the data centre is the use of HDLC link modules. Since so much data is to be transmitted it was decided to employ a point-to-point communication network between each data centre and its main computer, (except for the DDC centres which are multi-drop connected to data centres; see Fig 3.38). This would allow each data centre exclusive use of its HDLC link to achieve optimum speed of transmission.

The communications system employed avoids utilising more software and processor capacity to check transmitted data, by virtue of on-board microprocessor intelligence.

A twisted pair cable link with high immunity to electrical interference (RS422 standard) was employed. If interference causes the voltage signal on the HDLC line to go out of balance, the data centre on the unit CPU will recognise the fact since sections of the message will be obliterated or degraded. The communications module has been developed with transformer coupling to improve integrity to power surges or short circuits which might occur if the line is damaged, as well as to protect it from interference on the plant input side of the communications module.

A cyclic redundancy check is provided on the communications module, this being an additional error detection feature which requests retransmission if errors are detected in the number of bits counted. A 'handshake' mechanism is also a feature of the communication module enabling the module to pass and accept signals between stations.

The foregoing features are essential to the reliable operation of the HDLC links and enable secure operation and data transfers up to 125k baud for the data centres and 38.4k baud for DDC centres at distances up to 500 metres over RS422 standard links.

Each of the 24 DDC centres carrying out the boiler and turbine control functions employs identical HDLC communications with similar modules.

24.5.4 *Location of Data Processing Equipment*

The major computer suites which comprise the overall scheme are located in specially designed individual unit computer rooms (to limit fire damage) which have controlled temperature and humidity conditions.

The majority of the data acquisition is by remote scanners located in switchrooms. These scanners are microprocessor controlled and setting the analogue and digital plant signals are marshalled in cubicles alongside the scanners.

Provision is made for a centralised record room in the central control room which houses all the printers for log sheets and alarm records.

24.6 Boiler Control (DDC Centres)

Current practice hitherto has been to make major use of 'analogue' controllers for fossil-fired boilers. This practice has not been adopted on the Completion units in view of the general line of development towards computer control. The number of control loops involved on one boiler can be appreciated from Table 3.5. The level of diversity and back-up provided by the allocation of loops to the DDC microprocessors (control centres) is typical of that shown in Table 3.5.

Fig 3.40 shows one of the eight microprocessors per generating unit (24 in all for three 660MW units). The inputs represent plant conditions to the controller and the outputs initiate changes to plant parameters. The processors are connected to the main CPU via the multidrop link described earlier. A 'stand-alone' design has as far as possible been adopted to protect the operator of the plant from the effects of computer failures and to lessen the number of occasions when control loop faults would impose on the operator an unacceptable extent of manual control.

The purpose of the HDLC links to the DDC centres is to enable plant alarm data which is detected within the control programs to be transmitted to the main CPU for display and recording. The operator will also be able to select control parameters that are held in the control centre for display.

The standard hardware interface to all plant actuators is by a stepper motor, whether the actuator is hydraulic or pneumatic.

There is inherent diversity between mills, back-up being unnecessary because the number of mills required for boiler CMR could be only eight out of ten mills depending on the coal supplies. There is standby loop software in centres 1, 4, 5 & 6 for feed control, combustion air control, unit load control and superheater and reheat steam temperature control respectively. Various options were considered in terms of numbers of centres, as few as five and as many as 20. More centres mean more maintenance and less mean lower auto availability. Eight centres are considered optimum on this project.

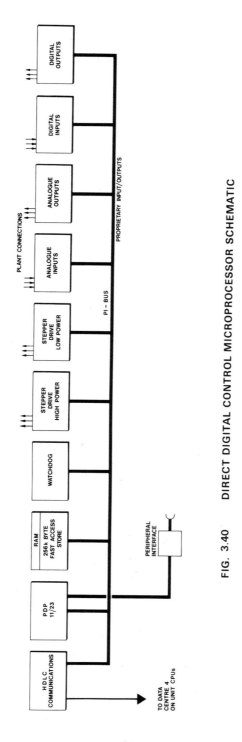

FIG. 3.40   DIRECT DIGITAL CONTROL MICROPROCESSOR SCHEMATIC

The DDC centres include data links to the main data processing system for supervisory control of the main and feed pump turbine run up. It is also possible to display certain formats particular to the DDC on VDUs at the unit desk. Links between centres are by DDC analogue outputs, hence reducing reliance upon communication links for the boiler control. Power supplies to equipment are duplicated but from a non-guaranteed source. Stringent testing has been carried out to ensure reliability and availability.

24.7 Data Processing Software

A decision was made to adopt the CEGB high level language CUTLASS (computer users technical language and software) for as much of the data processing and control software as possible. It uses engineer-orientated language subsets for applications programs. Similar generating plant can use similar CUTLASS programs even on different types of computer.

The major aims of CUTLASS are as follows:

(1) To meet an agreed set of user requirements.
(2) To achieve machine independence by compiling into a high level language Coral 66, which is available on a range of machines over a range of machine architectures.
(3) To use engineering staff who are familiar with the plant requirements rather than specialist computer programmers.
(4) To provide a robust system where software failure is minimised.
(5) To achieve configurability, ie to use only those facilities required for any given system.

CUTLASS has several subsets most of which have been used at Drax. These are as follows:

(a) Display subset — VDU picture generation and graphs.
(b) DDC subset — automatic control (mainly boiler).
(c) Sequence subset — plant startup and shutdown.
(d) General purpose subset — general data processing needs.

In addition some proprietary software is used for diagnostics, etc.

24.7.1 General Description of Software Facilities

The major features of the software facilities are as follows:

(1) Alarm display (and back-up).
(2) Alarm and event recording.
(3) Monitoring.
(4) Data display.
(5) Post incident recording.
(6) Efficiency data collection.
(7) Routine data logging.
(8) Main and feed pump turbine run up - supervisory control.
(9) Performance of calculation necessary for main turbine and feed pump turbine run-up and turbine loading.
(10) Boiler header life factor measurement, calculation, display and logging.

24.7.2 Alarm Display

All alarm messages are normally displayed on VDUs in the order of detection. The appearance of the alarm message is accompanied by an identifier, as part of the message. When accepted by the operator, the identifier stops flashing. Indication is given when an individual alarm condition clears and can be reset. Urgent alarms are displayed in a clearly identifiable fashion.

An audible alarm is also provided for each unit and sounds when initiated, until it is accepted.

A limited amount of alarm suppression is provided to suppress standing alarms from plant that is shutdown or on standby duties.

One VDU per unit desk is permanently allocated to alarm display. It should be noted that from the 1200 alarms in the data processing system (DPS), facilities are made available for back-up display on the unit desk for up to 400 alarms. These are grouped to ensure full unit cover in the event of catastrophic failure of the computers (ie prolonged outage of a unit computer).

24.7.3 Alarm and Event Recording

All alarms are recorded in order of detection and printed out with plain English messages and the time of detection.

24.7.4 Monitoring

Measured information from the plant is scanned and compared with pre-determined alarm limits. These limits are set by the computer program and may be high or low, or a combination of high and low.

24.7.5 Data Display

Plant measurements, calculated values and control program information are available, for display on data formats; capacity for 100 different formats per unit is provided. Three VDUs of the total of four per unit are available for display of formats, with one permanently held to display alarm information.

24.7.6 Post Incident Recording

The computer continuously monitors and stores a number of plant measurements and contact states so that, in the event of any one of several previously defined plant incidents, the preceding 25 minutes of information will be automatically printed together with the subsequent 5 minutes of information.

Provision is made for a subsequent incident, so that no information is lost during the printing out of information arising from the first incident.

24.7.7 Efficiency Data Collection

A separate log is made, containing all main unit plant data required for completion of monthly operating data returns. This log accumulates hourly, daily, weekly and monthly totals of the required parameters.

24.7.8 *Routine Data Logging*

Selected measurements are printed at regular intervals. The selection of measurements is made for specific purposes and provision is made for the selection to be varied, as required, by changing circumstances of plant operation.

In addition, the operator is able to compose a log of selected measurements for print-out on request.

24.7.9 *Supervisory Control Program for Main Turbine and Boiler Feed Pump Turbine*

The unit DPS, as required by the unit operator, will run supervisory control programs that will effect run-up of the main turbine and the feed pump turbine. These programs utilise a substantial number of plant inputs for turbine supervisory measurements and are therefore resident in the main DPS. The unit DPS uses the data links shown on Fig 3.38 to control speed ramp programs which are held in DDC centres 7 & 8. Hold, advise trip or raise/lower speed signals are sent to DDC centre 7 or 8 to control the ramp programs in accordance with the supervisory control program. Using the DDC centre and the unit DPS in this manner avoids the need to duplicate the supervisory inputs into the DDC centre (as well as the DPS) and is economic compared with the possible degradation of automatic run-up and loading availability.

Pre-start checks are carried out and optimum rates calculated from the plant thermal conditions. The operator selects the run-up rate and initiates the run-up, following which the run-up proceeds automatically, with the steam supply to the set being controlled to achieve the required rate, until control is taken over by the turbine speed governor.

During the run-up the computer monitors the plant and automatically modifies the run-up program when necessary, to ensure that plant stresses are kept within iimits.

Provision is made on the unit control desk for back-up manual control of run-up, when the automatic equipment is not available.

Separate equipment is provided for automatically synchronising and connecting the turbine generator to the external system. This equipment is entirely independent of the unit computer, and its operation is manually initiated by the operator after the completion of run-up.

The steam-driven feed pump is automatically run-up to speed in a similar way to the turbine generator. The run-up terminates with the speed control set point wound up to become an overspeed protection level and the boiler feedwater control programme regulating the feed pump turbine speed over its normal range, as required by feed demand of the boiler under 3-element control.

24.8 Auxiliary Plant Startup and Shutdown

On the Completion, the decision was made to use microprocessor control for startup and shutdown of those auxiliaries which are regularly used for unit hot startup and shutdown. Discrepancy switches are provided on the unit desks and panels in the central control room to be used by the operator as necessary during plant startup. The software used is the CUTLASS sequence sub-set.

The allocation of these auxiliary plant sequences to processors is as shown in Table 3.6. The diversity achieved is due largely to the number of mill group sequences required and also by the need to ensure that a single processor fault does not prevent the unit achieving 60% CMR during startup. Loss of a centre does not directly affect plant availability.

The use of software-based sequence control schemes represents the current best practice on a fossil-fired unit.

The procedure for preparation of the software involves establishing the plant startup/shutdown and protection operations. These operations are converted into a software preparation diagram called an algorithmic state machine diagram from which the coding of the tasks and schemes is carried out using the target microprocessor. Following testing of the compiled software using a test box in conjunction with a host processor, the software is taken to site and commissioned using 'dry runs' on the plant with the target processor, in the conventional manner as would have been the case with hard-wired equipment.

24.9   Monitoring of Solid and Gaseous Chimney Emissions

Sufficient instrumentation is provided to enable the station staff to meet the best achievable emission levels as agreed with HM Alkali Inspector.

The following are provided as measurements on each unit panel/desk:
Smoke density.
Coarse grit monitor.
Fine dust monitor.
Carbon monoxide at induced draught fan inlet.
Oxygen at boiler outlet.

Additionally at the supervisor's desk a closed circuit television monitors the chimney stack plume.

24.10   Plant Mounted Modulating Control Loops

Some mechanical sub-systems require close control of temperature and pressures. Conventional practice is to install, on the turbine and its auxiliary system, a number of modulating control loops to regulate various temperatures and pressures critical to the operation of the auxiliary plant concerned. These have been, on earlier stations and the First Half, mainly pneumatic controllers employing 'flapper and nozzle' controllers. On the Completion units, electronic controllers are employed, locally mounted in suitable enclosures. A list of the controllers is given in Table 3.7.

24.11   General Measurements

24.11.1   *Signal Transmission*

The standard adopted for transmitting signals to the central control room is 4mA to 20mA dc. This has the advantage of being able to share a transmitter for several functions. It also provides an inherent loop failure detection feature.

Temperatures are nearly always measured using thermocouples and these are led direct to the DPS or to the transmitters for presentation by indicators or recorders.

TABLE 3.6 SEQUENCE CENTRES

Centre	Sequence	Location
S1	Mill group A Mill group H ID fan A FD fan A	
S2	Mill group B Mill group G ID fan B FD fan B	Boiler house switch rooms
S3	Mill group C Mill group K Air heater A	G/H row
S4	Mill group D Mill group J Air heater B	
S5	Mill group E Mill group F	
S6	Feed suction pump A Start/standby feed pump A	Turbine house
S7	Feed suction pump B Start/standby feed pump B	annexe switch rooms
S8	Feed suction pump C Gland seal/vacuum raising	B/C row
S9	Dust hopper emptying	Coal plant control block annexe
S10	Auto chemical monitoring	Chemical monitoring room operating floor
S11 S12	CW pump 4 & 5 CW pump 6 & 8	CW pump house switch room

TABLE 3.7 LOCALLY MOUNTED CONTROLLERS

Title	Location
Fuel oil pressure	Fuel oil pumphouse
Turbine power fluid temp	Turbine HP area
Turbine lubricating oil temp	Turbine LP area
Hydrogen temp	Hydrogen cubicle (turbine)
Stator coolant temp	Hydrogen cubicle (turbine)
Turbine generator seal oil temp	Hydrogen cubicle (turbine)
Gland steam pressure	Local to regulating valve
Gland steam temp	Local to regulating valve

24.11.2 *Boiler Acoustic Leak Detection*

A system of microphones and amplifiers is employed to assist with the location of steam leaks on the boiler tubes. These are in 'dead gas spaces' and the information is used to speed up identification and thereby reduce outage time.

24.11.3 *Furnace Implosion and Loss-of-Flame Detection*

Under certain draught conditions in the furnace and boiler ducting it has become apparent that a furnace implosion can occur. Protection to detect the onset of these conditions and take remedial action is provided on the Completion boilers.

Additionally, loss-of-flame protection is provided by the 'mini puff' scheme which uses measurement of furnace pressure fluctuations to detect loss of furnace flame.

24.11.4 *Milling Plant Protection*

Current safety rules have required certain protection schemes for milling plant particularly for fire detection. Mill carbon monoxide content is sampled by a probe inserted at classifier level and the analyser signal is put onto a recorder to monitor the trend. Cycle time is about 5 seconds. Controls are provided on the unit desk to enable auto/manual recycling and putting any of the 10 mill signals to a recorder.

Mill infra-red fire detection is provided by monitoring the infra-red spectrum of the mill inlet belt (at three points 120° apart) and in the inlet primary air duct. The sensor outputs are analysed, amplified and compared to reference values to give an alarm in the central control room when the set reference is exceeded.

24.11.5 *Air Heater Fire Detection*

The rotary airheaters are provided with thermocouples to monitor the temperature at the exit of the gas passes. These are scanned once every 20 seconds. The output signal from the scanner is passed to an alarm comparator and is also displayed on a recorder to monitor the trend.

24.11.6 *Windbox Fire Detection*

Thermocouples are provided at the bottom of each windbox to detect fire that may be caused by oil leaks from the lighting up oil burners. The thermocouple outputs are compared with reference values in trip amplifiers which initiate alarms if fire is detected.

24.11.7 *Drum Level Measurement*

Drum level is monitored by manometric type transmitters for control purposes, but for protection purposes the Hydrastep method utilising four vessels for greater security is employed.

25. POST COMMISSIONING EXPERIENCE

The three generating units on the First Half of the station had been synchronised by 1974. After initial problems, mainly associated with machine vibration, the units settled down to a steady pattern of operation. Over a period of years the station electrical output has been consistently high.

Average overall thermal efficiency has approached 37%. This has contributed to a low cost of generation which has placed the station high in the merit order of stations selected for operation.

The first unit in the Completion (unit 4) was initially synchronised to the grid in December 1983 and unit 5 at the end of November 1984.

Over a typical three year period, the following availability figures were attained.

Year	Winter Peak Availability	Average Annual Availability
1981–82	91%	62%*
1982–83	91%	81%
1983–84	93.6%	79%

*Unit 1 out of service for nine months.

26. MAIN CONTRACTORS, MANUFACTURERS AND PLANT SUPPLIERS - DRAX COMPLETION

Mechanical

Main and auxiliary boilers	Babcock Power Ltd
Turbine-generators	NEI Parsons Ltd
HP and LP feed heaters and deaerators	GEC Turbine Generators Ltd
Starting and standby feed pumps	Weir Pumps Ltd
Feed suction pumps	Sulzer Bros (UK) Ltd
Gas turbines	GEC Gas Turbines Ltd
Turbine house crane	A.B. Cranes Ltd
Cooling water pumps	Sulzer Bros (UK) Ltd
Sodium hypochlorite plant	NEI John Thompson Ltd
Coal handling plant	GEC Mechanical Handling Ltd
Ash and dust handling plant	Babcock Hydro-Pneumatics Ltd
	Fletcher Sutcliffe Wild Ltd
Pipework and valves	Aiton and Co Ltd
Vacuum cleaning plant	Sturtevant Eng Co Ltd
Hydrogen plant	NEI John Thompson Ltd
Water treatment plant	NEI John Thompson Kennicott Ltd
Fire protection plant	Mather and Platt Ltd
Boiler fuel oil tanks	Copper-Neill International Ltd
Chemical sampling plant	George E. Low Ltd

Electrical

Generator transformers	GEC Power Transformers Ltd
Unit and auxiliary transformers	Hawker Siddeley Power Transformers Ltd
Gas turbine unit transformers	NEI Peebles Ltd
11kV and 3.3kV switchgear	NEI Reyrolle Ltd
415V and LV switchgear	Electro Mechanical Manufacturing Co Ltd
Batteries and chargers	Tungstone Batteries Ltd
Main connections	Balfour Beatty Power Construction Ltd
Main and auxiliary cables	N.G. Bailey and Co Ltd
Telecommunications	Reliance Systems Ltd
Radio equipment	Burndept Electronics Ltd
Site electrical supplies (installation and maintenance)	T.W. Broadbent

329

4.

Dinorwig Pumped Storage Power Station

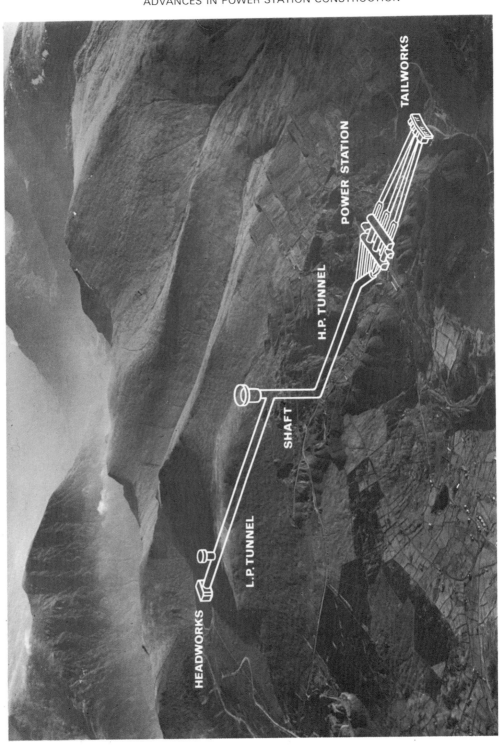

FIG. 4.1   AERIAL VIEW OF LLANBERIS PASS AND DINORWIG SITE

1. INTRODUCTION

The CEGB has a statutory obligation to maintain system frequency between 49.5Hz and 50.5Hz. However, the CEGB's own target is to maintain system frequency between 49.8Hz and 50.2Hz. Maintaining a reliable supply within these limits and at the specified voltage is operationally demanding.

It is not feasible to store electricity directly. Therefore, because absolute accuracy in forecasting is not achievable and to cope with conceivable generation/transmission failures, more generators than are strictly necessary to meet exactly the load demand at any time are kept running. By this means sudden changes in demand or supply are met by plant already connected to the transmission system.

Typically, with fossil-fired plant, a partly-loaded generator can increase output by 30% within 3 to 5 minutes of demand with half of this occuring within the first few seconds. This method of operation is uneconomic because of the higher production costs of the additional plant brought into operation, and the cost of running plant at loads lower than their optimum rating.

Gas turbines can be started and loaded in 2 minutes but this is not fast enough to meet all the sudden changes in demand.

The inherent ability for rapid loading of hydro-plant and its flexibility in changing from one operating mode to another makes it ideally suited as a means of meeting this kind of demand. Unfortunately most viable hydro-electric sites in England and Wales are being used, and therefore pumped storage schemes are being developed. Pumped storage plants are able to carry out the emergency response function and in addition they can be used economically for peak load generation. In March 1972, the CEGB decided to seek authorisation to proceed with the development of a six 300MW unit pumped storage station at Dinorwig in North Wales.

In contrast to this large pumped storage scheme, the CEGB is also engaged in the recovery of energy from water being discharged from storage reservoirs. This can be achieved using small simple hydro-stations. Several of these schemes have been evaluated and one such station has been built in partnership with the Northumbrian Water Authority at its Kielder Dam. At this station a 5.5MW unit recovers the energy from the water being discharged from the reservoir into the River Tyne for water regulation purposes whilst a smaller 0.5MW unit similarly deals with the river compensating flow. The power station project was started in 1981 and began operating in the summer of 1984.

2. DINORWIG PUMPED STORAGE STATION

2.1 Site Selection

In May 1970 investigations were started to assess the feasibility of pumped storage development at three sites in North Wales, near Dolwyddelan (Bowydd), Croesor and Llanberis (Dinorwig).

The topography of all three sites was favourable for the construction of large scale pumped storage plant. Of the three, Croesor offered the most favourable ratio of operating head to conduit length, but was the site at which topography would limit the maximum capacity which could be developed, whilst at Llanberis a working head of 519m was available.

Slate is the predominant rock form at each of the three sites, and it is known to be stable both in open quarries and in underground galleries with large unsupported roof spans. All three sites were geologically suitable for the excavation of the large underground caverns and tunnels associated with a pumped storage scheme.

There were no major problems at any site relating to the construction of dams of the size required to enclose the necessary volume of water storage. The reservoir basins could be made satisfactorily watertight and there was no reason to believe that they would not be stable.

The Dinorwig site is in Gwynedd, close to Llanberis, and to the northern boundary of the Snowdonia National Park. The power station and tunnel system is underground below the disused Dinorwig slate quarry. The site, being close to Llanberis and lying at the foot of Llanberis Pass, is of high public amenity value. See Fig 4.1.

An existing lake, Llyn Peris, was used as the lower reservoir but was enlarged by building an embankment dam at its northern end and by excavating material from the bed. The river Afon Nant Peris, which previously flowed into Llyn Peris was diverted through a tunnel along the western side of the lake and via a new channel into another lake, Llyn Padarn.

The upper reservoir was formed by enlarging a small water supply reservoir, Llyn Marchlyn Mawr, by means of a rock embankment dam.

Amenity consideration dictated that the access tunnels were driven from the existing quarry face and also precluded the possibility of providing an above ground switchyard, thus necessitating the construction of an underground high voltage transformer and switchgear compound.

The capacity of 1800MW was dictated by the maximum pumping power which could be transmitted to Dinorwig without additional reinforcement of the grid system.

In the case of Llanberis, the use of Llyn Padarn in conjunction with Llyn Peris as a lower reservoir was examined. This showed a marginally cheaper capital cost than the use of Llyn Peris alone, but both the operational difficulties in the control of floods and the interference with fishing and amenity interests were found to be very much greater. For these reasons it was decided that Llyn Padarn should not be utilised as part of the lower reservoir of the Llanberis scheme.

In considering the feasibility of the project it was necessary to take into account the following requirements:

(1) There must be two adequate reservoirs with minimum horizontal separation, but with maximum vertical separation. If the capacity of the reservoirs was not equal, then dams must be built to match their capacities. A high head and close horizontal proximity between the two reservoirs reduces capital costs and maximises output.

(2) The site must be close to the grid system, reducing installation and operating costs of transmission plant.

(3) There must be good access to the site because of the need to transport raw and finished materials to and from site.

(4) There should be favourable geological conditions; the ideal being a homogeneous rock, free from faults, to enable tunnelling and underground caverns to be formed.

(5) The site should preferably have low amenity value. Unfortunately, suitable sites in England and Wales are rare and often associated with areas of great natural beauty, such as Dinorwig. Hence, careful design must be undertaken to reduce the visual impact.

The final report, based on these considerations, concluded that the Dinorwig scheme would be the most economic and that it would have less impact on the environment than either of the other two sites. The CEGB accepted the recommendation that statutory powers should be sought to construct a pumped storage station at Dinorwig of about 1800MW nominal plant rating.In May 1972 an application was made to promote a Bill in the 1972/73 Parliamentary Session for authorisation of the scheme, and on 19 December 1973 the Royal Assent was given to the Act.

2.2 System Considerations

Dinorwig power station is one of the most advanced pumped storage systems in the world; the entire station output from the six 300MW units can be increased to full load from no load in 10 seconds, faster than any other station on the CEGB system.

The advanced technological nature of the station arises from the many functions that the station can carry out as part of the grid system. The station is designed so that two of the units can be used to provide the total standby capacity in case a major fossil-fired or nuclear unit trips out, and four units can be used to control the system frequency. This may require starting and stopping the Dinorwig units up to typically 40 times a day throughout the 40 years designed plant life. This requirement, combined with the high head and power output, has demanded a refinement in plant design far in excess of that normally required for a hydro station.

Traditionally, pumped storage plants have been used for load smoothing, ie providing night-time demand and generating during peak daytime periods. Dinorwig will also fulfil this function.

The economic deployment of pumped storage plant for peak lopping depends upon there being large differences in the costs of generation on the remaining types of plant. These wide differences can arise from having a large proportion of nuclear plant, and differences in thermal efficiencies or fuel costs. As the CEGB presently has an excess of generating capacity, operation of the high efficiency plant available rules against using pumped storage units for peak lopping, except when maximum demand is unusually high.

On the CEGB system therefore, pumped storage plant can be best employed by utilising its inherent flexibility. This flexibility arises from the speed with which electricity can be supplied to or taken from the grid and from the capability of the plant to do this repeatedly with high reliability and availability.

The time taken to increase generation safely to full output depends on the combined characteristics of the pump-turbine and hydraulic system.

At Dinorwig, the hydraulic system has been increased in size to cope with the rapid changes in flow which occur when the station output or input is changed.

The other flexible feature of pumped storage, ie repeated starting/stopping, is not normally exploited but at Dinorwig this will be utilised to provide fine frequency control for the entire system. When operating on standby the generator-motor/pump-turbines will normally be driven from the grid system with pump-turbine runners spinning at full speed in air. Air is pumped into the pump-turbine casing after closing the main water inlet valve and it pushes the water level down below the runner to reduce drag and therefore the power required for spinning.

The closing of the main water inlet valve and the injection of compressed air reduces the pressure in the machine casing and the pipework connecting it to the inlet valve. The pressure changes between 600m and 60m head every time the valves are operated, which during the machine's life is approximately 300,000. The use of the machine in this way therefore introduces into the design the need to consider fatigue and this is not normally required on pump storage plants.

At Dinorwig, the volume of the hydraulic system between the top reservoir and the pump-turbine inlet valves is greater than the volume of the underground caverns. A failure of the main inlet pipe would therefore lead to the complete flooding of the station if the failure coincided with a failure to close the inlet valve. These considerations have necessitated a fracture mechanics approach to the design, and manufacturing quality control of a particularly high standard.

The design has also had to take into account accessibility to critical highly stressed areas, to facilitate non-destructive examination periodically throughout the plant's life. This will ensure that fabrication defects can be monitored to provide assurance that they are not growing to critical dimensions as a result of the intensive cyclic stressing.

The use of pumped storage plant for both emergency reserve generation and peak lopping creates a difficult situation for the designer. The former role requires a very high degree of reliability and this may be endangered by repeated usage of the plant for peak lopping. At Dinorwig, special measures have been taken to ensure high plant reliability.

The design of the electrical system has also had to take account of the arduous operating regime and the need to meet a requirement for a reliability of 99% and an availability of 95%. The generating voltage of 18kV was kept as low as possible to reduce the risk of winding insulation failure. Further reduction would have meant a higher current flow for the same power output and this would have increased the 18kV air blast circuit breaker duty beyond existing capability.

A great deal of effort has been put into the full appraisal of the hydraulic design problems. Waterhammer calculations relating to the hydraulic system and its associated surge pond were carried out. Transient fluid conditions which could be expected during machine startup, shutdown and emergency trip conditions were analysed. Full scale comparison of the results were carried out at North of Scotland Board Cruachan Pump Storage Power Station which gave a full validation of the method.

It was absolutely essential that the design of the steel sections of the hydraulic system was such that they were able to withstand all conceivable dynamic loads. Following detailed analysis of the design information available, particularly regarding North Sea oil rig design, it was decided to design the free standing sections to the ASME Boiler and Pressure Vessel Code Section VIII Division 2.

Fracture mechanics fatigue assessment methods have been used extensively both for the penstock sections and also the spiral casing, the need being to ensure absolute safety, taking into account the onerous operating regime specified for the station, together with the adoption of a spinning-in-air mode of pump-turbine operation.

Stringent material fabrication standards, welding procedures, heat treatment and non-destructive test procedures were established and vigorously enforced during the whole of the manufacture and erection stages. The plant was designed to give good access for inspection of all the major welds both during erection and in service.

Component reliability studies have been carried out to ascertain the areas of risk. High safety integrity has been ensured by designing those areas of high risk to nuclear station standards.

2.3 Operational Roles

The required operational capabilities of Dinorwig were specified in terms of response rates, operating roles, modes of operation and number of mode changes. A detailed operating regime was not specified as this is best determined from operating experience and would probably evolve with time. Required operating roles were as follows:
(1) Generation — generation by any number of units in the load range zero to full load.
(2) Pumping — pumping by any number of units at full load.
(3) Emergency Reserve — emergency reserve is required to provide dynamic response up to the full station output capacity (also called spinning reserve).
(4) Frequency Regulation — frequency regulation by any of the units.
In the event of a system collapse, it is a design requirement that Dinorwig could be started to re-establish the system, so diesel generators and batteries are provided.

The designed operating times for the various operations are given in Table 4.1.

2.4 Availability

The economic importance of Dinorwig's regulation and reserve duties means that the main and auxiliary plant is designed to provide average availabilities better than 95% excluding overhauls. Overhaul periods will not exceed 6% of the time, averaged over a number of years.

3. STATION SITE LAYOUT (FIG 4.2)

The pumped storage scheme utilises a difference in level of 519m between the upper lake, Marchlyn Mawr, and the lower lake, Llyn Peris. Both lakes have 7 million cubic metres of water storage capacity and each was created by the enlargement of an existing lake.

The underground power station is a complex of nine caverns, the largest of which houses six 300MW reversible pump-turbines. As the avoidance of cavitation required a minimum submergence of 60m for the pump-turbines, the choice of an underground power station was the only practical possibility.

TABLE 4.1. MODE CHANGE OPERATING TIMES

Description of Mode Change	Station Times	Unit Times
Standstill to no-load generation synchronised (watered).	90s	90s
No-load generation synchronised (watered) to 1320MW output.	10s	10s
No-load generation synchronised (in air) to 1320MW.	11s	11s
Full-load generation to no-load generation synchronised (watered).	10s	10s
No-load generation to standstill.	6min	6min
Standstill to pump synchronised in air (de-watered start).	20min	9min
Pump synchronised in air to full pump discharge (full-load).	220s	90s
Full pump discharge (primed pump running at shut-off).	10s	10s
Standstill to full pump discharge (watered back-to-back start).	—	120s
Primed pump running at shut-off to pump synchronised in air.	15s	15s
Shut-off pump to standstill.	6min	6min
Full-load generation to full pump discharge (full-load).	37min	17min
Full pump discharge (full-load) to full load generation.		
Normal	8min	8min
Emergency	90s	90s
Pump synchronised in air to no-load generation synchronised (water).		
Normal	8min	8min
Emergency	150s	150s

DINORWIG PUMPED STORAGE POWER STATION

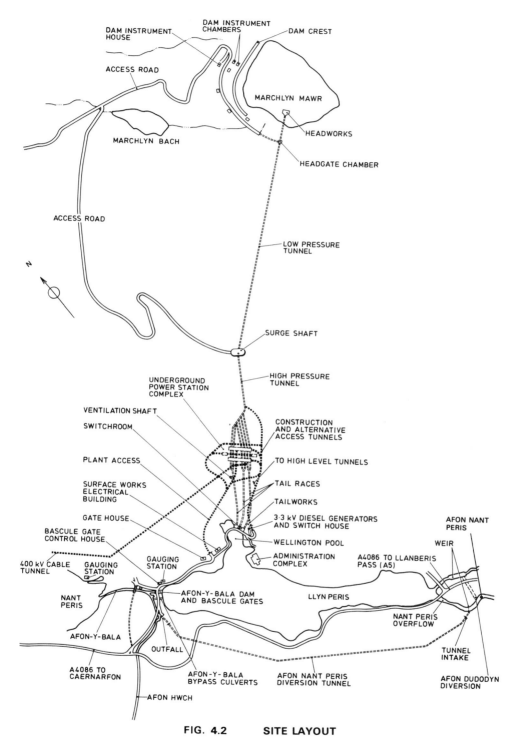

FIG. 4.2   **SITE LAYOUT**

To facilitate access and expedite construction, the underground power station had to be reasonably close to Llyn Peris. On the other hand there was a strong incentive to keep the high pressure system as short as possible. The compromise adopted was to make the tailrace tunnels as long as was practicable (about 500m) and not resort to surge shafts. This length also allowed acceptable gradients in direct access tunnels. Topography of the steeply sloping quarry area fortuitously permitted the water inlet surge shaft to be sited only 670m upstream of the power station.

Two access tunnels to accommodate the passage of plant and personnel to the main cavern complex were built. Two tunnels, rather than one, were required to provide security of access and egress. They also permitted rapid removal of rock spoil from the main underground excavations.

Siting of Marchlyn dam was virtually predetermined by the shape of the hanging valley in which the existing lake was situated. There was a natural moraine dam across the valley. Using this as a base, heightened by rockfill to the desired height, was cheaper than a comparable concrete dam. Dam height was determined by the storage volume requirement.

The provision of sufficient water storage in a lower reservoir was less available. Two existing lakes, Llyn Peris (36ha) and Llyn Padarn (100ha) were separated only by a barrier of glacial and alluvial deposits. Using the combined area, a fluctuation in level of 5m in both lakes would have achieved the necessary 7 million cubic metres. However, the resulting increase in the hazard of flooding the immediately adjacent village of Llanberis precluded the use of Llyn Padarn, leaving only Llyn Peris to form the reservoir.

The water conduits (see Fig 4.3) consist of a single low pressure (LP) tunnel, a high pressure(HP) shaft, and a tunnel with a manifold to the six steel-lined penstock tunnels serving the machines. On the Llyn Peris side of the station six steel-lined draft tubes lead into concrete-lined sections which are paired into three tailrace tunnels.

The whole of the downstream side of the machine hall is taken up by the 7 − storey switchgear annexe. This is a steel-framed structure with reinforced concrete floors and blockwork walls containing switchgear, cable flats and the control room.

3.1 Headworks

The installation consists of a main gate and stop gate located in an underground chamber alongside the Marchlyn Mawr (upper) reservoir. When lowered into position, the gates effectively isolate the upper reservoir from the low pressure tunnel. The upper floor of the chamber houses winching equipment for lowering and raising each gate and the lower floor provides a gate stowage area.

Location of the headworks installation relative to the reservoirs, power station complex and other gate installations is shown in Fig 4.2 and 4.3.

3.2 Low Pressure Tunnel

A single 10.5m internal diameter concrete-lined tunnel approximately 1.7km long, connects the upper reservoir at Marchlyn Mawr to the surge chamber and HP shaft situated on the upper edge of the Dinorwig quarries. A screened intake/outfall structure is provided at the entrance to the tunnel and fitted with stop logs for isolation purposes. The maximum design flow rate of water from the upper reservoir is $420m^3/s$.

3.3   Surge Chamber

The surge chamber consists of a vertical circular shaft 30m in diameter and 65m deep under an open headpond which is approximately 73m by 38m and 14m deep.

3.4   High Pressure Penstocks

A single vertical concrete-lined HP shaft of 10m diameter and 443m deep is constructed at the end of the high level tunnel below the surge chamber. From the base of the shaft a straight but slightly inclined tunnel of 9.5m diameter extends for approximately 446m towards the generating station, before curving into a manifold system of six tunnels of 3.8m internal diameter. These tunnels in turn connect with the six HP, steel-lined, penstock tunnels of 3.3m diameter at a distance of approximately 170m from the centre of the machine hall. Each of these penstock tunnels reduces to a 2.5m diameter just before entering the main inlet valve gallery at a distance of approximately 34m from the machine hall. After the main inlet valves the tunnels continue at the same diameter until just before the machine hall, where they taper to the spiral casing inlet of 2.3m diameter for connection to the spiral casing inlet pipes.

3.5   Tailrace

On the low pressure side of the machine the six draft tubes continue as 3.75m diameter steel-lined tunnels with concrete surround for a distance of approximately 38m to the centre of the draft tube valve gallery, in which the valves are installed whereby each machine can be isolated from the water in Llyn Peris when required for maintenance purposes. From the draft tube valves the tunnels are 3.75m diameter and steel-lined for 69m and change from steel-lined to concrete-lined where the diameter increases to 5.8m. The tunnels then lead into bifurcations where they are joined together with the three 8.25m diameter concrete-lined low level tunnels for the remaining 380m to 400m to the tailworks, located at Wellington Pool, where they join the lower reservoir via a channel.At the reservoir end of the three low level tunnels, a screen and gate structure is provided. The tailworks contain three cooling water and two fire pump houses, main and stop gates. The area is serviced by a 15t (radio controlled) Goliath crane.

The bascule gates installation is situated in the Afon-Y-Bala Dam at the north western end of the Llyn Peris (lower) reservoir. Temporary maintenance access to an individual bascule gate is afforded by a set of stop logs when these are lowered into position on the upstream and downstream sides of the gate. Handling the stop logs is by beams of a mobile crane and lifting beam. With the gates lowered into position the interspaces can be dewatered. Location of the bascule gates relative to the reservoirs, power station complex and other gate installations is shown in Fig 4.2.

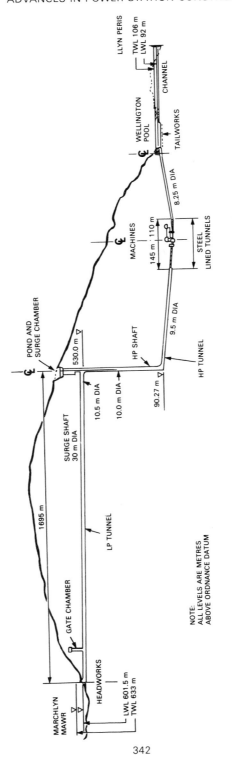

NOTE:
ALL LEVELS ARE METRES
ABOVE ORDNANCE DATUM

FIG. 4.3   STATION SCHEMATIC

3.6 Station Complex (Fig 4.4)

The original design criterion required that the excavated volume be kept to a minimum because of the high cost per cubic metre. This, plus plant requirements, dictated cavern dimensions.

Within the complex of nine caverns, the power station consists of the machine hall (Fig 4.5), with the main inlet valve (MIV) gallery (Fig 4.6) upstream, and the transformer hall on the downstream side with the draft tube valve (DTV) gallery lying between them. There are small galleries interconnecting the MIV gallery with the pump-turbine and generator-motor floors in the machine hall for routing pipework and control cables and for man access. There are three major busbar galleries, one for each pair of generator-motors, interconnecting the machine hall with the transformer hall and providing accommodation for the generator-motor main connections and electrical switchgear. At each end of the DTV gallery is a chamber housing the starting equipment for the generator-motors. Above the main complex is a system of high level tunnels used for ventilation and for monitoring rock stability. Access to the high level tunnels is via the construction access tunnel.

3.7 Machine Hall

The machine hall, 179.25m by 23.5m and 51.3m deep, contains six reversible pump-turbine/generator-motor units with units 1 and 2 on one side of the loading and assembly bay and units 3 to 6 on the other side. This arrangement was dictated by the need to have an access area during the construction period and at the same time maintain the location of units in pairs which was compatible with the arrangement of the busbar gallery equipment serving two adjacent units. The hall is divided into four major areas, with three floors housing the generator-motor/pump-turbine units, a 7-storey switchgear annexe, a 5-storey workshop and a pipe gallery. (See Fig 4.7).

The three main floors are the pump-turbine floor at the 34.80m Ordnance Datum (OD) level, the generator-motor floor at the 42.50m OD level and the machine floor at the 50m OD level. At this level is a loading and assembly bay area which leads into the main plant access tunnel. Hatches are provided between these floor levels for access and maintenance purposes.

Housed in the switchgear annexe is the station control room, operations room, station and unit switchrooms, control and instrumentation rooms, data logger room and, at the upper levels, a battery room and heating and ventilating equipment. All the necessary facilities for the operational staff and offices are also provided in the annexe.

The workshop is a 5-storey structure and has a loading bay and storage area, electrical and instrument workshops, mechanical maintenance workshops and all necessary maintenance staff facilities and offices.

Two 250t overhead (radio controlled) cranes are provided in the machine hall which can lift, when tandem operated, a maximum load of 460t (equivalent to the rotor and poles of a generator-motor). A further cab-driven 10t auxiliary crane is also provided together with a 4t goods lift and four personnel lifts.

Located at the lowest level of the machine hall is a pipe gallery which houses the dewatering pumps and associated system pipework. In the base of the gallery are the main drainage sumps, submersible pumping sets and oil interceptors.

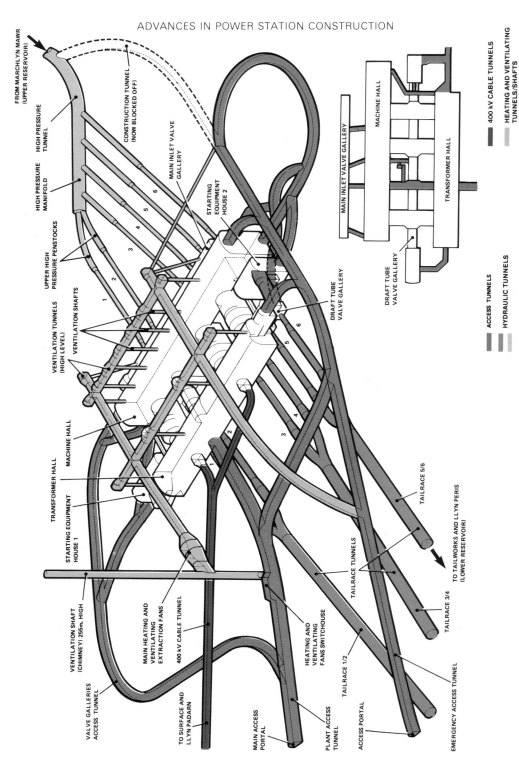

FROM MARCHLYN MAWR
(UPPER RESERVOIR)

HIGH PRESSURE
TUNNEL

HIGH PRESSURE
TUNNEL

CONSTRUCTION TUNNEL
(NOW BLOCKED OFF)

HIGH PRESSURE
MANIFOLD

MAIN INLET VALVE
GALLERY

STARTING
EQUIPMENT
HOUSE 2

UPPER HIGH
PRESSURE PENSTOCKS

VENTILATION TUNNELS
(HIGH LEVEL)

VENTILATION SHAFTS

DRAFT TUBE
VALVE GALLERY

TRANSFORMER HALL

MACHINE HALL

STARTING EQUIPMENT
HOUSE 1

VENTILATION SHAFT
(CHIMNEY) 255m. HIGH

MAIN HEATING AND
VENTILATING
EXTRACTION FANS

400 kV CABLE TUNNEL

VALVE GALLERIES
ACCESS TUNNEL

TO SURFACE AND
LLYN PADARN

HEATING AND
VENTILATING
FANS SWITCHHOUSE

TAILRACE TUNNELS

MAIN ACCESS
PORTAL

PLANT ACCESS
TUNNEL

TAILRACE 1/2

ACCESS PORTAL

EMERGENCY ACCESS TUNNEL

TO TAILWORKS AND LLYN PERIS
(LOWER RESERVOIR)

TAILRACE 3/4

TAILRACE 5/6

MACHINE HALL

TRANSFORMER HALL

MAIN INLET VALVE GALLERY

DRAFT TUBE
VALVE GALLERY

400 kV CABLE TUNNELS

HEATING AND VENTILATING
TUNNELS/SHAFTS

ACCESS TUNNELS

HYDRAULIC TUNNELS

FIG 44   CAVERNS AND TUNNELS

FIG. 4.8   INTERMEDIATE SHAFTS

3.10 Pipe Gallery

The pipe gallery, at the 19.20m OD level of the machine hall incorporates the dewatering and drainage system pumps and associated pipework, together with the cooling water system and standby mains, these systems being linked by culverts in the inverts of the draft tube and tailrace tunnels to the tailworks. Also leading into the gallery is the 600mm standby fire main which supplies the various fire protection systems in the gallery and on the upper floor levels. A 45m^3 capacity tank is installed adjacent to the drainage pumps for receiving, in the event of a fire, an oil/water mixture from the governor recesses and oil tank bund at the 34.80m OD level.

A brief description of the major systems is given as follows:

3.10.1 *Cooling Water System*

This system supplies the necessary cooling water for the six pump-turbine generator-motor units, associated generator-motor transformers, blowdown air compressors and the station heating and ventilating system.

After passing through the various heat exchangers of the units, associated transformers, air compressors and the station heating and ventilating system, the water is returned to the tailworks via the 1m diameter cooling water discharge culvert.

3.10.2 *Dewatering System*

This system is used when it is necessary to enter parts or the whole of the hydraulic system for inspection and maintenance purposes. The dewatering system pumps water out of the draft tubes either to dewater a pump-turbine only, between the MIV and DTV or to empty the whole of the hydraulic system. Each draft tube has two drainage sumps at its lowest point which are connected to a dewatering main which runs along the whole length of the pipe gallery.

3.10.3 *Drainage System*

The drainage system is used to pump away from the cavern and tunnel complex natural seepage and incident leakage water, this water entering two interconnected sumps in the gallery floor either directly via pipework and culverts or indirectly via oil interceptors.

3.10.4 *Standby Main*

A standby main is also incorporated in the pipe gallery which can be used for either the cooling water or drainage/dewatering systems and is interconnected with the tailworks via the 1m diameter standby culvert. At the tailworks the changeover is effected by means of interchangeable spool pieces and in the gallery by interconnecting pipework and valves.

3.11 Main Inlet Valve Gallery (Fig 4.6)

The main inlet valve (MIV) gallery, 147m by 8m and 18.6m deep, houses the six MIVs which are plinth mounted at a common floor level of 28m OD. Also located at this level are: the associated oil pumping sets (one per MIV) used for opening the MIV and for operation of the various control valves, two compressors located between MIV 2 and MIV 3 for supplying, via a busmain, the control air for each MIV, and two further skid-mounted auxiliary pumping sets for the provision of 'high integrity isolation' of the MIV. High integrity isolation is used when it is necessary for personnel to enter the hydraulic system downstream of the MIV, ie the pump-turbine suction cone or spiral casing.

A 16t radio controlled overhead crane is also provided, together with a vehicle loading bay at MIV 1 end.

3.12   Draft Tube Valve Gallery (Fig 4.9)

The draft tube valve (DTV) gallery, 172.7m by 9.2m and 19.7m deep, houses the six DTVs which are plinth mounted at a common floor level of 20.83m OD. Located between DTV 1 and DTV 2 is a blind sump tank designed to take an oil/water mixture in the event of a generator-motor transformer fire.

Also located at this level adjacent to each DTV is a hydraulic/electric control cabinet which controls valve operation via local or remote initiating signals. At the 37.2m OD level of the gallery are the 18kV starting busbars. These interconnect the two starting equipment houses, one at each end of the gallery, to the main 18kV busbars.

An 80t radio controlled overhead crane is also provided, together with a vehicle loading bay at DTV 1 end.

3.13   Transformer Hall

This is 161m by 23.5m and 17m deep and contains at the 25m OD (ground floor) level six generator transformers each serving a machine, three station transformers, excitation and other auxiliary transformers. Also at this level is a SF6 switchgear gas handling room, a battery room and an amenity area, all of which are adjacent to the main access tunnel entrance. SF6 gas-filled switchgear was selected because of its compactness.

Above this floor at the 60.5m OD level is the 400kV substation housing the 420kV metal-clad SF6 switchgear and associated circuit breakers and isolators. Logging and relay rooms are also at this level together with a 10t overhead crane.

The high voltage side of each generator transformer is connected via the SF6 switchgear to two sets of 400kV cables (Pentir 1 and Pentir 2) to the grid system at Pentir substation 10km distant.

3.14   Above Ground Buildings

All of these buildings have contact type foundations, ie the loads are transferred directly to the ground without the use of piles. See Fig 4.2 for the location of buildings.

3.14.1   *Administration Block*

This is a 2-storey building with a pitched slate roof. It has load bearing walls, the outer skin of the external wall is of slate masonry and the inner skin of blockwork. Because the building is in a very exposed location on a promontory of rock on the side of Llyn Peris, the cavity face of the inner skin is sprayed with silicone sealant to prevent penetration of damp.

FIG. 4.9   **DRAFT TUBE GALLERY**

3.14.2 *Diesel Generator and Switchgear House*

This is located adjacent to the portal of the emergency access tunnel. It is the only masonry clad building with a steel frame at Dinorwig. It has a pitched slate roof, cavity external walls, with the outer skin of slate masonry and the inner skin of blockwork. Two 2.5MW diesel generators for emergency supply in the event of loss of grid supplies are located in this building, together with their auxiliaries.

To preserve integrity still further the generators are housed in separate rooms, each of which is protected by a Halon gas system in the event of fire. The switchhouse contains on two floors, batteries, auxiliary switchgear, transformers and cabling marshalling facilities associated with both diesel generators.

3.14.3 *Surface Works Electrical Building*

This is located adjacent to the portal of the main plant access tunnel. It is of single storey construction with a pitched slate roof, cavity external walls, with the outer skin of slate masonry and the inner skin of blockwork. The building contains transformer and auxiliary switchgear for local distribution; also the terminations for the station public telephone system and a cabling marshalling facility.

3.14.4 *Security Lodge*

This is located on the opposite side of the main plant access tunnel portal to the surface works electricity building. It is a steel-framed timber studded building carried on a grid of steel beams supported on a reinforced concrete base slab. Lightweight construction on a steel grid was chosen because of the possibility of foundation settlement continuing for a considerable period. Provision has been made for jacking the building through the perimeter steelwork to offset any settlement of the base slab. The building has a proprietary light weight roof at a 6° pitch, except at the overhanging eaves where the pitch is 25°. The gatehouse portion of the building has toughened and tinted glass with a steel plate wall lining beneath this glazing.

3.14.5 *Bascule Gate Control House*

This is located adjacent to Afon-y-Bala Dam. It is of single storey construction with a pitched slate roof, cavity external walls, with the outer skin of slate masonry and the inner skin of blockwork. The hydraulic power packs for remote operation of the bascule gates are located in this building.

4. RESERVOIRS AND TUNNELS CIVIL WORKS

4.1 Marchlyn Mawr Reservoir and Dam (Fig 4.10)

Marchlyn Mawr, which forms the upper reservoir, is located in a steep sided corrie on the north side of the Elidir mountain. The reservoir has a usable storage capacity of 7 million cubic metres of water between the normal operating levels of 600m and 633m OD.

The dam is formed of graded rockfill material and the dam has an impermeable upstream surface membrane of asphalt. Below the asphalt layer, a compacted layer of graded rockfill is provided to allow any small amounts of water passing through the asphalt layer to migrate down to the concrete toe structure which has an integral inspection gallery and a drainage channel. Drainage from the rockfill passes into drainage pipes, which in turn, feed into the drainage channel formed in the toe gallery structure. To ensure water tightness below the toe structure and at each side of the dam, extensive multi-stage rock grouting was carried out.

A single LP 10.5m internal diameter tunnel, approximately 1.7km long, connects Marchlyn Mawr to the surge shaft, situated on the outer edge of the Dinorwig quarries. A screened intake/outfall structure leads from the LP tunnel via a sloping concrete apron into the reservoir. Coarse screens have been installed at the intake to prevent large solid objects entering the hydraulic system.

Various types of instruments were built into the dam and its foundations, during construction, for monitoring of the following:

(1) Behaviour of the foundation and structure during construction.
(2) Behaviour of the foundation and dam during impounding (initial filling of the reservoir).
(3) Behaviour of the foundation and dam during use.

4.1.1 Rockfill Embankment

An embankment dam was considered the most suitable form and rockfill was available from disused quarry workings in the area. A total of 1.5 million cubic metres of rockfill was brought by dump trucks up a haul road nearly 2km in length at a gradient of about 1:10.

Weather conditions at Marchlyn Mawr were such that at times all construction activities could become extremely difficult as a result of high rainfall, severe winds, mist and snow experienced at high altitude.

4.1.2 Asphaltic Deck

The total deck thickness of 140mm consists of a binder layer with a minimum thickness of 60mm which acts as a regulating course on the rockfill surface. This is covered by an 80mm thickness of dense impermeable asphalt and a sealing coat.

Satisfactory weather for laying asphalt exists at Marchlyn Mawr only between April and October and weather records showed that on average only 137 days during that period would be suitable for this work. Nevertheless, the whole of the asphalt work was carried out in one season, in 1979.

4.1.3 Landscaping

Some of the structurally unsuitable material excavated during construction of the dam, which included peat, was stockpiled and used for landscaping the downstream face of the dam and obliterating the temporary works by re-establishing vegetation appropriate to the area. In spite of the magnitude of the construction it does not intrude on this wild countryside.

4.2 Hydraulic Tunnels

The design of the LP tunnel and the tailrace tunnels incorporates pressure relief holes to limit the external hydrostatic loads which would be experienced if the tunnels were dewatered; generally the rock cover here is less than 100m.

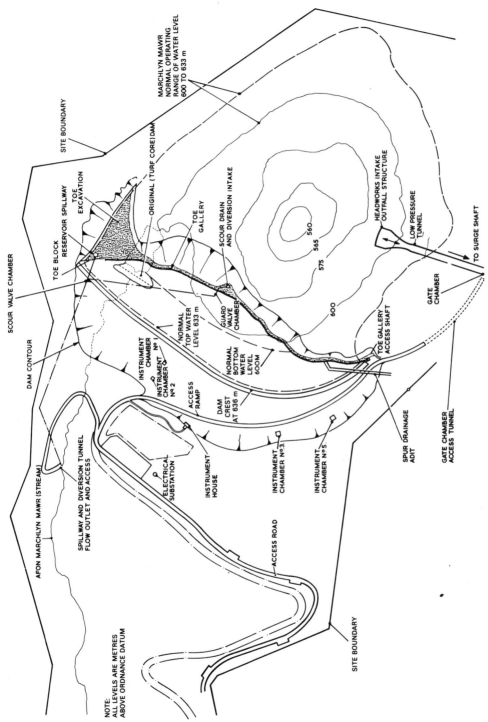

FIG. 4.10   MARCHLYN MAWR RESERVOIR AND DAM

In the HP tunnel the rock cover is less than the internal hydrostatic head and the lining was designed to be as impervious as possible to reduce any leakage from the tunnel to the surrounding ground. No pressure relief holes, therefore, were incorporated here.

Drill and blast methods were used for excavating the hydraulic tunnels. The penstocks were driven full face whereas the larger hydraulic tunnels were excavated as a top heading 8m high and full diameter leaving a flat invert. When the top heading had been excavated over its full length, the bottom segment was excavated in the reverse direction. This method enabled the plant to operate on a flat base.

Rock support in the tunnels comprised 25mm diameter high yield steel fully resin bonded rock bolts. Immediate shotcrete (sprayed concrete) protection was provided subsequent to scaling and additional tunnel support was occasionally required by steel arch ribs, mainly at the tunnel portals.

The concrete lining of the LP tunnel is nominally 300mm thick and was poured in two stages using tracked and adjustable steel shutters. With the invert complete the sides and crown were formed in a single pour. The concrete was not pumped into the ends of the 10m long shutter unit but placed through numerous side hatches. In the difficult crown section, pump hoses were plugged into one of a dozen ports and external vibrators ensured compaction. The HP concrete tunnels were poured in a similar manner and the linings are generally 1m thick. The concrete linings are reinforced only at tunnel junctions and small radius bends. Before being taken underground for erection the linings were coated internally with an isocyanate cured coal tar/epoxy resin based coating. Health and safety requirements dictated that this was carried out above ground. The suitability of the coating material for adhesion and resistance to erosion was determined in a CEGB laboratory by pressure cycling an internally-coated tube and by water impingement tests.

The steel lining sections were transported on a very low chassis trailer down the access tunnels to within 25mm of final position, blocked up and jacked clear of the trailer. Each lining section was winched into its final position on skates and welded up to a previously installed section. Concrete was then placed around each section of lining, with restraint jacks stopping any displacement. Finally, grouting was carried out to the areas behind both the concrete and the steel-lined tunnels. Cavity grouting to a pressure of 1.5 bar was applied throughout and where necessary, mainly where geological conditions are poor, fissure grouting to considerably higher pressures.

All concrete linings for hydraulic tunnels contain pulverised fuel ash (PFA) as a partial cement replacement. The main purpose of PFA is to provide greater chemical resistance to aggressive water. It also improves the workability of the concrete and is a cheaper constituent than cement.

4.3 Hydraulic Shafts

The HP shaft is surmounted by a surge shaft and surge pond (Fig 4.3). Sudden shutdown of the machines results in a transient pressure rise in the system and the water level consequently rises in the surge shaft and surge pond. The top of the pond is above the maximum water level of Marchlyn Mawr reservoir and was determined by mathematical modelling to ensure it would not overflow during tripping of the machines. The capacity of the hydraulic system above the junction of the LP tunnel with the HP shaft is sufficient to prevent air from becoming entrained in the hydraulic system during maximum downsurge.

The method of construction of the vertical HP shaft was essentially as follows: a 280mm diameter pilot hole was drilled down to the HP tunnel from the location chosen for the shaft. This hole was increased to 2.4m diameter by means of the drilling machine which remained at the top of the shaft and reamed upwards from the HP tunnel.

The drilling was carried out from a special sinking stage which had four heavy duty drilling booms mounted at its base. After blasting, the debris was raked down the hole by two centrally mounted JCB telescopic backactors. While men were on the shaft floor, inflatable bladders were installed in the mucking hole for safety. The sinking stage had three decks and shotcreting and rock bolting, using 3.7m long resin anchored bolts, could be carried out from any of the decks. The time taken to sink the shaft was six months.

4.4 Lyn Peris and Diversion Works (Fig 4.2)

Because the rivers Afon Nant Peris and Afon Dudodyn flow into the lower reservoir, Llyn Peris, and a pumped storage scheme requires a constant volume of water in the system, it was necessary to divert the normal flows from these rivers around Llyn Peris so that they now discharge immediately downstream of Lyn Peris. The diverted flows are carried in a tunnel on the south side of Llyn Peris.

The size of the diversion tunnel was based on a study of the hydrology of the area. It is horseshoe in section and is 6m high and 5m wide and has a maximum flow capacity of $55m^3/s$. When there is a higher flow than this in the Afon Nant Peris and the Afon Dudodyn the excess is discharged into Llyn Peris in order to reduce the risk of flooding downstream at Llanberis village. The overflow water is retained in Llyn Peris until lower river flows occur when it is discharged by gravity into Llyn Padarn.

The diversion tunnel passes through some areas of poor ground and through more varied strata than was encountered in the main power station area. In addition to driving this tunnel from each end it was decided in order to meet the construction period, to build a central adit to approximately midway along the length of the diversion tunnel so that four faces could be driven simultaneously. One of the environmental requirements of the scheme for the lower reservoir was to limit the difference between top and bottom water levels to 14m. In order to do this and provide 7 million cubic metres of live storage the surface area of Llyn Peris had to be enlarged to approximately its original size, ie before slate had been tipped into it during previous quarrying operations. So that the lake bed could be reprofiled in the dry and the lake area increased, the diversion tunnel had to be commissioned and the lake pumped out. The lake was pumped out into Llyn Padarn, either directly or by discharging into the diversion tunnel central adit. The water level in Llyn Peris was lowered by over 20m and this was achieved by electrically-driven pumps carried on pontoons and having a total capacity of $4m^3/s$. The lake enlargement involved moving 4 million cubic metres of material and had to be completed before any of the hydraulic sections of the power station could be filled with water.

The lake bed was shaped to reduce water velocities sufficiently to prevent excessive quantities of silt being taken into suspension and drawn through the system.

4.5 Integrity of Civil Engineering Works

The Marchlyn Mawr and Llyn Peris reservoirs come within the provisions of the Reservoirs Act. Instrumentation is built into the dams and the works have to be inspected at specified intervals and in accordance with the requirements of the independent engineer appointed and qualified within the terms of the Act.

Whilst there are no specific statutory requirements for inspecting the caverns or tunnels with regard to stability, and it is considered that all underground rock faces have been adequately reinforced and supported, a series of instruments and datum points were built into the works so that any rock movements can be detected.

With regard to the quarry benches and margins of Llyn Peris a comprehensive system of inspection and monitoring has been instituted.

The caverns were excavated by conventional drill and blast methods using drilling rigs which had up to five booms. For the 24m wide caverns (machine and transformer halls) a central roof heading was driven 8m wide. This was opened to 16m and then finally to full width. After each opening the roof support was put in. The shot holes were drilled to a depth of 3.6m and the average pull, involving some 300m^3 of rock and using about 300kg of explosives, was 3.3m.

Initial roof support for the caverns consisted of shotcrete having a minimum thickness of 30mm and normally incorporating an accelerator to give early strength. This was followed by installing 3.7m long resin anchored rock bolts at 2m centres. This is also typical of the rock support for all tunnels up to 8m wide. In critical areas the shotcrete was reinforced by a steel mesh pinned to the rock and subsequent layers of shotcrete were applied without the use of the accelerator. In the caverns the primary support normally consisted of the 36mm dia anchors, which have an ultimate tensile strength of 120t at 6m centres, though the support was modified where necessary to suit geological conditions.

The rock anchors provide overall stability for the caverns, the rock bolts secure medium size blocks of rock, whilst shotcrete and mesh secure the surface.

The method of opening the roofs of the main caverns and determining the spacing of the rock anchors was based on a trial enlargement carried out in the machine hall. An 8m wide heading was first driven the full length of the crown. A relatively narrow transverse cut was made from this heading, initially to give a span of 16m and then to the final span of 24m with roof support being put in as the span was increased.

Because the transverse cut was only 3m wide the roof was not truly spanning the full 24m. The width of the transverse cut was gradually increased in stages until it could be considered that a single way span of 24m existed. During each stage of opening out the roof movement was monitored and also the load on the rock anchors.

The total volume of underground excavation was 1,350,000m^3, of which about 400,000m^3 was from caverns. The majority of the excavated material was disposed of in quarry holes in the side of the mountain left from previous quarrying operations.

5. HYDRAULIC SYSTEM

5.1 Headworks (Fig. 4.11)

The hydraulic system can be isolated from Marchlyn Mawr (upper) reservoir by the operation of the headgates. The headgate installation consists of a single main gate and single stop gate which are located in an underground chamber alongside Marchlyn Mawr. When lowered into position, the gates provide double isolation for carrying out inspection and maintenance.

The gates permitted initial impounding of the upper reservoir and the provision of bypass valves in the gates facilitate controlled filling of the hydraulic system following a station dewatering. The stop gate is also used to isolate the upper reservoir from the main gate to enable the main gate roller paths to be inspected. The gates are designed to seal against a head of 54m, corresponding to the maximum level in Marchlyn Mawr on the upstream side with a dewatered LP tunnel on the other.

5.1.1 *Main Gate (Fig 4.12)*

The main gate which is of the fixed roller type with sealing on the downstream face is retained in the raised position by hydraulically-operated latches. The gate is held suspended by a 4-fall rope system from an electrically-driven winch unit, which is used only to raise the gate. When the gate is lowered, the winch motor is disengaged and the gate lowered under its own weight, a duplicated hydraulic braking system ensuring the gate is lowered at a safe speed.

The gate is of a mild steel fabricated structure made up of four sections. Each section comprises a flat skinplate stiffened and braced by pre-fabricated beams and struts. The sections are bolted together with rubber seals along the length of each skinplate joint, forming a structure 10.75m high by 8.2m wide and weighing 160t. Roller and guide wheels engage in tracks located in side grooves formed in the concrete and running from the gate storage area to the floor of the tunnel.

For maintenance purposes, the gate can be mechanically locked in a maintenance position.

5.1.2 *Stop Gate (Fig. 4.12)*

The stop gate is of the sliding bulk head type with sealing on the downstream face. Lowering and raising of the gate is by means of an electrically-driven winch unit via a 4-fall rope system. In the fully raised position the gate is mechanically secured.

The gate is of a mild steel fabricated structure similar to that of the main gate, except that the rollers and guide wheels are replaced by guide brackets with phosphor bronze bearing faces operating in side groove slide channels. The gate is 10.75m high, 8.67m wide and weighs 96t.

5.1.3 *Gate Seals*

Extruded neoprene rubber seals of d-shaped section are fitted along the top and sides of the gates. Rectangular section seals of similar material are fitted along the bottom of the gates. Maximum sealing is effected by the hydraulic force of the water in

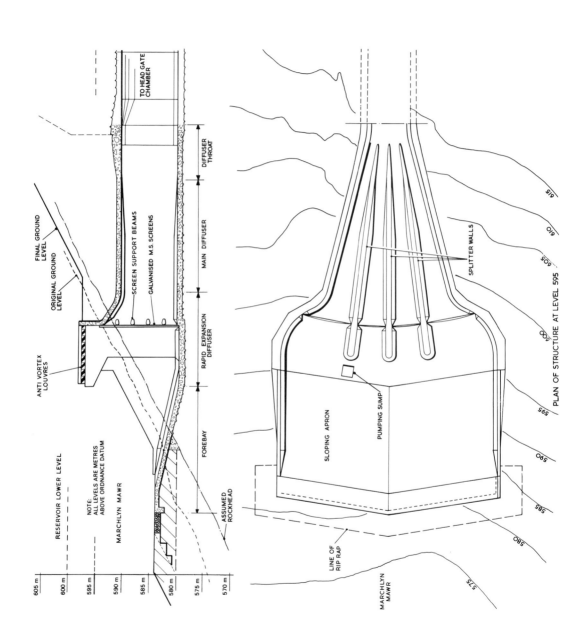

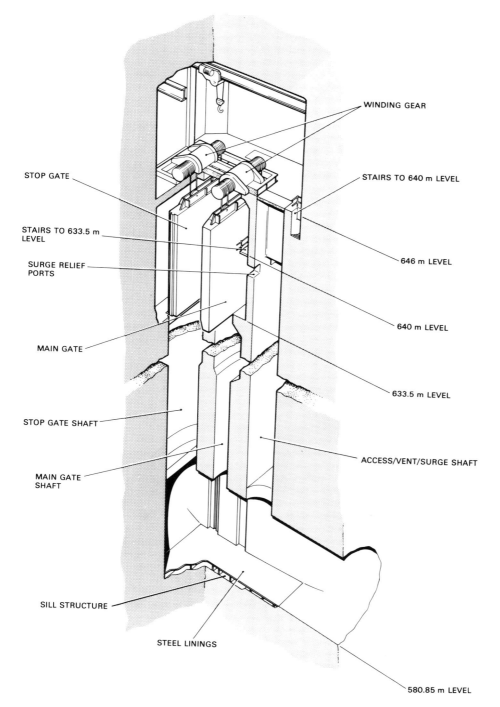

WINDING GEAR

STOP GATE

STAIRS TO 640 m LEVEL

STAIRS TO 633.5 m LEVEL

SURGE RELIEF PORTS

646 m LEVEL

640 m LEVEL

MAIN GATE

633.5 m LEVEL

STOP GATE SHAFT

ACCESS/VENT/SURGE SHAFT

MAIN GATE SHAFT

SILL STRUCTURE

STEEL LININGS

580.85 m LEVEL

FIG. 4.12   HEADWORKS GATE INSTALLATION

the upper reservoir acting on the gates, whilst sealing along the bottom of the gates between the rubber seals and embedded sill plates is effected by the compressive load of the gate structure.

5.1.4 *Bypass Valves*

Both gates are provided with a sliding blade valve system mechanically linked with the gate lifting pulleys for opening and closing operations. The valves are also biased towards the closed position by hydraulic cylinders supplied from a pressurised closed circuit oil system which, in turn, employs a bag-type accumulator charged with nitrogen as a constant pressure source. The main gate is provided with one bypass valve whilst the stop gate has four valves. An external bypass arrangement was considered but found to be too costly.

5.1.5 *Operation*

Control facilities for both gates are located in the gate winch chamber, although closure of the main gate can be initiated from the station central control room where there is also indication of the position of both gates.

The main gate is designed to close in 20min to 30min against a transient reversing flow resulting from a machine trip. The effect of gate closure on surge levels has been studied and reports on mathematical modelling plus model tests confirm that the closure time against these flow conditions introduces no unacceptable operational conditions. Raising of the main gate takes 3h with a differential head of 2m on the upstream side.

Raising and lowering of the stop gate takes 3h in either direction with a water pressure approximately balanced on both sides. Stop gate bypass valves are used to flood the interspace between both gates.

Interlocks prevent gate operation if more than one pump-turbine is in operation and inhibit turbine operation if both gates are not fully raised; stop gate operation is prevented if the main gate is not lowered.

Loss of electrical supply to the headworks installation is catered for by the provision of an emergency diesel generator at the headworks, which can either be started from the station central control room or arranged to start automatically on supply failure. With the gates in the normal raised position, however, operation of the emergency generator is not necessary to ensure gate closure, since all essential control and instrumentation supplies are provided from a continuously charged battery installation.

The headworks gates are not emergency gates, since the volume of water contained within the hydraulic system between the gates and main inlet valve is sufficient to flood the station. Thus gate closure in the event of gross failure in the hydraulic system would not prevent the station being flooded.

5.2 Tailworks (Fig 4.13)

At the tailworks, each of the three tailrace tunnels is fitted with a main gate to allow initial impounding of the lower reservoir and to isolate individual tailrace tunnels and associated pairs of pump-turbines for maintenance and inspection purposes. Splitter walls divide each tailrace tunnel into two outfalls at the final entry into Llyn Peris,

DINORWIG PUMPED STORAGE POWER STATION

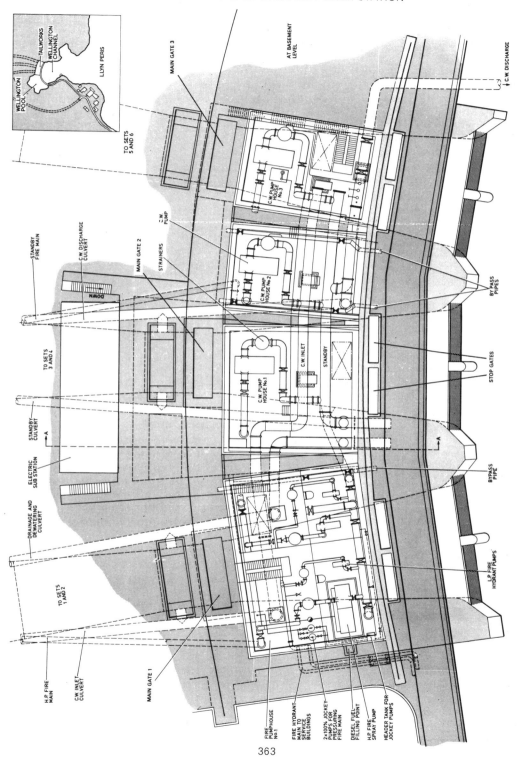

FIG. 4.13   TAILWORKS - LOWER RESERVOIR

363

where pairs of stop gates can be inserted to provide double isolation during dewatering of the tailraces. During normal station operation, removable coarse screens are placed in the stop gate positions to prevent large solid objects entering the hydraulic system.

The gates are designed to seal against a head of 32m corresponding to the maximum level in Llyn Peris on the downstream side with a dewatered tailrace tunnel on the other. The main gate is provided with an access door in the bottom section to allow entry into the interspace between the main and stop gates under dewatered conditions.

5.2.1 *Main Gates*

The main gates, which are of the fixed roller type with sealing on the upstream (station side) face are supported in the raised position by mechanical latches. Controlled lowering and raising of a gate is by means of a single hydraulic cylinder operated from an independant power pack. Hydraulic cylinders from the power pack also operate the mechanical retaining latches. A hand-operated pump facility for gate and latch operation is also provided on the power pack. Because of its position relative to the water level in the lower reservoir, the gate, when raised, is housed in a reinforced concrete bonnet fitted with a steel pressure cap. An air vent system maintains atmospheric pressure within the cap under all operating conditions.

The gates are of mild steel fabricated construction made up of three sections bolted together, the construction being similar to that of the headworks main gates. Ease of gate movement is achieved by rollers and guide wheels fitted to the gate side and running in vertical tracks built into side grooves formed in the concrete. Each gate is 8.6m high and 7.37m wide and weighs 60t.

5.2.2 *Stop Gates*

The stop gates are of the sliding bulkhead type made up of three sections with sealing on the upstream faces. The sections are stored ready for use at two locations on the tailworks gates installation complex. Lowering and lifting the sections into position is by means of an overhead electric crane and an automatic coupling type lifting beam. A complete gate consists of one top section and two lower sections, the latter being interchangeable. Sufficient sections are provided to isolate two complete tailrace tunnels simultaneously.

Each gate section is 3.54m high by 4.922m wide and is a mild steel fabricated structure similar to the main gates except that location is achieved by guide shoes rather than by rollers and guide wheels. The top gate section weighs 7.25t with the lower sections 6.65t each.

5.2.3 *Gate Seals*

The seal arrangement on the tailworks gates is identical to that of the headworks gates.

5.2.4 *Operation*

Control facilities for the tailworks main gates are provided locally, but closure of the main gates can be initiated from the station central control room where the positions of the gates are also displayed.

Although the tailworks main gates will normally only be operated in zero flow conditions, they are designed to be closed in 10min against a flow of $10m^3/s$ from the lower reservoir. Raising of and lowering the tailworks main gate is accomplished in 10min.

There are no bypass valves fitted in the tailworks gates but bypass pipes are included in each tailrace which allow controlled filling of the tailraces and equalising of water levels.

Interlocks are provided to trip the relevant pair of pump-turbine units if a tailworks main gate starts to close and operation of the MIV, DTV and associated pair of pump-turbines are prevented if the tailworks main gate is closed.

5.3   Bascule Gates Installation

To control both the rate of water discharge from the lower reservoir and its operating level to within the limits set by the local water authority, a gates installation is incorporated in the Afon-y-Bala dam embankment at the north western end of the reservoir (Fig 4.2). The gate structure is designed to be as unobtrusive as possible and not to protrude above the clean lines of the dam crest. This requirement precluded the more conventional method of gate operation from above the water level and a bottom pivoted bascule gate design was adopted. Stop gates inserted on the upstream and downstream sides of the gate allow maintenance access to individual bascule gates.

The bascule gates are designed to close from the fully open position against a water level 2.4m above the pivot centre on the upstream side with atmospheric pressure on the downstream side. From the fully closed position, the gates can open with an upstream level of 4.4m above the pivot centre.

5.3.1   *Bascule Gates*

The three bascule gates are of the bottom hinged-flap type mounted between the pier walls and incorporating sealing along the side and bottom edges. Controlled lowering and raising of each gate to maintain the required reservoir level is by means of a single hydraulic cylinder operated from an independent power pack. An electrical control system provides the necessary control facilities for operating the gate.

Each gate is a welded mild steel fabrication consisting of a rectangular, flat skin, face plate stiffened and braced longitudinally and transversely on the downstream side by T-section rolled steel beams and flat plate. The complete gate is 5.24m wide, 2.75m high and weighs 9t.

5.3.2   *Stop Gates*

The stop gates are of the sliding bulkhead type made up of a number of sections with sealing arranged along the sides and bottom of each section. Storage racks for the gate sections are provided adjacent to the gate installation, handling being by means of a mobile crane with automatic coupling type lifting beam.

The upstream stop gate comprises five sections each being 5.325m wide, 0.86m high and weighing 1.1t. The downstream stop gate is made up of two sections each being 5.325m wide and 2.4m high with a weight of 3t. Sufficient sections are provided to isolate one bascule gate.

The stop gate sections are welded mild steel fabrications consisting of a flat skin plate stiffened and braced by rolled steel members. Location is by means of guide shoes bolted to the side plates of each section.

5.3.3 *Gate Seals*

The sides of the bascule gates are fitted with pairs of extruded L-section neoprene rubber seals, arranged to effect automatically a seal against embedded side staunching plates when subjected to the hydraulic water loads from the upstream and downstream directions.

The staunching side plates, which are built into the pier walls, provide the seating faces for the gate seals. A heating system is incorporated in each staunching structure to prevent possible freeze-up between the seals and plate.

The stop gate side seals are of extruded neoprene rubber d-shaped section, an effective seal being formed by the hydraulic force of the water. Rectangular section seals of similar material are fitted along the bottom edge of each section, the compressive load of the structure forming the seal.

5.3.4 *Operation*

Control facilities are available on local control panels in the gate control house and remotely in the station central control room. The system can be placed on remote automatic control in which gate movement automatically follows the upwards or downwards movement of the lower reservoir level to maintain a pre-determined water flow from the reservoir. Visual indication of the particular function selected and the operational state of the gates is given at the control points. Closing time for the gates is 8 minutes.

6. MAIN PLANT

6.1 Pump-Turbine

The main plant consists of six pump-turbines, together with main inlet valves and draft tube valves which isolate them from the upstream and downstream water reservoirs. The units are each designed to produce 287MW at the minimum net head of 494m. As a structure the pressure containment components are tested against a pressure head of 120 bar.

The turbine is of the reversible Francis design which will pump when driven in one direction by grid system power, and generate power as a turbine when driven by hydraulic pressure from the upper reservoir in the other direction.

During intervals between generating periods, compressed air is injected into the runner chamber, in order to depress the water level down below the runner and reduce hydraulic drag and hence power consumption when operating in the spinning reserve mode.

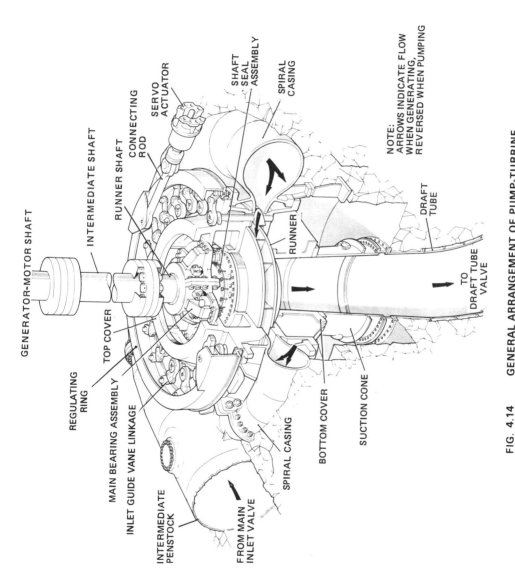

GENERATOR-MOTOR SHAFT

INTERMEDIATE SHAFT

RUNNER SHAFT

CONNECTING ROD

SERVO ACTUATOR

SHAFT SEAL ASSEMBLY

SPIRAL CASING

NOTE:
ARROWS INDICATE FLOW
WHEN GENERATING,
REVERSED WHEN PUMPING

DRAFT TUBE

RUNNER

TO DRAFT TUBE VALVE

REGULATING RING

TOP COVER

MAIN BEARING ASSEMBLY

INLET GUIDE VANE LINKAGE

INTERMEDIATE PENSTOCK

FROM MAIN INLET VALVE

SPIRAL CASING

BOTTOM COVER

SUCTION CONE

FIG. 4.14   GENERAL ARRANGEMENT OF PUMP-TURBINE

6.1.1 *Design Features (Fig 4.14)*

The pump-turbines operate at a speed of 500rev/min. The power output is controlled by 24 inlet guide vanes (IGV). If generation is optimised, a firm output of 1675MW can be maintained for 5 hours. The corresponding time for pumping the water back to the upper reservoir, Marchlyn Mawr, is approximately 6 hours. The average head is approximately 520m OD and the quantity of live water used is 7 million cubic metres. These machines have a maximum efficiency of 93.2% as a turbine and 92.5% as a pump. The spiral casing is manufactured from carbon manganese steel plate which is 76mm thick and is welded to a cast steel stayring in the factory. For handling and transport convenience it was built in four quadrants which were bolted together in situ at site.

The runner is cast in a weldable stainless steel alloy in order to facilitate repairs. The top and bottom covers are single castings which weigh in the order of 70t and house the bushes for the guide vane stems.

Much of the pump-turbine structure cannot be seen on site since it is embedded within the foundation concrete in order to physically restrain movement due to hydraulic forces and also to keep the noise levels down to an acceptable level. At the turbine gallery level access is provided to the regulating ring which is driven by the two governor servomotors to alter the gap/angle of the IGV and hence the power output of the machine as demanded by load requirements. Each guide vane link is attached to the regulating ring by means of a friction device which slips at a pre-set torque such that if any debris jams between two guide vanes, then the remaining guide vanes are able to close normally and shutdown the machine safely. Below the spiral casing which forms the main body of the machine, is the suction cone which is where the water level is monitored during spinning-in-air operation. It also provides access for runner inspection for cavitation damage in service, by means of manholes. Below the suction cone is the draft tube liner. This is a right-angled bend which streamlines the flow into and out of the machine under flow conditions. Since it is the lowest part of the hydraulic system, drainage connections are built into its design for tunnel dewatering purposes. As a precaution to minimise cavitation damage the station excavation was designed to arrange for a runner elevation 60m below the lowest level of the lower reservoir.

6.1.2 *Hydraulic Performance*

Early in the contract period a scale model of the turbine was built and tested under laboratory conditions in order to verify the performance parameters of generated output power, pump input power, turbine efficiency, pump efficiency and cavitation, ahead of manufacture for the machine itself. Additional information regarding runner side loads and axial thrusts during varied plant operation was also obtained.

6.1.3 *Pressure Containment Integrity*

The spiral casing and main inlet valve (MIV) will be subjected to cyclic pressure loading whenever the MIV is closed. This will happen frequently bearing in mind the spinning-in-air duties. Over 60% of the design plant life of 40 years is consumed by this mode of operation. This fatigue loading has been used as a design criterion for the pressure parts affected at Dinorwig. It is believed that this is the first time that the fracture mechanics technique has been used in the design of hydro-electric plant.

Traditional high-head plant would normally have taken advantage of a 23% reduction in tube wall thickness offered by high strength steels. However, the need for thicker sections to reduce stress levels and hence control the critical defect size and crack growth rates to within the capability of reliable NDT monitoring equipment, has turned the emphasis towards higher strength steels with their added advantage of better welding capability. Accordingly, carbon manganese steel was selected.

With the need to ensure the continuing integrity of the plant it was essential that the inspection techniques were capable of establishing that there were no defects which could cause failure within the design life.

The use of fracture mechanics in design takes account of the fact that metal structures contain defects produced during their manufacture and assesses the rate at which these defects grow during an operational loading regime before fracture would occur.

Such an assessment requires an accurate knowledge of the plant loading, a detailed understanding of component stresses together with the size and location of inherent defects, data on fracture toughness of materials and crack growth rates.

The Dinorwig machine life was defined as 300,000 cycles and the stress levels were predicted by photo-elastic models and finite element analysis and final confirmation obtained by strain gauge testing during the site pressure test.

Throughout manufacture, all measurable defects in fabrication welds and castings were sized and positioned using ultrasonic examination, recorded and assessed by fracture mechanics design life calculations, and unacceptable defects excavated and weld repaired.

In service, the assessment is used to determine the minimum inspection periods, to allow for safe operation and to enable strategic planning of any preventive maintenance work.

6.2 Main Inlet Valves (Fig 4.15)

Each of the six units has a rotary type MIV located in a separate gallery, which is adjacent to the machine hall. The MIV isolates the pump-turbine from the HP tunnel system, both when the unit is shutdown in normal operation and for maintenance purposes. Normally, the MIV operates in a balanced condition, ie zero flow with the pump-turbine guide vanes closed, but in an emergency, it is capable of closing against a full turbine discharge of $70m^3/s$. The MIV is closed whenever its associated pump-turbine is either shutdown or is spinning-in-air. The MIV is opened by twin oil pressure operated servomotors in 5s and is closed by weights with the servomotors acting as dash pots in 20s. In each case the operation is part of the operating sequence of the pump-turbine. The MIV functions very similarly to a conventional ball valve, except that there is a clearance between the moving valve rotor and the housing. These clearances are closed by steel rings. The ring at the downstream end is called the service seal and operates automatically as part of the normal operating sequence. The ring at the upstream end is called the maintenance seal and usually remains open, then only being closed for isolating purposes.

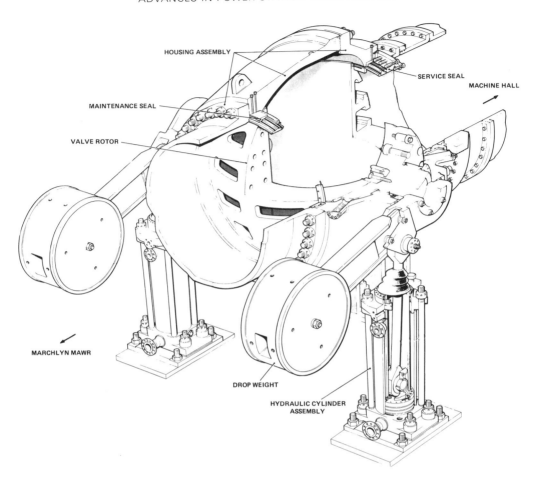

FIG. 4.15   GENERAL ARRANGEMENT OF MAIN INLET VALVE

In order to avoid damage, these seals must never be closed unless the rotor is closed; conversely the rotor must never be opened unless both seals are open (retracted) first. The MIV is rigidly connected to the end of the penstock and can slide axially on its foundations to accommodate changes in the length of the penstock when the MIV opens or closes. Adjacent to each MIV is its own high pressure oil pumping unit, a local control panel and two separate panels carrying the associated oil and water controls. The operation of the rotor and service seal in the opening and closing modes is normally initiated remotely by the sequence control equipment. There are also facilities for remote manual operation of the rotor, the service seal and the maintenance seal from the unit control desk. In addition, means are provided for independently opening the service seal under conditions of penstock resonance. Local manual operation of the MIV and its seals can be selected by operation of a switch at the local control panel. For normal isolation of the pump-turbine from penstock water, the

MIV is closed and locked off. In this condition the maintenance seals and the service seal are applied, the locking pins are inserted and pressure oil for opening the MIV is isolated. Control water for keeping the seals closed is drawn from the penstock and/or from its auxiliary seal water pumping sets depending on whether the penstock is full or not. Where a specially high level of security is required, eg when a manhole is removed and personnel are working inside the pump-turbine suction cone or spiral casing, the MIV is put into a condition called 'high integrity isolation'. This condition, which can be reached only after dewatering the pump-turbine with the MIV in the locked-off condition involves the fitting of a deflector plate to the upstream section of the dismantling joint after removal of the downstream section. Furthermore, should the service seal closing pressure be inadvertently lost, its opening movement is restricted by the movable flange and the escape water rigidly controlled, within the capacity of the station drainage system.

6.3 Draft Tube Valves (Fig 4.16)

Six draft tube valves (DTV) are located in a separate gallery downstream of the draft tube. These are of the lattice blade, butterfly type and serve as isolating valves for tunnel dewatering purposes and also as a safety valve to prevent pressurising of the draft tube from the upper lake. Each DTV is opened by two oil-operated hydraulic cylinders in 120s and is closed by drop weights with the hydraulic cylinders acting as dash pots in 30s. Its operation plays no part in the operating role for the station. The valve has a bore size of 3.75m, weighs 80t and is designed for 60 bar water pressure. Two bypass pipes, each fitted with an electrically-operated isolating valve which is mounted between two manually-operated guard valves, are mounted on each DTV. The bypasses are for filling the draft tube tunnel, pump-turbine and intermediate penstock, to equalise the pressure on both sides of the DTV before opening. In addition, drain points are positioned between the isolating valve and guard valve for draining any leakage water which escapes through the stop gates at the tailworks when the DTV is closed. The DTV is rigidly connected at the downstream end and via a dismantling joint at the upstream end to the hydraulic tunnels. The DTV is mounted on sliding feet which can move axially to accommodate movement in the LP pipework due to water pressure. Adjacent to each DTV is its own hydraulic/electric control cabinet which supplies the pressure oil, together with an electric switch assembly which controls the operating sequence.

6.4 Generator-Motor

6.4.1 *Design Basis*

(1) Design Objectives

The design objectives were to ensure reliable operation of the generator-motors, with particular reference to the frequent starting, stopping and changing of modes and the high availability necessary to meet the CEGB's system requirements. Under normal system conditions the machines will probably be started and synchronised three times a day in either the generating or pumping modes. System requirements may increase this to six times a day. During the life of the station the daily number of operations is unlikely to fall below these figures, but

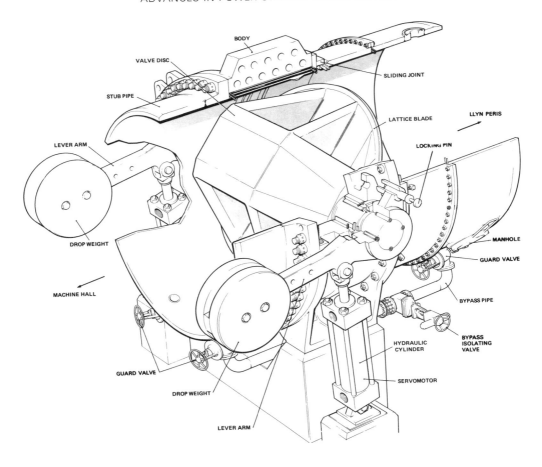

FIG. 4.16   GENERAL ARRANGEMENT OF DRAFT TUBE VALVE

rather increase due to the flexible role ascribed to the pumped storage plant. This will result in the maximum number of starts per machine per year ranging between 1250 and 2500.

High engineering standards were to be applied to avoid the necessity of maintenance between annual overhauls and still achieve a starting reliability of not less than 99% for both generation and pumping.

Additionally, to meet the frequency and mode control requirements specified, it is estimated that each generator-motor will be subject to 15,000 mode changes annually.

(2) Scheme Parameters — Choice of Basic Machine Rating

The output from Dinorwig was determined from hydro-geological considerations and from the capacity of the uprated existing transmission line. (Further overhead lines were ruled out on amenity grounds). These factors gave an output of 1800MW.

This output left four practical possibilities for machine size:

(a) 8 machines of 225MW at 600rev/min
(b) 6 machines of 300MW at 500rev/min
(c) 4 machines of 450MW at 375rev/min
(d) 4 machines of 450MW at 428rev/min

The specific speed of the pump-turbines dictates the generator-motor speed.

(3) Cost

In cost, (a) was the highest with (c) and (d) the lowest, the higher cost of the larger number of smaller machines being attributable mainly to civil engineering works. Plant costs showed little reduction in specific cost with increase in unit rating. This is consistent with the reduction in optimum speed of the pump-turbines as the size increases.

(4) Cooling and Reliability

Air-cooling appeared to be just about possible for 450MW ratings stretching all parameters to the limit, but it was considered that water-cooling for both the stator and rotor would be necessary to produce a design with an acceptable degree of extrapolation in critical design areas.

The water-cooled machine was considered unlikely to meet the project reliability requirements for the following reasons:

(a) The CEGB's long standing problems with leakage from water-cooled stators.
(b) No international experience was available with reversible water-cooled sets.
(c) Since existing and proposed designs for reversible sets did not include high air pressures within the machine, water leaks would be critical and cause damage to stator windings.
(d) The large number of load cycles and stress cycles specified for the project would give rise to accelerated thermal fatigue problems on water cooling pipes.
(e) There was no experience of reversible pump-turbines at outputs as great as 450MW with a high head of 500m.
(f) There was no point in reducing the size of the rotor by employing water-cooling, if additional inertia had to be built-in to maintain the required inertia constant.
(g) The 300MW and 225MW machines are within acceptable parameters for air-cooling.
(h) The lack of experience of any set running at 600rev/min above 135MW made the 500rev/min machine preferable.
(i) Experience with fully water-cooled hydro-machines was limited to a few single rotation units with minimal numbers of cycles per year. The only British manufacturer had no experience in the manufacture of water cooled hydro-machines. On the other hand, 375rev/min air-cooled machines of about 300MW were being designed for pumped storage operation.

(5) Adopted Design

As compared with the scheme with four 450MW machines, the scheme with six 300MW machines had distinct advantages in the assessment of complexity, problems in manufacture, reliability, maintenance requirements and loss due to outage. The scheme with eight 225MW machines showed no further improvement in reliability, but incurred a substantial cost penalty.

It was concluded that the maximum size of generators which should be considered, based on a limited extrapolation of existing designs was six sets of 300MW at 500rev/min.

6.4.2 *General Design (Fig 4.17)*

Because the six 330MVA 18kV 500rev/min generator-motor units are used for frequency control as well as the traditional roles of peak lopping and standby generation, the sets are involved in a large number of mode changes, and each machine has been designed to withstand up to 40 mode changes a day throughout its 40 year life.

Many of these mode changes include a reversal in the direction of rotation, subjecting the machines to severe electrical and mechanical stresses during repeated cycling between standstill and full speed at rated load.

The specified sequence times for mode changes include a requirement for the output of the station to be increased from zero to 1320MW in 10s with six units operating, whilst maximum permitted times for starting the machines to the generating or pumping modes are 90s and 540s respectively. By using electrical braking on the machine, the units are brought to a standstill in 360s. In addition, each machine is capable of transferring from full-load pump to full-load generate in an emergency in 90s.

In spite of these very demanding operational requirements, the generator-motors and their auxiliary systems have been designed to operate continuously for 12 month periods without any off load maintenance and to achieve a high startup and mode change reliability.

Each generator-motor has its own independent excitation system, automatic voltage regulator, HP oil system, braking and jacking system and brake dust extraction system.

Two sets of variable frequency starting equipment are provided, each of which is capable of starting the six units sequentially.

The generator has an upper thrust bearing together with bottom and top guide bearings. The thrust bearing is supported in a housing which is integral with a top bracket comprising two deep bridge-type girders. A long stator core length and the specified requirement that the critical speed shall be not less than 20% above the maximum overspeed of 763rev/min, led to the top guide bearing being placed below the thrust bearing, so as to keep the distance between guide bearing centres to a minimum.

The rotor design comprises a friction-type rim mounted on a fabricated mild steel spider. The latter is in two parts, a lower and an upper part, because the unusual length would make a one part spider awkward to machine. Laminated poles are secured to the outer periphery of the rim by six 'T-headed' projections on the pole engaging with similar shaped slots in the rim. Round copper bars forming the damper winding are

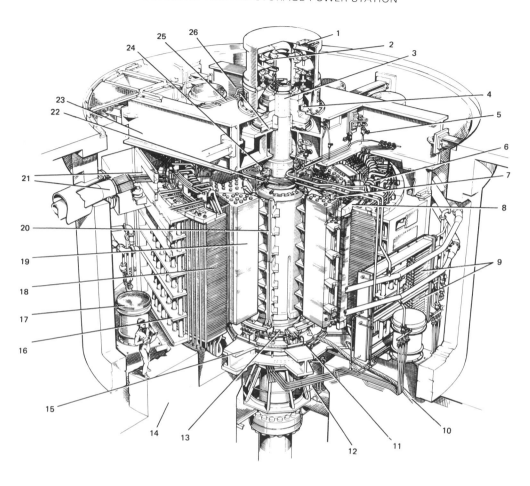

1	AIR FILTER	14	BOTTOM BRACKET
2	EXCITATION BRUSHGEAR	15	MAIN FRAME
3	ROTOR SHAFT	16	STATOR SEGMENT FRAMES
4	SHAFT CURRENT EARTHING BRUSH	17	MOTOR DRIVEN COOLING AIR FAN
5	THRUST BEARING OIL COOLER	18	STATOR CORE ASSEMBLY
	WATER SUPPLY	19	ROTOR CORE ASSEMBLY
6	STATOR WINDINGS	20	SPIDER
7	STATOR WINDING CONNECTOR	21	STATOR WINDING BUSBAR
	(TO BRUSHGEAR)		CONNECTION (ONE PHASE)
8	POLE WINDING 4b POLE FACE	22	TOP GUIDE BEARING OIL COOLER
9	AIR/WATER COOLERS	23	MAIN FRAME · TOP BRACKET
10	BRAKE DUST EXTRACTOR	24	TOP GUIDE BEARING ASSEMBLY
11	BRAKE ASSEMBLY	25	THRUST BEARING JACKING
12	BOTTOM GUIDE BEARING OIL COOLER		OIL SUPPLY
13	BOTTOM GUIDE BEARING ASSEMBLY	26	THRUST BEARING ASSEMBLY

FIG. 4.17   GENERAL ARRANGEMENT OF GENERATOR MOTOR

fitted into semi-closed slots in the pole face and shorted at the ends by copper laminations sandwiched between the steel laminations to give adequate support to the ends of the bars.

The stator winding is formed from single-turn bars placed as a double layer in open parallel slots and lap connected. Top bars are subdivided into a larger number of strands than are the bottom bars. This gives the advantage of balancing the self-induced strand loss between top and bottom bars. The bar-to-bar joints are formed by butting together top and bottom bar strands in a sufficient number of groups to permit a very reliable brazed joint to be achieved. The precise technique was developed after several very searching tests on prototype joints.

A Roebel transposition to the high voltage bar stator winding has been used with a view to eliminating the circulating current losses. The stator winding insulation is, essentially, a system based on epoxy-resin-rich, mica-paper material with which there has now been an appreciable number of years service experience. Although the insulation complies fully with Class F standards, the machines have been designed conservatively to have temperatures not exceeding those permissible for Class B insulation.

Ripple spring separators, which are always maintained under compression, are provided in the slot depth of the stator winding to ensure that the stator bars are held securely in the slots to prevent mechanical vibration. The slot wedges are of glass-mat material which does not shrink in service and so helps maintain the general tightness of the bars in the slot.

The machine ratings are as follows:

Motor -
    Continuous output (50Hz)   292MW/312MVA
    Power factor   0.95

Generator -
    Continuous output   313.5MW/330MVA
    Power factor   0.95

6.4.3 *Excitation System*

The excitation of each generator-motor is provided by a thyristor rectifier which is controlled by a fast-acting automatic voltage regulator.

The main power loop comprises a 3-phase excitation transformer, an excitation triple pole ac circuit breaker, a thyristor rectifier, a field suppression circuit breaker and the dc connections to the machine brush gear and then to the rotor.

The excitation transformer is supplied directly from the 18kV system so that under all starting conditions, in both the generate and pumping modes full excitation is available.

Following total station shutdown and when the 18kV supply is not available, the machine is excited by energising the field from the 240V dc unit battery to enable generation to commence.

The excitation rectifier consists of a full wave 6-pulse arrangement of thyristors supplied from the secondary winding of the excitation transformer. Duplicate firing circuits are provided and a fault in the operational firing circuit results in an automatic changeover to the standby firing circuit. Auxiliary power supplies for the thyristor excitation equipment and automatic voltage regulator (AVR) are derived from a tertiary winding on the excitation transformer.

The circuit breaker between the excitation transformer and the thyristor cubicle provides protection in the event of a short circuit on the dc output of the rectifier. It also permits isolation of the excitation equipment from the excitation transformer (which is continuously energised) for maintenance purposes.

A conventional arrangement of field suppression circuit breaker and discharge resistor is used for the protection of the generator-motor field in the event of a stator fault. It also provides isolation of the field before the application of a 3-phase shorting switch to the generator-motor stator in preparation for a dynamic braking sequence.

The excitation equipment for each generator-motor is housed in a suite of cubicles located near the unit on the 42.50m level.

The equipment has a rated continuous output of 357V dc 1700A; while field forcing, the permissible output is 766V dc 3830A for 5s.

There is one thyristor in series per arm and six thyristors in parallel, thus giving a total of 36 and the equipment will provide its full rating with up to two thyristors per phase arm out of circuit.

6.4.4 *Automatic Voltage Regulator*

The continuing increase in unit size of generating plant together with the trend towards lower short-circuit ratio generators has resulted in a greater dependence on the excitation control system as a whole to maintain the power system stability. The generator motors at Dinorwig have dual channel automatic voltage regulators.

The AVR determines by means of a dc control signal the output of the thyristor excitation rectifier which supplies the machine field, thereby regulating the power factor of the machine. The AVR holds the machine terminal voltage constant over the range no-load to full-load and is active over the whole range with no dead band.

The dual channel AVR consists of two identical electronic control circuits plus a common voltage matching, fixed excitation, output selection and relay circuit. Usually both the main and standby channels are energised and functional. However, only the output of the main channel is switched by the output selection circuit to control the converter. If this channel fails then control is automatically switched to the standby channel.

The normal mode of operation is auto-control but fixed excitation and voltage matching are available for machine starting and dynamic braking. Additionally, manual control of the excitation is available for testing and setting up purposes or in the unlikely event of failure of the auto-control of both channels.

The auto-control circuits consists of two interdependent feed back control loops (voltage and current) with limiting circuits for under-excited MVAR, rotor heating, current and power system stabilizers. Amplifiers in the voltage control circuits compare the voltage feedback signal, representing stator volts, with the voltage reference signal, which represents the required volts. The difference signal is the auto-control signal, and amplifiers in the current control loop compare the current feedback signal representing the excitation current with the current reference signal, to give the thyristor control signal.

Signals produced by monitoring stator voltage and current levels are introduced into the voltage control loop to provide MVAR limit and power system damping, ie rotor. Rotor heating current limit is provided by introducing a signal derived from the current feedback into the current control loop as necessary.

A joint control scheme maintains an even distribution of the required reactive load between machines which are switched into the scheme. It also corrects the station reactive load to a new level when so required by the operator. Any number of machines may be switched into the scheme by the operator provided the associated AVR is in auto-control and the main generator circuit breaker is closed.

The joint control circuits monitor the reactive MVAR load of all generators in the scheme. The total station reactive load is determined by summing the individual machine load signals and the required average machine load can then be calculated. Comparator circuits sum the required average machine load with the actual machine loads and activate relay circuits to raise or lower the relevant unit excitations as required until the difference is reduced to zero. When a change in station demand is required the control potentiometer is set to a new level and a station error signal is produced and fed to the comparator circuits which initiate the appropriate raise/lower action.

6.4.5 *Starting Equipment*

For the Dinorwig pumped storage project various methods of starting the machines in the pumping mode were considered, and finally a static variable frequency starting system was adopted as the main starting method with the alternative of back-to-back starting using a second machine if the static equipment was out of service.

In order to be able to start all six machines in the pumping direction in approximately 30 minutes, two starting equipments have been supplied and these achieve a starting time of 9min per machine. In the event of failure of one starting equipment, then the other is able to start all six machines in approximately 60 minutes.

Using the variable frequency starting equipment, starting is divided into two stages. During the first stage the converter operates with forced commutation because the voltages obtained from the motor are too low to commutate the inverter. The frequency for this forced commutated system is controlled by rotor position transmitters. On the machine shaft there is a tooth wheel and surrounding this are three sensors mounted on a suitable bracket. The sensors control the pulses of the converter in such a way that the frequency of the ac voltage from the converter follows the frequency (speed) of the machine. This method is used up to 5Hz.

When a frequency of 5Hz is reached, the control of the converter is changed so that the commutation is accomplished with the aid of the voltage generated by the motor. The power connection is altered at the same time so that the converter is connected to the motor via a transformer. The purpose of this is to match the higher motor voltage during this stage with a level suitable for the converter. The motor is now accelerated with almost constant current up to overspeed (105%) and the converter is disconnected and as the motor slows down it is synchronised to the network at 100% synchronous speed.

During this second stage of the starting sequence, the field current is maintained constant, thus the voltage on the motor increases linearly with speed.

A dc current limiting device has been provided in order to protect the inverter against short circuits when the rectifier fuses would be too slow to operate. A cast-in-concrete smoothing reactor rated 1.6mH is also provided in the positive dc interconnections between the rectifiers and inverters.

6.4.6 Cooling System

The generator-motor is air-cooled by a closed ventilation system comprising a closed air circuit and air/water heat exchangers. Since the reverse rotation in the pumping mode precludes the use of efficient rotor-mounted fans, the cooling air is circulated within the machine by 12 separate motor-driven aerofoil type cooling fan units. Six of these units are located at the top of the generator and the other six are located at the bottom over ducts in the floor of the concrete enclosure and between the stator frame and the enclosure wall.

Both sets of fans draw cooled air from the annulus space between the stator frame and the enclosure wall and circulate it into the ends of the machine. This air passes over the stator coil windings, the rotor poles and field coils and then via the air-gap enters the air ventilation ducts in the stator core. The salient poles of the rotor assist this circulation of air.

The air flowing through the ventilation slots in the stator core cools the core and stator windings. From the back of the core the air passes through eight air/water heat exchangers mounted on the periphery of the stator frame, and into the annulus space between the stator and the enclosure wall. The eight air/water coolers are mounted in four groups of two coolers each, the two coolers in each group being mounted one above the other on the stator frame.

Each of the eight parallel water circuits for the air/water coolers has a control valve in the inlet side to the coolers, whilst a flow indicator and a water stop-valve are located in the outlet side of each circuit. In addition there are two flow relays in the common outlet from the eight cooler circuits, one flow relay being a standby to the other.

In the air circuit of each air/water cooler there are two thermocouples, one on the inlet side measuring hot air temperature and the other on the outlet side measuring cold air temperature. Signals are taken from these thermocouples to the data logger located in the station control room.

In addition to the thermocouples there are two thermometers in the air circuit measuring stator 'hot' air temperature and the other measuring stator 'cold' air temperature. These are dial type thermometers with the usual mercury in steel bulb type sensors and each thermometer has two sets of normally open contacts for alarm and trip purposes. In addition to providing alarm and trip signals, these thermometer contacts are used to override, if necessary, the control of the fans during running in the spin-generate or spin-pump modes.

6.4.7 HP Oil System

The HP oil system provides lubricating oil to the generator-motor thrust bearing pads during starting and stopping of the machine when the rotational speed is insufficient for hydro-dynamic lubrication.

Before the generator-motor is started, the HP oil system establishes an oil film on the face of the thrust pads and provides lubrication in this manner until the set reaches 85% of synchronous speed (425rev/min). Similarly, when the machine is being shutdown, as the speed falls below 85% of synchronous speed, the HP oil system is brought into operation and continues to supply oil to the thrust pads until the set is stationary.

Whilst the machine is above 85% synchronous speed, lubrication is obtained automatically by hydro-dynamic means whereby the speed of the machine is sufficient to generate a wedge-shaped oil film between the faces of the thrust bearing pads and the runner disc.

The HP oil system comprises one ac motor-driven main pump and one dc motor-driven standby pump together with associated valves, LP and HP filters.

The motor-driven pumps, filters, valves and pressure gauge for the HP oil system are assembled onto a fabricated steel housing which is located on the outside wall of the concrete enclosure on the upstream side of the motor-generator.

The sequence control equipment ensures that the generator-motor cannot be started until correct HP oil pressure is available, and also primes initiation of the emergency dc motor driven pump on failures of HP oil pressure.

6.4.8 *Thrust Bearing*

The thrust bearing consists of a highly polished steel thrust face on the generator-motor rotor, which bears on a ring of 10 segmental tilting pads, each supported on a spring mattress. The thrust collar is shrunk on to a tapered seating on the rotor shaft and a split ring key holds the collar in position.

The thrust pads are of stress-relieved mild steel faced with a 3mm thickness of lead-based white metal. The face of the white metal has a circular channel machined in it for the high pressure oil and each pad is water-cooled by two cooling circuits. Each circuit consists of two holes drilled radially in the pad with their inboard ends connected by cross-drilling. The outboard ends of the four holes are tapped for the cooling water connections. Cooling water for the pads is taken from a ring main within the bearing housing which is in turn supplied from the main generator-motor cooling water system.

The pads are not all identical since some are provided with drillings for thermocouples and thermometer sensing units for temperature monitoring. Each thrust pad is supported on a spring mattress comprising 580 compression springs. During manufacture the dimensions of the springs and the thrust pads are carefully controlled to ensure a uniform distribution of the thrust load over all of the 10 pads. The springs are mounted on a spring plate fixed to the floor of the bearing housing.

The thrust bearing is immersed in oil contained in the bearing housing and oil circulation is induced by the rotation of the thrust collar fixed to the generator-motor shaft. The oil is cooled by two oil/water heat exchangers mounted in the bearing housing, one on each side of the bearing assembly. Oil for the thrust bearing housing is supplied from the main lubricating oil system.

The thrust bearing housing is fabricated from steel plate and is an integral part of the generator-motor top bracket with bolted-on covers on the housing walls giving access to the thrust bearing. The top cover of the housing is electrically-insulated from the housing and carries the stationary components of the vapour seal and the 'earthing' brush assembly.

Post commissioning experience exposed problems with the thrust bearing, which necessitated profiling the bearing surfaces to an extremely high accuracy to prevent failure (see Section 19.1).

Instrumentation is provided to monitor the oil level, oil temperature, thrust pad temperatures, flow of cooling water to the pads and flow of cooling water to the oil coolers.

Insulation for the thrust bearing is located at the joint between the runner disc and the thrust collar.

6.4.9 *Braking and Jacking System*

A specially-designed dual-function arrangement provides for braking or jacking of the generator-motor.

Six cylindrical-shaped pressure units complete with pistons are fitted to the bottom bracket of the machine and when charged with air at 7.5 bar, they provide the braking effect. This is automatically applied when the machine speed falls to 25rev/min, and is maintained until the rotor comes to rest. The brakes are maintained for 10min after the unit has come to rest.

An independent air supply provides compressed air for the braking system for all six generator-motors and a dust extraction system is provided to collect any brake dust and prevent it entering the generator-motor ventilation system.

The friction pads of the brakes act on a steel brake track bolted to the lower end of the rotor, so braking is applied to the complete shaft assembly, ie rotor shaft, intermediate shaft and turbine runner.

These friction brakes are controlled through the general and auxiliary control relay equipment. When these six pressure units are charged with high pressure oil at 140 bar, they act as a rotor jacking system, and are capable of lifting the complete shaft assembly comprising the rotor shaft, intermediate shaft and pump-turbine runner. The free distance available is some 14mm.

Each generator-motor has its own self contained HP oil pack for supplying HP oil to the jacking units. The control valves in the braking and the jacking systems are mechanically intercoupled so that the compressed air is isolated from the HP jacking oil.

The brake/jacking control panel associated with each unit housing the control valves and pressure gauges is located at the 34.80m level and the panels of two adjacent machines are mounted on a common support structure positioned between the two units.

7. MECHANICAL AUXILIARY SYSTEMS

7.1 Cooling Water System (Fig 4.18)

To keep the power plant in continuous operation, an essential supply of cooling water (CW) must be supplied to cool the oil in the machine bearings, the oil in the transformers, the air which circulates around the generator-motor windings and the pump-turbine shaft seal.

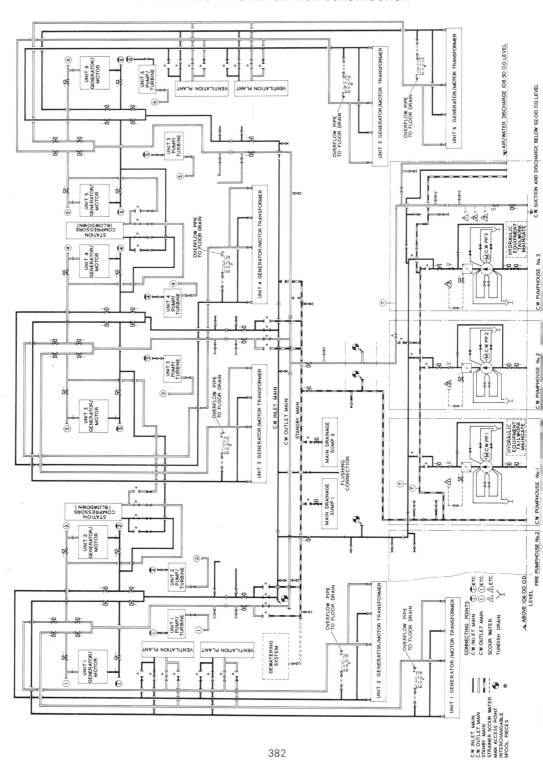

Additional supplies are needed for the ancillary plant such as air compressors and cavern air conditioning equipment.

Three pumps are located in the tailworks structure, any two of which normally supply the total needs of a full operating station with the third as an idle standby unit. Each pump is of the horizontal split casing centrifugal design and is capable of a duty flow of 1000 litre/s. The pumps are housed in separate compartments such that a pipe failure would only lead to flooding of one pump unit.

The pumps draw water from the lower lake and transfer it down to the underground works via a 1m diameter culvert buried within the tailrace excavation, through the station coolers and back to the surface works by means of a similar culvert which finally discharges into the neck of Wellington Channel, away from the pump intakes, in order to avoid recirculation of warm water. Both ends of the system are below the lower lake level of 92m OD so that the pump only has to generate pipe friction head and no static head.

The pump discharges through an outlet valve which closes automatically in the event of a pump failure, in order to avoid reverse flow from system pressure. Flow from each pump is screened by a rotary automatic self-cleaning strainer before it enters the supply culvert to the underground works.

A similar autoclose valve is fitted to the station outlet main, such that a pipe failure within the underground works would not lead to a back flow from the lower lake and that a system shutdown would seal both ends of the pipework. Valve closure times have been set to keep the water hammer stresses within the structural capability of the pipework and valve materials.

7.2 Dewatering System

To drain the hydraulic tunnel system, gates can be seated to seal off the top lake and allow the station turbines to transfer water into the bottom lake until lower lake level exists throughout. At this point a further gate can be seated at the tailworks and remaining water pumped out separately.

From each pump-turbine draft tube a system of pipework is provided to feed the suction of two dewatering pumps, discharging the remaining tunnel water into the station drainage main which constitutes a 1m diameter pipe following the tailrace tunnel up to the surface works, with an outfall into the lower lake at a level high above top water level. Each pump is of the horizontal split casing centrifugal design with a duty flow of 600 litre/s.

7.3 Drainage System

It is necessary to maintain the underground caverns in a dry and acceptable environment. In view of the natural seepage water, incident leakage water and condensation, the civil works design incorporates various pipes and gulleys which ultimately lead to a drainage gulley within the pipe gallery from which drainage water enters the main drain sumps after passing through a multi-pass oil separator. There are two main sumps each equipped with three normal drain pumps and two emergency drain pumps all of which discharge into the station drainage main. These pumps are of a fully submersible pump-motor design such that flood levels above floor level can be tolerated without plant failure.

Normally one pump would be in service at a time but in an emergency flood, high level overspill at pipe gallery floor level would bring the second pump into automatic operation. All pumps are operated on a sequence dictated by individual water level detection switches for starting and stopping. The emergency pumping capacity is designed to cope with the fracture of any pipe which falls outside the main isolating area of the hydraulic system formed by the main inlet valve and the draft tube valve. The normal pumps have a duty flow of 20 litre/s and the emergency pumps 400 litre/s and are all of proven commercial design.

7.4 Blowdown Air System

To provide the large quantity of air needed to depress the water level down against the lower lake head to a level below the runner for spinning-in-air operation, four high pressure station compressors are installed. Each compressor is of the 3-stage 3-cylinder reciprocating design and can deliver a minimum of 28.4m^3 of free air.

This bank of compressors has to charge the six spherical units blowdown receivers up to a storage pressure of 31 bar and recharge following a 6-unit simultaneous blowdown within 15 minutes. The receiver discharge system can blow the pump-turbine down in 15s and can maintain the water level within defined limits by means of automatic control of a topping-up valve by water level monitoring equipment.

7.5 Air Admission System

During part load generation, flow behaviours tend to be noisy and this can be effectively suppressed by the injection of a metered flow of compressed air into the runner area drawn from the blowdown air receivers.

7.6 Maintenance Air Supply

In view of the large quantity of available storage capacity provided by the blowdown air receivers, and the ability of each operating receiver to serve two adjacent units during operation, a maintenance air supply at a pressure of 7 bar is taken from the blowdown air system, via a pressure reducing valve.

The distribution pipework extends to provide connection points throughout the main plant areas in the underground complex including the workshops.

7.7 Oil Handling System

To minimise the possibility of oil spillage in plant areas around the station, a central oil storage, purification and handling system has been provided. Pipework enables oil to be delivered by road tanker and pumped into a station tank where it can be raised to the correct standard of cleanliness before being pumped into the machine bearings, or governor system. Dirty oil can be drained back to the central tank for cleaning or disposal as also required. Portable oil purifiers can be connected at selected locations to facilitate on-load purification should this be required.

8. FIRE PROTECTION AND DETECTION

8.1 Fire Protection

In comparison with a thermal power station, the fire risk in the underground works is considered to be low because the pumped storage process fluid is water at about room temperature; nevertheless the use of inflammable materials has been kept to a minimum, and the structure is non-combustible and fire resisting. The emphasis, therefore, has been to identify the fire risk areas in the station and ensure the rapid detection and extinguishing of any fires that might occur, in order to minimise the production of smoke.

8.2 Water Systems

The principal fire risks arise from the use of oil, which is required for cooling the larger transformers as well as for lubrication and hydraulic servo-systems, and from certain items of electrical equipment which use combustible materials in their insulation. Fixed fire protection has been provided over all oil-filled transformers and other major oil concentrations. This is mainly in the form of high velocity waterspray initiated by a quartzoid glass bulb detection system and controlled by automatic deluge valves located away from the item protected. Manual operation of the deluge valves is possible.

The oil-filled cables which are used for the 400kV connections enter the transformer hall via cable tunnels. Since the cables are buried in the floor of the tunnels, any fire risk is confined to the sealing ends where high velocity waterspray is provided.

Although cables used throughout the station are of the flame retardant type, the principal areas of concentrated cabling in the switchgear annexe, busbar galleries and unit cable shafts are protected by a quick-acting waterspray system. Additionally, segregation of cables has been adopted to ensure that any fire damage would be confined to one unit only.

A large number of fire hydrants is provided throughout the complex, most hydrant points being supplied with foam making equipment.

Water discharge onto the largest single fire risk, a generator-motor transformer, is limited to a period of 5min by timers in order to control the quantity of water discharged in the underground works.

Water supplies for fire fighting purposes are provided from a fixed pumping installation in the fire pumphouse at the tailworks. An electrically-driven pump with a standby diesel-driven pump supplies water to the automatic waterspray system via a 600mm diameter duty or 600mm diameter standby main (Fig 4.13).

Duplicate electrically-driven hydrant pumps supply water to the hydrants installed in the access tunnels and the surface hydrant system.

A pumping-in point is arranged at the Afon-y-Bala channel where a constant source of water will allow the local fire authority to connect their appliances and pump into the hydrant system in the event of a failure of the fixed hydrant pumps.

In the fire pumphouse, the fire pumps are located in segregated compartments to protect against loss due to flooding of the pumphouse.

8.3 Gas Systems

Automatic bromotrifluoromethane (BTM) gas protection has been provided for the six generator motors, the six marshalling cubicle rooms, the data logger room and the input and output transformers. The systems are designed on the basis of totally flooding the enclosures with a 6% concentration of BTM gas, each enclosure being provided with two gas cylinders, one acting as duty with the other as standby.

BTM gas protection is also provided on the emergency diesel generator installation which is located in an above ground building at the tailworks.

8.4 Foam Systems

Certain areas of the station are too remote to be served by the fixed water spray installation. These are the above ground emergency diesel generator transformers at the tailworks and the emergency diesel generator together with its associated transformer located at the headworks. A pre-mix foam system is installed to protect these risks since they are, in two of the three cases, outdoor installations and gas protection is not appropriate.

8.5 Heat and Smoke Detection Systems

The heat and smoke detection system covers 33 areas throughout the power station and is interfaced with the fire control panel located in the central control room. Smoke and heat detectors are employed to provide early warning of fire in the protected areas.

The automatic system is based on a fail-safe self-policing philosophy with the lines of the field devices being continuously monitored for open and short circuit conditions. A separate control panel for each area is located in an accessible position. The zone alarm signals are fed to the fire control panel and warning devices. The control equipment is supplied from the station 110V ac power supply.

8.6 Heat and Smoke Optical Type Detection System

Heat and smoke optical detection systems are fitted to the tunnels, transformer hall, 400kV substation valve galleries, busbar gallery and main machine hall. These zones are large areas which require both heat and smoke detection to provide alarm facilities.

Each detection device comprises an emitter unit and receiver unit, the emitter producing infra-red light pulses which are detected by the receiver unit, the signal being analysed for the presence of heat or smoke.

8.7 Smoke Detection Systems

In addition to the areas requiring both heat and smoke detection systems there are further areas requiring only smoke detection facilities. These areas consist mainly of individual rooms throughout the power station.

The detection devices fitted to these systems are ionisation and optical type smoke detectors that monitor the protected area for smoke conditions produced by combustion of materials. The detectors are installed in 2-wire circuits with the monitoring and control carried out by the local panels in each area.

8.8 Ultra Violet Detection System

One ultra violet detector is installed above the electric fire pump in the tailworks building. This detector has a fast response to ultra violet radiation produced by flames and is suitable in an unsupervised area.

8.9 Maximum Temperature Heat Detection System

Heat detectors are fitted above the cooker units in the mess rooms at various locations in the power station and administration building. The detectors operate on the principle of fusion of an alloy, producing a short circuit across two conductors and initiating an alarm.

The alarm contacts of the detectors initiate the common alarm on the local control panel plus the appropriate alarm on the central control room alarm panel.

8.10 Break Glass System

Break glass units are installed throughout the power station to provide manual alarm facilities in addition to the automatic detection/alarm system.

8.11 Heat Detecting Cable System

A heat detecting cable system of the analogue type employing the change in characteristics with temperature of a specially formulated coaxial cable, is installed in the cable flats of the switchgear annexe and busbar galleries and also in each unit marshalling room. Its purpose is to provide initiation of the fixed fire fighting equipment in those areas.

The cable flats are protected by a conventional waterspray system in which the sprinkler heads perform the dual role of detecting and extinguishing a fire. In addition each sprinkler head is fitted with a 'Metron' actuator. This comprises an electric fuse requiring a given current to flow for a specific time in order to ignite a chemical charge, which generates gas to drive the piston of the actuator so that it bursts the quartzoid bulb of the sprinkler head.

By using the detection signal from a heat detecting cable arranged in close proximity to the potential hazard, the electrical actuator devices on otherwise conventional sprinkler heads may be operated at an early stage, whilst maintaining the inherent ability of the sprinkler head to operate normally, thus giving early detection with rapid response of actuation for the fixed fire fighting equipment.

The cable racking in each designated area is broken down into zones, each zone comprising a small area covered by a group of sprinkler heads. Heat detection is applied to the cable racking contained within the zone by suspending a length of heat detecting cable above each rack and below the bottom rack. Operation of an individual sprinkler head or the heat detecting cable will cause an alarm on the fire control panel in the control room.

In the case of the unit marshalling rooms the fixed fire protection is in the form of a Halon gas extinguishing system. This system is initiated by either coincident operation of two smoke detectors or by the operation of the heat detecting cable system in that area.

9. HEATING AND VENTILATION EQUIPMENT

The heating and ventilation (H&V) system, which is designed to remove plant heat emissions and to maintain a good environment, briefly consists of a battery of air extract fans and a number of air circulation plants. Air extracted from the roof arches of the machine and transformer halls is replaced by fresh air brought down to two access tunnels and through the air circulation plant rooms where it is treated and distributed in ducting to the various areas of the works. Control of the system under normal conditions of operation is from local control cubicles.

The extract system is also the means by which smoke will be removed from underground in the event of fire.

Under fire conditions, the H&V system will be controlled from the fire panel in the central control room on which a separate switch is provided for each area of the system. Each switch has three positions, ie shutdown, normal and smoke clearance.

The shutdown position is basically intended to shutdown the air handling plant concerned in order to avoid the spread of smoke-laden air. The smoke clearance position will close all air handling plant recirculation dampers, open all dampers for full fresh air flow and in addition bring into service the maximum extract fan capacity in order to purge the area concerned of smoke.

A single similar switch having the same functions is provided in the above ground security lodge. The shutdown position of this switch duplicates the shutdown mode of all area switches whilst the smoke clearance mode places all H&V systems in the smoke clearance mode.

10. STATION AND UNIT AUXILIARY ELECTRICAL SYSTEMS

10.1 Introduction

The six 300MW generator-motors are connected in pairs to the 400kV system.

There are no suitable Area Board supplies available to the station and therefore it has been necessary to derive all station supplies from the generator voltage side of the generator-motor transformer.

10.2 Operating Criteria

In addition to the overall system needs the auxiliary system was required to meet the following requirements:
(1) The auxiliary system design must be such that significant changes in electrical loadings, as the main plant design was developed, could be accommodated without changing the basic arrangement upon which decisions on cavern design and layout were made.
(2) In the absence of grid and any Area Board supplies the source of supply required for a dead station start must also be rated to supply vital services at the lower works, headworks and cavern areas.
(3) The system must allow for the provision of separate and segregated supplies to main and standby plant.

388

(4) Provision should be made for the disconnection of a starting or station transformer and return of the associated generator-motor unit to service in a period of time not exceeding one shift of 8 hours.

(5) Since the majority of the electrical equipment would be located in the cavern, the auxiliary system design must recognise that space is at a premium.

(6) The permanent auxiliary system supplies should be available for the commercial operation of each unit and standby supplies should be available as soon as possible afterwards, consistent with the total project programme.

10.3 Design Considerations

In addition to the 'needs' as defined above, the following factors have been recognised in the selection of the electrical auxiliary system for Dinorwig:

(1) The starting equipment must derive its supply in such a way that it can be adequately protected from all internal faults, thus ensuring that a starting equipment fault does not result in the loss of a pumping or generating unit.

(2) All the 'unit' auxiliaries are of such a rating that it was necessary to establish a unit board at one voltage level only, ie 415V.

(3) As is usual practice on power station sites, where auxiliary system supplies are provided to the transmission substation, the system phase relationships have been arranged so that the 415V levels are in phase with each other and with the 400kV system. This is to ensure that primary injection testing of the grid system protection can be undertaken from the Dinorwig end without the need for special equipment.

(4) Each starting equipment transformer has been rated to supply the load of one starting equipment. It is considered unnecessary to provide permanent standby facilities, since the failure of a starting transformer does not constitute a threat to the availability of the Dinorwig machines in the pumping mode, since a second starting transformer and starting equipment is available with 'back-to-back' starting as a further alternative. However, in order to cover for the possible combination of a starting transformer 1 fault coincident with a starting equipment 2 fault (or vice versa) an interconnection at 11kV is provided.

(5) For security of supply reasons a minimum of two separate cable routes are required between the cavern and the lower works (diesel generators, CW pumps, etc) and the equipment must be so rated that one route may be out of service without degrading the normal station operation.

(6) Two 100% diesel generators are required in order that the design duty can be met when one diesel is out of service for maintenance or fails to start.

(7) The supply to the headworks is essential for the safe and efficient operation of the power station and, although a short break can be tolerated, it must be available at all times. In order to meet this requirement an alternative supply is necessary.

Six alternative auxiliary system designs which met the aforementioned criteria were considered and analysed and the scheme which was finally selected gave the following advantages without any disadvantages:

(1) Lowest acceptable capital cost scheme.

(2) Faults on the input to the starting equipment are protected by a circuit breaker and do not affect the availability of the main machines.

(3) Each machine has a completely individual unitised system.

(4) The harmonic distortion associated with the operation of the starting equipments are significantly reduced because the point of common coupling is at 18kV for equipment connected to units 1 and 3 and at 400kV for equipment connected to units 2, 4, 5 and 6.

The complete electrical system is shown in Fig 4.19 and 4.20.

10.4   400kV System

Each pair of generator-motor units and their associated generator-motor transformers and 400kV switch isolators are connected to a new single busbar 400kV switching station within the cavern. The connections at 400kV consist of two section circuit breakers between the generator-motor circuits and two circuit breakers for the circuits to Pentir substation. This arrangement is shown in Fig 4.19.

In order to meet the stringent space limitations within the cavern, the switching station is constructed using SF6 metal clad switchgear throughout.

The switching station is connected to the existing Pentir 400kV substation, the connection being made by two 400kV circuits, one of cable throughout and the other part-cable and part-overhead line, the overhead line being a section of the existing Pentir-Trawsfynydd 400kV circuit between Pentir and Penisarwaun.

10.5   18kV System

10.5.1   *Generator-Motor Main Connections*

The equipment described is located in the main busbar gallery units shown in Fig 4.21.

Six sets of phase-isolated busbars and tee-offs are provided to connect the generator-motors to their respective generator-motor transformers and auxiliary equipment via phase-isolated switchgear as shown in Fig 4.22. One set of phase isolated connections is provided between the starting equipment output isolators and the tee-offs from the six sets of main busbars. The connections and tee-offs are pressurised with dry air to a pressure of 12.5mbar and the main connections supplier also provides the compressed air system which includes compressor, air receiver, filters and air dryers, control panel and associated alarms, all in the plant rooms.

10.5.2   *18kV Switchgear and Associated Systems (Fig 4.22)*

The generator-motor switchgear is either of the airblast or motor-driven type. The six generator-motors, GM1 to GM6 are connected to the common 3-phase, phase-isolated 18kV starting busbar via motor-driven isolators or to the grid via air blast circuit breakers 1M0 to 6M0 and the generator-motor transformers. The braking switches 1A3 to 6A3 are used during the shutting down operation of a machine and are air operated isolators. The motor-driven heavy current isolators are used in the following applications:

(1) Busbar section isolators 1S4 to 5S4.

(2) Starting equipment output isolators 1Y3 and 2Y3.

(3) Generator-motor starting isolators 1M7 to 6M7 and 1M2 to 6M2.

(4) Generator-motor changeover isolators 1M3G to 6M3G and 1M3PP to 6M3PP.

A different type of motor-driven equipment is used for earthing switches 1M1 to 6M1 and 1S1 to 6S1 which are used to earth the busbars during maintenance periods.

The necessary compressed air system to drive the air operated equipment comprises three separate ring main systems, one for each of two generator-motor units. Under abnormal conditions the systems can be cross connected so that one system can feed four units. The air is stored in receivers at a nominal 147 bar and pressure reduced to a nominal 28 bar, which is the operating pressure for the circuit breakers. The operating pressure for the braking switches is nominally 16 bar and is reduced and stored at this pressure in local, wall mounted, air receivers. The system is fully automatic and will maintain a supply of clean dry air of sufficient quantity to carry out all switching operations.

Associated with the switchgear is a mechanical key interlock system which enables the equipment to be isolated and earthed for maintenance purposes. The locks are mounted on the switchgear but the keys to operate them are inserted in local exchange boxes mounted on the walls adjacent to the switchgear. The local keys are obtained from the main key exchange system after following a prescribed sequence of events that finally proves that the plant is isolated and earthed and safe to work on.

The operational electrical interlocks scheme for the 18kV equipment ensures that only the following switching combinations are possible:

(1) One 400kV/18kV transformer is connected to its own generator-motor.
(2) One set of starting equipment is connected to one generator-motor.
(3) One generator-motor is connected to another for a back-to-back start.

There may be up to six of these combinations in operation at any given time, each involving the use of its circuit breaker and isolators. Switching of plant is divided into distinct priority levels to ensure that whichever combination is used, the plant is safe to operate.

The interlocking scheme comprises three identical logic channels and a channel disparity monitoring and alarm channel. Contact inputs indicating the open and closed stage from each phase of an item of switchgear are series connected and fed as a single pair to each channel. The contacts are interconnected within each channel to comply with the logic and priority switching levels and finally an output contact is fed to the switchgear items to enable them to operate as required. At each switchgear control circuit the output contacts from each channel are connected in a two-from-three logic configuration in the 110V control supply line, so completing the final stage of safety interlocking. Contacts from each stage of each channel are also fed to the channel disparity circuits where they too are connected in a two-from-three logic configuration to monitor any channel disparity. If a disparity is detected an alarm is raised in the control room and a flag is operated in the disparity cubicle.

10.6   11kV System

Unlike normal large thermal and nuclear power stations, at Dinorwig there is no necessity for a large number of auxiliary motor drives and therefore 3.3kV was chosen as the voltage level for the main distribution voltage.

However, the rating of the two starting equipments, each some 14kVA, indicated that a higher voltage system was required to provide a satisfactory source of supply for these equipments and 11kV was chosen for this purpose.

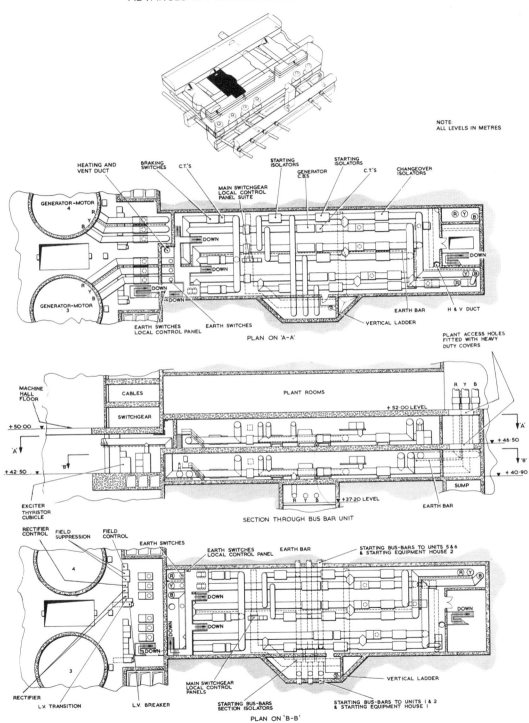

NOTE:
ALL LEVELS IN METRES

PLAN ON 'A-A'

SECTION THROUGH BUS BAR UNIT

PLAN ON 'B-B'

FIG. 4.21   MAIN BUSBAR GALLERY UNITS

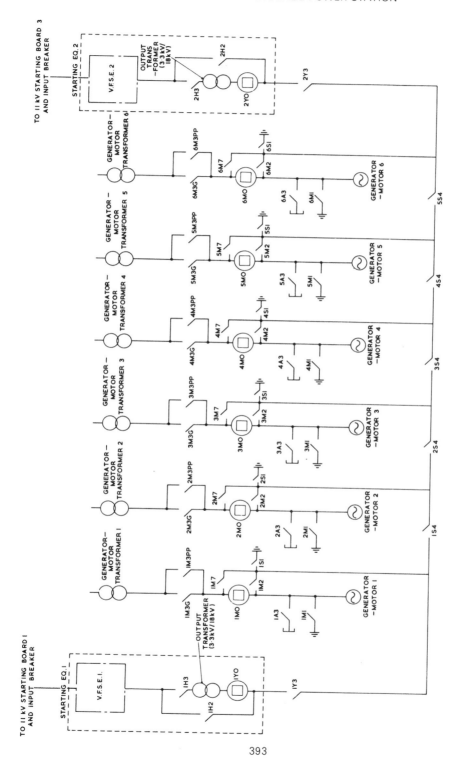

FIG. 4.22   SCHEMATIC OF ELECTRICAL SYSTEM

The 11kV system derives its supply from two 16MVA starting transformers which are teed-off from the 18kV main connections to generator-motors 1 and 3 respectively (Fig 4.19).

Each starting transformer is rated to supply one starting equipment and is connected to its respective starting board via an 11kV 1200A 750MVA breaking capacity circuit breaker. The starting equipments are connected to the starting boards via off-load isolators.

If either of the starting transformers should fail, both switchboards can be fed from the remaining transformer, but only one starting equipment must be energised to prevent overloading of the starting transformers.

Interconnection between the two starting boards is made by an air break circuit breaker which is normally open. The two incoming circuit breakers are interlocked with the interconnecting circuit breaker so that it is not possible for the two incoming supplies to be in parallel.

10.7   3.3kV System

A 3.3kV auxiliary system has been adopted to cater for motor ratings in excess of 220kW and individual loads greater than 1MVA.

The 3.3kV station auxiliary power system derives its supply from three 10MVA 18kV/3.45kV station transformers teed-off from the 18kV main connections to generator-motors 1, 3 and 5 respectively. (Fig 4.19).

Each station transformer has been rated to meet the following simultaneous loads:
(1) Supply half the 3.3kV station service drives.
(2) Supply half the 415V station services.
(3) Supply the headworks services.
(4) Supply the lower works services.
(5) Supply the 400kV cable cooling plant (this is an alternative supply).

In this way the total station services requirements can be met, using only two station transformers with the third station transformer acting as a standby and this would normally be station transformer 3.

The 3.3kV station auxiliaries are located in the power station complex and comprise dewatering pumps, emergency drainage pumps and blowdown air compressors and are divided evenly between the two main 3.3kV station boards 1 and 5. The feeder providing an alternative supply to the 400kV cable cooling equipment is connected to 3.3kV station board 1.

The feeds to the various loads at the lower works are derived from the 3.3kV diesel house board which is connected to the station boards by two 3.3kV cable feeders each on a separate route. One route is via the plant access tunnel and the second route is via the construction access tunnel.

Diesel generators 1 and 2 have been provided and located above ground as a source of standby supplies so that in the absence of grid supplies, these will maintain essential services in the main power station complex such as lighting, heating and ventilation plant to the personnel areas, battery charging and normal drainage. They will also provide supplies to enable a generator-motor to be started even if the grid supply is lost.

Since it is necessary to establish ac supplies in the vicinity of the headworks, and the cable route distance involved (5.5km) is such that the voltage regulation would be outside acceptable limits at 415V, a 3.3kV supply has been provided, fed from 3.3kV station board 1 and this is stepped down to 415V by a 500kVA headworks services transformer 3.3kV/433V. A standby diesel generator has been installed at the headworks to supply essential loads if the main supply is lost.

10.8 415V System

A 415V auxiliary system has been adopted to cater for motor ratings up to 150kW and individual loads 2MVA and below.

The auxiliary drives for the pump-turbine and generator-motor are all of such a rating that a single voltage level of 415V is the most economic and therefore a single 415V unit services board has been provided for each unit, energised from the 18kV generator-motor supply via an 18kV/433V unit services transformer (Fig 4.19).

In order to reduce the amount of equipment connected at generator-motor voltage, the winding arrangement for the unit services transformer has been made star/interstar so that each 18kV system can be earthed via the neutral of this transformer, thus avoiding the need for separate earthing transformers.

An alternative supply has been arranged to each unit services board from a station services board and this is used in an emergency to allow essential unit auxiliaries to be energised in the event of loss of grid supplies. During such an emergency, the station services board will obtain its supply from the standby diesel generators.

The station 415V auxiliary services are fed from three station services transformers, 3.3kV/433V and each transformer is rated to supply half the cavern services load; thus the third transformer acts as a standby to the other two.

It is not possible to operate the auxiliary system with any of the transformers connected in parallel, since unacceptable circulating current would result should the associated generator-motors be operating in opposite modes, ie one generating and one pumping. The transformer impedances, therefore, have not been chosen with parallel operation in mind and loading transfers between transformers are undertaken 'off-load'.

The tailworks area of the site obtains its 415V supply from the 415V diesel house board and this switchboard is located in an annexe adjacent to the diesel generator building at the entrance to the construction access tunnel. During normal operation the diesel house board is connected to the station auxiliary power system at 3.3kV via two separate interconnectors.

A 415V supply has also been established at Marchlyn Mawr, the upper dam, and since the distance from the power station is some 6km, the voltage regulation would be outside acceptable limits if 415V were used and the interconnector thus operates at 3.3kV with a 500kVA 3.3kV/415V step-down headworks services transformer. A standby diesel generator is located here to supply the essential loads in case of failure of supply from the power station.

10.9   Transformers

10.9.1   *Generator-Motor Transformers*

The generator-motor transformers convert the generator terminal voltage of 18kV to the nominal transmission system voltage of 420kV. The transformers are located underground in the transformer hall of the power station and are each rated at 340MVA.

The transformer is cooled by forced oil circulation through an oil/water heat exchanger and the intermediate cooling water is further cooled by a water/water heat exchanger with the coolant being the station raw water supply. The rating of 340MVA is obtained when both oil/water and water/water coolers are in operation.

The winding arrangement is HV star and LV delta, with a star-connected primary and a delta-connected secondary.

Tappings at the neutral end of the high voltage winding connect to an on-load tap changer which can vary the system voltage by $+5.5\%$ to -14.43% in 18 equal steps.

Each transformer is provided with three oil/water heat exchangers and two water/water heat exchangers and under normal load conditions two of the 50% duty oil/water coolers and one of the 100% water/water coolers are sufficient to cool the transformers; the other oil/water and water/water cooler acts as standby.

The intermediate water circuit is designed to ensure that water pressure in the oil/water coolers is always below that of the oil, thereby eliminating any possibility of ingress of water into the oil system in the event of a fault or leak developing in the coolers.

10.9.2   *Unit Services Transformers*

The six unit services transformers, each rated at 1.6MVA 18kV/433V have been installed to step-down the 18kV supply from the generator-motor to 415V to supply the miscellaneous auxiliary plant which is necessary to operate each generator-motor.

The transformers are located underground in the transformer hall and are naturally oil-cooled, the tanks being fitted with integral cooling radiators.

It should be noted that the winding arrangement for the unit services transformers is star/interstar and this has been done so that the 18kV system associated with each generator-motor can be earthed via the unit services transformer neutral, thus avoiding the use of a separate earthing transformer.

To limit the earth fault current, a neutral resistance type earthing arrangement has been used.

10.9.3   *Works System Transformers*

The works system transformers at Dinorwig can conveniently be split into three categories:
(1) The 18kV/3.45kV station transformers.
(2) Miscellaneous 3.3kV/433V auxiliary transformers naturally oil-cooled.
(3) Miscellaneous 3.3kV/433V auxiliary transformers naturally air-cooled.

Generally where auxiliary transformers are located underground, they are of the naturally air-cooled type with Class C insulation to minimise fire risk and to avoid providing extensions to the waterspray fire protection system. Where transformers are not located underground they have been designed as normal naturally oil-cooled units.

The detailed information on the three transformer arrangements is as follows:

(1) Station Transformers

Three station transformers rated at 10MVA are located in the transformer hall and convert the 18kV voltage associated with the generator-motor to 3.3kV for supplying the three 3.3kV station switchboards.

(2) Naturally Oil-Cooled Auxiliary Transformers

These comprise six 1MVA and one 500kVA transformers located at various areas on site, external to the underground works, and are delta/star connected 3.3kV/433V.

(3) Naturally Air-Cooled Auxiliary Transformers

These comprise five 2MVA and three 1MVA 3.3kV/415V transformers which are all located underground within the power station. They have Class C insulation and are housed within sheet-steel naturally-ventilated enclosures having interlocked doors so that access can only be obtained when the associated circuit breakers are isolated.

10.10 Earthing

The main generator-motor transformers and the associated 400kV substation employing metal clad switchgear are sited below ground, adjacent to the machine hall. A substantial rise of earth potential was anticipated as a consequence of an earth fault on the high voltage equipment and this strongly influenced the design of the station earthing system.

To provide for safety of personnel the aim is to ensure that in both normal and abnormal conditions no dangerous voltages can appear at any point to which personnel have unrestricted access. This voltage may appear as a voltage rise in a plant area, or as a potential difference between two items of adjacent plant.

The general land mass at Dinorwig comprises mainly slate, and as such has a much higher resistivity than is normally associated with power station sites.

Fault studies indicated that an earth fault on the Dinorwig station 400kV busbars would produce the highest rise of earth potential and the relevant projected value of earth fault current applicable during the lifetime of the station would be 22,000A. Soil resistivity values varied greatly over the site but using a deduced value of 3500 ohm metres the earth electrode resistance of the underground station alone was calculated to be about 8 ohms.

The value of earth electrode resistance has a significant effect on the distribution of the return fault current to Pentir between cable sheaths and ground, the latter current being responsible for generating the rise of earth potential on the site.

An appreciable reduction in the rise of earth potential could only be achieved by a significant reduction in the station resistance.

Since the effective mass of the electrode earthing system is associated with the foundations of the civil works and installed plant, a cavern earth electrode system was established making maximum use of inherent earth contact of plant and civil works in the station construction, such as:

(1) High pressure penstocks and main inlet valves.

(2) Draught tube valves and draught tube tunnels.

(3) Spiral casings of pump-turbines and associated foundations.

(4) Structural steelwork, eg columns, crane rails and framework of the electrical annexe.

Above-ground electrodes were established on a similar basis using the steelwork incorporated in the main access tunnel portal and construction tunnel portal, the tailworks and cooling water pump house structures, diesel house area and bascule gates control house at Afon-y-Bala.

Since all the aforementioned electrodes were associated with high soil resistivity it was considered necessary to augment the system with a further electrode located in lower soil resistivity. Geological surveys indicated that a favourable location was in an extensive wedge of low resistivity silt existing below the bed of the lower reservoir. Although this permitted the installation of an effective electrode, it nevertheless extended the ground voltage profile nearer to the village of Llanberis. However, the availability of such an electrode to distribute current at high density into such a volume of low resistivity material, before transmitting it at a very much lower density into the surrounding high resistivity bedrock was considered an advantage which out-weighed the other possible penalties. The electrode adopted consisted of a group of deep-driven piles.

Within the electrode areas all exposed plant was bonded together except for those items exposed to touch potentials or subject to general maintenance which extend outside the major electrode area. These items were provided with isolation at the interface edge of the electrode area and at selected locations between electrode areas.

The sections of reinforcing mesh used in the floors of the caverns were welded together at intervals and bonded into the main earthing system to ensure an equipotential surface throughout the station. Because of the very large void that the cavern presents in the mountain, additional mesh screen was installed up the walls and across the roof of the cavern and this too was bonded into the earthing system to guard against the possible effects from lightning. Thus the earth electrode of the underground cavern became roughly spherical in shape.

The determination of the values of rise of earth potential, step, touch and transformer potentials and the location of the 650V contour line for such a dispersed electrode system, situated in an area of widely varying soil resistivity was too complex to achieve reliable results from calculations alone. A series of tests were therefore carried out with injection along the power conductor of the 400kV circuit 1 (all cable) and circuit 2 (line/cable) respectively, to obtain more reliably the distribution of the fault current, the rise of earth potential of the site as a whole, the ground voltage profile radiating away from the site, the location of the 650V contour and the levels of step, touch and transfer potentials about the site. A representative model of the site electrode system was developed from the measurements obtained.

As a result of implementing the various special design features discussed, the rise of earth potential of the power station has been held to 2kV, and the 650V equal potential line lies only just outside the perimeter of the power station site. The maximum step/touch potential within the power station is estimated at 310V, and whilst this value is extrapolated from measured results and cannot be considered absolute it is far below the designed 650V maximum and therefore does not present a problem.

10.11 DC Equipment

DC systems at three voltage levels of 50V, 110V and 240V dc have been provided. Each system comprises a number of battery chargers and floating battery with the facility to boost charge the battery off-load. The battery sizes have been based on a standby period of half an hour and it is assumed that by this time ac supplies will be available from the standby diesel generators.

In the case of the 50V batteries for operating the transmission equipment and telecommunications equipment a longer standby period has been incorporated.

10.12 240V DC Equipment

Within the power station underground complex there are two 240V dc switchboards. Each has its own float/boost charger, battery and fused output circuits. There is an interconnector between these two switchboards so that one battery/charger can supply the two boards in an emergency whilst the second battery is on boost charge.

A similar arrangement has been included in the tailworks area of the site for two diesel house switchboards. These are located in the annexe adjacent to the diesel generator building and again an interconnector is included between the two switchboards.

The 240V dc system operates with the centre point earthed through a high resistance, and an earth fault relay is provided to check that the system remains earth free and to give an alarm if the earth current exceeds a certain value.

10.13 110V DC Equipment

Since the station 110V dc loads are only small, batteries and chargers have not been provided for the station system, but instead there are three 110V dc station switchboards designated 1, 3 and 5, each fed from an appropriate 110V dc unit board.

For the 110V dc unit system in the station there are six 110V dc unit boards, one being associated with each unit and each board is complete with its own battery and battery charger.

To provide facility for off-load boost charging and standby supplies, unit boards are connected in pairs.

As previously mentioned, the station boards are supplied from an appropriate unit board and thus each charger and battery for a unit board is rated to carry the load of two units plus one half of the station 110V dc load, plus the trickle charging of two 110V dc unit batteries.

The 110V dc system operates in a similar manner to the 240V dc in that the centre point is earthed through a high resistance and an earth fault relay is provided to check that the system remains earth free and to give an alarm if the earth current exceeds a certain value.

In addition to the 110V dc unit and station boards located underground, there are two 110V dc systems located above ground, one for the tailworks area the other for the headworks area.

Again, each of these systems has two charger/batteries so that the off-load boost charging facility is available and each charger/battery can act as a standby to the other.

10.14 50V DC Equipment

50V dc supplies are required for sequence control equipment alarms and indications, telephones and other telecommunications equipment.

Since the station 50V dc loads are quite low, a separate station 50V dc system could not be justified and, as is the case for the 110V dc system, there are three 50V dc station boards fed from the 50V dc unit boards.

There are six 50V dc unit boards which are interconnected in a similar manner to the six 110V dc unit boards, to provide off-load boost charging and standby facilities.

In addition to the 50V dc supplies required for the unit and station boards, there are 50V dc systems for telecommunications, auxiliary telecommunications, diesel house services and headworks.

The telecommunications and headworks 50V dc systems employ two batteries and chargers per system in a similar manner to the unit 50V dc system in that off-load boost charging and standby facilities are included. However, for the auxiliary telecommunications and diesel house services, only one battery and charger is employed since these loads are not considered quite so essential as the other 50V dc systems.

11. TELECOMMUNICATIONS

The following communication systems have been provided:
(1) Private automatic telephone exchange.
(2) British Telecom private automatic branch telephone exchange.
(3) 400kV tunnel sound powered telephones.
(4) Emergency sound powered telephone system.
(5) Senior executive intercommunication system.
(6) Maintenance and commissioning telephone jack system.
(7) Station sirens and siren control panels.
(8) General indication telephone system.
(9) Concentrator panel and direct wire telephone service.
(10) Staff location system.
(11) Grid control emergency radio system.

11.1   Private Automatic Telephone Exchange

The private automatic telephone exchange (PAX) is located underground and is based on current practice, involving crossbar type equipment using 500 switching principles such as mass marking, self-steering control and second attempt working. It is a 200-extension system with a 4-extension numbering scheme.

Access is provided into the staff location system using a one digit code plus a predetermined code for calling and a fixed code for answering. One call at a time can be established.

The PAX employs loop-disconnect signalling throughout. Incoming calls except for the staff location circuit are via extension line circuits. The PAX system operates from a 50V negative dc source within the limits of 46V to 54V.

Due to the slate terrain at the power station site and the inherent difficulty of providing a low resistance earth path, electrical isolation equipment has been fitted on certain telecommunication cables in order to provide isolation in the event of a system fault on the 400kV circuits causing differences in earth potentials between various locations. The isolation equipment has been fitted as follows:
(1) Between the upper works and the underground works.
(2) Between the lower works and the underground works.
(3) Between locations within the lower works.

Where telecommunication equipment cannot work through isolation transformers, provision has been made to use multi-channel voice-frequency equipment.

11.2   British Telecom Private Automatic Branch Telephone Exchange

The British Telecom private automatic branch telephone exchange (PABX) is located in the Post Office cable room in the administration block at the lower works.

11.3   400kV Tunnel Sound-Powered Telephone System

This system incorporates sound-powered telephones to enable it to function under normal and emergency working. The telephones are therefore dual-purpose being able to work with or without a power supply. The system enables communication to be maintained at all times between required locations in the 400kV tunnel, the District Engineer's office and the communication desk in the power station control room located underground.

11.4   Emergency Sound-Powered Telephone System

This system has been provided to enable communication to be maintained between the gatehouse, and all or part of the underground works during an emergency caused by the failure of the PAX and radio supplies. Signalling is provided by a hand-operated magneto-generator fitted to each telephone set and this enables an audible signal to be given at all telephone stations. Each station has its call code.

The master station is situated in the gatehouse and has a capability of calling any one of the sound-powered telephones (one at a time) by means of individually associated press buttons.

11.5 Senior Executive Intercommunication System

This is a solid state system providing loudspeaking, 'hands-free' communication for 10 stations initially and 20 stations ultimately. The 10 stations comprise eight master stations, one executive station (Station Manager) and one secretarial station. Each station is equipped with a handset allowing loudspeaker operation to be dispensed with and communication to take place in privacy.

11.6 Maintenance and Commissioning Telephone Jack System

This system enables point-to-point communication between equipment, relay panels, station control room, etc, for operational, maintenance and/or commissioning purposes. Jack sockets are provided on all equipment where necessary and are wired in such a manner as to give access to the telephone jack system as required.

11.7 Station Sirens and Siren Control Panels

This system enables a state of emergency existing on the station to be communicated to all site personnel.

Sirens are located throughout the underground works and controlled from one of two panels, one being on the communication desk and one in the gatehouse.

The sirens operate from a 240V dc supply and provide an audio output of 100dB $\pm$ 10dB at 1800m. Provision is made for monitoring the supply voltage to the sirens and a supply failure condition causes an alarm on the station services panel in the control room.

11.8 General Indication Telephone System

Telephones have been provided in the power station which are connected directly to the CEGB's National General Indication Telephone System to provide communications with the Grid Control Centre.

11.9 Concentrator Panel and Direct Wire Telephone Service

This provides radio cover within the Dinorwig power station complex; portable transmitter/receivers are used both in the hand-held and body-worn mode for personal use, together with mobile type units for use in vehicles.

A combination of radiating and co-axial cable and conventional antennae has been incorporated to provide the required coverage above and below ground. Heated antennae are provided for the upper works and grid control base stations to minimise icing problems. The routing of the radiating cables has been designed to provide the optimum coverage with minimal signal attenuation. Duplexers and transmitter combiners have been provided where required. Care has been taken to minimise the dangers due to voltage build-up in the radiating cables under power line fault conditions by the insertion of filters with high voltage (10kV) blocking capability at suitable points in the cable runs.

11.10   Staff Location System

This system is overlaid onto two channels of the personal radio system and uses the same base stations and antenna systems (apart from the 400kV tunnel equipment). Both 'individual' and 'group' calling is incorporated within the system. Separate pages are used for the two systems giving different alerting tones to allow individual and group calls to be distinguished when both types of pager are carried.

11.11   Grid Control and Emergency Radio System

This is a single channel link to Wylfa Power Station.

12.   CONTROL AND INSTRUMENTATION

Manning at Dinorwig for operation purposes had been defined as two engineers plus three roving attendants per shift.

One engineer (Shift Charge Engineer) is involved in normal power station management, ie permit for work routines, switching, etc and therefore the layout of the central control room is designed to enable normal and emergency operation to be carried out by one engineer with assistance from his colleague during abnormal operating conditions.

The operational requirements of the power station naturally divides into the following basic operational working areas within the control room:
(1) Central control and communications.
(2) Unit control, alarm and indications.
(3) 400kV/18kV transmission and machine voltage.
(4) 3.3kV/415V auxiliary electrical services.
(5) Hydraulic system.
(6) Cooling water general services.
(7) Fire alarms.

The arrangement of the working areas are designed such that if necessary two operators can work satisfactorily within the same or in adjacent areas.

The single operator at Dinorwig has to cover all six units plus the station services controls and indications. He therefore needs all the minute-to-minute controls and indications grouped around him such that he can operate them from one control position. The controls and indications in the working areas that require less frequent operation are arranged such that the operator, when situated at the minute-to-minute control position, can easily view and have good access to them.

12.1   Station Operation Requirements

The station is designed to perform the following operational requirements:
(1) Generation as instructed by the Grid Control Centre.
(2) Pumping at full load as instructed by the Grid Control Centre.
(3) Immediate response to cover for a sudden loss of generation on the grid system.
(4) Frequency regulation to correct system frequency deviations.

The operator at Dinorwig carries out the operational roles with the following facilities:

(1) Generation — this role involves the operator in adjusting the output of individual units as instructed by the Grid Control Centre. The operator adjusts the individual generator outputs by using the governor load control equipment and the sequence equipment should a mode change be required.

(2) Pumping — this role involves the operator in initiating a mode change using the sequence equipment.

(3) Immediate response — the operator selects two units to run in the spinning reserve mode using the sequence equipment and then selects the immediate response under-frequency relays for those units.

(4) Frequency regulation — the operator selects 2, 3 or 4 units to operate in the spinning reserve or generate mode using the sequence equipment. He then regulates the total station generated output by using the joint load control equipment. To maintain a high generation efficiency all units operate in the 60% to 100% load range. The operator manually initiates individual spinning reserve to generate, and generate to spinning reserve mode changes, depending on the station generated output demanded by the system frequency deviations.

In addition to the operating roles the operator carries out routine control and supervision of the unit and the station services controls and indications during normal and abnormal operating conditions. This involves the operator in dealing with alarms, electrical switching, changeover of standby plant, communications, monitoring of analogue indications, etc.

The operator may also have to control manually the units during abnormal conditions, eg back-to-back pump starting, dead station start, etc.

12.2 Central Control Room Facilities

The layout of the desks and panels within the central control room is shown on Fig 4.23.

The central control and communications desk provides the operator with all the necessary communication facilities within and outside the station. It also provides the operator with the means to initiate the six units into any of their operating modes and also to apply joint load and/or joint MVA control to any number of selected units.

The unit control desk contains the alarms, control and indications for all units in six identical sections. Each unit section contains the facilities to enable the operator to run a unit in any operating mode including the abnormal situations of dead station start and back-to-back pump starting.

The unit controls required on the central control and communications desk are duplicated from the unit desk where the point of control is selected. The unit desk is designed to enable two operators to work on adjacent unit sections.

12.3 *Electrical and Other Station Services*

The remaining electrical and other station services are provided on five vertical control panels:

(1) 400kV/18kV system.
(2) 3.3kV/415V electrical auxiliaries system.

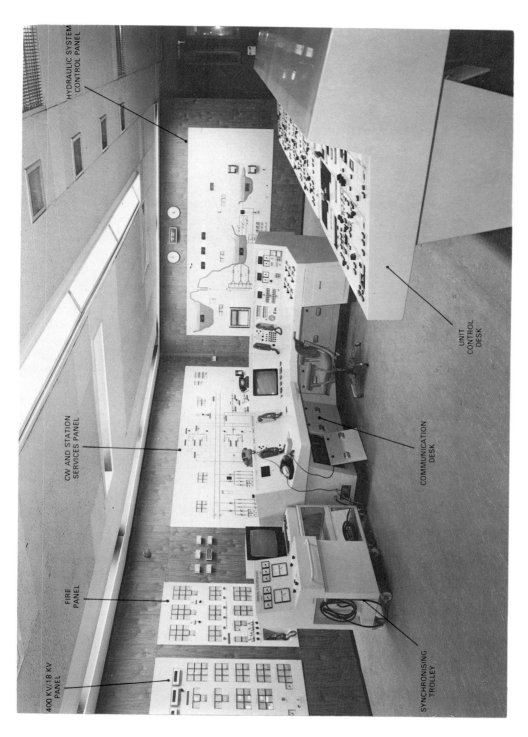

HYDRAULIC SYSTEM
CONTROL PANEL

CW AND STATION
SERVICES PANEL

FIRE
PANEL

400 KV/18 KV
PANEL

UNIT
CONTROL
DESK

COMMUNICATION
DESK

SYNCHRONISING
TROLLEY

FIG. 4.23   CENTRAL CONTROL ROOM

(3) Hydraulic system.
(4) Cooling water and station services.
(5) Fire alarms.

12.3.1 *400kV/18kV System Control Panel*

The 400kV/18kV system control panel is flush-mounted within the control room wall. The panel is positioned in such a manner that the unit control desk operator can identify the various displays and indications.

The panel front is made from standard modules and a mosaic in the form of a mimic diagram which displays visually the complete 400kV/18kV electrical system at Dinorwig, together with the 11kV starting equipment. Included in this diagram are control and indication facilities for circuit breakers and isolators together with analogue and digital indications for the various electrical parameters (amps, volts, MW, MVAR, frequency). 400kV synchronising is carried out from the controls on this panel in conjunction with the synchronising trolley.

12.3.2 *Electrical Auxiliaries System Control Panel*

The electrical auxiliaries system control panel is flush-mounted within the control room wall. The panel is positioned in such a manner that the unit control desk operator can identify the various displays and indications.

The panel front is made from standard modules and a mosaic in the form of a mimic diagram which displays all the main feeder circuits for the station 3.3kV and 415V electrical systems with sufficient controls and indications to be able to configure all main electrical supplies in the station. Only the major and essential circuits are shown in full on the mimic diagram, the less important circuits having 3.3kV circuit breaker indication only. Facilities on the mimic diagram include control and indication of circuit breakers and isolators, and analogue indications of amps and volts in the various parts of the system, together with digital indication of voltages on all external feeds into the system on the mimic diagram. Remote control facilities for the 3.3kV diesel generators 1 and 2, and 415V headworks diesel generator 3 are also included.

3.3kV synchronising of certain designated circuits is carried out from the control facilities on the electrical auxiliaries system control panel, in conjunction with the synchronising trolley. A clearly identified, separate synchronising socket is provided for each such circuit.

12.3.3 *Hydraulic System Control Panel*

The hydraulic system control panel is flush-mounted in the control room wall and in such a position that the unit control desk operator can identify the various displays and indications.

The panel front is made from standard modules and a mosaic in the form of a mimic diagram which displays the overall water management system for the power station in the form of such items as:
(1) Reservoir and river levels.
(2) Station MWh capacity as a function of water quantities in the upper and lower reservoirs.
(3) Surplus water control (bascule gates).
(4) Power station water isolating facilities (headworks and tailworks gate closure).

In addition, telemetry system selection is provided on the panel.

12.3.4 *Cooling Water and Station Services Control Panel*

The cooling water and station services panel is flush-mounted within the control room wall and in such a position that the unit control desk operator can identify the various displays and indications.

The panel front is made from standard modules and a mosaic in the form of a mimic diagram/control panel which displays, control and facilities for the selection and control of the three cooling water pumps, the valve controls for isolating sections of the cooling water system, indications of cooling water system leakage and drainage sump levels.

12.3.5 *Fire Alarms Control Panel*

The fire alarms control panel is flush-mounted in the control room wall and is positioned such that the unit control desk operator can identify the various displays and indications.

The panel front is made from standard modules and a mosaic, and has alarm modules mounted in separate fascia groups, each group representing a specific geographical area within the station. Individual alarms within the modules are all associated with fire/smoke detection, together with the monitoring of equipment located within the specific area. Other facilities provided are controls and indicators concerned with the main diesel and electric fixed waterspray fire pumps and the fire hydrant water pumps. In addition, facilities are available to switch the heating and ventilation plant from normal operation to shutdown, or smoke clearance only, in the various plant areas. In the event of a generator-motor transformer fire, the transformer oil pumps can be tripped to limit the spread of oil.

12.4 Water Management System

The water management system monitors and controls the amounts of water available for use by the station to within designed limits. As water levels play a critical role within the station operating procedures, it is extremely important that accurate high resolution monitoring facilities are employed within the water management system to register small changes to water quantities. The significance of this required accuracy and high resolution can be seen from the fact that 2cm at the upper reservoir (ie the incremental step in a normal working range of 0-33m) or 1cm at the lower reservoir (ie the incremental step in normal working range of 0-14m) can represent up to 8MWh of power station energy capacity.

Monitoring the two reservoirs and their associated rivers and lake takes the form of level indications on the hydraulic system control panel combined with alarms should abnormal (high or low) levels be met.

The control function is provided by the bascule gates on the lower reservoir and their prime function is to release surplus water from the upper/lower reservoir system to maintain maximum power station MWh capacity.

13. SEQUENCE, INTERLOCKING AND DISCREPANCY CONTROL EQUIPMENT

The equipment can be separated into four areas:

(1) A sequence control scheme which provides for semi-automatic transition between any of the operating modes, ie after the desired operating mode has been selected, the transition from the existing mode to the new mode is completed without further manual intervention.

(2) General and auxiliary relay control equipment which provides a second level of control to the sequence control scheme. The equipment also caters for control functions requiring continual surveillance as opposed to those uniquely associated with mode changing.

(3) Trip and interlock relay equipment which provides conditioning logic for plant signals as part of the permanent protection of major items of plant and which is completely independent of control actions generated by other equipment, eg sequence control and general and auxiliary relay control equipments.

(4) Discrepancy control and indication equipment which, in conjunction with an external discrepancy control switch, provides discrepancy and alarm indication of a circuit breaker, or isolator or similar device. Control of plant devices is direct from the discrepancy control switch.

13.1 Sequence Control Equipment

This is a micro-computer system developed by its manufacturer as programmable equipment for relaying and measurement (PERM) applications. In order to carry out its required function as a plant sequence controller, PERM has at its centre a large scale integrated circuit microprocessor and memory units.

The equipment is programmed to control the plant in the sequences of startup preparation, standstill to generate, standstill to pump, generate to standstill, pump to standstill, emergency generate and transition to shutdown.

13.2 General and Auxiliary Relay Control Equipment

This equipment provides a second level of control to the sequence equipment. It enables certain plant items to be controlled manually for commissioning purposes and, in particular, the equipment also caters for control functions requiring continual surveillance as opposed to those uniquely associated with mode changing.

13.3 Trip and Interlock Relay Equipment

The trip and interlock equipment provides a signal conditioning facility for the permanent protection of major items of plant completely independent of control actions from other equipments, eg sequence control, and general and auxiliary control equipments. The protective function covered by this equipment can be summed up in three basic areas as follows:

(1) Monitoring of and performing logic on those plant parameters which require indirectly to initiate a unit trip.

(2) Providing protection to major auxiliary items of plant and their associated periphery plant items, eg cooling water pumps.

(3) Preserving safety of the hydraulic system by appropriate interlocking of the main valves and gates.

The trip and interlock equipment is housed in sheet-steel cubicles and is based on relay logic using high quality, heavy duty, plug-in relays. The integrity and reliability of the equipment is based upon comprehensive subfusing of the equipment, dc supply voltages and, where necessary, circuit redundancy.

14.   SYNCHRONISATION

Synchronising facilities are provided for the 400kV, 18kV, 3.3kV and 415V services at the station. These facilities may be detailed further as follows:

(1) Automatic synchronising at 18kV.

(2) Manual synchronising at 18kV and 3.3kV.

(3) Manual synchronising at 400kV.

(4) Manual synchronising at 415V for the headworks diesel generator only.

14.1   Automatic Synchronising at 18kV

Automatic synchronising is provided for each machine for the closure of the 18kV generator circuit breakers and is available in the pumping and generating modes of machine operation.

For the pumping mode the machine speed is raised by the sequence starting equipment to a level which is above the normal system frequency and then the starting equipment is switched out. The machine, will then reduce speed and allow the automatic synchronising to take place when the appropriate conditions are satisfied. During this period of time the speed of the machine cannot be adjusted and automatically falls from 52.5Hz to 49Hz in a set time of 40s.

In the generating mode, synchronising takes place while the machine speed rises. All six machines may be synchronised at approximately the same time using individual automatic synchronising equipment under the sequence starting control.

14.2   Manual Synchronising at 18kV and 3.3kV

Manual synchronising at 18kV and 3.3kV is provided in the form of a synchronising trolley which utilises a common set of relays for the operating logic and check synchronising and houses one set of instruments for all circuits at these voltages.

Manual synchronising at 18kV on the generator-motor unit is limited in practice to the generating mode only, as it is impracticable in the pumping mode since the availability of the synchronising relay signal is extremely short.

14.3   Manual Synchronising at 400kV

In the case of the 400kV switchgear, the relays for the operating logic and check synchronising are fixed and do not use those mounted in the trolley. However, instruments for this voltage are fitted to the synchronising trolley.

14.4 Manual Synchronising at 415V for Headworks Diesel Generator 3

This diesel generator normally operates only on loss of ac supplies and synchronising is not required. However, for testing purposes, the diesel generator is loaded onto the grid system and for this mode of operation manual synchronising has been provided.

When the 'test' condition is selected, a pilot light indication is given locally when correct synchronising conditions exist. The operator then closes the solenoid-operated diesel generator circuit breaker to permit the diesel generator to be loaded onto the 415V auxiliary system and hence the grid.

15. DATA LOGGING EQUIPMENT

A data logger has been installed to carry out the function of logging and recording essential parameters associated with the units, station services and transmission. This is to relieve the operator of having to keep a manual log and also to provide a source of chronological information after an emergency condition.

The data logger does not carry out any control functions or provide essential displays of information since this is the function of the sequence, alarm annunciator and analogue or digital display equipment. However, it is fair to state that resulting from commissioning experience more use is made of the visual display unit (VDU) format displays (pump-turbine/generator-motor parameters) than was originally envisaged.

Whilst the data logger has a main system and a standby system, the main system does not have processor redundancy or input/output redundancy. The standby system provides a line processor and memory which is available to enable a manual changeover during any data logger equipment faults.

Data presentation facilities consist of six line printers and one video display monitor. The printers are located in a separate printer room and the video display and associated system keyboard are fitted in the communication control desk.

16. ALARM ANNUNCIATOR SYSTEM

The alarm equipment is located in the station control room and unit equipment rooms and gives visual and audible indications to the operator of the alarm conditions, due to malfunctioning equipment in the plant. The operator is thus able to take suitable action for the safety and continued running of the plant.

The facia display equipment is the prime method of annunciating alarms to the operator with the VDU of the data logger equipment as a supporting facility.

The alarm input signals from the plant equipment are used to initiate both the alarm annunciator system and the data logger but fire alarm signals initiate only the alarm annunciator system; fire alarms are classified as reportable incidents and are expected to be manually recorded in the shift log.

17. FLOOD DETECTION AND CONTROL

17.1 Detection

To inform the control room operator of the incidence and extent of flooding, a comprehensive flood detection system is installed throughout the underground complex.

All surface water drainage flows via drainage channels to the two sumps in the pipe gallery. Excessive drainage from the main inlet valve (MIV) and draft tube valve (DTV) galleries causes an alarm to sound.

The operation of the normal drain pumps is not an alarm condition but indication of drain pumps running is given in the central control room. However, if the water in the sump continues to rise, a high level alarm is given. A further rise in water level brings into operation the first emergency drain pump followed by the second pump if the water level is still increasing. The running of the emergency drain pumps is indicated audibly and visually in the central control room. Should water level in the sump still continue to rise, an extra high level alarm is sounded.

A number of float switches suspended at increasing heights from the walls of the pipe gallery, MIV and DTV galleries and turbine floor raise alarms in the control room of the level that flooding has reached in those respective areas.

The height to which flood water has risen above the floor of the main drainage sumps situated below the pipe gallery is continuously indicated in the central control room by means of gauges operated by pressure transducers mounted in each sump. This will indicate floodwater level up to about turbine floor height, or some 20m above the sump floor.

The cooling water system total flow into and out from the underground works are compared and any discrepancy in these two figures causes an alarm to sound in the central control room. Additionally, flow to and from pairs of units are compared with any discrepancy alarmed. All of these system alarms are arranged to indicate two stages of discrepancy, either a small difference or gross difference. Comparison between flow and return quantities can be determined at any time by means of an indicator and selector switch on the panel in the central control room.

17.2 Flood Control

The control of the ingress of floodwater is an operator action and will be based upon interpretation of the alarm raised by the detection system described. The initial indication will be whether the drainage system is handling the water ingress or if water is continuing to rise at an uncontrolled rate. If the drainage system is coping with the situation, the operator will despatch a plant attendant to determine the location and cause of the flooding and will then implement appropriate isolation measures.

Flooding caused by leakage from the cooling water system will have been detected in sufficient detail to enable any necessary isolation to be implemented from the central control room.

Floodwater ingress at a rate in excess of the capacity of the drainage system will cause alarms to sound in relation to the location and severity of the failure. Whilst major unit leakage caused by manhole failures etc, can be isolated by closure of the

MIV and DTV, this would depend upon the operator being able to identify the unit involved. If this was possible, then the leakage could be contained by closing down one unit only. If, however, conditions are such that the affected unit cannot be identified, then the station would have to be shutdown by individually tripping each unit followed by the cooling water system and finally initiating closure of the headworks gates and tailworks gates.

By further analysis of the alarms and flood water level indicators, the operator can determine if the unit or station trip has been effective in containing the leakage or whether station evacuation is desirable.

Flooding is not a hazard to station staff, in view of the relatively long time taken to flood the station, and the ample provision of alternative escape routes. Emergency lighting, operated by batteries and switchgear at 62.30m OD, is provided and the lifts have a special control system and permanent escape facilities to ensure the safety of lift passengers during a flood.

17.3 Station Safety

As the whole of the main underground works are located below the levels of both Marchlyn Mawr and Llyn Peris the potential exists for complete flooding of the station. Although a comprehensive flood detection system is provided to detect rapidly and locate unwanted water ingress, the primary consideration has been to design and install high integrity plant and systems throughout. The principal source of water which could cause flooding is the main hydraulic system. Gates are provided in the water tunnel systems to provide isolation from the reservoirs for station maintenance requirements, but, even if they were fast acting, they provide no safeguard against pressure system failure within the station, as the water quantities in the tunnel system alone are sufficient to cause serious flooding. Within the station itself, the plant is isolated from the high pressure penstock by the main inlet rotary valves and on the low pressure side by the draft tube butterfly valves. Failure of any plant between these valves is protected by the valves themselves, but the integrity of the valves and the sections of steel tunnels linking them into the buried tunnel system has to be assured throughout the station life.

The integrity of the system has been ensured by obtaining an accurate and detailed knowledge of the loadings imposed, by using proven techniques to produce a conservative design with materials in the construction suitably selected to obtain the required mechanical properties. Where necessary additional test work was commissioned to provide data on mechanical properties required.

Estimates of the operational pressure transient conditions were obtained from computer simulations of the hydraulic network. The number of mode changes giving rise to such transients will be up to 15,000 per unit per annum over the design life of the plant.

From the design point of view, the critical regions of the pressure envelope are the A-section of the upper penstock, the MIV, the intermediate penstock, the pump turbine inlet pipe and the pump-turbine itself. The tendered design for each of these components was assessed in detail. The route followed in the assessment was to establish the operational loading conditions and then through an appropriate stress analysis technique to derive operational stresses for code comparison purposes. For complex regions of the MIV and pump-turbine it was necessary to resort to scale photoelastic modelling and stress freezing followed by confirmatory strain gauging during proof pressurisation. In the analysis, particular attention was directed to regions of stress concentration and possible manufacturing difficulty.

Consideration of the high cyclic life resulting from operational mode changes assumes a dominant role in the design of the pressure boundary. Fatigue loading generally causes a defect to grow in size until it finally reaches the size which would grow in an uncontrolled manner under the imposed load.

Conventional methods of design assessment were therefore supplemented by the use of fracture mechanics techniques, which enable the lifetime of a structure containing cracks or crack-like defects to be predicted from a knowledge of specific material properties and operational stresses. Since all components contain fabrication defects of one type or another, it is essential to establish that the size of the defects which will remain is in fact compatible with the achievement of the design life.

During manufacture, fracture mechanics was also used to assess defects which lay beyond the manufacturing quality control standards and allowed concessions to be granted in a logical and safe manner.

The DTV and suction cone have been designed to withstand an internal pressure corresponding to approximately that of the top reservoir level, which far exceeds the maximum static head encountered in service. It is not necessary to consider fatigue as a design component for these items since, unlike the critical regions of the high pressure envelope referred to, this part of the system does not experience the pressure transients associated with operating mode changes.

In order to ensure that the pressure-containing plant items are of the highest possible integrity, the CEGB placed contracts imposing formal quality assurance requirements on its contractors. Each major contractor was evaluated by the CEGB against its own declared quality system and was required to correct any deficiency or non-conformity. Regular monitoring of the quality system was carried out during the contract period.

Contractors were required to submit for approval quality plans for all items or services provided under the contract prior to commencement of manufacture and to obtain quality plans from his subcontractor. The control to be exercised over the subcontractor was determined by the main contractor and agreed with the CEGB.

Each contractor was required to submit for approval to the CEGB those procedures or method statements necessary to ensure that all production operations or assembly sequences were accomplished under controlled conditions and to carry out his declared inspection and test policy as set out in the approved quality plan. The CEGB carried out independent inspection and tests to ensure that the declared standards were attained and that the product met the design specification.

To record the implementation of specified manufacturing quality control and provide a datum against which future in-service inspections could be carried out, case history records were compiled. Main contractors were made responsible for providing the necessary case history records which were supplemented by relevant inspection reports.

In-service inspection is the final safeguard in a chain of quality control measures applied throughout the design and construction of the plant. Linked with the assessment of defect significance by fracture mechanics, it protects against a fracture arising from service-induced fatigue damage. It also guards against fracture occurring in-service as a result of defects produced in the manufacturing process which may have not been detected by the earlier quality control measures.

18. CONSTRUCTION METHODS

After receiving the Royal Assent in December 1973, the first major contract placed was the Exploratory and Access Tunnel Contract. This contract entailed the driving of two major tunnels, one for exploratory access (plant) and the other for plant/construction access.

In February 1974 work started on establishing the portal for the exploratory tunnel in the Dinorwig Quarry. The rock face and bench into which the tunnel was being driven, was heavily jointed and exposed to climatic conditions and as a result the complete face collapsed when only a few metres of portal length had been driven. This major failure resulted in the examination of the design already prescribed. As a result of the re-examination, a decision was made to continue with the exploratory tunnel to the original design, but introducing a massive support system to the rock for the first 40m of length before re-starting the exploratory tunnel drive. Encompassed in this decision was the need to drive the exploratory tunnel at a very slow rate to minimise disturbance to the rock prestressing that had been introduced. This portion was also concrete-lined and buttressed as excavation progressed.

In the original design, the access tunnel portal was to be established in the same bench approximately 33 metres to the north. As a result of the exploratory tunnel portal failure, a more stable area for the location of the access tunnel portal was found and work on this tunnel drive started in September 1974.

The problem of stabilising newly-blasted rock was dealt with by introducing, for the first time in the UK, the use of shotcrete (sprayed concrete) to provide immediate support to newly-blasted excavations. This system of support with rock bolting and mesh, where necessary, proved highly successful.

To assist the permanent rock support designers for what was to be the main cavern excavation later in the programme, a matrix of small tunnels was established 30m above the intended main excavations. This allowed for geological conditions to be determined by instrumentation and thus give some assurance on the permanent support design requirements for the machine hall and transformer hall roofs. Thus it was intended that some programme time would be saved. However these small exploratory tunnels subsequently fulfilled a more important role in assisting with the construction and also permanent ventilation of the underground works.

Methods of working and means by which time and money could be saved were investigated. The main efforts were concentrated on considering methods of excavation and equipment that should be employed. It had been determined very early in the project planning that the most critical area of the work would be the machine hall through from excavation to completion of the units. Various methods of excavation were considered and it was firmly established that a system of headings with ramps and shafts would produce the quickest results.

FIG. 4.24   GENERATOR-MOTOR UNDER CONSTRUCTION

FIG. 4.25   PUMP-TURBINE UNDER CONSTRUCTION

Techniques for building structures within the main caverns were considered and because of their varying profile the use of mobile equipment was rejected. Therefore a case was made for the installation of overhead cranes as early as possible in the excavation programme. These cranes were capable of covering the full length, width and depth of the machine hall excavation.

Priority was given to the provision of large supplies of fresh air to all the underground workings in order to remove all traces of slate dust and other noxious gases.

A complex system of ventilation tunnels was constructed above the main cavern workings and connected via a battery of fans to a ventilation shaft exhausting above ground level. Additional measures used to minimise pollution underground was the use of water jets at the working face of the rock to collect dust created by the rock drills.

The first equipment to be erected underground was the MIV gallery crane. The method of erection was by means of lifting points anchored in the roof of the loading area. Problems were experienced in the load testing of the lifting points but nevertheless this system was proved reliable and was used later for the transformer hall crane installation.

The machine hall has three permanent cranes, two 250t and one 10t underslung cranes. The need to have these cranes available for the heaviest tandem lift (460t assembled rotor) necessitated an unusual method of erection of cranes of this size. Again the designed system of erection of the main cranes was via roof lifting points. This equipment required extensive testing and erection prior to use and also required a certain amount of strip-down between lifts. A quicker method was devised using two heavy duty mobile cranes and a special support column for erecting the 250t cranes.

As a result of the extreme size and weight of the generator-motors, it was decided that for machine integrity, it was advantageous to manufacture the stator on site where it would be possible to wind it in one section and so avoid unnecessary connections. Also to ease logistic problems it was decided to build the main rotor underground. Fig 4.24 shows the generator-motor under construction and Fig 4.25 shows the pump-turbine under construction.

Both these decisions demanded that factory-like conditions should be established underground, and a dust-free atmosphere established. Initially it was proposed that the on-site work should be carried out within an air tent, but the high volume ventilation equipment installed for the excavation phase of the project proved highly successful, such that the air tents were dispensed with.

19. POST COMMISSIONING EXPERIENCE

19.1 Plant Commissioning and Performance

Early operating experience of the plant has shown that in spite of a number of difficulties encountered during commissioning the design intent has been met. In particular, it was found that following the commissioning of the first unit at the end of 1981, problems occurred with the operation of the thrust bearing. Initially this was considered to be caused by the large thrust loadings to which the bearing was being subjected. Investigations showed that the down thrust due to the hydraulic performance of the turbine was greatly in excess of that predicted from the scale model studies done in the early design work. Modifications were made to adjust the pressure differential across the turbine runner and the loading on the thrust bearing reduced to the design figure.

Further troubles with the thrust bearings then developed and the bearings again failed in use. After much study the profile of the bearing pads has been found to be very critical to the satisfactory operation of the bearing interface oil film. A form of profile has subsequently been determined which is proving satisfactory in service.

During the early testing of the generator it became clear that the machine was overheating. Modifications have been necessary to adjust the generator air-cooling flows and also the machine winding design to reduce the effects of stray circulating currents.

The peripheral drainage pipework of the turbines had to be completely reconfigured to eliminate the resonant frequency of the system as vibrations, sympathetic to those being operated by the turbine rotation, were destroying it through fatigue failure.

The pump-turbine shaft seal was found to overheat under certain operational conditions and modifications were made to its cooling arrangements to overcome this.

Collectively these phenomena caused delays to the station commissioning programme, some of which were compensated for by altering the sequence in which the machine commissioning was done, units 3 and 4 coming ahead of units 1 and 2.

With the common plant and services, some difficulties have been experienced with the distribution and monitoring of cooling water flows. Modifications are in hand to improve matters.

The 18kV mechanical interlocks for maintenance access whilst being safe, were cumbersome and time-consuming to operate. A revised scheme has been designed and is being put into operation.

Of the automatic control systems, it has been found that, with a few minor exceptions, the operational experience of the equipment has shown it to be excellent. This is despite the fact that during commissioning about 70 modifications were necessary to the operating sequences in order for the plant to achieve its design criteria. The complexity of the changes varied from minor wiring changes to complete software rewrites for a particular mode transition. Much of the latter work was eased by the programmable nature of the sequence controller.

The commissioning and early operation has validated the design intent of the station. The required performance of the plant is:

(1) Main and auxiliary plant shall achieve average annual availabilities of better than 95%, excluding annual overhauls.
(2) The automatic equipment shall have a starting reliability of not less than 99% for both generation and pumping.
(3) Each unit shall be capable of up to 40 mode changes per day.

During the early operating period and following the modifications to the plant, the performance has been such as to give confidence in the station being able to meet these performance criteria. For example, for the month of August 1984 the figures were as follows:

Unit Number	Availability		Starting Reliability
1	98.95		99.42
2	98.51		99.27
3	}	Out of service due to	
4		plant modifications	
5	99.20		99.59
6	98.67		99.27
6	Average number of mode changes per day = 20.5		

19.2 Operating Roles

The operational requirements for Dinorwig, as defined at the time of its concept, are summarised as follows:

(1) To provide additional generating capacity of 1800MW to meet the growing system demand.

(2) To provide a rapid response system reserve of up to 1300MW in 10s to cater for the (then) envisaged new breed of generator design of up to 1320MW unit size.

(3) To improve system frequency control (frequency regulation).

However, electrical power system requirements change as circumstances change, and consumer demand on the electricity supply system actually declined over the period 1975 to 1983, with only a very slow growth expected over the following decade. Because of this, the largest generator size on the system has remained at 660MW and, with the odd exception, will continue at this level for some time to come.

Whilst bearing these facts in mind and with due consideration of the present economic circumstances, it is incumbent on the CEGB to exploit the valuable assets of the Dinorwig pumped storage scheme to maximum advantage.

The overall capacity and capability of the station is such that a great flexibility of operating roles is both possible and desirable. These roles, in order of economic/technical benefit to the CEGB are summarised as follows:

(1) System reserve or emergency reserve — these are the terms used to describe the spare capacity held on standby ready to respond to sudden losses of generating plant already supplying the grid, or to compensate for rapid increases in demand (peak lopping). For this role Dinorwig will have two or more turbine-generator units spinning, in air, throughout the day (but exclusive of the overnight pumping period) in readiness for picking up quickly to full-load, either by operator or automatically initiated action. This capability displaces the more expensive method of maintaining spinning system reserve on partially-loaded thermal plant.

(2) Economic generating — in this role Dinorwig supplies the grid system when the cost of its electricity production is competitive with other types of generating plant. The major cost of production at Dinorwig is that of pumping the water from the lower to the upper reservoir during the overnight off-peak period, suitably adjusted for any losses incurred in the process. Based on the current fuel prices, using the pump storage plant is cheaper than using oil-fired generating plant and

much cheaper than gas turbine plant. The special attraction of using Dinorwig under these circumstances is its ability to deal with short duration peaks without the attendant startup and standby costs incurred by other types of plant.

(3) Pumping — during the course of each day the system reserve and economic generating roles use up most (if not all) of the water stored in the upper reservoir. This water is restored during a 6h overnight off-peak period by using the pump-turbine/generator-motor units in the pumping mode. This provides an additional load for otherwise under-utilised, cheap production cost, coal-fired generating plant, thus helping to level out the night time demand trough. An additional benefit accrues in the pumping mode as it is possible, following the sudden loss of another large generating unit, to interrupt pumping on one or more Dinorwig machines (either manually or automatically) and thereby gain instant relief for the system.

In the existing economic circumstances the allocation of Dinorwig generating plant to the role of frequency regulation is not contemplated. This role continues to be filled more cost effectively by the conventional thermal plant operating on the system.

19.3   Operating Regime

The allocation of pump-turbine/generator-motor units to a particular role constitutes the station operating regime. Currently the amount of Dinorwig plant required to cover the catastrophic loss of generating units on the system has been reduced from 1320MW to 660MW, whilst the value of energy savings in the system reserve roles has increased and the full utilisation of the 6h overnight pumping period is now most important.

The most significant feature of the new operating regime is its flexibility. Whereas in the past a specific number of units have been allocated to a particular role, the new regime is designed to respond more cost effectively to the day-to-day variations in electricity supply and demand, ie to make maximum use of the station's potential and further increase the economic merit of the scheme.

20.   KIELDER RESERVOIR AND DAM - SMALL HYDRO PROJECT (FIG 4.26)

The Kielder Reservoir and Dam was developed by the Northumbrian Water Authority (NWA) in order to provide the North East conurbation area with adequate supplies of water. Water is discharged from the reservoir into a stilling basin from which it flows into the river North Tyne and then into the distribution system.

In partnership with the NWA, the CEGB has installed two turbine generators at the base of the dam to produce electricity which will be fed into the Area Board's supply systems.

One 5.5MW turbine generator absorbs the potential energy in water released for regulation of the main rivers, and a 500kW turbine generator is powered by release of compensating water. That is the minimum daily requirement of water into the river to safeguard environmental considerations.

FIG. 4.26   KIELDER RESERVOIR AND DAM

The power station is required to operate as an unattended station, remotely started and stopped from the NWA control centre some 50 miles away, where the regulation of water release is of paramount importance and the generation of electricity and its associated revenue is of secondary importance.

The station will be loaded as required to match the water flow regulating requirements from the Kielder reservoir to provide the water needs of the NWA water management. Within these water management needs the operation of the station will be such as to optimise the revenue available from the supply of electricity. This suggests that the compensating set will run during the night and the main machine during the day.

During construction of the dam, the NWA installed foundations for a power house. It was therefore necessary to design the power station, consisting of two turbine generators together with transformers and high and low voltage switchgear together with all their ancillary equipment, into an existing foundation slab.

20.1 Station Description (Fig. 4.27)

The power station consisting of two hydro-electric turbine generators, utilises water from the reservoir to produce up to approximately 6MW of electricity for the national grid.

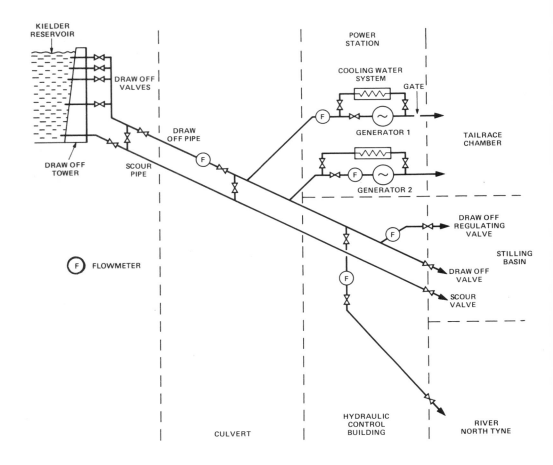

FIG. 4.27   KIELDER HYDRAULIC SYSTEM

Water releases from the reservoir will vary from a minimum of 56,625m³ per day (winter compensation water flows, December to March inclusive), or 113,250m³ per day (for April to November inclusive), to river regulation releases up to a maximum of 1,220,000m³ per day. The design aim is to utilise the generating sets to cover this range of flow as far as possible, the power produced varying between 0.2MW and 6.2MW.

Turbine generator 1 (synchronous) produces 5.7MW taking advantage of the water release for regulation of the main rivers, and turbine generator 2 (compensating) using the release of compensating water which is the minimum release from the reservoir.

Construction of the Kielder Reservoir and Dam started in Spring 1976, impounding commenced in Autumn 1979. The normal crest level of this reservoir is at 185.2m OD, with a maximum overflow level of 186.2m OD during flood conditions. When full the reservoir contains 188 million cubic metres of water.

A 2m dia draw-off pipe approximately 400m long, together with a scour pipe 2.9m diameter in parallel, with interconnecting branches and valves, are contained within a culvert from the reservoir to the power station and supply the two generating sets. The turbine draft tubes discharge into a stilling basin at the foot of the spillway channel adjacent to the power station.

Turbine generator 1 generates at 11kV and is connected to the 33kV system through a transformer and two circuit breakers. Turbine generator 2 generates at 415V and is connected to the main 11kV system via two circuit breakers, contained within a switchboard, and a transformer. The output of the station is connected via a 33kV overhead line to the Electricity Area Board substation at Spadeadam. Station auxiliary supplies are derived from the 415V switchboard and a standby 415V incoming supply is taken from the Electricity Area Board local system. The electrical system is shown in Fig 4.28.

The sets are started, stopped and loaded by local/remote auto control, with provision for auto/manual synchronising. Emergency local manual control is also provided.

The station normally runs unmanned and is remotely controlled by the NWA from their Howdon control centre situated in Newcastle-upon-Tyne, some 50 miles distant from Kielder. The plant is designed to give a starting reliability of not less than 99%, reliability being defined as the number of times a successful start is achieved within the given time limits, expressed as a fraction of the total number of attempts.

20.2   Plant Layout

The layout of plant was dictated by the existence of the power station concrete perimeter walls which were established during the construction of the dam and culvert.

Within the power station building and at ground level (144.61m level) but accessible externally, are located the 33kV/11kV generator transformer, 11kV/415V auxiliary transformer and the 33kV Spadeadam feeder earthing transformer together with the neutral earthing equipment associated with the 33kV/11kV generator transformer and 11kV generator 1. The turbine 1 tailgate is located adjacent to these transformer compounds and is operated during maintenance periods by the use of a portable crane.

The mechanical and electrical plant within the power station building is located on various floors as follows:

(1) Tank Floor 149.46m level.

Located over the switchgear room and containing the transformers' fire protection pre-mix foam tank.

(2) Operating Floor (ground level) 144.61m level.

The 30t pendant-controlled crane is located above this level and is used for erection and maintenance of the plant. A loading bay at one end of the building, together with access openings throughout all floors, are provided to facilitate this. The station control and mimic panels are located adjacent to the loading bay, with the 33kV and 11kV circuit breakers located in an adjoining switchroom. The transformers' fire protection equipment is also located on this floor.

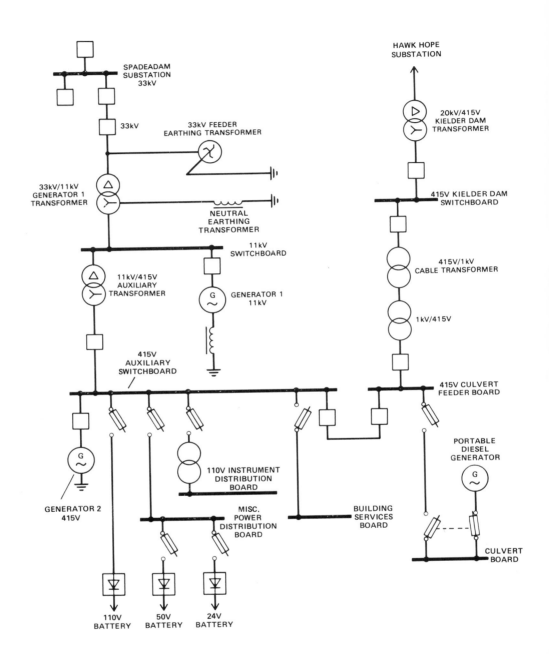

FIG. 4.28   KIELDER ELECTRICAL SYSTEM

(3) Services Floor 141.11m level.

Situated on this floor are the 415V auxiliary board, station data logging/generator sequencing equipment cubicle, main protection relay panel, auxiliary motors control panel, telemetry equipment, battery and charger rooms.

(4) Generator Floor 137.61m level.

Situated on this floor are 11kV generator 1, turbine generator 2 with magnetic flow meter and auxiliaries, 415V building services board, 415V culvert/power station feeder board and associated P1 feeder transformer, stores and workshop area.

(5) Turbine Floor 131.25m, 132.50m and 134.60m levels.

Situated on this floor are the turbine 1 auxiliaries comprising governor control equipment, hydraulic equipment, main inlet valve and magnetic flow meter.

(6) Basement 128.60m level.

Situated on this floor are the drainage and dewatering pumps, housed in the station sump, and the basement floor drain sump. Access is provided at this level into the turbine 1 draft tube for maintenance purposes.

20.3 Main Turbine

A vertical shaft Kaplan turbine 1 is provided having a gross output of 5.88MW under a gross head of 47.35m and a flow of 14.1m^3/s. The machine which is fed by a 2m diameter inlet pipe, has an embedded spiral casing of fabricated steel pipe construction, the profile of which has been developed using a standard computer programme.

The 20 cast stainless steel guide vanes, are controlled by hydraulic servomotors and the 6-bladed runner is similarly controlled, optimum relationship between guide vane opening and runner blade angle being maintained by a combination linkage and cam arrangement. Guide vane closure times have been selected to ensure that during trip conditions, unacceptable pressure surges or water spillage in the valve tower are avoided.

Water flow to the turbine is measured by a 2m diameter magnetic flow meter upstream of the main inlet valve.

An oil-lubricated water-cooled guide bearing carries a 356mm diameter forged steel shaft driving the direct coupled generator at a speed of 500rev/min.

20.3.1 *Main Inlet Valve*

A 2m diameter main inlet valve of the hydraulically-operated butterfly type is installed at the spiral casing inlet. The valve is held open hydraulically and closed by the action of a counterweight by releasing oil pressure.

Valve closure time has been selected having in mind the need to close as quickly as possible, whilst at the same time ensuring no unacceptable pressure surges occur in the reservoir draw-off pipe and that water spillage from the draw-off pipe vent in the valve tower does not take place.

The valve is provided with a bypass arrangement to permit flooding of the spiral casing following dewatering of the machine.

20.3.2 *Draft Tube*

A fabricated steel draft tube and bend embedded in concrete conveys water into the stilling basin. A draft tube valve is not provided but the machine can be isolated from the tailwater by insertion of a gate of the sliding bulkhead type.

20.3.3 *Control System*

The control system is the conventional hydraulic system employing a continuously charged air/oil vessel. Duplicate oil pumps and compressors are provided and the system has sufficient capacity for two complete strokes of the guide vane servomotors when the oil pressure falls to that when emergency trip is initiated. The control system for the MIV is similar to, but independent of, the main turbine control system.

20.4 Compensation Turbine

A horizontal shaft Francis turbine 2 is provided, mounted on a combined bedplate with the generator. The turbine has a gross output of 533.8kW under a gross head of 47.7m and a flow of $1.325m^3/s$. The main inlet pipe to the machine is 600mm diameter.

The spiral casing is of steel plate whilst the 16 guide vanes and runner are aluminium bronze castings. Guide vane control is by hydraulic servomotor and the hydraulic system is supplied by the air/oil system of the main turbine.

The turbine is directly coupled to the generator and operates at a speed of 1000rev/min.

Water flow to the turbine is measured by a 600mm diameter magnetic flow meter downstream of the main inlet valve.

20.4.1 *Main Inlet Valve*

A 600mm diameter main inlet valve of the hydraulically-operated butterfly type is installed at the spiral casing inlet. The hydraulic system is supplied by the air/oil system associated with the main turbine and the valve is held open hydraulically and closed by counterweight action on release of oil pressure.

20.4.2 *Draft Tube*

A fabricated steel draft tube embedded in concrete directs water into the stilling basin. Since the compensation turbine is located above all but the extreme flood tailwater level, neither draft tube valve nor tailwater gate is provided on this machine. Should the machine be under maintenance however, a blank can be fitted to the power station end of the draft tube in safety since the water level in the stilling basin is always known.

20.5 Main Generator

A vertical shaft, salient pole, continuous maximum rated, closed air circuit water-cooled, brushless ac generator is provided designed to operate in conjunction with a Kaplan type water turbine. The generator has a continuous maximum rating of 6200kVA at 11,000V, 50Hz (nominal) and power factor range of 0.9 over-excited to 0.9 under-excited.

The generator mounting skirt is a fabricated steel structure capable of supporting the weight of the generator and turbine rotating parts, and of withstanding the turbine hydraulic thrust and fault conditions including short circuits. The exciter stator pack is underslung on the top bearing bracket. The permanent magnet generator stator packs and enclosures are supported by the non-drive end bearing assembly.

The exciter is located between the rotor body and the non-drive end bearing with the rotating excitation diodes and heat sinks mounted on the rotor body. The leads are secured to the surface of the generator shaft. The excitation permanent magnet generator is mounted above the non-drive end bearing.

The excitation system of the machine comprises a brushless revolving armature stationary field ac exciter, the output of which is rectified in a revolving diode bridge mounted on the rotor and fed directly to the field of the ac generator.

20.6   Compensation Generator

A closed air-circuit air-cooled, continuous maximum rating horizontal shaft, foot mounted induction generator is provided designed to be operated with a Francis water turbine. The generator has a continuous maximum rating of 510kW at 415V 50Hz (nominal).

The stator core structure is mounted on a bed plate and supported by feet welded onto the core end plates. The stator bed plate is mounted on a common turbine and generator skid assembly.

20.7   Control and Instrumentation

20.7.1   *Principles*

Whilst the station is designed for remote operation it must also be possible to operate the generating units locally, at Kielder site, for the purposes of commissioning and recommissioning following a maintenance period.

The modes of operation for the two generating units and associated station services have been defined as follows:

(1) Remote operation from Howdon for the automatic startup, loading and shutdown following manual initiation.
(2) Local operation at Kielder for the automatic startup, loading and shutdown following manual initiation.
(3) Local operation at Kielder for the manual control of startup, loading and shutdown.

To enable local manual operation (3) to be carried out, a control panel is provided at Kielder incorporating all the controls and indications necessary for that purpose and covering both generating units; the auxiliary electrical system within the station and the hydraulic system comprising the reservoir outflow management.

20.7.2   *Local/Remote Automatic Operation Selection*

The selection between Kielder or Howdon as the location for the initiation of automatic operation is also made at the Kielder control panel. This is a fairly obvious choice since it must be established that the plant is capable of being operated unattended before its control is passed to a remote point. Likewise the man on the spot, ie Kielder, must always be able to gain control of the plant should circumstances dictate.

Local/remote control selection is made on an individual system basis, the systems comprising generator 1 operation, generator 2 operation, hydraulic system in the form of individual valve control, electrical auxiliary system giving control of the 33kV Spadeadam feeder circuit breaker and the two alternative 415V feeders to the auxiliary board.

In selecting the generating systems to remote operation it is essential that all subsystems which should be in automatic mode are so selected. In order that the station is not left unattended without this condition being satisfied, alarms are provided at Kielder.

20.7.3 *Remote Control Facilities (Howdon)*

A telemetry link between Kielder and Howdon is the means through which the station will be controlled remotely and the following controls and indications will be provided for the Howdon controller.

The controls for turbine generator 1 will be initiation of startup and shutdown via the automatic sequence, load level adjustment in terms of water flow, the selection of loading and unloading rates and the initiation of emergency shutdown, should that become necessary.

The indications for turbine generator 1 will be status signals of whether the unit is available, whether it is in operation or has tripped. There will be plant parameter levels (for format display) of speed, water flow, power, reactive power, stator voltage, stator current, turbine inlet water pressure and guide vane (load level) set point.

The output of the station through the 33kV feeder will also be signalled to Howdon in the form of power output (MW), reactive power (MVA), export/import energy (MVA), field current and voltage.

With regard to the electrical auxiliary system, interlocks are arranged to ensure incorrect operation cannot occur either remotely or locally.

20.7.4 *Local Control Facilities (Kielder)*

The operational facilities at Kielder are incorporated on a control panel which provides all the remote control facilities, plus the selection switches for local/remote and automatic/manual control.

Manual control also requires additional facilities, eg excitation controls and indications, synchroscope facilities and generator circuit breaker control in the case of generator 1.

The main function of the this control panel is related to the normal operation of the plant. Where it is necessary for maintenance/commissioning engineers to carry out individual plant item checks and testing, local manual control facilities at the plant item are provided. However, where such manual control can be achieved at the switchgear controlling the plant item, control stations local to it are not provided.

20.7.5 *Automatic Control Equipment/Alarm Monitoring*

The means by which either or both of the two generators will be automatically started, loaded and shutdown will be through a programmable logic controller/data logger (PLC/DL).

In addition to automatic sequence control this equipment monitors and logs plant alarm conditions. The equipment also serves another function, particularly with regard to the generating plant; in the course of monitoring alarm conditions, logic is applied such that if, individually or in combination, certain inputs reach dangerous levels, trips are applied to isolate and shutdown the plant. This latter function, ie protection of the generating units, is not covered entirely by the PLC/DL equipment, in fact there is a separate protection relay system which monitors the whole electrical system and generating plant for unsafe conditions.

The PLC/DL equipment is a microprocessor system, which combines the flexibility necessary when designing control and indication systems for plant, with the reliability being proved by such equipments. Because of the multi-function role the PLC/DL performs a key feature is its self-diagnostic routines. Its normal power supply is backed-up by an inverter operating from the dc battery system and this, together with the self diagnostic routines, ensures that in the automatic mode the plant always operates in a safe and controlled manner.

The alarm conditions are recorded in hardcopy form on a printer associated with the data logger. In addition to the alarm function this record lists the condition in time and event order, which is advantageous to engineers arriving at the unattended station to correct any abnormal situation which has occurred.

21. MAIN CONTRACTORS, MANUFACTURERS, AND PLANT SUPPLIERS - DINORWIG

Mechanical

Pump-turbines, main inlet valves, draft tube valves	Boving and Co. Ltd
Heating and ventilation plant	Haden Young Ltd
Cranes	Herbert Morris Ltd
Turbine and valve manufacture	Markham & Co. Ltd
Fire protection plant	Mather & Platt Ltd
Headgates, tailgates, bascule gates	Newton Chambers
HP tunnel steel linings	Whessoe Ltd
Lifts	W.M. Wadsworth & Sons Ltd

Electrical

Generator-motors	GEC Large Machines Ltd
Starting equipment	A.S.A.E. (GB) Ltd
Control equipment	Babcock Bristol Ltd
18kV busbar equipment	Balfour Beatty Power Construction Ltd.

18kV switchgear	British Brown Boveri Ltd
Batteries and chargers	Chloride Standby Systems Ltd
Alarm annunciators	D. Robinson & Co. Ltd
Metering	Ferranti Ltd
Data logger and sequence equipment	Fisher Controls Ltd
Auxiliary transformers	NEI Parsons Peebles
	Distribution Transformers Ltd
	Bonar Long Ltd
415V switchgear	Scott & Electromotors Ltd
11kV and 3.3kV switchgear	NEI Reyrolle Ltd
Generator-motor transformers	NEI Parsons Peebles Power
	Transformers Ltd
Telephone systems	Plessey Communications Systems Ltd
Radio communications equipment	Pye Telecommunications Ltd
Diesel generators	Ruston Diesel Ltd
Cabling	T.W. Broadbent
400kV SF6 circuit breakers	GEC High Voltage Switchgear Ltd
400kV SF6 busbars	NEI Reyrolle Power Switchgear
400kV Cables	Pirelli General Cables Works Ltd

Civil

Marchlyn reservoir	Gleeson Civil Engineering Ltd
Main civil works	McAlpine-Brand-Zschokke
Exploratory and access tunnels	Mowlem (Scotland) Ltd
Ventilation and surge shafts	Thyssen (GB) Ltd
Road works	A.P.W. Construction Ltd

Consultants

Civil engineering	James Williamson & Partners
	Binnie & Partners
Pump-turbine/generator-motor	Merz & McLellan

Aerial Photography

Aerial Photography	Airways (Manchester) Ltd

5.

Heysham 2 – AGR Nuclear Power Station

1. INTRODUCTION

In January 1978, the Government authorised the CEGB to begin work with a view to ordering a further AGR station (Heysham 2) as soon as possible. This decision was taken in the light of the encouraging operating experience from Hinkley Point B (and from the similar station, owned and operated by the South of Scotland Electricity Board, at Hunterston B) which had, by that time, been accumulating over a period of some two years.

Experience with the earlier AGR stations had left no doubt that to start the construction of a nuclear plant in advance of adequately detailed design was a certain route to problems, delays and additional costs. This experience, in fact, only confirmed what had already become abundantly clear from the CEGB's earlier experience of commissioning their very large programme of 500MW coal and oil-fired units during the late 1960s and early 1970s. It lent further emphasis to the policy of replicating proven designs as the standard means of meeting load growth whilst proceeding separately at an optimum pace with the evolution and introduction of more advanced plants.

It was therefore decided to base Heysham 2 as closely as possible on the Hinkley Point B design. The only design changes which would be permitted were those necessary to meet the enhanced safety requirements which had been adopted during the decade since the Hinkley Point B station had been ordered and to remedy a number of deficiencies revealed during the early operation of the Hinkley Point B and Hunterston B stations.

1.1 Design Changes Relative to Hinkley Point B

1.1.1 *Main Parameters*

The main parameters of Heysham 2 are shown, together with those of the earlier AGRs, on Table 5-1.

Parametrically, Heysham 2 is little different from Hinkley Point B. There is an increase in the number of channels from 308 to 332 in order to avoid the shortfall in output which was experienced at the earlier station. There is a small increase in circulator outlet pressure from 42.4 bar to 43.3 bar which improves the circulator operating point by reducing volumetric flow. Additional operating margin is also provided by a 4% increase in the heat transfer surface area of the boiler module.

Furthermore, the strengthened fuel element adopted for Heysham 2 during 1984, which is described in Section 10, has a lower resistance to flow than that in use at Hinkley Point B. This allows full design output to be achieved with a large circulator margin and provides the potential for increasing reactor output in the longer term.

1.1.2 *Safety Requirements*

Safety requirements underwent considerable evolution during the building of the earlier AGRs.

TABLE 5.1. AGR MAIN PARAMETERS

		Dungeness B	Hinkley Point B and Hunterston B	Hartlepool and Heysham 1	Heysham 2
Net electrical output per unit	MW	606	620	625	625
Reactor heat rating	MW	1453	1500	1507	1556
Mass of uranium per reactor	t	122.1	105.3	110.8	113.5
Fuel channels per reactor		408	308	324	332
Fuel elements per channel		7	8	8	8
Nom. max. fuel element rating	MW/tU	19	23	22	22
Bulk carbon dioxide gas outlet temperature	°C	675	650	640	634
Carbon dioxide gas pressure at circulator outlet	bar	31.0	42.4	41.3	43.3
Gas circulators per reactor		4	8	8	8
Power input per circulator	MW	11.2	4.6	4.6	4.8
Turbine stop valve steam conditions	bar/°C	159.6/566	159.6/538	159.6/543	159.6/538

Shutdown Systems

Secondary and tertiary shutdown systems were introduced in order to provide diverse back-up to the gravity-operated control rods which constitute the primary shutdown system. This diversity was provided for the first time at Hartlepool and Heysham 1 and similar arrangements have been made for Heysham 2. (See Section 8).

Post-Trip Heat Removal

An important change as compared with Hinkley Point B is the arrangement of the essential electrical supplies for the post-trip heat removal systems as independent trains. Each of the four quadrants into which the boiler system of each reactor is divided, has associated with it, two independent trains of electrical equipment supplying diverse methods of heat removal. Eight diesel generators provide a diverse source of electrical power for the station if grid supplies are lost.

The post-trip heat removal systems are described in more detail in Section 14.

Protection Against Internal Hazards

Improved facilities are now provided in the shape of restraints, barriers and venting routes and of a greater degree of segregation and separation of essential auxiliary equipment and cables. These minimise the possible effects of potential internal hazards such as the release of hot carbon dioxide as a result of a depressurisation incident, the release of steam following the fracture of a steam main, missiles generated by failed components and fires.

Protection Against External Hazards

The external hazards which now have to be considered include seismic disturbances, high winds and aircraft crashes.

The probability of a damaging earthquake at any particular location in the UK is such that the economic risk implied by not designing for continued operation during or after such a disturbance can be accepted. It is therefore necessary to consider only the

434

safety implications of a seismic event. Essentially, the integrity of the reactor plant and shutdown systems must be maintained so that the reactor can be shutdown safely and an adequate post-trip heat removal capability must remain operational. The relevant components and systems are therefore designed to withstand a so-called 'safe shutdown earthquake' (SSE).

In the design process, it is convenient to introduce the concept of a 'design basis earthquake' (DBE) and ensure that when subjected to a disturbance of this magnitude, all components which are required to have a seismic capability, will retain their elastic behaviour. This procedure provides a large margin of capacity for absorbing energy in the inelastic regime. It is considered that this process will ensure adequate response at the SSE level.

A similar approach is taken to counter high wind loading. Additional conservatism beyond that used in conventional station structures is applied to the selection of the factors covering speed, ground conditions and return period built into the British Standard Codes of Practice CP3 Chapter V Part 2. This corresponds to a criterion that the probability of a high wind causing a hazard in any one year should not exceed one in ten thousand.

Heysham is not situated in an area of high aircraft activity and the probability of an aircraft crashing on the site is very small. However the segregation and separation of essential equipment which is being provided as a protection against potential internal hazards would afford some measure of protection in such a remote eventuality.

1.1.3 *Irradiated Fuel Disposal Facilities*

Operational experience at Hinkley Point B and Hunterston B indicated that any new AGR could, with advantage, be provided with a greater margin of capacity at certain points in the irradiated fuel disposal (IFD) facilities.

At Heysham 2, the active maintenance facility, in which the re-usable plug units are serviced, has duplicate plug unit maintenance positions as well as a separate control rod assembly cell so that plug unit and control rod assembly and maintenance can proceed in parallel.

In order to ease the IFD cell cooling duty, Heysham 2 has a buffer store designed on the basis of a 28 days storage period. All irradiated fuel is first routed to the buffer store where it is cooled by the natural convection of carbon dioxide in pressure tubes before being moved into the IFD cell. To secure these ends, Heysham 2 has 32 buffer storage tubes as compared with a total of 22 at Hinkley Point B where the original provision was as few as eight such tubes.

The refuelling facilities are described in more detail in Section 11.

1.1.4 *Man Access to the Reactor Vault*

When the prestressed concrete pressure vessel was adopted and with it, the concept of containing the whole primary circuit within the vault, it was thought that man access would be severely limited both as to duration and position. However, it has been demonstrated in practice that regular entries for inspection and minor repairs are quite practicable.

Whilst this has not been allowed to affect the extent of the facilities provided for remote viewing — where camera coverage has, in fact, been greatly increased — the arrangements for man access to the reactor vault have been improved. In particular, man access to the spaces below the boilers and the diagrid is possible without the need to withdraw a circulator, as is necessary at Hinkley Point B.

In addition, the vault dimensions have been increased so that there is enough space for men in protective suits to reach all surfaces of the boiler annulus. There is also enough space around each boiler module to permit repair work should this ever be necessary.

2. MAIN PLANT DATA

Station

No. of reactors	2
Electrical output (net)	1250MW
Electrical output (gross)	1320MW
Thermal efficiency	40%

Reactor

Reactor heat rating	1556MW
Reactor coolant	CO_2
Reactor gas inlet temperature	299°C
Mean fuel channel gas inlet temperature	334°C
Mean fuel channel gas outlet temperature	635°C
Total gas mass flow through fuel channels	3909kg/s
Weight of uranium	113.5t
Mean fuel rating	13.65MW/tU
Mean fuel discharge irradiation	18,000MWd/tU
Mean fuel channel rating	4.67MW

Core

Moderator	Graphite
Number of fuel channels	332
Lattice pitch (square)	460mm
Number of control rods	89
Number of fuel elements per channel	8
Active core mean diameter	9458mm
Active core height	8296mm

Fuel Elements

No. of elements	2656
Fuel	Enriched uranium dioxide ceramic
No. of fuel pins	36
Element length	1039mm
Element weight	83.52kg

Gas Baffle

Construction	Dome headed vertical cylinder
Material	Fabricated steel plate
Dome inside diameter	13.85m
Dome (top cap) radius	9.5m
Plate thickness	67mm
Cylinder inside diameter	13.85m
Cylinder wall thickness	35mm
Cooling gas nozzles	24 (two rows of 12)
Dome design pressure/temp	3.1 bar at 375°C
Cylinder design pressure/temp	3.2 bar at 375°C

Pressure Vessel

Material	Concrete, helical prestressed, steel lined Internal diameter
Internal diameter	20.25m
Internal height	21.87m
External diameter	31.86m
Wall thickness	5.76m
Pile cap thickness	5.427/7.347m
Base slab thickness	7.484m
Design pressure	46.65 bar

Gas Circulators

Type	Centrifugal, single stage
Number per reactor	8
Drive	Constant speed induction motor
Flow control	Variable inlet guide vanes
Speed	2970rev/min
Outlet gas pressure	43.3 bar
Gas flow (8 circulators)	4270kg/s
Pressure rise	2896mbar
Outlet gas temperature	299°C
Motor input power/reactor	42.6MWe

Boilers

Type	2 start, once through serpentine platen rectangular unit
Number of boilers/reactor	4
Number of units per boiler	3
Boiler inlet gas flow	4202kg/s
Gas inlet temperature to reheater	615°C
Gas outlet temperature	290°C
Heat transferred to steam	1577MW
Feed flow	500kg/s
Superheater outlet header pressure	166 bar
Superheater outlet temperature	541°C
Steam generation	500kg/s
Reheater outlet manifold pressure	40.7 bar
Reheater outlet temperature	539°C
Reheater flow	455kg/s

Reactivity Control

Number of black rods	44
Lift 1	22
Lift 2	22
Number of black rods as sensors	7
Number of grey rods	45
Regulating/safety	16
Regulating	29
Secondary shutdown system	Nitrogen and boron glass beads

Diesel Generators

	X Train	Y Train
Number of sets	4	4
Generator output voltage	3.3kV	3.3kV
Generator rating	5.2MW	6.735MW
Power factor	0.85	0.85

Turbine Generators

Number of sets	2
Type	5 cylinder single shaft
CMR	660MW
Steam pressure — HP inlet	159.6 bar
Steam temp — HP inlet	538°C
Steam pressure — IP inlet	39 bar
Steam temp — IP inlet	538°C

Speed	3000rev/min
Steam consumption at CMR	500kg/s
Generator voltage	23.5kV
Generator current	19,619A
Generator CMR	776MVA
Generator power factor	0.85

2.1 Operational Characteristics

AGR stations up to and including Heysham 2 have been designed for base-load duty over an operating life of 30 years. They do not, therefore, necessarily have characteristics suited to more flexible modes of operation and the Heysham 2 control scheme is of the 'turbine-follows-reactor' type.

Although different dynamic characteristics could be designed for, the capability of AGR in this respect is likely to be limited by fuel behaviour. However the extent to which the present forms of fuel element may be able to withstand regular load cycling will only be determined definitively by operational experience.

The joint UKAEA/CEGB fuel development programme may lead to improvements in fuel endurance which could result in the prospect of a greater degree of flexibility.

2.2 Capability

As compared with the Magnox units, the AGR is highly rated, employs steam conditions identical with those of conventional practice and was built at the 660MW unit size without the advantage of a semi-commercial scale precursor of any magnitude or of close similarity.

Inevitably, a considerable degree of uncertainty attached to the forecasts of the level of reliability which might be achieved with the novel and complex mass of plant and equipment subjected to high temperatures, high dynamic forces and high noise fields within the primary circuit.

It can be seen from Table 5.2 that, during the early years of operation of Hinkley Point B, these grounds for caution manifested themselves in substantial loss of capability due to planned overhauls whilst the loss due to actual breakdowns has been at an encouragingly low level since 1979.

TABLE 5.2. Hinkley Point B Reactor Units Loss of Capability

Reason for Loss of Capability	1978/79	1979/80	1980/81	1981/82	1982/83	1983/84
	%	%	%	%	%	%
Reactor-unit breakdowns	15.19	1.00	4.37	0.92	4.10	1.08
Reactor refuelling	—	—	—	—	8.20	4.35
Overhaul programme	33.04	23.16	63.23	15.33	12.57	11.31
Total breakdowns, refuelling and overhaul programme	48.23	24.16	67.60	16.25	24.87	16.74

3. THE SITE

3.1 Topography and Geology

The Heysham site is bounded on the north by Heysham harbour and on the west by the coastline. It comprises land reclaimed from the sea at about the turn of the century by the former London and North Western Railway Company who built a sea-wall and filled the enclosed area to about 8m OD.

The site investigation revealed the existence of a geological fault running approximately north-south and bisecting the useful area of the site. Triassic sandstone exists to the west of the fault and is suitable for the support of power station loads. A complex sequence of Namurian mudstones, sandstones and siltstones, which are not suitable for heavy ground loadings, exists to the east of the fault.

The upper surface of the Triassic sandstone varies in level from about 3m OD in the south to about −7m OD in the north and Heysham 1 power station is situated at the northern end of this area, adjacent to the harbour.

3.2 Site Layout (Fig 5.1)

The lines of the sea wall and the geological fault converge towards the south of the site and thus create, to the south of Heysham 1, a roughly triangular area on which Heysham 2 could be located.

At the time that planning permission was sought for Heysham 1, the stage 2 development was envisaged and shown on the planning application as a mirror image of Heysham 1. Although Heysham 2 could not, in the event, conform with that concept, the implied commitment has been recognised to the extent of maintaining parallelism of the station centre-lines in the north-south direction.

The arrangement of the main plant was governed by the need to keep the cooling water (CW) culverts and the 400kV cables as short as possible. The CW pumphouse is located in the harbour alongside the Heysham 1 pumphouse and the Heysham 2 400kV transmission lines must leave the station area in a north easterly direction. It was therefore economically advantageous to site the turbine house and generator transformers at the northern end of the available area and the nuclear island to the south of the turbine house.

This arrangement led to the location of the active effluent treatment plant and the solid waste store in a position to the south of the reactor building with access from the nuclear island by means of a bridge.

The essential supplies buildings are strategically positioned around the main building envelope to satisfy the safety requirements.

The block containing the administration, welfare and workshop facilities is located so as to satisfy a number of conditions, namely:

(1) To fit into a rational pattern of personnel movement.

(2) To occupy land unsuitable for other uses.

(3) To permit the movement of station personnel and visitors to and from the block with little or no contact with other station activities.

HEYSHAM 2 – AGR NUCLEAR POWER STATION

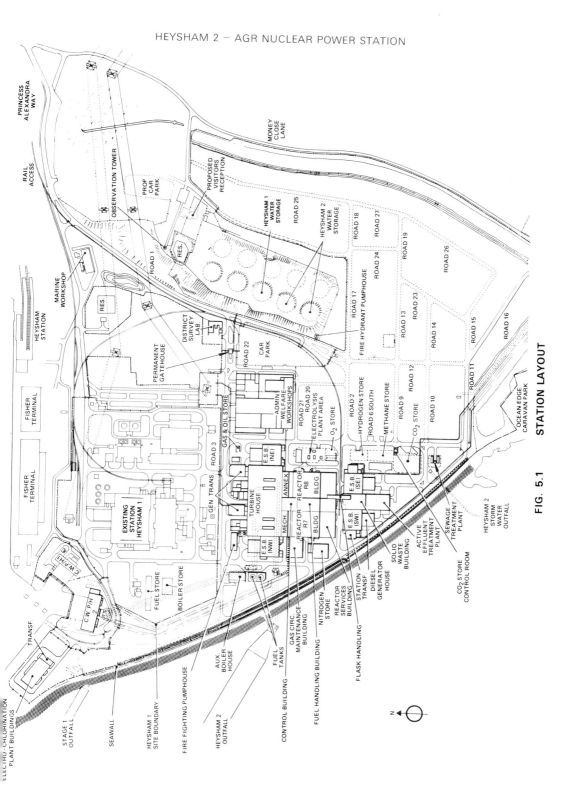

FIG. 5.1   STATION LAYOUT

Sea water for cooling purposes is drawn from Heysham harbour and is discharged in a westerly direction from the turbine hall through culverts and tunnels to an outfall in Morecambe Bay.

3.3   Flood Protection

The ground level in the area occupied by the Heysham 2 station has been raised, by using excavated material, from the previous level of about 8m to 9m OD. This may be compared with a predicted highest tide level at Heysham due to gravitational and topographical effects under average meteorological conditions of 5.8m OD and with the highest recorded water level in the harbour of 6.93m OD on 11 November 1977.

The sea frontage of the site is protected by a bound masonry wall at a minimum height of 7.8m OD and a super-imposed wave reflection wall which gives a further 1.7m of height. Thus, the lowest level of protection along the sea frontage is 9.5m OD giving a margin of 2.57m over the highest recorded still-water level. Drainage facilities behind the sea wall accommodate wind blown spray and any waves which may overlap the sea wall.

To the south, a westerly extension of an existing escarpment is included in the site landscaping works so that, to the south and east, the site is fully protected against flood water ingress by natural or artificial banks.

3.4   Seismology

Study of all the available data, on both local and national scales, led to the definition by probabilistic analysis, of a safe shutdown earthquake (SSE) which is employed in the design process. The SSE corresponds to two recordings of the Parkfield (USA) 1966 earthquake modified to give a peak horizontal ground acceleration of about 0.25g. The north-south fault across the site is considered to be 'incapable'; that is to say, it is not capable of surface rupture. It does not, therefore, constitute an additional potential hazard over and above that represented by the probabilistically derived design basis.

3.5   Power Lines

The Heysham 400kV and 132kV substation is located about 0.6km east of the power station site and is separated from it by a golf course. Two parallel double-circuit 400kV overhead lines carry the outputs of Heysham 1 and 2 to the substation by a route which leaves from the northeast corner of the site whilst, 132kV supplies are cabled to the station transformers by a southerly route.

3.6   Facilities for Receiving Heavy Loads

Some 15 loads of up to 400t were delivered by sea-going vessels and handled by existing harbour facilities.

For Heysham 2, a further temporary docking facility was constructed at the western end of the harbour in order to be able to off-load four fabricated assemblies each of which, with its road transport trailer, exceeded 1000t. These factory-built items comprised the two reactor gas-baffle assemblies, each weighing 950t (1150t with trailer) and the two reactor pressure vessel roof liner and standpipe assemblies, each weighing 1000t (1300t with trailer). All four were transported to Heysham on sea-going barges.

4. PLANT LAYOUT

The buildings which house the main plant occupy an area of 51,400m^2 or some 16% of the useful site area. Their total volume is 1,380,000m^3 which corresponds to a figure of just over 1m^3 per kilowatt generated.

Site constraints were not such as to enforce departure from established layout principles which include:

(1) The transverse arrangement of turbine generators.
(2) The location of reactors and turbine generators on a common centre-line.
(3) The symmetrical arrangement of the main steam and reheat pipework for each unit with respect to the common centre-line of that unit.
(4) Identical arrangements of plant and other pipework for each unit so far as possible.
(5) The selection of the turbine room basement level so that pumping costs are minimised by the syphonic recovery of CW pressure head.

4.1 Layout of the Nuclear Island (Fig. 5.2)

The nuclear island complex comprises two buildings which accommodate the reactor units and their support systems, a common fuel handling building which is located centrally between them, a control annexe to the north and reactor services annexes which are positioned on the south side of each reactor. A central control room is situated in the control annexe on the transverse centre-line of the station.

The charge hall extends the full width of the nuclear island and spans both reactor pressure vessels and the centrally positioned fuel handling facilities.

The turbine house constitutes the most northerly block of the main building complex and is separated from the control annexe by a mechanical annexe.

The central location of the fuel handling building leads inevitably to the 'handed' location of the rooms which house the items of reactor ancillary plant such as the primary coolant processing plant and the decay-heat condensing plant. Slight variations of layout of plant within the ancillary plant rooms are necessary in order to accommodate the 'handed' access into these rooms but so far as possible, identical ancillary plant arrangements are provided for both reactor units. By this means, operator familiarity gained on one reactor unit is rendered, in large measure, applicable to the other.

The central location of the fuel handling building also leads to some assymetry, with respect to the north-south centre line, in the reactor services annexes where it is necessary to adopt offset locations for certain of the fuel handling facilities, such as the pond water treatment plant, new fuel reception and the switchgear associated with the fuel handling equipment.

A unique feature of the Heysham 2 layout is the provision of wholly internal routes from the gas circulator withdrawal positions in the reactor buildings to the gas. circulator maintenance workshop. Care has been taken to ensure that there are no open

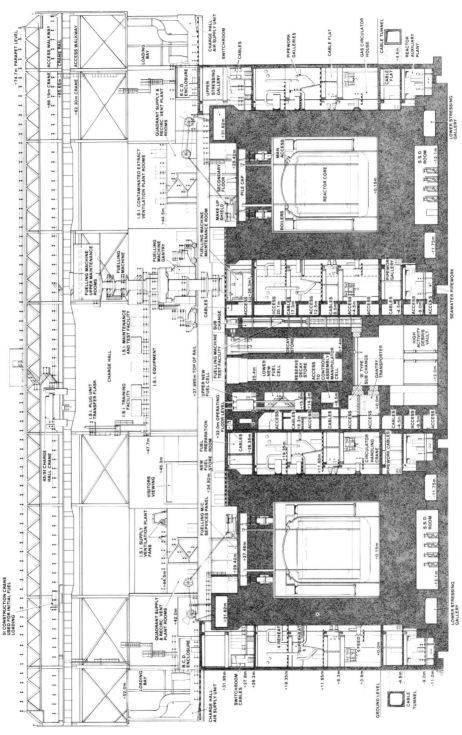

FIG. 5.2   ELEVATION THROUGH REACTOR AND FUEL HANDLING BUILDING

gullies or discontinuities in the floor finish of these routes along which the machines are moved on specially designed air flotation transporters. Any loose contamination which may be shed by the circulators during movement will therefore be confined to spaces internal to the reactor buildings, having surfaces from which it may easily be removed.

4.2 Segregation of Reactor Auxiliaries and Their Power Supplies

The essential switchgear, diesel generators and lead acid batteries associated with the reactor auxiliaries are segregated and situated at four widely spaced locations.

The switchgear and lead acid batteries associated with two quadrants of the first reactor primary coolant system are located in an essential supplies building situated south of the reactor services annexe and on the west side of the solid waste building. The corresponding equipment for the second reactor is located on the east side of the same building. The comparable equipment for the other two quadrants of the first reactor is located to the west of the turbine house and that for the corresponding quadrants of the second reactor is positioned to the east of the turbine house.

Contiguous with each essential supplies building is a diesel generator house containing two machines, so that a total of four physically separated pairs of diesel generators are provided to support the 'short break' supplies of which the most important are those associated with the reactor post-trip cooling. As mentioned, two independent and diverse methods are provided to each quadrant for post-trip cooling. These are known as X and Y trains and are described in Section 14. Each separately-located pair of diesel generators supplies the 'short break' essential power to one quadrant in both reactor units with the X trains in both of these quadrants being supplied by one of that pair of diesel generators and the corresponding Y trains by the other.

In this way, the risk of a major disruption of the essential auxiliary supplies by any one incident such as a fire or an aircraft crash on the site, is eliminated.

As explained in Section 12, the boiler modules and gas circulators of each reactor are divided functionally into four quadrants. Externally to its pressure vessel, each reactor building is also divided into four quadrants which house auxiliary equipment associated with the boilers and circulators in the corresponding quadrants of the boiler annulus. The quadrants external to a reactor pressure vessel are separated from each other by barriers. Each quadrant of a north-south pair is separated by an east-west barrier capable of giving protection against propagation of a fire for at least one hour. Each of the two north-south pairs of quadrants for each reactor is separated by a north-south barrier assessed as being capable of resisting a fire for four hours or long enough for the combustible material on the other side to burn itself out.

4.3 Effects of Seismic Design on Layout

The need to design nuclear safety-related systems for possible seismic disturbance has had a number of effects on the layout, size and nature of structures.

The need to accommodate a seismic allowance in the pipework design stress levels leads to a need for space to accommodate large expansion loops with substantial steel supporting frames and devices (snubbers) to dampen or reduce unwanted motions.

A major consequence of the need to provide structural stability in the event of a seismic incident, is the replacement of structural steel by reinforced concrete as the main structural element. This has had a marked effect on the reactor building which is designed as a massive double concrete box rising to a height of 52m surmounted by a seismically qualified structural steel roof. The roof of the turbine house, however is designed as a conventional structure because turbine generator integrity is not essential to the safety case.

Seismic considerations also led to the development of separate foundations for the various components of the main building structure, as described in Section 4.6.

4.4 Layout of Turbine House (Fig 5.3)

The two 660MW turbine generators are arranged transversely with each machine sharing a common centreline with the associated reactor unit. This gives a separation of 78m between centrelines which provides space within the turbine house for a central loading bay capable of accommodating a low-loader and also of providing sufficient space for laying down the turbine components during the complete overhaul of one machine.

The water treatment plant occupies the space below the loading bay and, in consequence, 'handed' pipework is necessary to and from the unit polishing plants and to the reserve feed water tanks.

4.5 Mechanical Annexe

The mechanical annexe is an important feature of the Heysham 2 layout and represents a considerable improvement upon that of Hinkley Point B. In addition to housing such unitised items of ancillary plant as deaerators, emergency feed pumps and reserve feedwater tanks, it also provides the space required for such facilities as the station computers, the unit simulator, the main changing rooms, the services for the central control room, etc.

4.6 Foundations

The walls of the main buildings have strip footings and the two concrete pressure vessels have individual foundation slabs. Columns within the building envelope are provided with pad footings and all such slabs and footings are founded directly upon the Triassic sandstone. The basement floors which span the spaces in between are sufficiently massive to resist the upward pressure of ground water. A movement joint is provided between each of the pressure vessel foundation slabs and the surrounding basement floors.

The work was executed within an extensive sheet-piled cofferdam, dewatered by wells installed both inside and around the perimeter until such times as the deadload of the structures substantially exceeded the potential uplift created by the natural water table of the site.

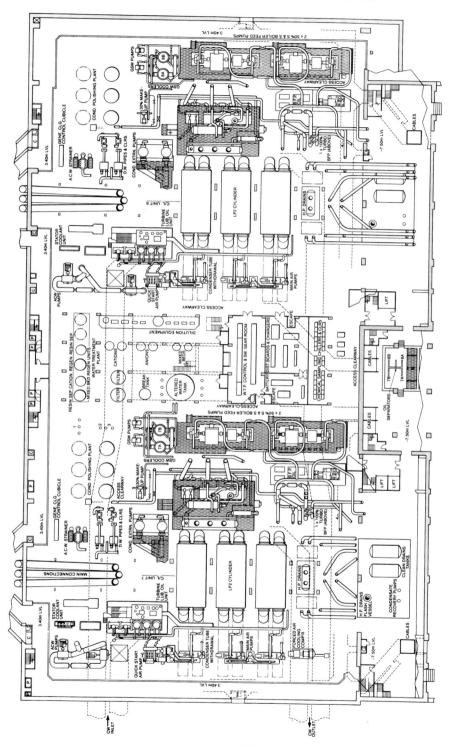

FIG. 5.3   LAYOUT OF TURBINE HOUSE AT -7.5M LEVEL

5. REACTOR PRESSURE VESSEL (FIG.5.4)

5.1 Functional and Safety Requirements

The reactor pressure vessel, which comprises a prestressed concrete structure, integral gas-tight liner and penetration closures, thermal shield and cooling system, is required to provide the pressure envelope of the primary coolant circuit. The integrity of the envelope must be maintained both during normal operations and following all credible faults or component failures including those which lead to the loss of the reactor pressure vessel (RPV) cooling system.

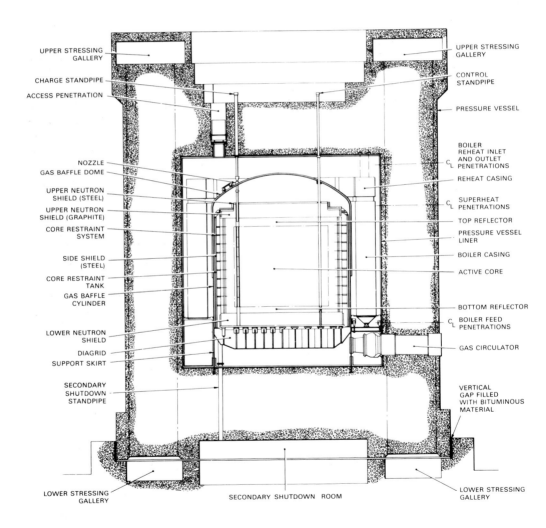

FIG. 5.4   CROSS SECTION OF REACTOR

The concrete structure is also required to support the weight of the reactor core and its supporting diagrid, the gas baffle and the boilers, all of which are located within the vessel. External items which must be supported are the charge face secondary floor which covers the standpipe closures and the fuelling machine when moving or positioned over the reactor.

5.2 Prestressed Concrete Structure

The prestressed concrete structure is a vertical cylinder, constructed of high strength concrete, with an internal diameter of 20.3m and an internal height of 21.9m. The thicknesses of the cylinder wall and of the bottom slab are 5.7m and 7.5m respectively whilst that of the roof slab varies from 5.427m to 7.347m.

The design pressure is 46.65 bar on all internal surfaces other than that part of the bottom slab within the periphery of the gas-baffle skirt where it is 49.62 bar.

The walls and end slabs are pre-stressed by some 3600 helical tendons which are anchored in annular stressing galleries at the top and bottom of the structure. The tendons may be re-tensioned or re-strung during the life of the reactor.

Although closely associated, in design terms, with the reactor/fuel handling buildings, the concrete vessel is a free-standing structure supported vertically on laminated elastomeric support pads. The pads are situated on two rings of concrete plinths on either side of the lower stressing gallery which is formed by the annular interspace left between the two rings of plinths. Horizontal seismic loads are transmitted by a shear upstand from the raft foundation around the periphery of the bottom slab of the vessel.

There are two large penetrations through the top slab for man access to the upper parts of the vault and 452 smaller penetrations of which 332 are fuelling standpipes. The other 120 comprise 104 for control and other services at interstitial positions of the core and a further 16 outside the core zone for reactor flux instrumentation and the remote in-service inspection facilities.

There are 10 large horizontal penetrations through the cylindrical walls, eight of which house the circulator motors and two provide man access to the vault spaces below the steam generators and the core diagrid. Other horizontal penetrations are provided for the steam and water pipework connections to the boilers and for instrumentation and remote inspection purposes.

The bottom slab of the vessel has 28 secondary shutdown penetrations through which nitrogen can be injected into the core from a high pressure main in the secondary shutdown room which is formed by a space between the bottom slab and the vessel foundation within the inner ring of vessel support plinths. Another 16 penetrations provide access for instrumentation and remote inspection purposes.

The penetrations are formed with permanent steel shutter tubes which become the gas-tight liners for the penetrations after being sealed by closures to complete the primary coolant envelope.

The weight of the reactor core, diagrid and gas baffle is transmitted through the gas baffle support skirt to the bottom slab of the vessel. The weight of the boilers is transmitted partly through the gas baffle skirt and partly through supports set into the cylindrical walls.

5.2.1 *In-Service Monitoring of the Concrete Structure*

In addition to the facility for monitoring the tension in the prestressing tendons, thermocouples and strain gauges are embedded in the concrete and thermocouples are attached to the vessel and penetration liners in order to monitor the thermal condition of the concrete.

A further monitoring facility comprises 16 unlined holes which extend the vertical length of the vessel. These are provided to facilitate the detection of any leakage of carbon dioxide through the vessel liner into the concrete. They are so located that any such leakage would be detected before a significant number of tendons could suffer thermal relaxation.

5.3 Pressure Vessel Liner

5.3.1 *Functional Requirements of the Liner*

A pressure vessel liner is installed in order to prevent erosion of the internal surfaces of the structure by hot carbon dioxide, to prevent its diffusion into the porous concrete and to provide a foundation for the thermal shield and vessel cooling pipes which protect the concrete from unacceptable temperatures and temperature gradients. It must also be capable, where necessary, of transferring loads from the plant items housed within the vault to the concrete structure.

The possibility of liner failure can however be accepted, from the point of view of nuclear safety, as the monitoring facilities described in Section 5.2.1 would indicate the occurrence of such failure well before the integrity of the prestressed concrete structure could be impaired.

5.3.2 *Design and Construction of the Liner*

The vessel liner, which is fabricated from fully-killed carbon steel, is made up of three main sections, apart from the penetration sections. The vertical barrel section and the floor liner have uniform thicknesses of 13mm and 16mm respectively whilst those of the roof liner are 13mm over the boiler annulus and 32mm in the central region which is penetrated by the fuelling standpipes.

As mentioned in Section 3.6, the roof liner and standpipes were fabricated off-site as a single assembly weighing 1000t and transported to Heysham by seagoing barge.

The junction of the vertical section and the roof liner consists of a ring of 171mm diameter round bar to which those two main sections of the liner are welded. The same arrangement is adopted at the junction of the vertical section to the floor liner where a similar ability to resist high compressive loads between the concrete and liner is required. Where dead weight loadings are transmitted through the liner from the gas baffle skirt to the bottom slab of the vessel, both skirt and floor liner are welded to a forging of cruciform section which in turn is welded to an extension of the skirt which terminates at an anchor ring within the bottom slab.

The liner is secured to the concrete by means of anchor studs and hooks. Anchor studs are used mainly to secure the floor liner and penetrate some 228mm into a grouting which fills a space between the bottom slab of the vessel and the underside of the floor liner. Anchor hooks are used mainly to secure the roof liner and the barrel section and penetrate some 400mm into the walls and top slab of the vessel. In addition, the liner sections are keyed into the concrete by vertical or radial panel stiffeners and by the network of 24mm square section cooling pipes which are fillet-welded to the concrete side of the liner at a pitch of about 100mm.

Where sections of penetration liners and closures are not fully supported by concrete, secondary restraint and containment systems are provided to prevent excessive gas leakage in the event of failure.

5.4 Pressure Vessel Thermal Shield (Fig 5.5)

5.4.1 Functional Requirements of the Thermal Shield

A thermal shield is provided in order to reduce heat loss from the primary coolant and, in combination with the pressure vessel cooling system described in Section 5.5, to enable the pressure vessel liner temperature to be held below about 75°C. This in turn ensures that excessive temperatures and temperature gradients are not imposed upon the concrete of the vessel structure. The permissible conductance varies from 2.5 to about $6W/m^2/°C$ depending upon the local gas temperature and other relevant factors.

The shield has also to withstand the loadings imposed upon it by acoustic vibration, thermal cycling and reactor depressurisation.

Loss of thermal shielding over large areas could result in damage to the vessel liner but, provided that the pressure vessel cooling system continues in service, concrete temperatures would remain acceptable and gas diffusion monitoring guards against loss of integrity of the vessel structure.

5.4.2 Design and Construction of the Thermal Shield

The insulation consists of layers of ceramic fibre interleaved with overlapping foils of type 310 stainless steel which reduce its permeability in the direction normal to the shielded surface. Mild steel cover plates are used for the floor area and lower part of the cylindrical wall where the gas is at reactor inlet temperature; elsewhere, the cover plates are of stainless steel.

Except in the roof area, each of the main cover plates is held in position by a centrally-positioned Incolloy 800 stud which is welded to the pressure vessel liner. Secondary cover plates contain and protect the fibre against the acoustic environment.

The edges of adjacent main cover plates are joined by flexible U-shaped ties which accommodate expansion of the plates in a plane parallel to the insulated surface but apply a constraint perpendicular to that plane. Thus, in the event of random stud failures, the cover plate affected remains secured in its position by the adjacent plates.

During installation, the cover plates are used to compress the insulation to a pre-determined extent in order to ensure that no significant gaps, in which gas circulation could occur during the life of the reactor, remain. The resulting nominal shield thicknesses are 51mm on the cylindrical wall and 76mm on the floor.

For the central area of the roof, the thickness is 96mm and the primary retention of the stainless steel cover plates is achieved by welding them to Incolloy 800 tubes which are in turn welded to the lower ends of the penetration liners. Over the boiler annulus, the assembly is similar to that on the floor and walls but additional support is given to the interleaved foils by means of short studs which do not extend as far as the hot face.

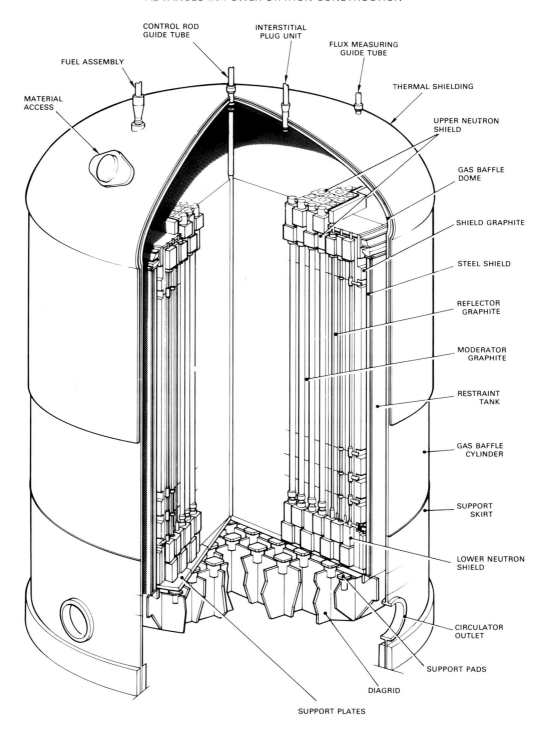

FIG. 5.5   **REACTOR ASSEMBLY**

The horizontal penetrations for the steam and water pipework connections to the boilers are insulated with ceramic fibre which is wrapped around its containment cylinder and compressed onto it by means of polypropylene tape. Once in position the tape is released by means of resistance heating wires to permit the ceramic fibre blanket to expand to the full bore of the penetration liner. The containment cylinder is attached at its inboard end to the penetration liner by 16 straps. This represents a 25% redundancy to guard against random failures of individual straps.

The fuelling and peripheral in-service inspection and flux measurement standpipes which penetrate the pressure vessel roof are normally closed at their lower ends by plug units. However, the fuelling standpipes are exposed to hot gas during on-load refuelling and, as a precaution against leakage past the plug units, the liners of these penetrations are insulated over virtually their full length. The liners of penetrations such as those for control rods which are enclosed in guide tubes and cooled by gas at reactor inlet temperature, are insulated over the lower third of their length.

5.4.3 *Monitoring the Condition of the Thermal Shield*

The performance of the thermal shield is monitored by thermocouples attached to the vessel and penetration liners. In addition, the physical condition of the shield on the walls and floor and over a part of the roof can be visually examined by means of the remote in-service inspection facilities.

5.5 Pressure Vessel Cooling System

For each pressure vessel, a cooling system comprising two 100% duty circuits is provided. The two circuits are arranged in parallel so that adjacent cooling pipes are not part of the same circuit. The 3.2mm thick, 24mm square section pipes are welded to the concrete side of the liner. Each circuit is provided with two 100% duty pumps to circulate treated water around the circuit and two 100% duty heat exchangers through which the heat is transferred to the reactor sea water system.

Each circuit is provided with degassing facilities since the water is subject to radiolytic decomposition into hydrogen and oxygen during its passage through the reactor vault.

The pressure vessel cooling systems also serve the central irradiated fuel disposal facilities which are described in Section 11.

6. REACTOR CORE

6.1 Functional and Safety Requirements

The reactor core, which consists of a stack of graphite bricks, peripheral neutron shields and restraint system, is required to act as neutron moderator and reflector and to provide channels to accommodate the fuel stringer assemblies, neutron sources and specimens and their coolant flow. It is also required to permit secure entry for the neutron absorbers which effect and maintain shutdown and which consist of gravity-operated rods, high-pressure nitrogen and boron glass beads.

The core and its restraint system must be highly redundant structures in order to ensure that random local failures will not result in the consequential failure of surrounding components or in gross channel distortion.

6.2 Design and Construction of the Core (Fig. 5.6)

Graphite, which is necessary as moderator and reflector, has a number of structural disadvantages. Its weakness in tension necessitates the use of steel, which has a coefficient of expansion some four times greater, both for the foundation and for the lateral location of the stacked columns of graphite bricks which form the fuel channels. Furthermore, graphite shrinks under irradiation and may suffer a radiolytic oxidation loss of up to 30% throughout a 30 years life with consequent loss of strength.

The graphite structure is 16-sided with an active core mean diameter of 9.46m and an active core height of 8.3m. The complete assembly comprises the active zone of moderator graphite containing 332 fuel channels on a 460mm square lattice pitch and an outer zone of neutron reflector and shield graphite.

The structure has a complex interlocked keying system by means of which each column of graphite bricks is located radially with respect to its neighbours by eight keys. These permit differential vertical movement between the vertical columns and transmit load radially outwards to the edge of the structure where it is linked to a steel restraint system. A total of some 25,000 graphite components are used to form the channels and the numerous other flow passages which are required to maintain the graphite at a temperature below 550°C, above which thermal oxidation would become significant.

The use of isotropic graphite is essential for AGR cores in order to limit shrinkage and avoid unacceptable internal stresses. The graphite for Heysham 2 was obtained from the source which has supplied Hartlepool and Heysham 1.

The permeability of the graphite is also important, and is not in itself sufficient to provide adequate means of access to the graphite pores for the inhibition of graphite corrosion by methane which is added to the coolant for that purpose. At Hinkley Point B, the rate of replenishment of methane within the brick structure was considerably improved by reducing, as far as practicable, the length of the graphite ligament to be traversed and this was achieved by drilling 70 small vertical holes through each moderator brick. For Heysham 2, the fact that the gas flows around the bricks for cooling purposes and creates a pressure difference between the outside and the bore, has been exploited. Diffusion of the methane into the bricks is enhanced by sealing the leakage gaps at the end of each brick and so increasing the pressure difference by a factor of about four. This greatly reduces the number of axial holes required.

To reduce the levels of radiation in accessible regions of the vault outside the core, shielding is provided by steel bricks at the top of the core and by steel plate at the bottom. Shielding at the sides is provided by a combination of steel rods placed in vertical holes within the graphite bricks used for the two outermost rings and steel plates located at the inner face of the restraint tank.

HEYSHAM 2 – AGR NUCLEAR POWER STATION

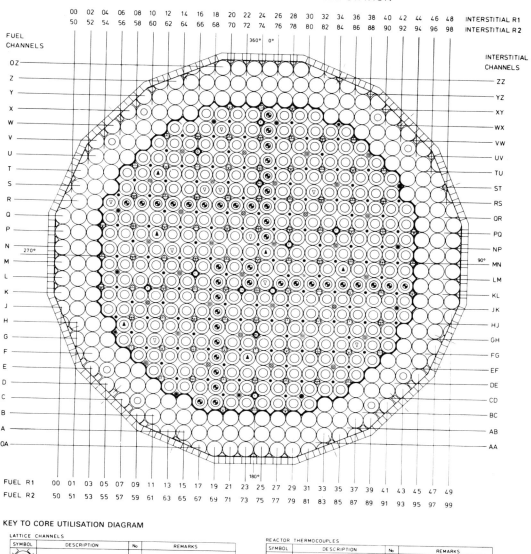

KEY TO CORE UTILISATION DIAGRAM

LATTICE CHANNELS

SYMBOL	DESCRIPTION	No	REMARKS
	FUEL CHANNELS	332	ALL WITH STANDPIPES (ON LATTICE)
	DESIGNATED TREPANNING CHANNELS	40	

INTERSTITIAL CHANNELS

SYMBOLS	DESCRIPTION	No	REMARKS
	BULK CONTROL ROD (GROUP 1)	15	
	BULK CONTROL SENSOR ROD (GROUP 1)	7	WITH INTERSTITIAL STANDPIPES
	BULK CONTROL ROD (GROUP 2)	22	WITH INTERSTITIAL STANDPIPES
	REGULATING / SAFETY	24	WITH INTERSTITIAL STANDPIPES
	REGULATING	21	WITH INTERSTITIAL STANDPIPES
	FLUX SCANNING CHANNEL (REDUNDANT)	6	WITH INTERSTITIAL STANDPIPES
	GRAPHITE SAMPLE CHANNEL	2	WITH INTERSTITIAL STANDPIPES
	C14 MONITORING CHANNEL (REDUNDANT)	1	WITH INTERSTITIAL STANDPIPES
	NEUTRON SOURCE CHANNEL	2	WITH INTERSTITIAL STANDPIPES

REACTOR THERMOCOUPLES

SYMBOL	DESCRIPTION	No	REMARKS
	UPPER NEUTRON SHIELD	6	
	LOWER NEUTRON SHIELD	6	
	MODERATOR COOLING GAS INLET	6	
	MODERATOR 5 LAYERS	6 x 5	LAYERS 2/3, 4/5, 6/7, 8/9, 10/11 INTERSECTION
	FUEL CHANNEL ENTRANT GAS	6	
	FUEL CHANNEL ENTRANT GAS	10	
	FUEL CHANNEL GAS OUTLET THERMOCOUPLES	996	3 THERMOCOUPLES IN EVERY FUEL ASSEMBLY

FLUX MEASURING CHANNELS

SYMBOL	DESCRIPTION	No	REMARKS
	FLUX MEASURING CHANNEL	8	WITH STANDPIPE ON LATTICE

SECONDARY SHUT DOWN (BOTTOM ENTRY)

SYMBOL	DESCRIPTION	No	REMARKS
	NITROGEN & BEAD INJECTION	32	S/PLATE CENTRE POSITION
	NITROGEN INJECTION	12	S/PLATE CENTRE POSITION
	NITROGEN INJECTION	119	S/PLATE EDGE POSITION

FIG. 5.6   CORE UTILIZATION DIAGRAM

6.3 Cooling Flow (Fig 5.7)

Cooled carbon dioxide gas is drawn from the bottom of the boilers by the gas circulators and is discharged into the space below the core. About half flows directly to the fuel channel inlets whilst the remainder, known as the re-entrant flow, passes up an annulus surrounding the core and returns downwards through the core in the passages between the graphite bricks to rejoin the main coolant at the bottom of the fuel channels. The re-entrant flow thus cools the gas baffle, the core restraint system and the graphite bricks. The combined flow passes upwards through the fuel channels and is led through the space above the top of the core and through the gas baffle in the fuel assembly plug unit and discharged into the area above the gas baffle which connects with the top of the boilers.

6.4 Core Restraint System (Fig 5.8)

At the core periphery, steel puller rods are located in recesses in selected bricks at each inter-layer position. The rods are in turn attached to restraint beams that form 16-sided polygons around the core at each interlayer position, the beam ends being loosely connected as a secondary restraint feature. Each beam is inset and supported by the graphite and attached to the cylindrical restraint tank by two ball-ended links which provide the means of load transfer from the core. This type of unit allows for vertical differential expansion between the graphite structure and the restraint tank.

The restraint tank is welded to the diagrid and is treated for construction purposes as a component of the gas baffle assembly.

7. GAS BAFFLE ASSEMBLY (FIG 5.9, 5.5 AND 5.7)

The gas baffle assembly, which was transported to site as a single 950t load comprises:
(1) The diagrid which supports the reactor core.
(2) The core restraint tank.
(3) The gas baffle itself.
(4) The skirt through which the dead weight is transmitted across the bottom gas space to the bottom slab of the concrete vessel.

7.1 Diagrid

The function of the diagrid is to support the reactor core without exceeding permissible limits of deflection. It is a circular grid of 1.96m deep box sections made of 38mm thick carbon steel plate. The sides of the box sections are 920mm long and are joined through cruciform-shaped extrusions. This form of fabrication permits the examination of the total volume of all major structural welds.

The load imposed by the reactor core is transmitted to the diagrid through core support plates which are themselves supported by levelling pads on the diagrid webs. The core restraint tank is welded to the top of the diagrid close to its periphery. The diagrid itself is supported at its periphery where its radial webs are butt welded to T-extrusions which form an integral part of the skirt.

The integrity of the diagrid is assured by structural redundancy which is such that a number of the box sections would need to fail before the remainder became overloaded.

7.2 Core Restraint Tank

Although welded to the diagrid to form part of the gas-baffle assembly before delivery to site, the core restraint tank is functionally a component of the core restraint system.

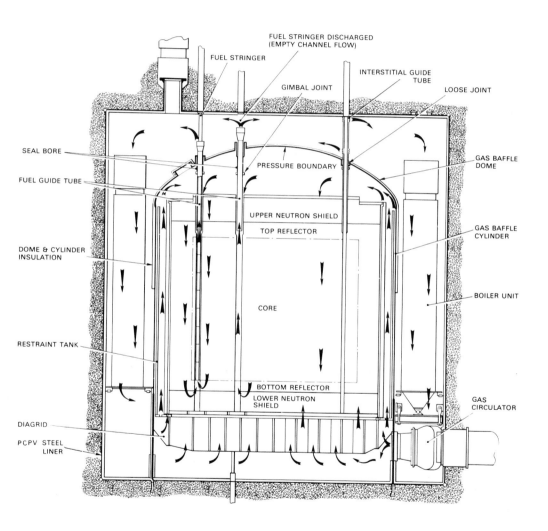

FIG. 5.7   REACTOR GAS FLOW

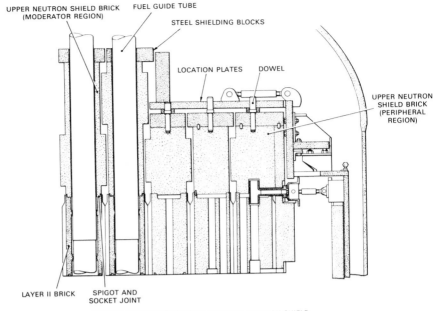

SECTION THROUGH PERIPHERY OF UPPER NEUTRON SHIELD

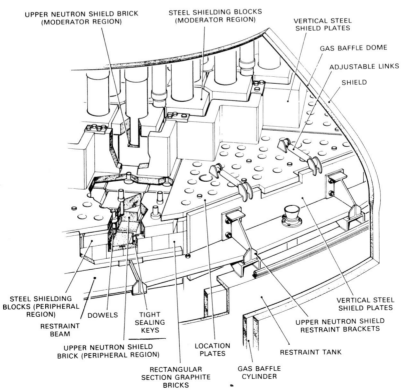

FIG. 5.8   **UPPER NEUTRON SHIELD, STEEL SHIELDING AND RESTRAINT SYSTEM**

7.3 Gas Baffle

The gas baffle constitutes the barrier between the spaces around the core which contain the relatively cool re-entrant gas flows and the spaces external to the baffle which contain the gas after it has been heated during its upwards passage through the fuel channels and ducted through the top dome of the baffle.

A gross disruptive failure of the baffle could cause damage to other reactor components, prevent entry of control rods and allow the coolant to bypass the fuel channels. The integrity of the baffle must be such therefore that the possibility of such a failure can be discounted. The material chosen for this structure is a fine-grained carbon steel. This is an easily worked and well understood type of material which has been used for the same purpose at Hinkley Point B. It is tough at all relevant temperatures and is considered to provide the best guarantee of achieving a sound structure which will give trouble-free service over the life of the reactor.

The fast neutron dose to which the baffle is subjected is less than that which would cause a significant change in the properties of the steel, and so irradiation does not have to be considered in the design or subsequent in-service assessments of the continued integrity of the baffle.

The dome is torispherical with an internal crown radius of 9.5m, a knuckle radius of 1.4m and a thickness of 67mm. The central section is perforated and fitted with 332 large nozzles to which the fuel channel guide tubes are attached and 104 small interstitial nozzles for the control rod guide tubes and other facilities. Core construction requires three large penetrations in the outer section, two of which have provisions for subsequent man access during the operational phase of the station. There are a further 16 nozzles in the outer section for in-service inspection and flux measuring purposes.

The dome is welded to the thinner cylindrical part of the baffle via a tapered transition piece. The cylindrical section is 35mm thick and has an inside diameter of 13.85m.

7.3.1 *Gas Baffle Thermal Shield*

Thermal shielding is provided which, in conjunction with the coolant gas flows, maintains the carbon steel of the upper parts of the baffle at temperatures below which thermal creep occurs. The extent of the insulation down the cylinder is arranged to ensure acceptable temperatures and thermal gradients local to its termination.

The thermal shield for the dome is nominally 90mm thick. It is similar to that for the pressure vessel liner except for the method of attachment in the central area where each cover plate has welded to it an upstanding cylinder which is attached to the upper end of a gas baffle fuel channel nozzle by six straps. The six straps constitute a 50% redundant design so that secondary retention features are not required. For the dome annulus region, and also for the cylindrical section where it is only 37mm thick, the arrangement of the heat shield is generally as described for the liner of the pressure vessel wall. Loss of insulation panels over small areas could be tolerated.

Each fuel channel guide tube is attached, below the dome, to a mounting sleeve which, at its upper end, is welded to the upper end of the corresponding gas baffle nozzle, which rises above the dome. This arrangement forms two annular spaces; that between the mounting sleeve and the guide tube is filled with insulation whilst that between the sleeve and the nozzle is used as a cooling passage through which gas is bled from the underside of the dome to the space above. The insulation is retained by the cylindrical upstand which forms the dome insulation cover plate attachment to that nozzle.

The interstitial guide tubes, which span the space between the gas baffle dome and the top slab of the pressure vessel, are insulated over that length. The insulation is contained by stainless steel cylinders which have a number of bellows units to take up differential expansion. In conjunction with an internal flow of gas for cooling purposes, this maintains the temperature and alignment of the control rods or other equipment contained within those tubes within safe working limits.

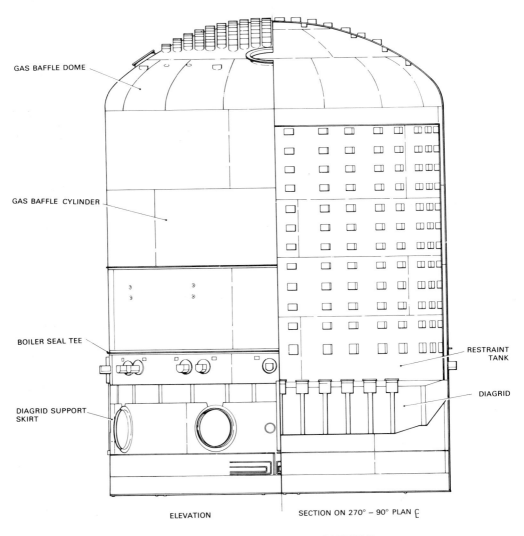

GAS BAFFLE DOME

GAS BAFFLE CYLINDER

BOILER SEAL TEE

RESTRAINT TANK

DIAGRID

DIAGRID SUPPORT SKIRT

ELEVATION   SECTION ON 270° – 90° PLAN

FIG. 5.9   GAS BAFFLE ASSEMBLY

7.4 Skirt

The skirt is 50mm thick and has the same internal diameter as the cylindrical section of the gas baffle to which it forms a lower extension. The diagrid is supported from its inner face as described in Section 7.1 whilst, on the other side, forgings are butt welded to the skirt in order to provide the inboard mountings for boiler support beams. The structural arrangements for transmitting the dead weight loads through the liner of the bottom slab of the pressure vessel are described in Section 5.3.2. Those loads exceed the upwards thrust on the gas baffle due to the differential gas pressure across it and the skirt is therefore always in compression.

Below the diagrid support level, there are eight large openings through which the gas circulators discharge coolant to the underside of the core and two man access openings.

Under operating conditions, the skirt is exposed to gas at a temperature of about 300°C. In order to reduce the temperature of the bottom end of the skirt to the pressure vessel liner operating temperature of around 40°C, the lower part is cooled by two independent systems of water cooled pipes which are welded to it and provided with a thermal shield on both sides. The cooling system is similar to, but separate from, the reactor pressure vessel cooling system.

8. REACTOR CONTROL AND SHUTDOWN DEVICES

Reactor power is directly proportional to neutron density which is controlled by neutron absorbers. Three separate systems of absorber are installed with the secondary and tertiary systems providing back-up against the remote possibility of a fault which would prevent a substantial number of the control rods, which constitute the primary system, from entering the core when required.

8.1 Primary Control and Shutdown Devices (Fig 5.10)

The primary system for control and shutdown comprises 89 absorber rods and drives housed in standpipes in the top slab of the concrete pressure vessel. Of these, 44 are 'black' rods, so called because they absorb most thermal neutrons impinging on their surface, and of these seven are also sensor rods, designed to detect, when being raised or lowered, any gradually developing guide tube misalignment or dome distortion that may occur.

Each black rod, other than the seven sensor rods, consists of eight cylindrical sections linked axially by articulating joints. Each of the lower six section consists of a 9% Cr 1% Mo steel sheath containing four tubular inserts of stainless steel alloyed with 4.25% by weight of boron. Inside the tubular inserts are two solid cylindrical graphite inserts to reduce fast neutron streaming. The upper two sections form part of the top reflector when fully inserted and the sheaths enclose solid graphite inserts only.

The seven sensor rods differ in that the third, fourth and fifth sections from the bottom are replaced by one long section in order to ensure interference if unexpected relative movement occurs.

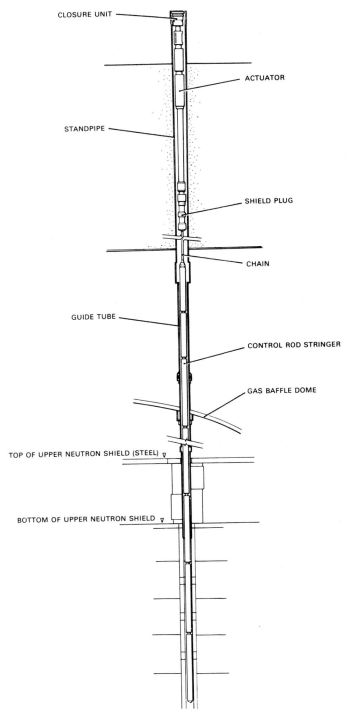

FIG. 5.10   CONTROL CHANNEL

The other 45 rods are termed 'grey' because they absorb neutrons less strongly and are used primarily as regulating rods. Of these, 24 are also used as a safety group by having them normally out of the core when the reactor is shutdown in order to provide a safeguard against an inadvertent criticality. The grey regulating rods are of similar design to the fully articulated black rods except that the stainless steel inserts in the lower six sections contain no boron.

A control rod assembly comprises a control rod, a plug unit, a control rod actuator and the standpipe closure unit. The complete assembly may be removed by the refuelling machine whether the reactor is operating or not and at any reactor pressure.

The plug unit, which is a steel plug with a central hole for the control rod suspension chain, reduces to acceptable levels the radiation streaming from the core through the standpipe penetrations of the top slab of the concrete pressure vessel.

Each control rod actuator is complete with motor-operated winding gear and suspension chain storage, an electromagnetic clutch, rod position indicator and limit switches. The actuator is designed for frequent small movements but in the event of a reactor trip, the clutch is de-energised to allow rod insertion under gravity. The rate of fall is controlled by a carbon disc brake governed by a centrifugal mechanism. For the final deceleration, the brake is actuated by an independent mechanism based upon rod position so as to limit the stresses in the chain suspension and gearing.

The housing for the penetration closure is screwed and seal-welded to the top of the standpipe and the closure unit is installed and removed remotely by the refuelling machine. The closure locking mechanism is provided with a pressure-sensitive interlock to prevent unlocking when the differential pressure across the closure is sufficient to eject the control rod assembly.

8.2 Secondary Shutdown System

The secondary shutdown system consists of a high pressure 2-stage nitrogen injection system. The nitrogen store, which is common to both reactors, is connected to solenoid-actuated pneumatically-operated trip valves located in the secondary shutdown rooms below the bottom slabs of the concrete pressure vessels. See Fig 5.4.

The first stage of operation ensures rapid shutdown of the reactor. A high initial rate of nitrogen flow purges a group of 165 interstitial channels of carbon dioxide and fills them with nitrogen which has a relatively high neutron absorption cross section. The second stage of operation follows up with a reduced rate of supply in order to provide the make-up to the interstitial channels as the nitrogen makes its way into the re-entrant flow passages and the fuel channels. The second stage gradually builds up the nitrogen concentration in the coolant gas circuit to the point at which it is sufficient to hold the pressurised core in the subcritical condition.

If it is intended to depressurise the reactor and insufficient control rods are available to hold down the activity of the core, then the tertiary system will be brought into play.

8.3 Tertiary System

The tertiary system uses a solid neutron absorber in the form of boron glass beads of 3mm diameter which are stored in hoppers in the secondary shutdown rooms. When needed, this system is activated by manual operation of valves adjacent to the hoppers and the beads are conveyed pneumatically by carbon dioxide to fill, simultaneously, 32 of the group of interstitial channels used for the secondary shutdown system.

The capacity of the glass beads for absorbing neutrons is sufficient to counteract the positive reactivity effect of the decay of fission product poisons and is capable of holding the reactor in the subcritical condition indefinitely. Provision is made for the beads to be recovered from the interstitial channels and collected in shielded flasks.

9. REACTOR PROTECTION SYSTEM

The reactor protection system guards against all credible faults and maintains the fuel parameters within acceptable limits. The system used at Heysham 2 is a logical development of the systems used for earlier AGRs and reflects the enhancement of safety requirements over the years. In particular, operating experience has led in recent years to a wider recognition of the risk of a common-mode failure; that is to say, the failure due to a common cause of all of some group of identical elements of a system designed to provide high integrity by means of component redundancy. This risk has strongly influenced the design of the system for Heysham 2. Considerable attention has also been paid to the reduction of the susceptibility of the electronic equipment to radio frequency interference.

The basic elements of the reactor protection system consist of multiple sensors connected to trip amplifiers which compare the detected value of the measured parameter with the trip setting. Where possible each identified reactor fault is detected by two diverse physical parameters using sensors and trip amplifiers of different designs. When the setting is exceeded the trip amplifiers initiate signals to what are known as the 'guardline logic networks'. Moreover, the trip amplifiers are of high integrity fail-safe designs in which the failure of an amplifier component would also produce a trip output.

The guardline logic networks will, on receipt of trip signals from two-out-of-four trip amplifiers, operate pairs of contactors so connected that whereas a trip output from one guardline only will be disregarded, the operation of any two of the pairs of contactors will initiate one or both of the shutdown systems described in Section 8. They will at the same time initiate the post-trip heat removal systems.

9.1 Main Guardline System

The main guardline system for Heysham 2 uses the highly reliable 'Laddic' logic system which was first adopted for the Magnox reactors at Oldbury and has subsequently been used for all AGRs. The Laddic system uses a ladder-like ferritic logic element incorporated in a dynamic pulse train circuit in such a way that failure of a component would interrupt the pulse train and cause the output circuitry to go into the trip state.

This system is widely regarded as having extremely high integrity particularly where, as at Heysham 2, four such guardlines are used per reactor and their output contactors are connected so that the de-activation of any two-out-of-four guardlines initiates operation of the primary shutdown system by removing the electrical supplies from the electromagnetic clutch, as referred to in Section 8.1.

The Heysham 2 Laddic system uses revised circuitry which permits the use of modern high frequency transistors and is engineered to make the best use of modern circuit and packaging techniques.

9.2 Diverse Guardline System

Even when allowance has been made for the known integrity of the main guardline system, account has still to be taken of the risk of a common-mode failure in meeting the very high reliability target set for the reactor protection system as a whole. There is consequently a need for a diverse guardline system to cover the more frequent reactor faults and a relay-based logic system has been adopted to satisfy this need. This form of logic system also offers a very high immunity from radio frequency and similar forms of interference.

The diverse system, which has three guardlines per reactor, receives signals from trip amplifiers associated with channel gas outlet thermocouples and neutron flux sensors. The guardline contactors are connected on a two-out-of-three basis and initiate both the primary and secondary shutdown systems.

The levels at which the diverse guardlines are de-activated are set higher than those for the main Laddic guardlines, so that de-activation of the diverse guardlines and initiation of the secondary shutdown system is avoided if both the main guardline system and the primary shutdown system operate effectively.

9.3 Auxiliary Guardline System

An auxiliary guardline system which uses relay-based logic, is provided to protect against possible fault conditions during reactor shutdown. Such conditions could arise as a result of fuel or control rod movements and the sensed parameter for this component of the reactor protection system is neutron flux.

10. FUEL ASSEMBLY (FIG 5.11)

Heysham 2 uses a design of fuel assembly which, in most respects, is the same as that in use at Hinkley Point B. The fuel element for Heysham 2 has however been strengthened as a consequence of operating experience at Hinkley Point B in order to permit on-load refuelling at higher levels of reactor power.

The fuel assemblies are complex, composite assemblies which include and extend from the closure units at the top of the fuel standpipe in the top slab of the concrete pressure vessel to their bottom supports which sit on support stools on the core support plates.

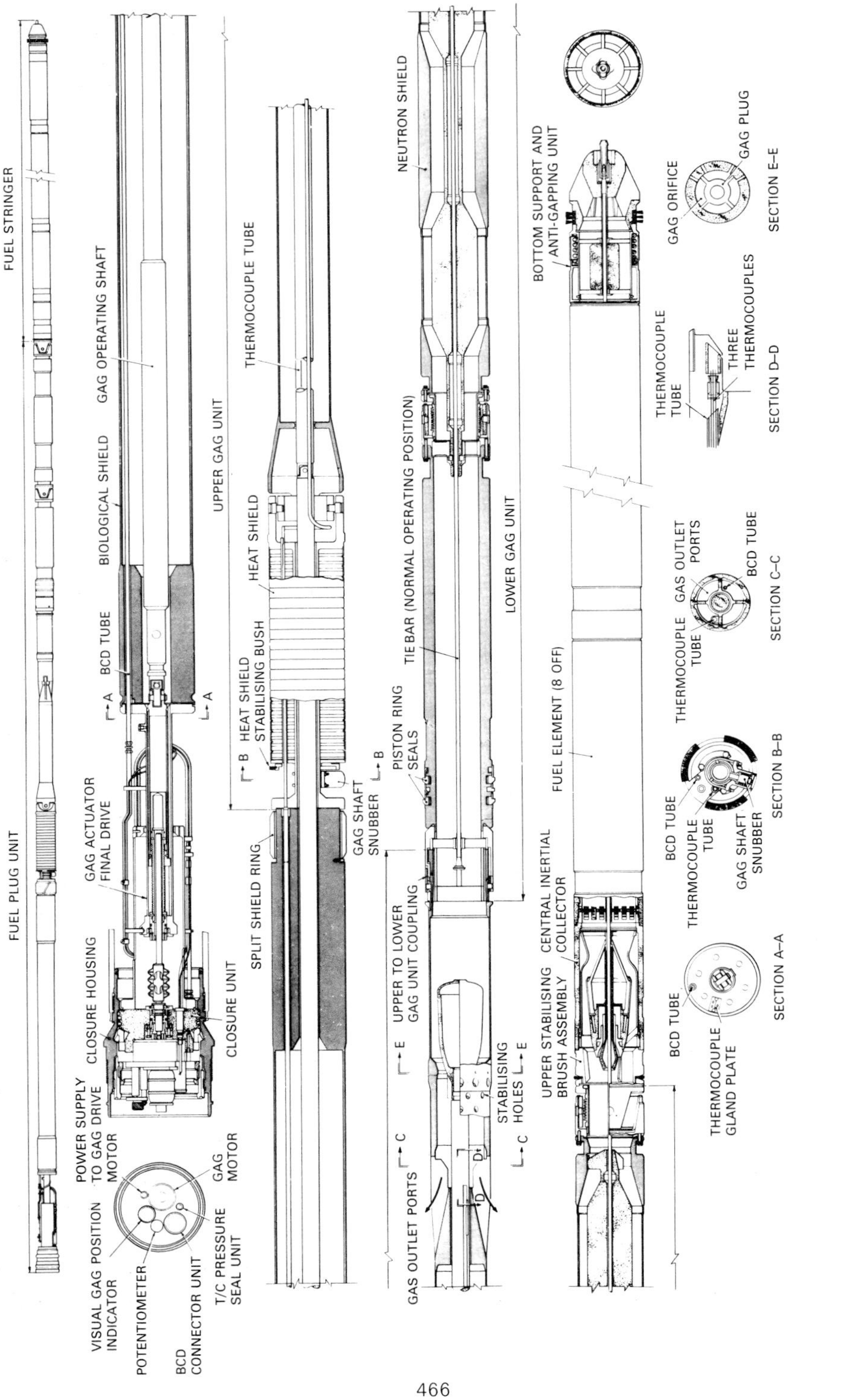

FIG. 5.11   FUEL ASSEMBLY

There is an important functional distinction between those parts of the assembly which constitute what is known as the 'fuel stringer' and those parts which constitute the so called 'plug unit'which is re-usable with replacement fuel stringers after having been serviced in the active maintenance facility (see Section 11).

10.1 Fuel Stringer

The fuel stringer comprises a bottom support and bottom reflector assembly surmounted by eight fuel elements which, in some channels, are in turn surmounted by a central inertial collector for removing, from the gas stream, oxide particles which may spall from the fuel pin cladding after prolonged operation at high temperature.

A central tie bar takes the weight of the fuel stringer during loading into and removal from the core and protrudes about a quarter of the way up the plug unit to the point at which that weight is transmitted to the latter unit.

10.1.1 *Fuel Element (Fig 5.12)*

The uranium oxide fuel is fabricated in the form of hollow ceramic cylindrical pellets of 14.51mm outside diameter. They are contained within sealed stainless steel cans known as cladding, to form fuel pins each of which is about 978mm long. 36 such pins are arranged in clusters within graphite sleeves to form a fuel element.

The graphite sleeves confine the hot cooling gas and thereby permit the re-entrant flows within the core to maintain the permanent core graphite at a temperature below 550°C as described in Section 6.2. It is the graphite sleeve which has been strengthened to permit on-load refuelling at higher reactor powers than can be permitted with the original design of fuel as used at Hinkley Point B, because the impacts received by the sleeves during refuelling are a function of the gas flow rate in the channel.

When in the reactor, the weight of almost the whole of the plug unit is carried through the fuel element sleeves to the bottom support. This ensures both the stability of the column in the gas flow and effective sealing between the ends of adjacent elements.

Two features are incorporated to enhance the rate of heat transfer from pellet to coolant gas. The space remaining within the can after the cladding has been compressed onto the pellets during manufacture is filled with helium which is both inert and a good heat conductor. Externally, the cladding is provided with small transverse ribs.

To avoid the deterioration of the fuel pin which would be caused by relative movements between the stack of pellets and the cladding during reactor operation, a number of pellets are grooved around the circumference so that the cladding becomes locked onto the stack by the compressive force exerted by the coolant gas.

Slightly enriched uranium is required for an AGR. The elements for the inner zone have an enrichment of 2.11% whilst those for the outer zone have the somewhat higher enrichment of 2.77%, in order to flatten the neutron flux across the core.

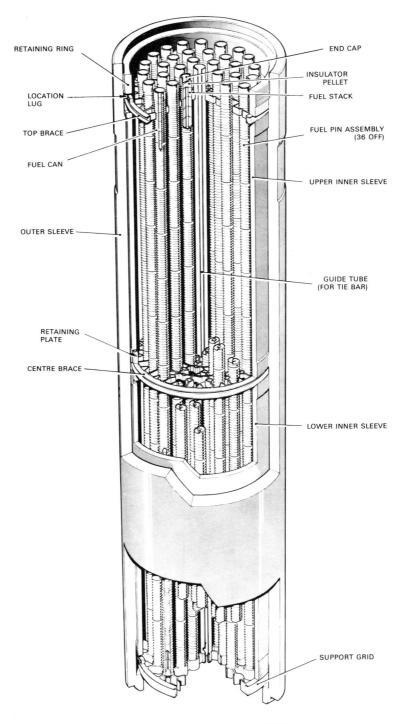

RETAINING RING

LOCATION LUG

TOP BRACE

FUEL CAN

OUTER SLEEVE

RETAINING PLATE

CENTRE BRACE

END CAP

INSULATOR PELLET

FUEL STACK

FUEL PIN ASSEMBLY (36 OFF)

UPPER INNER SLEEVE

GUIDE TUBE (FOR TIE BAR)

LOWER INNER SLEEVE

SUPPORT GRID

FIG. 5.12   FUEL ELEMENT

10.2 Plug Unit

The lower part of the plug unit serves to channel the hot gas from the top of the fuel stringer to the gas outlet ports of the plug unit which communicate with the gas space above the gas baffle dome. At its lower end it incorporates a neutron shield to prevent fast neutrons from streaming up the plug unit. Where it passes through the guide tube in the gas baffle dome, it is encircled by a pair of segmented graphite piston rings to limit the leakage of cool re-entrant gas through the annulus between the plug unit and its guide tube.

Between the level of the piston ring seals and the outlet ports, the hot gas passes through a flow control valve, known as a gag, which permits the adjustment of coolant flow through individual channels. It is operated by a shaft which passes up the centre of the plug unit to a remotely-controlled motor-driven actuator located in the closure assembly at its upper end. A gas sampling line and a thermocouple tube are similarly accommodated.

The upper part of the section between the gas outlet ports and the bottom of the standpipe in the top slab of the concrete pressure vessel is insulated in order to reduce the heat flow into the standpipe section.

The lower part of the standpipe section consists of two mild steel blocks joined by a mild steel tube which attenuates the radiation up the standpipe. The upper end has a telescopic section which permits the separation of the dead weight of the closure assembly, which is taken by the standpipe closure housing, from that of the rest of the plug unit which is transmitted downwards through the fuel elements for the reasons described in Section 10.1.1.

The housing for the closure assembly is screwed and seal-welded to the standpipe. The closure locking mechanism is provided with a pressure sensitive interlock to prevent unlocking when the differential pressure across the closure is sufficient to eject the fuel assembly. The primary lock is operated by the grab of the fuelling machine which handles the fuel assembly as one unit. A secondary lock, also operated from the fuelling machine but by a different mechanism, provides a safeguard against failure of the primary lock.

The plug unit is articulated so as to accommodate any slight misalignment between the standpipe and its associated gas baffle guide tube.

11. REFUELLING FACILITIES

Fuel routes within the plant embody extensive control systems to ensure that all fuelling operations are carried out safely. Redundancy and diversity of equipment is provided in the monitoring systems which are capable of initiating remedial actions if anything goes wrong.

At Hinkley Point B these systems incorporate hard-wired, plug-in, transistor logic modules. For Heysham 2, two microprocessors backed-up, where appropriate, by a relay-based system provide greater flexibility and permit a more extensive monitoring of the state of the plant.

11.1 New Fuel (Fig 5.13)

Fuel elements are delivered by road in steel boxes that conform with the relevant regulations of the International Atomic Energy Authority. These regulations specify the maximum dose rates allowed at the outer surfaces of the boxes and other design parameters which ensure their integrity. To avoid any possibility of accidental criticality during transport, the amount of uranium in each box and the number of boxes per vehicle are limited.

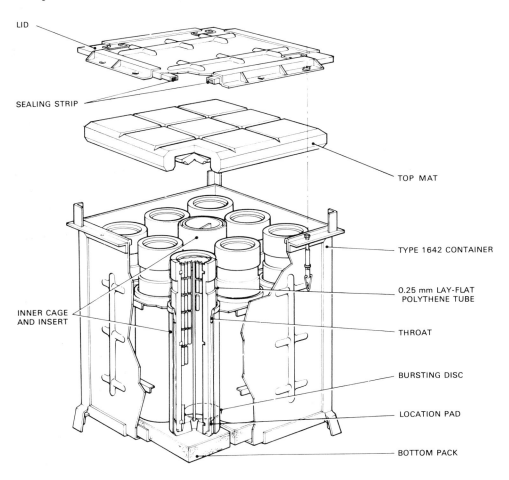

LID

SEALING STRIP

TOP MAT

TYPE 1642 CONTAINER

0.25 mm LAY-FLAT POLYTHENE TUBE

INNER CAGE AND INSERT

THROAT

BURSTING DISC

LOCATION PAD

BOTTOM PACK

FIG. 5.13   NEW FUEL TRANSIT CONTAINER

Within the station, similar attention is given to the design of the new fuel route for safety, as for example, by ensuring that the new fuel stores can never be flooded with water which would act as a moderator.

New fuel elements and other fuel stringer components are received in the reactor services building from whence the tiebars are taken to the charge hall for storage. As required, a new tiebar is taken to one of the plug unit maintenance positions in the central fuel handling facilities where it is inserted into a plug unit which has been serviced and is ready for re-use.

New fuel elements are taken from store in the reactor services building and made up with their top and bottom assemblies into fuel stringer stacks in the new fuel cell. This is also located in the central fuel handling facilities and is accessible by the fuelling machine which inserts a serviced plug unit complete with new tiebar into the cell. After the tiebar has been connected to the bottom assembly of the fuel stringer stack, the completed fuel assembly is withdrawn into the fuelling machine for insertion into one or other of the reactors.

11.2 Fuelling Machine (Fig 5.14)

A single fuelling machine weighing some 1000t serves both reactors and is capable of refuelling 240 channels a year. It is carried on the saddle of a gantry which spans the width of the charge hall. The transverse travel of the saddle across the gantry together with the longitudinal travel of the gantry along the length of the hall, permit the machine to be aligned with any fuel channel or interstitial standpipe or with any of the connecting points on the operating floor of the central fuel handling building.

The main body of the machine is a shielded pressure vessel containing a 3-compartment rotatable turret of sufficient height to accommodate a complete fuel or control rod assembly. Because the reactor charge faces are below the level of the reactor pressure vessel pre-stressing galleries, over which the machine must pass, a removable length of shielded pressure tube is provided and secured to the bottom of the machine for operations on the charge faces. To avoid gaps, this extension piece known as the 'make-up shield' has a telescopic snout which connects with an extension sleeve which is fitted during refuelling, in order to extend the standpipes to the level of the charge face floor. The machine itself has a similar telescopic snout at its lower end.

The topmost section of the pressure vessel above the turret contains the hoist drive shaft which passes through seals to an external motor. The grab is raised and lowered on twin roller chains, the excess length of which is tensioned and anchored in a vertical chain locker tube external to, but integral with, the vessel. The hoist drive incorporates a spring-operated, chain tension control which, in association with a load cell in the grab and other monitoring equipment, provides protection against both overloads and underloads. It also limits shock loading on the suspension system and the fuel or control rod assemblies during handling operations. The jaws of the grab are gravity operated and spring-loaded to engage and solenoid-operated to disengage; moreover, mechanical interaction and a spring-loaded cross lock ensure that the jaws cannot be disengaged when loaded. The electric cables to the grab are carried between the two lines of roller chains.

For a refuelling operation, two compartments of the turret are loaded in the fuel handling building, one with a replacement fuel assembly and the other with a shield plug and closure unit which can be inserted temporarily into the standpipe if, for any reason, the operation has to be interrupted after removal of the irradiated fuel assembly from the reactor.

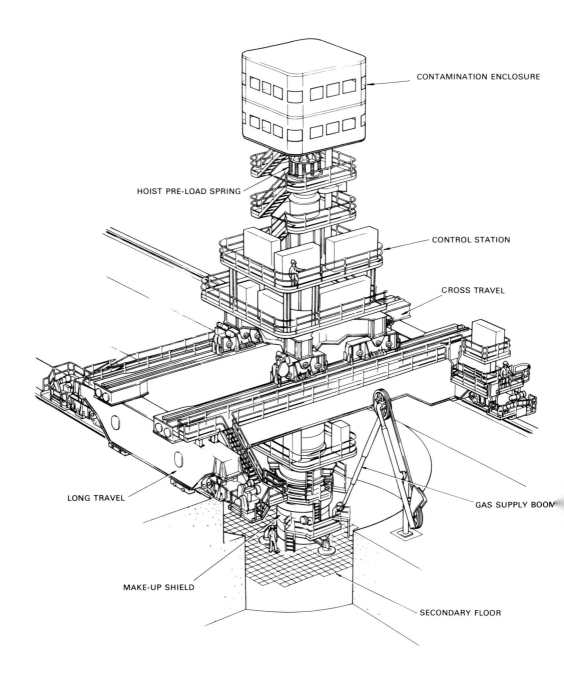

CONTAMINATION ENCLOSURE

HOIST PRE-LOAD SPRING

CONTROL STATION

CROSS TRAVEL

LONG TRAVEL

GAS SUPPLY BOOM

MAKE-UP SHIELD

SECONDARY FLOOR

FIG. 5.14   FUELLING MACHINE

The primary lock of the plug unit closure is released by the fuelling machine grab whilst the secondary lock is operated by a drive mechanism built into the standpipe extension sleeve. The grab also has electrical contacts which connect the channel gas outlet temperature thermocouples on the fuel stringers to the external instrumentation so as to maintain continuous monitoring of the condition of the channel during movement of the assemblies.

When withdrawn into the remaining compartment of the turret, the irradiated fuel assembly is supported on turret latches thus permitting the grab to be disengaged and withdrawn from that compartment.

Both during the withdrawal of an irradiated fuel assembly from the reactor and during the insertion of a new fuel assembly, a sliding sleeve inside the fuelling machine pressure vessel is arranged to seal with the bottom of the turret tube in use. The combination of this feature and a relatively small flow of carbon dioxide, which is injected into the fuelling machine to flow down over the grab and fuel assembly, prevents the coolant which is being forced up the fuel stringer by the gas circulators, from spreading generally into the fuelling machine pressure vessel. Instead, the stringer coolant joins with the injected carbon dioxide to flow down the annulus between the stringer and the standpipe to the relatively low pressure region above the gas baffle dome where it rejoins the reactor coolant circuit. This arrangement obviates the need for a recirculating cooling system on the fuelling machine.

Once an irradiated fuel assembly is within the fuelling machine and whilst the machine remains at pressure, natural circulation of the carbon dioxide within the machine transfers sufficient heat to the vessel walls to maintain a satisfactory fuel stringer temperature. However, when the machine is blown down to atmospheric pressure in order to transfer the fuel assembly into the irradiated fuel disposal cell, forced cooling may have to be employed, depending on the amount of decay heat the stringer is generating at the time.

The problem of ensuring that the 1300t of fuelling machine and gantry remain sealed to the reactor at 43.3 bar during a seismic disturbance is met by providing hydraulically-operated clamping devices which fix the machine relative to the gantry and the gantry relative to the building structure. The make-up shield pressure tube incorporates a large spherically-mounted section to accommodate relative movement between the fuelling machine pressure vessel and the standpipe.

11.3 Irradiated Fuel Disposal Facilities (Fig 5.15)

11.3.1 *Buffer Storage*

A buffer store, located in the central fuel handling building, accepts freshly discharged fuel assemblies whilst their fission product heat generation decays to a rate compatible with the cooling arrangements at the irradiated fuel disposal (IFD) cell. It consists of a number of water-cooled pressure tubes, known as buffer tubes, within which the fuel assemblies are placed and cooled by the natural convection of heat to the pressure tube walls by pressurised carbon dioxide.

Only eight buffer tubes were provided initially at Hinkley Point B but a further 14 were subsequently added. The decision to extend the facility resulted from operational experience of part load batch refuelling which had not been originally envisaged as a

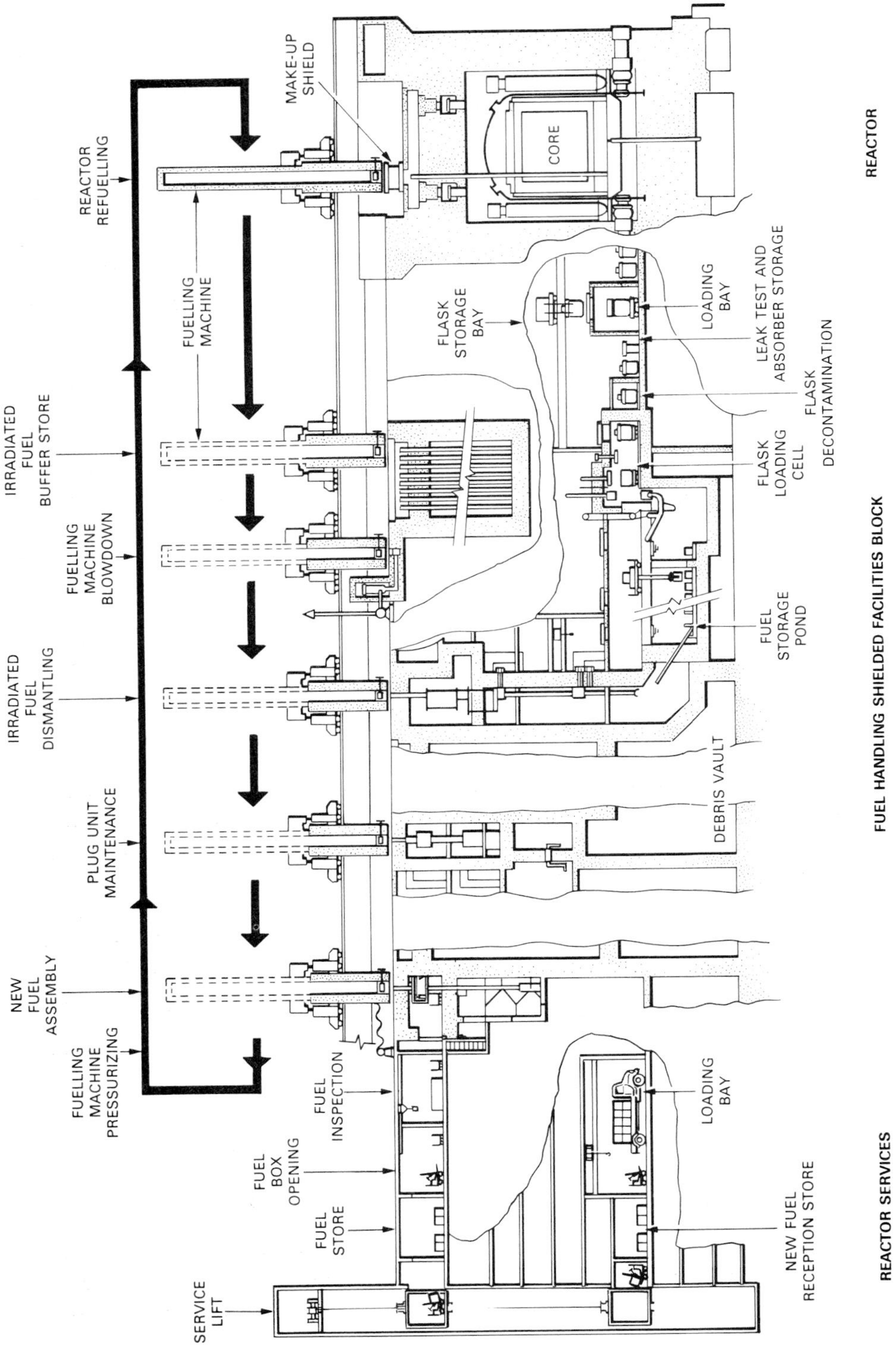

FUEL HANDLING SHIELDED FACILITIES BLOCK

FIG. 5.15   FUEL HANDLING FLOW CYCLE

normal operational route. At Heysham 2, the fuel route was designed from the outset to permit a 28 day cooling period for all fuel assemblies prior to transfer to the irradiated fuel disposal cell, as a means of easing the duty on the cooling system for that cell. In consequence, the buffer store at Heysham 2 contains 32 buffer tubes.

11.3.2 *Irradiated Fuel Disposal Cell (Fig 5.16)*

From the buffer store, the fuelling machine transfers each irradiated fuel assembly to the IFD cell. The tie bar bottom end fitting is first removed, after which the fuelling machine raises the plug unit so that the tie bar may be cropped and disposed of. The fuelling machine then transfers the plug unit to the maintenance facility for servicing prior to reuse.

Components of the fuel stringer other than the fuel elements themselves are discharged into an adjacent debris vault and the fuel elements are passed down a discharge tube into the fuel pond.

At Hinkley Point B, operations within the cell are viewed directly through shield windows and master/slave manipulators are installed for the purpose of rectifying any process faults. At Heysham 2, windows have been eliminated for seismic reasons and replaced by closed circuit television. The various tools used for in-cell operations are attached to removable shielding plugs so that, normally, they can be withdrawn complete for rectification or servicing with active components still in the cell. A robotic arm facilitates recovery from fault situations.

When it is desired to store a fuel element in the dry condition for subsequent post irradiation examination, the element is loaded into a reusable bottle prior to transfer to the pond.

11.3.3 *Fuel Pond and Flask Transfer Bay*

The fuel elements are removed under water from the end of the IFD cell discharge tube by a pond fuel manipulator and placed in skips in which they are stored in the pond. After a further cooling period of about 70 days, loaded skips are transferred into flasks for transport by rail to the irradiated fuel processing plant.

At Hinkley Point B, the 55t transport flask is lowered into the pond for loading. With this arrangement, the possibility that the flask may be dropped during the loading operation has to be considered and guarded against. At Heysham 2, the skip is raised out of the pond and placed in the flask within a shielded cell immediately adjacent to the pond. The loaded flask is transferred out of the pond area through a shield door on a motorised trolley and is then lifted to the decontamination and storage bays.

The operations within the pond and flask loading cells are remotely controlled from a pond control room and viewed by closed circuit TV.

11.4 Active Maintenance Facility

The provision of only one active maintenance position at Hinkley Point B proved to be a limitation. In consequence, there are, at Heysham 2, duplicate cells for inspection and maintenance of plug units and a separate cell for dealing with control rod and other assemblies.

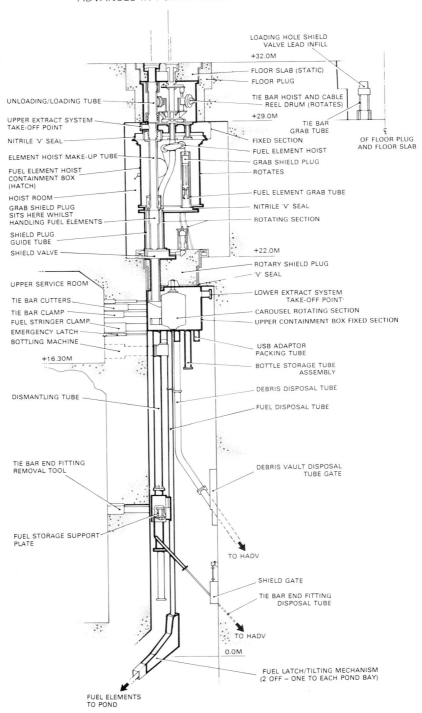

FIG. 5.16   IFD CELL

476

The cells provide for operations to be carried out at various levels. At the lower levels, the components to be inspected and maintained are highly active and the operations are carried out remotely. At the upper levels, facilities are provided for removing the more complex mechanisms such as the control rod actuators and the gag actuators of the plug units and for transferring them to an adjacent active maintenance workshop. The workshop is designed to facilitate the installation of glove boxes should those mechanisms prove in practice to become contaminated with active dust.

12. BOILER SYSTEM

The boiler system uses 600km of boiler tubing and accounts for some 10% of the capital cost of the station. It is located in a boiler annulus formed between the gas baffle and the wall of the reactor pressure vessel. As described in Section 1.1.4, the annulus has been increased in size as compared with that at Hinkley Point B, in order to permit man access for repair should this ever be necessary.

The annulus is divided into four quadrants, each of which contains three boiler modules and two circulators. The circulators are situated under the boiler modules and draw hot carbon dioxide from the space above the gas-baffle dome down through the banks of boiler tubing before discharging it into the space below the reactor core.

The main function of the boiler system is to generate superheated steam with conditions of 166 bar and 541°C at the superheater outlet header, for expansion in the HP cylinder of the turbine and subsequently to reheat it to 539°C at 40.7 bar prior to admission to the IP cylinder. The evaporator/superheater sections are of the once-through type in order to minimise the number of pressure vessel penetrations.

The three modules in each quadrant are controlled as a single unit and any one quadrant may be taken out of service while the reactor is on load. Each module includes, below the main evaporator section, separate banks of tubes which form the decay-heat boiler system. These banks are connected to form one decay-heat boiler per quadrant. Both the main boiler units and the decay-heat boilers play major roles in the reactor post trip heat removal systems.

12.1 Boiler Module (Fig 5.17)

Each of the 12 modules is 16m high, weighs 120t and was transported to site as a single unit. Although located in an annulus, the module has a rectangular cross section which permits the use of straight horizontal tubing in the direction giving the greater length of tube between vertical serpentine return bends.

The serpentine tubes which make up the decay heat boiler and the main evaporator/superheater section are assembled in pairs by means of welded spacers to form vertical platens. 44 such platens hang side-by-side in the module casing and are supported at regular intervals across and down the module by links attached to beams which span the casing. That part of the module which comprises these platens and their stainless steel casing is supported at its base on mild steel beams. The beams are slung from flexible links anchored to the gas baffle skirt on the one side and to the reactor pressure vessel wall on the other. The load is transferred from the module casing to the beam through a pin joint. This permits the anchor points of the beam to

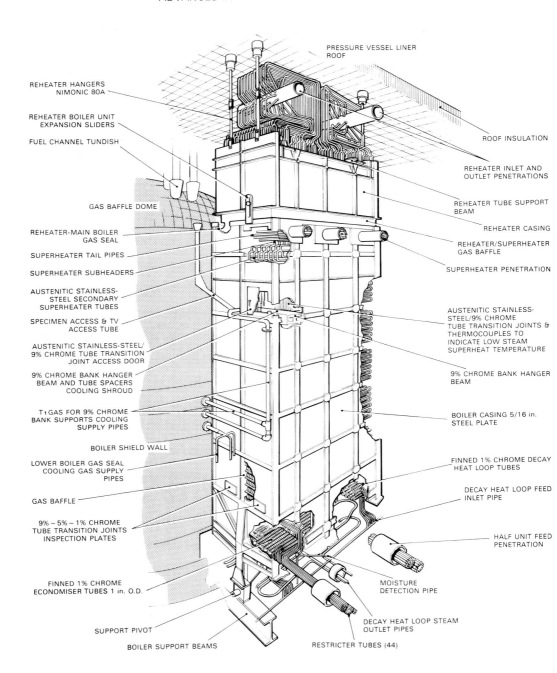

REHEATER HANGERS
NIMONIC 80A

REHEATER BOILER UNIT
EXPANSION SLIDERS

FUEL CHANNEL TUNDISH

GAS BAFFLE DOME

REHEATER-MAIN BOILER
GAS SEAL

SUPERHEATER TAIL PIPES

SUPERHEATER SUBHEADERS

AUSTENITIC STAINLESS-
STEEL SECONDARY
SUPERHEATER TUBES

SPECIMEN ACCESS & TV
ACCESS TUBE

AUSTENITIC STAINLESS-STEEL/
9% CHROME TUBE TRANSITION
JOINT ACCESS DOOR

9% CHROME BANK HANGER
BEAM AND TUBE SPACERS
COOLING SHROUD

T_1 GAS FOR 9% CHROME
BANK SUPPORTS COOLING
SUPPLY PIPES

BOILER SHIELD WALL

LOWER BOILER GAS SEAL
COOLING GAS SUPPLY
PIPES

GAS BAFFLE

9% – 5% – 1% CHROME
TUBE TRANSITION JOINTS
INSPECTION PLATES

FINNED 1% CHROME
ECONOMISER TUBES 1 in. O.D.

SUPPORT PIVOT

BOILER SUPPORT BEAMS

PRESSURE VESSEL LINER
ROOF

ROOF INSULATION

REHEATER INLET AND
OUTLET PENETRATIONS

REHEATER TUBE SUPPORT
BEAM

REHEATER CASING

REHEATER/SUPERHEATER
GAS BAFFLE

SUPERHEATER PENETRATION

AUSTENITIC STAINLESS-
STEEL/9% CHROME
TUBE TRANSITION JOINTS &
THERMOCOUPLES TO
INDICATE LOW STEAM
SUPERHEAT TEMPERATURE

9% CHROME BANK HANGER
BEAM

BOILER CASING 5/16 in.
STEEL PLATE

FINNED 1% CHROME DECAY
HEAT LOOP TUBES

DECAY HEAT LOOP FEED
INLET PIPE

HALF UNIT FEED
PENETRATION

MOISTURE
DETECTION PIPE

DECAY HEAT LOOP STEAM
OUTLET PIPES

RESTRICTER TUBES (44)

FIG. 5.17   BOILER MODULE

478

move relative to one another without themselves causing the module to tilt, whilst at the same time accommodating the slight radial tilt which is produced by thermal expansion of the superheater penetrations.

A major manufacturing innovation for Heysham 2 was the use of automatic welding robots to carry out the four million spacer welds necessary to form the platens for the 24 modules required.

The reheater section of the module is located above the superheater section and is hung from the top slab of the reactor pressure vessel. The tubing is again arranged in the form of platens but in this case there are 36 platens each of which consists of four parallel steam paths in the same vertical plane. The casing can move relative to that of the lower part of the module and a gas-tight seal is provided by means of a flexible foil.

12.2 Boiler Tubing

The tube materials are chosen to avoid excessive gas or waterside corrosion or erosion. They are, therefore, graded from 1%Cr 0.5%Mo steel for the feed penetrations and the low temperature sections, through 9%Cr 1%Mo, to type 316H austenitic steel for the top of the superheater and for the reheater where gas temperatures are highest.

The point of transition from the 9%Cr 1%Mo ferritic steel to the austenitic material is particularly sensitive because it is necessary to ensure both that the gas temperature does not exceed 550°C, in order to avoid excessive gas side oxidation of the ferritic steel, and that the steam is superheated to a temperature which is sufficiently above saturation temperature to avoid any risk of stress corrosion in the austenitic steel. A superheat margin of 70°C was adopted as the design value for Heysham 2. The junction between the two materials is achieved with a short transition section of Inconel 600 which has a coefficient of expansion roughly halfway between those of the ferritic and austenitic tubing materials. It was found, during the manufacture of the boilers for Hinkley Point B, that closure welds between these dissimilar materials were difficult to make. In consequence, the transition pieces for Heysham 2 were manufactured and fully inspected with a stub of 9%Cr 1%Mo tubing at one end and of type 316H at the other thus allowing the closure welds to be made between similar materials.

Tubing of 1%Cr 0.5%Mo ferritic steel is used throughout the decay-heat boiler sections, so that operation with water of lower quality than that specified for the main boilers can be tolerated.

Both the decay-heat boilers and the low temperature sections of the main steam generators have finned tubes in order to improve their heat transfer performance.

12.3 Seismic Restraints

Seismic forces would, in large measure, be accommodated and boiler movement restrained by features of the boiler system such as the support beams and the sheath tubes which surround the superheater outlet tailpipes and connect the upper ends of the main casings to the wall of the concrete pressure vessel. To restrain circumferential movement, specific seismic restraints are provided at two levels which link the casings to anchors set in the pressure vessel walls.

12.4 Design Analysis

The design analysis assessed the capacity of the platens to withstand the loads due to pressure, aerodynamic drag, self weight and thermal expansion for defined operating conditions. Platen dynamics were analysed to determine natural frequencies both in and out-of-plane, and were then compared with the excitation frequencies which could be generated in the gas circuit. These analyses were supported by tests in which fullsize sections of platens and their tailpipes were subjected to the full range of aerodynamic operating conditions in pressurised test rigs.

In addition, the behaviour of the platens as a consequence of incremental plastic deformation and the creep fatigue damage which could result from the predicted loading patterns of the reactor units, was assessed in order to be sure of their continued integrity over a 30 year life.

13. GAS CIRCULATORS (FIG 5.18)

The first commercial AGR to be committed by CEGB, Dungeness B, employs four circulators per reactor with variable speed drives situated outside the pressure envelope. As the design of that station progressed, the advantages of increasing the number of circulators became apparent and Hinkley Point B is provided with eight circulators per reactor. This implies a mass flow of around 530kg/s and the resulting machine size made it possible to adopt an 'encapsulated' design which eliminates the need for a shaft seal by enclosing the motor within the pressure boundary. Flow control by means of inlet guide vanes was adopted and the vane drive mechanism was also housed within the motor enclosure.

An intensive materials and component development programme was undertaken which culminated in the testing of a complete prototype circulator unit under reactor operating conditions in a pressurised test rig. Subsequent operating experience at Hinkley Point B confirmed the suitability of the design in general but also identified a number of areas in which detailed improvement could be made. The circulators at Heysham 2 are, therefore, based upon, but improved versions of, the original Hinkley Point B design.

The machine is a single-stage centrifugal compressor giving a pressure rise of 2896mbar and operates with an outlet temperature of around 300°C. It has a horizontal shaft supported on two journal bearings and an overhung impeller. Because the original Hinkley Point B design of machine had a resonant frequency nearer to running speed than had been anticipated, the mass distribution was adjusted for Hinkley Point B and altered for Heysham 2 so as to raise the natural frequency of vibration to a value appreciably above the normal operating speed of 2970rev/min.

Each machine has its own forced lubrication system and interrelated ventilation system, the design being arranged to minimise oil ingress into the reactor coolant circuit. The labyrinth and its associated barrier plate also limit the interchange of gas between the primary coolant circuit and the motor compartment. In addition, they would constitute a secondary containment in the event of failure of the penetration closure plate which constitutes the outboard end of the machine capsule.

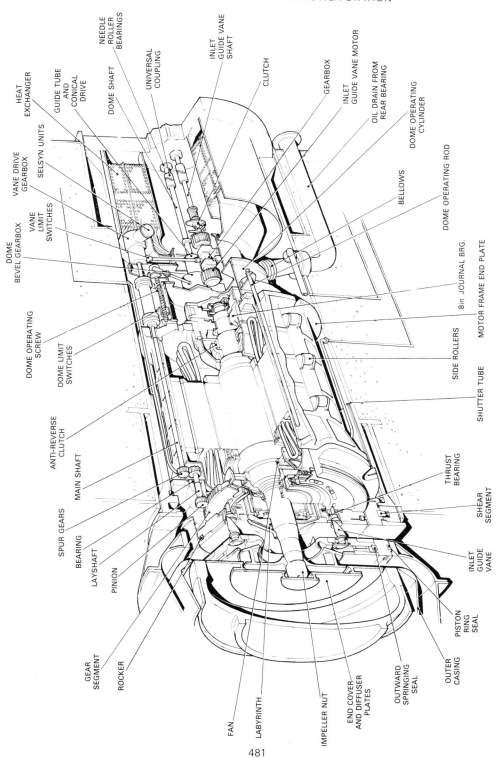

NEEDLE ROLLER BEARINGS

INLET GUIDE VANE SHAFT

GUIDE TUBE AND CONICAL DRIVE

UNIVERSAL COUPLING

DOME SHAFT

CLUTCH

HEAT EXCHANGER

GEARBOX

SELSYN UNITS

INLET GUIDE VANE MOTOR

VANE DRIVE GEARBOX

OIL DRAIN FROM REAR BEARING

DOME BEVEL GEARBOX

VANE LIMIT SWITCHES

DOME OPERATING CYLINDER

DOME OPERATING ROD

DOME OPERATING SCREW

DOME LIMIT SWITCHES

8in JOURNAL BRG.

BELLOWS

MOTOR FRAME END PLATE

SIDE ROLLERS

SHUTTER TUBE

ANTI-REVERSE CLUTCH

MAIN SHAFT

THRUST BEARING

SPUR GEARS

BEARING

LAYSHAFT

PINION

SHEAR SEGMENT

INLET GUIDE VANE

GEAR SEGMENT

ROCKER

FAN

LABYRINTH

IMPELLER NUT

END COVER AND DIFFUSER PLATES

OUTWARD SPRINGING SEAL

PISTON RING SEAL

OUTER CASING

FIG. 5.18   GAS CIRCULATOR

A sealing dome is provided to enable the motor compartment to be isolated from the reactor gas circuit when it is required to remove a circulator for maintenance.

Inlet guide vanes vary the gas flows when the machine is running and control the reverse flow when the machine is idle. At Heysham 2, the vane actuator drive motor and gearbox are located on the outer face of the penetration closure in order to permit ready access for maintenance purposes. The inlet guide vanes, inlet and outlet fairings and casings are attached to the penetration and are situated immediately below the boilers. Gas from the impeller outlet is discharged through a vaned radial diffuser and an annular casing diffuser system into a nozzle situated in the gas-baffle skirt. A seal accommodates thermal movement between the nozzle and the circulator casing and prevents leakage of gas from outlet to inlet.

The impeller is driven by an induction motor with a rating of 5220kW when supplied at 50Hz for normal operation with the reactor on load. A forced-cooling system transfers heat from the motor to recirculating carbon dioxide which is, in turn, cooled in gas/water heat exchangers mounted within the motor compartment on the inside face of the primary closure. Two 100% duty heat exchangers are provided and the heat is normally rejected through a further pair of 100% duty heat exchangers to the reactor sea water system. A circulator auxiliaries diverse cooling system with an alternative heat sink is available for use in post trip conditions. See Section 14.

For post trip operation, a variable-frequency electric supply is provided. This enables the circulator to operate over the range from 100% down to 15% of the normal speed according to operating conditions. These arrangements also are described in more detail in Section 14.

14. POST TRIP HEAT REMOVAL SYSTEMS (FIG 5.19)

Post trip heat removal systems are automatically initiated by a reactor trip and completely controlled by post trip sequencing equipment so that no operator action is required during the first 30 minutes after the trip.

Since the arrangements for removing heat following a trip must be highly reliable, independent and diverse groups of plant systems are provided. These are arbitrarily designated as X systems and Y systems. The X systems, which constitute the primary means of heat removal when the reactor remains pressurised, are seismically qualified and are therefore able to ensure adequate cooling following the occurrence of seismic disturbances of severity up to or equal to that of the safe shutdown earthquake.

Both X and Y plant systems employ conventional and well-tried plant, incorporate redundancy and are protected against internal and external hazards by segregation and civil engineering design features.

14.1 Arrangement of the X and Y Systems

Each reactor is provided with four groups of X systems and four groups of Y systems each of which is known as a 'train'. Each reactor quadrant is served by one X train and by one Y train.

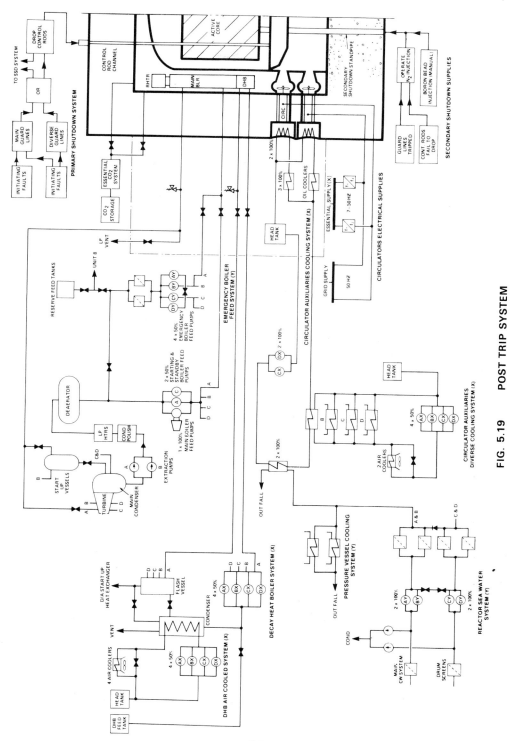

FIG. 5.19   POST TRIP SYSTEM

Each train comprises automatic post trip sequencing equipment, essential auxiliaries switchboards and associated electrical distribution systems, selected mechanical and electrical plant items and necessary controls and instrumentation. Each train is further identified by one of the letters A to D in order to relate it to the quadrant of the reactor with which it is associated. Some mechanical equipment however is common to all the trains of a particular category; for example, the steam flash vessel and atmospheric condenser of the decay-heat boiler system described in Section 14.2.3 are common to all four X trains of the reactor.

The preferred source of power for both the X and Y trains is the national grid system. The trains for the A and B quadrants are supplied via the 11kV station boards and those for the C and D quadrants via the 11kV unit boards, to which further reference is made in Section 20. Diesel generators are started automatically after a reactor trip and, in the event of loss of grid supplies, are connected to the 3.3kV essential auxiliaries boards and loaded automatically by the post trip sequencing equipment. The location of the eight diesel generators provided at Heysham 2 and their allocation to reactor quadrants is described in Section 4.2.

The X train sequencing equipments automatically introduce the X systems into service following a reactor trip. At the same time, the Y train sequencing equipments make the Y systems ready for duty and automatically introduce them into service should X system failures occur or should the reactor depressurise when certain systems in both categories are required.

Diversity of automatic sequencing equipment is provided by using a microprocessor-based logic system for the X trains and a relay-based logic system for the Y trains.

14.1.1 *Pressurised Faults*

For pressurised faults, the X and Y systems provide alternative diverse methods of removing the heat generated after a reactor trip.

The X systems are regarded as the primary means of heat removal and include the gas circulators operating at 15% of their normal speed together with a decay heat boiler system. These systems are redundant to the extent that one of the four X trains provided per reactor is capable of satisfying the post trip duty. The X systems also include the gas circulator auxiliaries cooling system and its diverse heat sink.

The Y systems include the emergency boiler feed system which operates in association with the main evaporator/superheater sections of the boilers, together with natural circulation of the carbon dioxide in the primary coolant circuit. Thus, the Y systems would take over in the unlikely event of the total loss of the circulators or the total failure of the decay-heat boiler system. The Y systems also include the reactor sea water system and provides one of the two alternative heat sinks for the gas circulator auxiliaries cooling system.

14.1.2 *Depressurisation Faults*

Depressurisation faults have a low probability of occurrence and it is not considered necessary to have a diverse, alternative means of heat removal. The single means provided makes use of both X and Y systems; the gas circulators (an X system) are operated in association with the emergency boiler feed system (a Y system). This single means of heat removal, however, has considerable redundancy. Adequate coolant can always be maintained by two-out-of-four quadrants and by only one-out-of-four if the reactor pressure remains above 2 bar.

14.1.3 *Boiler Protection*

When a reactor trip occurs it is necessary to protect the boilers against sudden changes of temperature. To this end, the gas circulators are tripped and run down prior to connection to a low frequency electrical supply and the rate at which feed water is supplied to the main evaporator/superheater sections is immediately reduced to about 10% of the normal flow. This is achieved by using the electrically-driven starting/standby feed pumps which then maintain that flow until such time as the gas temperature has fallen sufficiently. When the danger of overheating the lower sections of the boiler has passed, in about 15 to 20min from the time of the trip, this flow is isolated and cooling of the carbon dioxide is effected solely by the decay heat boiler system which has been started up as an X system. The steam generated in the main boilers during this time is normally discharged to atmosphere via a low pressure vent system which reduces the boiler pressure to 80 bar although the starting/standby feed pumps can, of course, supply feedwater at pressures up to that at which the pressure relief valves operate.

Since the starting/standby feed pumps are not available following the loss of grid supplies, the reduced rate of feed flow for boiler protection purposes would in these circumstances be provided by the emergency boiler feed pumps which would have been made ready as a Y system.

14.2 Plant Systems

As indicated, the X and Y systems embrace a variety of electrical and mechanical plant systems which include essential electrical systems. The latter are a part of the station electrical system and are described in Section 21.

14.2.1 *Forced Gas Circulation System*

The mass flow required in the pressurised primary circuit for decay heat removal is achieved by operating the circulators at 15% of their normal speed with the inlet guide vanes fully open. For this mode of operation, the circulator motors are supplied at 7.5Hz by means of solid state variable frequency convertors. Each circulator has its own independent variable frequency convertor to which it is connected automatically following a reactor trip.

For a depressurisation fault, higher circulator speeds are required and the frequency convertors automatically increase the frequency of the circulator supply as the reactor gas pressure falls.

14.2.2 *Gas Circulator Auxiliaries Cooling Systems*

Each gas circulator is provided with three 100% lubricating oil coolers and two 100% gas/water heat exchangers for recooling the carbon dioxide which recirculates through the circulator motor. The coolers and heat exchangers for each pair of circulators form part of the independent circulator auxiliaries cooling system provided for each quadrant.

Each system has two 100% pumps and two 100% heat exchangers through which the heat from the auxiliaries of the pair of circulators is normally transferred to the reactor seawater system. This latter system, however, is a Y system and is not therefore seismically qualified. In consequence, each circulator auxiliaries cooling system also includes a further heat exchanger in order to connect it to the circulator auxiliaries diverse cooling system which provides a diverse and seismically qualified heat sink as an alternative to the reactor seawater system.

The gas circulator auxiliaries diverse cooling system is an X system. It consists of a single loop per reactor with four 50% pumps and two forced draught air coolers each of which has two fans and rejects the heat to atmosphere. This system is brought into service by the automatic sequencing equipment and has a capacity which is sufficient only for removing the heat generated by the circulator auxiliaries when operating at post trip loadings.

14.2.3 *Decay-Heat Boiler System*

The decay-heat boilers are isolated on the water/steam side and kept dry during normal operation. The decay-heat boiler system is an X system and is therefore automatically introduced following a reactor trip.

This system comprises, for each reactor, a feed tank, four 50% duty feed pumps, four decay heat boilers, a flash vessel, a condenser and a recirculating condenser cooling sub-system. The feed pumps draw water from the condenser well via a common inlet header and discharge to a common outlet header. Individual feed lines run from the outlet header to the decay heat boiler in each quadrant. Saturated steam is generated at 35 bar and is directed via a flash vessel to a condenser operating at atmospheric pressure. In the early stages of a post trip operation, steam is vented to atmosphere until the reactor heat generation reduces to about 30MW, at which level the condenser cooling duty matches the heat load and allows the feed system to become fully recirculating.

The recirculating condenser cooling sub-system comprises four 50% duty circulating pumps with common inlet and outlet headers and four forced draught air coolers each of which has two fans and rejects the heat to atmosphere.

14.2.4 *Emergency Boiler Feed System*

The emergency boiler feed system is capable of supplying feedwater to the main boilers at the rate of 10% of the full load flow. There are four emergency feed pumps per reactor which draw water from the reserve feed water tanks via a common suction header. The discharge is split on a half reactor basis by a normally-closed interconnecting valve and each pair of pumps is normally connected through electrically-operated isolating valves in the individual feed lines to the boilers of two quadrants. Each pump is rated at 50% of the reactor duty so that one pump of each pair has the capacity to provide the whole of the emergency feed flow required by two quadrants.

The steam generated is discharged through vent valves, which maintain the boiler pressure at 80 bar, into flash vessels from which it is discharged to atmosphere.

14.2.5 *Reactor Seawater System*

Each reactor has its own seawater system. They are connected to the intake and outfall of the main cooling water system, which is described in Section 20. They are Y systems and each is split into two loops on a half reactor basis with normally-closed interconnecting valves between them. Each loop has two 100% duty pumps.

In addition to constituting the normal heat sink for the gas circulator auxiliaries cooling systems, the reactor seawater systems also cope with the heat removed from the reactor pressure vessel and the fuel pond in order to maintain them at acceptable working temperatures.

14.3 Sequence of Post Trip Operations

The post trip sequencing equipments initiate the same basic succession of events after every reactor trip and modify it only if grid supplies are lost or if the reactor pressure falls below 28 bar.

The basic sequence introduces the X systems into service and, at the same time, makes the Y systems ready for duty. Appropriate Y systems are then automatically introduced into service on a quadrant basis should X system failures occur, or in all working quadrants if the reactor is depressurising.

All diesel generators are automatically started up after every reactor trip but the diesel auxiliary switchboards are not connected to the station electrical system unless the grid supplies are lost.

14.3.1 *X Systems Operations*

Immediately after a reactor trip, the following occurs:
(1) Normal quadrant boiler feed valves are closed and the normally-closed valves in the post trip bypass circuits are opened.
(2) The starting/standby feed pumps are started and supply feedwater through the bypass circuit at about 10% of the full flow rate in order to protect the low temperature ends of the main evaporator/superheater sections of the boiler system.
(3) The gas circulator inlet guide vanes are driven to the fully open position.
    Subsequently, after appropriate intervals of time:
(4) The gas circulators are tripped and allowed to run down to about 10% of their normal speed before being connected to the variable frequency supply which is set to supply them at 7.5Hz and thereby drive them at a constant 15% of normal speed.
(5) The feed and other pumps and the air coolers of the decay heat boiler system are started and the valves operated in sequence to introduce these systems into service.
(6) The gas circulator auxiliaries diverse cooling system is brought into service.

All these operations are normally completed within about 1.5min of the reactor trip. After about 15 to 20min, on receipt of a signal indicating that gas temperatures have fallen sufficiently, the feed flows to the main evaporator/superheater sections of the boiler system are terminated. Within a period of a few hours, the decay heat boiler system becomes fully recirculating and the venting of steam to atmosphere ceases.

14.3.2 *Y Systems Operations*

Immediately after a reactor trip, the normally-open valves in the unit HP steam balancing header are closed so that the steam mains are split on a half reactor basis in line with the arrangement of the emergency boiler feed system.

The LP vent system which becomes operational after about 35s, reduces the main boiler pressure to 80 bar at a rate of 0.5 bar/s. Subsequently, the emergency boiler feed pumps are started.

The reactor seawater system is normally operational with two of its four pumps in service but after a reactor trip, the other two pumps are started up.

If reactor pressure is maintained, the emergency boiler feed valves will remain closed unless one of the following X system failures is detected:

(1) Loss of 10% feed flow from the starting/standby feed pumps whilst gas temperatures are still high.
(2) Loss of the decay-heat boiler system after the 10% feed flow from the starting/standby feed pumps has been terminated.
(3) Loss of all circulators, when natural circulation of the reactor coolant must be relied upon to maintain the fuel cladding temperature and the reactor gas outlet temperature below their permissible limits.

Once introduced, emergency boiler feed will continue until terminated by operator action. For long term use, a recirculatory route, making use of the boiler startup vessels and the condensers of the 660MW turbines, can be established and the discharge of steam to atmosphere through the LP vent terminated.

14.3.3 *Loss of Grid Supplies*

If the reactor trip is occasioned by the loss of the grid connections, all plant is first cleared from the switchboards, the diesel generators are run up and connected to the 3.3kV essential auxiliaries boards and the post trip sequencing equipments follow, in general, the basic succession of events described in Section 14.3.1 and Section 14.3.2. In fact, the timing of the sequence of events is compatible with the time required for the diesel generators to be run up and connected and with the provision of a satisfactory loading schedule. However, it is not practicable to start the large starting/standby feed pumps on the diesel generator supply, so the basic sequence is modified to introduce the emergency boiler feed pumps for boiler protection purposes.

If a post trip loss of grid supplies occurs, all plant is first cleared from the switchboards, the diesel generators, which have already been run up, are connected and the plant is restarted in the same sequence as before.

14.3.4 *Depressurisation Faults*

Reactor depressurisation is indicated by a post trip reactor gas pressure of less than 28 bar. The basic sequence of events is then modified in two ways. Firstly, the emergency boiler feed pumps are introduced into service since the main evaporator/superheater sections of the boiler system are required in these circumstances; secondly the output frequency of the supply from the frequency converters is increased as the reactor gas pressure falls so that the gas circulators reach their normal speed when the reactor pressure has fallen to about 3 bar.

Operator action is required to inject carbon dioxide from the storage and distribution system in order to prevent air ingress for at least 72 hours following the reactor trip or, in the case of a minor failure in the pressure envelope, to provide what is termed 'pressure support' and maintain the reactor pressure above 2 bar so as to retain a greater degree of redundancy in the post trip cooling provisions.

15. CARBON DIOXIDE AUXILIARY SYSTEMS (FIG 5.20)

15.1 Storage and Distribution

Normal operational requirements for carbon dioxide include reactor coolant make-up, supplies for cooling the fuelling machine and for refilling the reactor from time to time after it has been blown down to permit access to the vault for inspection and maintenance. In addition, supplies must be available to perform certain safety functions such as the provision of pressure support and prevention of air ingress in the event of a loss-of-coolant accident and the operation of the tertiary shutdown system.

Carbon dioxide is stored on site as liquid under pressure in refrigerated tanks and is passed through steam heated vaporisers into ring mains. Some 750t is stored at 25 bar and is available for distribution through the low pressure ring main whilst a further 78t stored at 59 bar, is connected to the high pressure ring main. The high pressure system is replenished from the low pressure storage tanks by transfer pumps having a capacity of 6t/h.

15.2 Gas Processing

The extent of the radiolytic oxidation of the core graphite depends upon the characteristics of the graphite itself, the coolant pressure, the radiation dose level and the presence of methane and carbon monoxide in the coolant. Methane inhibits graphite corrosion but its presence in high concentrations can lead to carbon deposition, particularly on the fuel stringers. The coolant composition must therefore be maintained within defined limits.

Some 0.7% of the reactor gas flow is passed continuously through a processing system, using the circulator pressure differential as the driving force.

Argon-free oxygen is generated by an electrolysis plant and is added to the coolant before it enters a recombination unit in which a platinum/aluminium catalyst effects the oxidation of carbon monoxide, produced by radiolysis in the primary circuit, to carbon dioxide. Dryers are situated downstream of the recombination unit to remove the moisture produced by the chemical reactions and to deal with possible water ingress into the primary circuit as a result of boiler tube failures. Methane is added in gaseous form to replace that lost by chemical reactions within the reactor.

Each reactor can be isolated from its gas processing plant by valves which are automatically closed when a reactor trip occurs, in the event of excess flow or reverse flow, or if carbon dioxide is detected in the processing plant area.

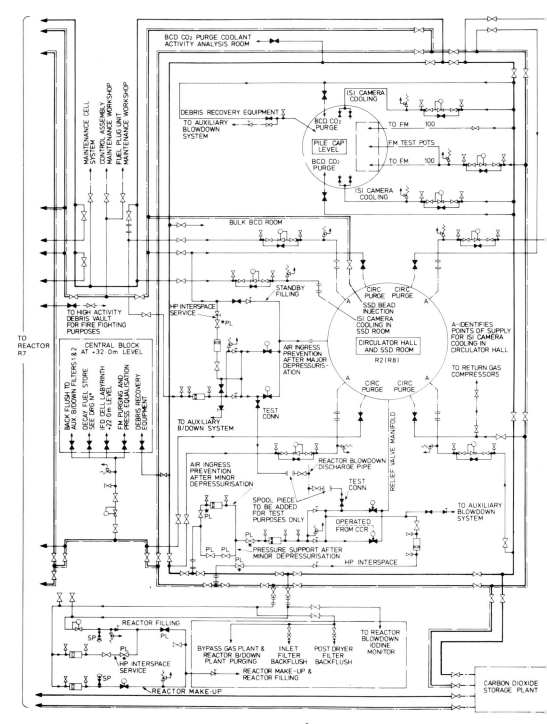

FIG. 5.20   CO2 DISTRIBUTION SYSTEM

15.3   Discharge System

The reactor coolant discharge system, which provides the means for the controlled discharge of contaminated gas from the reactor and associated equipment, is described as Section 18.

16.   REMOTE IN-SERVICE INSPECTION EQUIPMENT

Remote visual inspection of plant and equipment within the reactor vault is performed with the aid of TV cameras which are introduced along predetermined access routes by means of various forms of manipulator.

16.1   The TRIUMPH TV Camera System

The TRIUMPH (acronym — Television Remote Inspection Unit Multi-Purpose Head) TV camera system has been developed by the CEGB and has undergone a continuous process of improvement and development during the course of the AGR building programme. It is a modular concept which divides the cylindrical camera assembly horizontally into a number of modules each of which has its own functions. See Fig 5.21. This permits the ready incorporation of improvements in technology and the interchangeability of modules between the various TRIUMPH models.

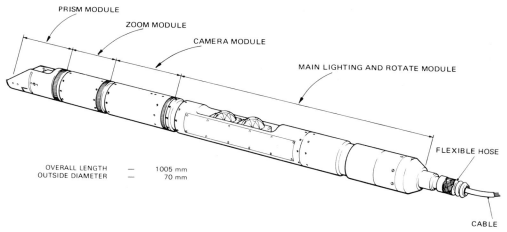

| OVERALL LENGTH | — | 1005 mm |
| OUTSIDE DIAMETER | — | 70 mm |

FIG. 5.21   TRIUMPH CAMERA

A shortened version and also an articulated version of the standard arrangement are available for use in physically-restricted situations and the choice of camera assembly depends upon the route which it has to follow and the functions required of it.

16.2 Charge Face Routes

Eight peripheral and four interstitial standpipes through the top slab of the concrete pressure vessel are provided specifically for the introduction of in-service inspection (ISI) equipment. Any fuel or control rod standpipe can also be used for ISI purposes once the fuel assembly or control rod assembly is removed.

The ISI standpipes are fitted with closure plugs of such length as to close also the corresponding penetration of the gas baffle dome. In addition, their plan positions have been chosen to maximise the inspection coverage of the boiler annulus and of the central region of the dome which is penetrated by the fuel and other guide tube nozzles.

16.2.1 *Charge Face Route Preparation*

The closure plug units of the ISI standpipes are removed by the ISI route preparation equipment. When a fuel or control rod channel is to be used, the assembly must first be removed by the fuelling machine which then inserts a temporary closure plug unit which can be handled by the ISI equipment.

The ISI route preparation equipment consists of an alignment frame which sits on the charge face floor and the plug unit transfer containment which is fitted to the alignment frame. The containment consists of a glove box and a flexible bellows within which a grab is lowered to engage with the plug unit. The charge face crane then lifts the bellows head and the plug unit is hoisted into the expanded bellows. A bung is fitted to the top of the standpipe and the containment, complete with the plug unit inside, is removed to the maintenance and storage facilities.

An appropriate ISI manipulator, which is also capable of removing and replacing the standpipe bung, is then placed on the alignment frame.

16.2.2 *Charge Face Manipulators*

Peripheral and Interstitial Manipulators

The peripheral manipulator at Heysham 2 is based largely upon the design of those provided at Hartlepool and Heysham 1 for remote visual inspection in the space between the gas baffle dome and the underside of the pressure vessel roof.

The camera assembly, which incorporates within itself the means of rotating its lighting and viewing modules, is positioned primarily by the use of hinged links which are free to move in one direction only. Thus, when extended, the chain constitutes a rigid arm except in the direction in which it can be reeled onto a drum in the charge face housing of the manipulator equipment. The camera assembly can, moreover, be deployed vertically from the end point of the chain of links by paying out its service hose.

The major increase in camera coverage provided at Heysham 2, as compared with Hartlepool and Heysham 1, and the complexity of some of the routes within the vault, has necessitated the introduction of computer control to allow the manipulator to follow automatically, predetermined sequences for each identified route.

The peripheral manipulator (Fig 5.22) can deploy the camera assembly inside the gas baffle, above the gas baffle dome and in the boiler annulus. It can operate from peripheral fuel channel standpipes as well as from the eight peripheral ISI standpipes and its chain of links can extend up to 7m from the bottom of the access standpipe.

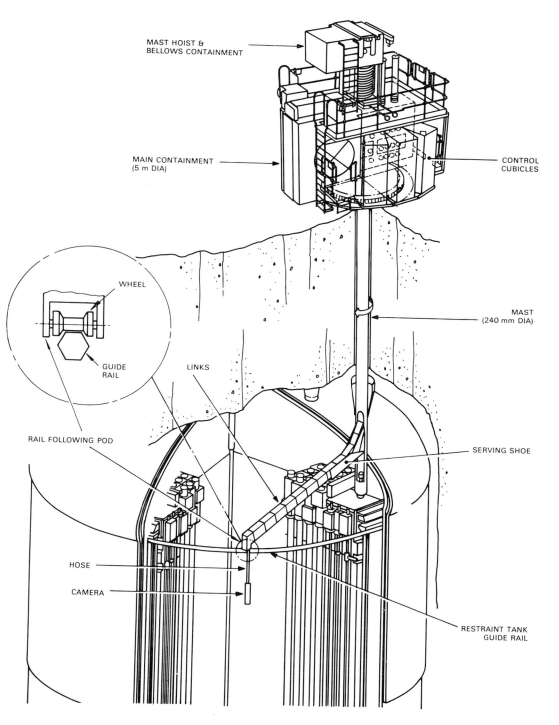

MAST HOIST &
BELLOWS CONTAINMENT

MAIN CONTAINMENT
(5 m DIA)

CONTROL
CUBICLES

WHEEL

GUIDE
RAIL

LINKS

MAST
(240 mm DIA)

RAIL FOLLOWING POD

SERVING SHOE

HOSE

CAMERA

RESTRAINT TANK
GUIDE RAIL

FIG. 5.22   PERIPHERAL MANIPULATOR

493

The interstitial manipulator is similar to the peripheral equipment but is simpler in that the camera assembly cannot be deployed from the end of the chain of links which can itself extend only up to 4m from the bottom of the standpipe in use. Restrictions of space prevent the reeling up of the chain so that it must be raised and lowered complete with its drives, by the charge face crane. It is intended primarily for use at the four interstitial ISI standpipes but can also be used at the peripheral ISI and fuel channel standpipes if required.

Hoist Unit

The hoist unit, which is based upon that provided at Hartlepool and Heysham 1 for viewing the core restraint system, consists essentially of a winch capable of paying out the camera assembly on its service hose. It is capable of lowering the camera down the full length of a fuel or control rod channel or to the bottom of the debris vault. It has also been adapted to recover graphite sample canisters from the reactor core.

16.3 Secondary Shutdown Room and Circulator Penetration Routes

72 ISI routes are provided in the boiler annulus and they were chosen to maximise the coverage of the pressure vessel heat shield, the external surface of the gas baffle cylinder and its heat shield and the boiler casings. Limited viewing of boiler platens also is possible. 16 of these routes can be used only by the articulated version of the camera assembly for which special guidance tubes are installed. Tundishes at each platform level provide uninterrupted vertical access down each route.

Within the gas baffle, 24 vertical routes are provided which permit viewing in the annulus between the core periphery and the core restraint tank. The top of the restraint tank is fitted with a hexagonal rail which, by providing location and guidance, enables the peripheral manipulator to be used for viewing within the 150mm wide annulus between the core restraint tank and the inner surface of the gas baffle cylinder.

16.2.3 *Routes in the Boiler Annulus and within the Gas Baffle*

12 ISI standpipes are provided in the bottom slab of the reactor pressure vessel between the secondary shutdown room and the lower end of the boiler annulus.

The lower boiler annulus manipulator also employs links which are stored on a drum and driven up the standpipes into the boiler annulus. The line of sight of the camera is controlled by the combined wrist and knuckle action of a drive head fitted to the end link.

Viewing within the space below the diagrid requires that a circulator be removed from its penetration and be replaced by a special access cylinder which isolates the gas circuit from the penetration for viewing purposes. A boom can then be driven through the access cylinder to position the camera assembly below the diagrid. As with the lower boiler annulus manipulator, the line of sight of the camera is controlled by the wrist and knuckle action of a drive head fitted, in this case, to the end of the boom.

16.4 Storage, Testing, Training and Maintenance Facilities

The ISI equipment is tested and operators are provided with realistic training, using fully representative models of the reactor components, in facilities located on the northern side of the charge hall. The equipment, including the ISI standpipe plug units, is also maintained and stored in this area.

As stated earlier, improved arrangements have been made for man access to the reactor and training of a different character is provided in the same facilities to prepare personnel for entry through the man access penetrations. Noise generators and air heaters are used to give a partial simulation of the environmental conditions which are encountered during such operations.

17. RADIATION AND CONTAMINATION CONTROL

Radiological control measures are incorporated into the station design to ensure that the station may be operated in accordance with the specified radiological criteria. These criteria require that all radiation exposures of station staff are as low as reasonably practicable and that all occupational radiation doses and dose rates conform to design target values.

The control measures include overall and detailed plant layout, personnel and plant access control, provision of shielding, suitable ventilation, attention to surface finishes to facilitate decontamination; also a comprehensive system of health physics control, including installed and portable instrumentation and monitoring facilities.

The station layout is such that adequate provision is made for the personnel access required for operation, inspection and maintenance of the plant. To facilitate radiological control on-site, a controlled area is identified, outside which there will be no significant radiological hazard arising from normal reactor operations.

Inside the controlled area are the two reactor buildings, charge hall, fuel handling building (containing the central shielded facilities block), reactor services annexe and certain ancillaries such as the active waste building.

Within the radiologically-controlled area, general layout and access control design principles have been followed for reducing radiation levels in inaccessible areas, reducing personnel access times required within the active areas and containing and controlling contamination. These principles include:

(1) The grouping together of active plant and equipment and their segregation from less or non-active plant by distance, containment barriers or shielding.

(2) The provision of personnel access routes minimising the time spent in transit through active areas.

(3) The use, wherever appropriate and practicable, of remotely-operated, high reliability, low maintenance, easily removable plant and associated services.

(4) The allowance of appropriate space (for radiological working) around plant unavoidably located within an active area and requiring local operation, inspection or maintenance.

(5) The provision of personnel change areas and plant hand-over facilities at the boundaries of potentially contaminated zones.

For areas subject to variable or intermittent high radiation or contamination levels, administration control is employed in conjunction with door locking and appropriate interlock schemes.

The contamination control design provisions, in terms of physical layout, segregation, ventilation, drainage, sub-change facilities, surface finishes suitable for decontamination and health physics instrumentation, are made in accordance with the relevant radiological zone classification for each plant area. Where the classification of an area is likely to change during the operation of the station, the provisions made are those necessary for the higher classification.

All personnel entering the radiologically-controlled area pass first through the main change room area. From the main change area, personnel access routes are provided to all regions within the controlled area. A lift and stairs in each reactor building give access to all levels. External staircases on the ends of the reactor buildings also provide access between main floor levels as well as emergency access to ground level. All these access routes are shielded from active plant to give general dose rates of less than 1μSv/h (0.1mrem/h) and are intended to be kept in an uncontaminated condition. No distinction is made between access routes to radiation and contamination areas. Barriers, shielding doors and labyrinth entrances are utilised where necessary at the boundaries of higher radiation areas, and sub-change facilities and plant hand-over bays are provided at the boundaries of potentially contaminated areas. In addition to the main radiologically-controlled routes, for general safety (including radiological safety), emergency exit routes are also provided in all buildings.

The main principle of contamination control is that all sources of contamination are contained as close to the source as practicable, or restricted to within a defined, controlled area. Also the spread of contamination within the potentially-contaminated zone and release or discharge to other areas or the environment is minimised. All potentially-contaminated areas where personnel access is required are ventilated in accordance with these principles, as well as being provided with appropriate standards of fresh air and temperature control, as described in Section 24.

Provision is made in the station design for decontamination of contaminated plant items and areas. To facilitate decontamination, attention is paid to the design of the plant or plant area and particularly to the final surface finishes. The provision of smooth surfaces and avoidance as far as practicable of potential traps for contamination (eg exposed mechanisms, sharp corners or partially covered surfaces) also has the primary benefit of reducing contamination accumulation.

A decontamination centre is provided in the central shielded facilities block in the fuel handling building for equipment such as pumps, valves, pipework, instrumentation, tools and equipment. Provision is also made for a large decontamination facility in the gas circulator workshop for the decontamination of larger plant items.

In potentially-contaminated plant rooms, ceilings and walls are generally finished in chlorinated rubber paint and all doors and skirtings, etc are finished in gloss paint for ease of decontamination. Epoxy-based paints or other more expensive alternatives are only utilised in areas where the radiation levels are such that their use is warranted for radiation damage reasons, or where other environmental conditions (eg presence of oil) favour their use.

18. RADIOACTIVE WASTE MANAGEMENT

Fission products comprise the major proportion of the radioactive wastes produced during nuclear power station operation. They are however retained, for the most part, within the fuel pins and are transferred, after cooling periods, to an off-site irradiated fuel reprocessing plant.

However, if defects develop in fuel pins during use, some fission products may escape into the carbon dioxide coolant or into the irradiated fuel pond and appear, along with various neutron activation products, in the power station wastes. These wastes arise in gaseous, liquid and solid forms and are managed in such a way as to protect both the station staff and the general public in accordance with the objectives laid down by the International Commission on Radiological Protection.

The wastes produced by AGR power stations are similar to those which arise at their Magnox precursors and the decisions made regarding the facilities for Heysham 2 are therefore founded upon some 20 years experience of the management of radioactive wastes arising at gas-cooled nuclear power stations. In addition, account has been taken of possible future developments in the arrangements for the disposal of radioactive waste products in the UK.

18.1 Policies and Principles

Gaseous and liquid wastes are, as a general rule, treated by filtration in order to reduce the activity of the fluid streams to the very low values which are acceptable for disposal by dispersion in the atmosphere or in the sea.

Solid wastes are, where practicable, disposed of at appropriate intervals rather than accumulated indefinitely at the station. Facilities at Heysham 2 for interim storage are capable of accommodating at least one year's accumulation. For wastes of a somewhat higher level of activity, such as filter sludges, the station storage is sufficient for at least five years' accumulation and plant is provided to process them into a form suitable for off-site disposal.

Long term storage is provided at the station for the highly active solid wastes, such as the fuel stringer components, recognising that the task of eventual disposal will be significantly eased by the fall in the level of radioactivity which will occur during such storage.

In the implementation of these policies, the fundamental principle followed is that the radiation exposures of individuals and the collective dose to the general public should be as low as reasonably achievable, taking into account both social and economic factors. Moreover, in view of the long timescale between design and the end of the station's operational life considerable flexibility is built into the waste management facilities in order to be able to cope readily with possible future increases in the quantity or activity of the wastes and with possible future changes in permissible discharges and off-site disposal policies and practices. At all times, discharges will be maintained as low as reasonably practicable.

The experience at Magnox power stations has confirmed the self-evident advantages of confining waste management activities, so far as practicable, within a single building and, so far as CEGB stations are concerned, this principle has been implemented for the first time at Heysham 2.

18.2   Gaseous Wastes

Gaseous effluents are basically either carbon dioxide released during various reactor and fuelling operations or ventilation air from contaminated or potentially-contaminated areas.

18.2.1   *Carbon Dioxide Effluents*

Fig 5.23 shows a typical schematic of the treatment methods adopted to process carbon dioxide released from the reactor. Reactor coolant is treated and then released to atmosphere whenever a reactor or the fuelling machine is depressurised. In addition, a proportion of the carbon dioxide which is circulated to cool the irradiated fuel disposal cells, is drawn off to maintain a sub-atmospheric pressure within the cells. Moreover, it is purged completely whenever man access is required.

Carbon dioxide coolant which has been exposed to a defective fuel pin may contain volatile radioactive iodine compounds and the central feature of the treatment of these effluents is the removal of such compounds by passing the gas through beds of activated charcoal. The beds, which remove more than 99.5% of the iodine compounds present, are maintained manually and it is important to ensure that radioactive particulate matter is reduced to negligible proportions before the gas stream enters the charcoal. For this purpose, resin-bonded fibre-glass filter elements are used which have a removal efficiency of 99.9% for particles down to 0.1 microns.

An initial filtration, using sintered stainless steel filter elements, takes place before the gas enters the charcoal bed pre-filters and post-filtration is also provided to remove any dust carried over from the charcoal beds themselves. The treated effluent is released to atmosphere at a height of 70m above ground level.

Facilities are incorporated which allow in-situ testing of the filters and the charcoal beds, for functional performance, at regular intervals. In addition, sampling equipment is provided to enable estimates to be made of the quantity of radioactive particulate and volatile material discharged to atmosphere at all times.

As with any large pressurised system, some leakage of gas takes place through glands and seals and constitutes a contaminant of the air used to ventilate the pertinent reactor areas.

18.2.2   *Ventilation Air*

The ventilation air from contaminated or potentially contaminated areas is passed through high efficiency filters, which have a removal efficiency of 99.9% for particles down to 0.3 microns, before being discharged to atmosphere at high level. Provision is made for regular in-situ testing. The spent filters constitute low activity solid waste.

During periods when man access is required, the reactor vessel is continuously purged with air. The extracted air, in addition to being passed through high-efficiency filters, is also passed through a charcoal bed to remove any iodine compounds which may be present.

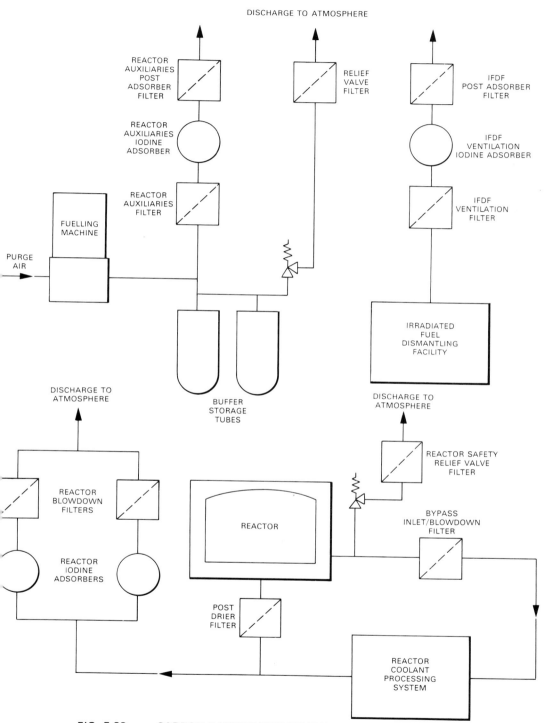

FIG. 5.23   CARBON DIOXIDE EFFLUENT TREATMENT SYSTEM

18.3 Liquid Wastes

The principal sources of liquid effluents are:
(1) The reactor coolant driers.
(2) The irradiated fuel pond.
(3) Filter sludge and spent ion exchange resin storage tanks.
(4) Drainage from change rooms, decontamination and other reactor areas.

18.3.1 *Treatment of Liquid Wastes*

The facilities for liquid waste management consist of active drainage systems, tanks, pumps, filters, ion exchange units and pipework through which treated effluent is discharged into the outlet culverts of the main cooling water system. Filtration and ion exchange are carried out as batch processes and the effluents are monitored before and after treatment. The possibility of leakage of radioactivity from these facilities is minimised by the appropriate choice of structural materials, the employment of double containment techniques, and the concentration of waste treatment plant in a single waste management building. Facilities for the detection of leakage are incorporated where appropriate as, for example, inside concrete cells which contain tanks of highly active material.

The techniques of treatment are based upon those proven in operation at the Magnox stations. The filters are of the precoat disc type which are easily maintained since the discs can be rotated at speed and the mixture of sludge and spent precoat removed by centrifugal action. Spent resin is removed hydraulically from the ion exchange units. Both forms of waste are stored under water in tanks to await solidification.

18.4 Solid Wastes

Solid wastes consist, primarily, of:
(1) Items irradiated in the reactor core which are not re-usable, such as fuel stringer components and flux measuring instruments.
(2) Spent ion exchange resins, filter sludges and spent precoat from the liquid effluent treatment plant.
(3) The miscellaneous collection of combustible and non-combustible items such as filters, protective clothing and laboratory equipment.

18.4.1 *Treatment of Solid Wastes*

Irradiated Components
Highly active items are stored in the central fuel handling building in the debris vault. It is not intended that this material will be disposed of until such time as the station is decommissioned, by which time the level of activity will have diminished significantly. For the first time at a CEGB station, the debris vault at Heysham 2 has been specifically designed to facilitate the eventual retrieval of its contents for disposal off-site.

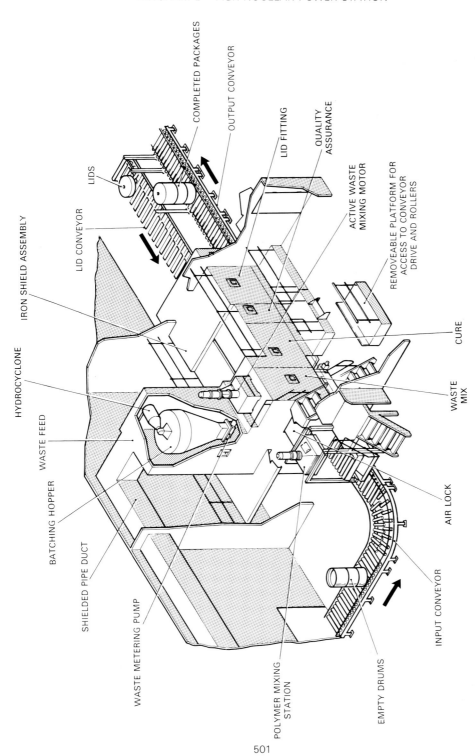

COMPLETED PACKAGES

OUTPUT CONVEYOR

LIDS

LID CONVEYOR

IRON SHIELD ASSEMBLY

HYDROCYCLONE

WASTE FEED

BATCHING HOPPER

SHIELDED PIPE DUCT

WASTE METERING PUMP

POLYMER MIXING STATION

EMPTY DRUMS

INPUT CONVEYOR

AIR LOCK

WASTE MIX

CURE

REMOVEABLE PLATFORM FOR ACCESS TO CONVEYOR DRIVE AND ROLLERS

ACTIVE WASTE MIXING MOTOR

QUALITY ASSURANCE

LID FITTING

FIG. 5.24   RESIN CONDITIONING PLANT

Wastes from the Liquid Effluent Treatment Plant

The sludges, spent precoat and spent ion exchange resins from the liquid effluent treatment plant are stored initially under water in stainless steel tanks within concrete cells. Each stream has its own tank which is sized to accommodate arisings over a minimum of five years.

Subsequent processing is carried out in plant similar to that previously developed by the CEGB and commissioned at Trawsfynydd Magnox power station during 1984. Fig 5.24 shows the principal features of the Trawsfynydd plant. The Heysham 2 plant is capable of solidifying these waste materials in a plastic matrix or in a cement mix according to the requirements of the disposal route. The matrix or mix is encapsulated in a drum which is made-up to provide shielding appropriate to the activity of the batch of waste being processed. A store is provided which can accommodate the drums for up to a year before they are transported off-site for disposal. The whole of the solidification process, including the transfer of material, is remotely controlled.

Miscellaneous Waste

The miscellaneous waste is generally of very low activity and is collected manually from the different locations within the power station in bags or drums and transferred to the waste management building where facilities are provided for sorting and monitoring. Smoke detection and automatic fire fighting equipment is installed in areas in which combustible material is accumulated.

Combustible waste is burnt in simple incinerators where the radioactive contaminants are retained in the low volume of ash. The ash, together with all the other non-combustible waste, is compacted hydraulically into steel drums which are then accumulated in a drum store where there is capacity for one year's arisings. Normally, however, such waste is removed from the station every six weeks or so.

19. POWER PLANT

The use, in the AGR, of uranium oxide fuel in stainless steel cladding permits the adoption of gas coolant temperatures which are capable of producing steam at the conditions of pressure and temperature employed in CEGB coal and oil-fired practice. Much of the power plant provided for Heysham 2 is therefore similar to, and in some respects identical to that installed for the Drax Completion coal-fired station.

Some of the differences between the Drax First Half and the Heysham 2 power plant and equipment arise for such reasons as the obsolescence of particular items and improvements in manufacturing processes. Others arise from the nature of the AGR as a steam supply system. The modular nature of the boiler system and its division into quadrants affects the arrangement of the main steam circuits. The economic optimisation of the plant, within the constraints imposed by material behaviour, leads to the omission of high pressure (HP) feed heaters from the boiler feed train. Thus, the final feed temperature at Heysham 2 is limited to 156°C. This compares with 254.4°C at Drax where HP steam is bled for the final stages of feed heating from the boiler feed pump turbine and the cold reheat pipework.

19.1 Main Steam Circuits (Fig 5.25)

The reheat steam cycle employed at Heysham 2 is similar to that used at the CEGB's 500MW and 660MW coal and oil-fired stations and at the earlier AGR stations. At Heysham 2, steam is superheated to 541°C at 166 bar at the boiler outlets and admitted to the HP turbine steam chests at 538°C and 159.6 bar. After expansion in the HP turbine cylinder, it is, for the most past, returned to the boilers for reheat to 539°C at 40.7 bar. Readmission to the turbine, at the intermediate pressure (IP) steam chests, is at 538°C and 39 bar. That portion of the steam which is exhausted from the HP cylinder and not returned to the boilers for reheating, provides the normal steam supply for the boiler feed pump turbine.

A boiler startup system bypasses the steam mains and the turbines and permits the discharge of up to 30% steam flow direct to the main condensing plant.

19.1.1 *HP Steam Circuits*

The HP steam pipework is fabricated in 0.5% Cr 0.5% Mo 0.25% V steel.

The smallbore superheater tailpipes from each boiler module are taken in groups through three reactor pressure vessel penetrations. Outside the pressure vessel, the tailpipes, from the total of nine penetrations in each quadrant, are connected to a quadrant header and the four quadrant headers of each unit are then connected by 267mm internal diameter 54mm thick pipes to the unit HP balancing header.

A normally-open manually-operated stop valve and a remotely-operated stop valve are located in series in the outlet from each quadrant header. Branches for the pressure relief valves are situated between the header and the manually-operated valve.

Since the unit balancing header penetrates the barrier wall which separates the north west quadrant from the north east quadrant, remotely-operated valves are located in the header on either side of the wall in order to be able to isolate the affected half unit in the event of a steam main fracture. These valves are also closed immediately after a reactor trip as explained in Section 14.3.2.

From the balancing header, four steam mains are routed as two pairs, one pair from each half of the balancing header, to the HP turbine steam chests. The temperature of the HP steam is controlled by adjustment of the regulating valves in the boiler feed system, as described in Section 19.3.3.

19.1.2 *Steam Reheat Circuits*

The cold reheat pipework is fabricated in carbon steel whilst that for the reheated steam is made of the same ferritic alloy as the HP pipework.

The arrangements for returning the partially-expanded steam from the HP turbine exhaust to the boiler and conveying the reheated steam from the boiler to the IP turbine steam chests are similar to those described for the main HP steam circuits. They differ in that there are three inlet and three outlet reheater penetrations for each quadrant and there are no isolating valves in either the cold reheat or the hot reheat unit balancing header.

A reheater bypass connects the cold and hot reheat headers of each quadrant to provide the means of attemperating the reheat outlet steam temperature. One pipe of each pair of cold reheat steam mains are interconnected and the normal supply of steam to the boiler feed pump turbine is taken from the interconnection.

19.2 Turbine Generators

The turbine generators are 660MW 5-cylinder single-shaft machines. Both the machine and its ancillary equipment are similar and in some respects identical to those used for the Drax Completion. The major differences arise as a consequence of the choice of LP cylinders with pannier condensers for the first three machines at Drax which were introduced into service between July 1974 and March 1976. Whilst the advantage of uniformity resulted in the adoption of pannier condensers for Drax Completion, new LP modules with underslung condensers, considered to have greater export potential and giving better access for turbine maintenance, had been developed and the first of this type were introduced into CEGB service in March 1982 as a component of the 4-cylinder machine at Littlebrook D. The designs of IP cylinders were also, of necessity, altered as a consequence of changes to the rotor systems dictated by the new LP modules.

19.2.1 *Steam Chests*

Both the HP and the IP steam chests differ in detail from those installed for Drax Completion. In the case of the HP chests, new non-destructive examination requirements result in some lengthening of the welded connections between sections. In the case of the IP chests, a dimensional change has been made to achieve compatibility between the type of steel foundation used for the turbine generators at Heysham 2 and the pipes which connect the IP steam chests to the IP turbine cylinder.

19.2.2 *HP, IP and LP Turbines*

The single-flow double-casing HP turbine is identical with that used for Drax Completion except for the blade heights which are sized to suit the Heysham 2 steam flow.

The IP section consists of a single double-flow machine with a double casing and is based upon the IP turbine used successfully by the South of Scotland Electricity Board at their Hunterston B AGR station. The exhausts from each end of the cylinder are modified as a consequence of the adoption of underslung condensers at Heysham 2 and four, instead of two, steam inlet connections are used. This reduces the diameter of the inlets and facilitates the keying of the inner casing.

The LP section consists of three double-flow double-casing LP turbines. They constitute a particular version of the LP modular concept which has already been used with success elsewhere. The outer casing of each module is supported at its ends and incorporates the journal bearings. The proportions of the casing are such that short connecting shafts are required between the adjacent LP rotors. They are, however, light in weight and for alignment purposes are treated as extensions of the LP 2 and LP 3 rotors. The inner casing is of fabricated design and all the diaphragms are kinematically supported. The LP rotors are only partially bored in order to provide an increased margin of safety against brittle fracture.

Some detailed improvements to this design of LP module are being introduced in the machines for Heysham 2; in particular, a Mk 2 design of the 914mm long last row blade is being used. Interblade ties provide tip restraint in the Mk 2 design and eliminate the maintenance task which arises from the limited life of the arch cover bands used in the Mk 1 design.

19.2.3 *Turbine Supervisory and Protection Equipments*

The turbine supervisory equipment, which monitors and records the conditions under which the rotors and bearings are operating at any time, is identical with that installed for Drax Completion.

The turbine protection equipment is essentially the same as that provided for Drax Completion but has added features related to the nuclear steam supply system; in particular, protection is provided against the possibility of water carry-over from the once-through evaporator/superheater sections of the Heysham 2 boiler system.

19.2.4 *Electro-Hydraulic Governing System*

Although the control scheme for Heysham 2 is different from that for Drax, the electronic governors, valve operating relays, trip system and on-load testing arrangements are identical with those provided for Drax Completion.

The storage tank pumps, coolers, filters and conditioning equipment for the fire resistant hydraulic fluid form a factory assembled packaged unit. This unit, which also serves the governing system of the boiler feed pump turbine, is similar to equipment already supplied outside the UK.

19.2.5 *Condensing Plant (Fig 5.26)*

Each LP turbine module has its own underslung condenser which is spring supported but rigidly attached to the turbine casing. The shell of each condenser contains two single pass parallel flow transverse water circuits having double tube plates, titanium tube nests and separate water boxes. Titanium tubing has virtually eliminated the problem of tube leaks but access to the water boxes is arranged so that tube inspection can take place whilst the adjacent water box remains in service.

Three 50% duty air extraction pumps are provided to maintain the vacuum in the condenser steam space under normal conditions of operation.

19.2.6 *Generator*

The Heysham 2 generators are electrically and mechanically interchangeable with those provided for Drax Completion and the differences between them are of a minor nature.

The rotor excitation system is similar to that provided for Drax Completion.

19.2.7 *Turbine Generator Foundation and Turbine Supports*

The steel foundation for the turbine generator is of the single-level 'table top' type.

The arrangements for axial location of the rotor assembly and for supporting and locating the HP turbine and the steam chest end of the IP turbine are the same as for Drax Completion. At the generator end of the IP turbine, however, the pedestal contains only the generator end IP turbine bearing since the LP turbine bearings are housed in the LP turbine casings. This pedestal, which is bolted to the foundation, supports and locates the IP turbine casing both axially and transversely and also locates the LP 1 turbine casing transversely.

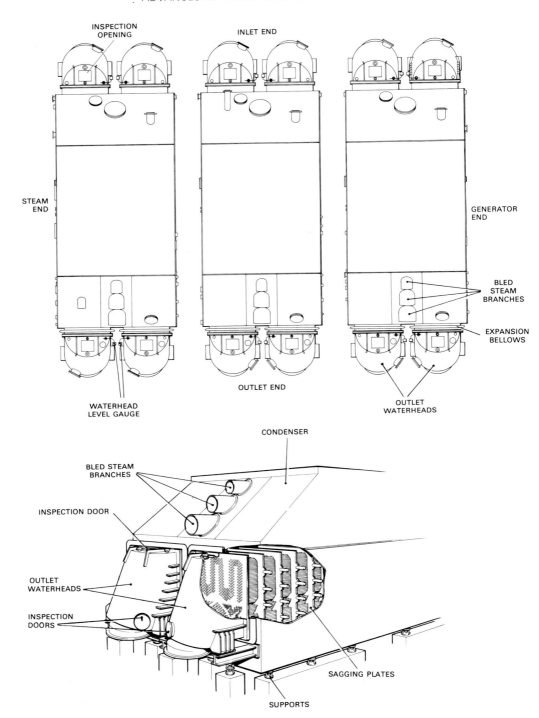

FIG. 5.26   CONDENSER ARRANGEMENT AND ASSEMBLY

The steam chest end of the LP 1 casing, like those of the LP 2 and LP 3 turbines, is supported and located axially on a key block which is dowelled and bolted to the steel foundation. Vertical keys on the centre line locate the LP 2 and LP 3 casings transversely. The generator end of each LP turbine is free to slide axially and is supported on plates which are dowelled and bolted to the foundation.

All three bearing pedestals may require adjustment during overhaul and are supported on wedge units to facilitate rapid re-alignment.

19.3 Main Boiler Feed System

As at Drax Completion, there are two 100% duty vertical-spindle caisson type condensate extraction pumps per unit and the condensate is first passed through the turbine gland steam condenser.

At Heysham 2, the whole of the condensate is then passed through a polishing plant in order to maintain continuous control of the quality of the water entering the boiler. The polishing plant, which processes the feed of each generating unit as a separate stream, consists of mixed bed units of ion exchange resins of sufficient capacity to allow regeneration to take place without impairing the performance of the system.

From the polishing plant, the condensate is passed through a triple cascade of horizontal LP feed heaters to a high level deaeration system and thence to the boiler feed pump suction main.

Each generating unit is provided with three reserve feedwater tanks. The take-off points for normal purposes are situated about half way up the tanks in order to ensure that an adequate reserve is always available for the emergency boiler feed system.

19.3.1 *Main Boiler Feed Pump*

The main, 100% duty boiler feed pump provided for each generating unit is steam turbine driven but, since it is not economic to have HP feedheating at AGR stations, the Heysham 2 boiler feed pump turbine is a simple back-pressure machine without extraction points. It takes its normal supply of steam from the cold reheat steam mains and normally exhausts to the steam pipes which connect the IP turbine exhausts to the LP turbine inlets on the 660MW machine. To augment the feed pump power when the 660MW machine is operating at low load, the boiler feed pump turbine is fitted with a single impulse stage which takes HP steam direct from the boiler from one of the pairs of HP steam mains.

The governor valves for the boiler feed pump turbine are mounted directly on the turbine casing. The electro-hydraulic governing system and the electronic overspeed trip system are similar to those employed for Drax Completion.

The HP pump set is of the high speed, horizontal-spindle, barrel casing design. It is of rugged design and construction and is capable of dry running. It is of cartridge construction to meet the requirements for rapid replacement of the internals. The suction section has a single-stage double-entry impeller which is driven through a gearbox from the steam inlet end of the boiler feed pump turbine. The pressure section is a 2-stage unit driven directly by the turbine at 6680rev/min.

19.3.2 *Starting/Standby Feed Pumps*

Two 50% duty electrically-driven pump sets are provided for each generating unit for starting and standby purposes. The sets are generally similar to the main pumpset. The suction section is driven directly from one end of an induction motor whilst the pressure section is coupled to the high speed output shaft of a combined fluid coupling and gearbox which is in turn driven from the other end of the induction motor.

The use of these pumps for boiler protection purposes following a reactor trip is described in Section 14.

19.3.3 *Main Feed Circuits (Fig 5.27)*

A non-return valve and an electrically-operated stop valve are positioned in each of the three feed pump discharges, after which they are commoned. A single feed main of niobium-treated carbon steel is run from the turbine house to the top of the pipewell in the mechanical annexe where it is bifurcated to the east and west pairs of quadrants. A further bifurcation takes place on the reactor east-west centre line to carry the feed into each quadrant.

Following the second bifurcation, quadrant feed valves are provided which control the feed flow during normal startup and during the period of post trip boiler protection by the starting/standby feed pumps.

The pipework on the discharge side of the quadrant valves receives the connection to that quadrant from the emergency boiler feed system before dividing into six branches which connect through two penetrations per boiler module, with the 1% Cr 0.5% Mo boiler inlet tail pipes.

For control of the feed flows during normal power operation, each of the six branches has a regulating valve which controls the feed flow to that half of the boiler module to which the branch is connected. The temperature of the superheated steam from each half module is controlled by trimming these valve positions and the speed of the main boiler feed pump is controlled so as to maintain a constant mean pressure drop across the total 24 half module regulating valves employed on each reactor.

19.4 Auxiliary Boilers

The auxiliary boiler system comprises four standard shell boilers, feedwater and fuel systems housed in a single building near the west end of the turbine hall. The fuel oil tanks are located in a bunded area immediately south of the boiler house.

The boilers are of the 'quick start' type and are capable of achieving the full load rating of 3.78kg/s at a discharge temperature and pressure of 206°C and 10 bar respectively, within one hour from a cold condition. They can be individually operated manually or automatically sequenced once having fired.

The main requirement for steam from the auxiliary boilers occurs when both reactors are off-load and the station demands cannot be met from the turbine bled steam source. The main requirements for auxiliary steam are heating and ventilating, deaerator heating, turbine gland sealing, decontamination and laundry facilities, nitrogen evaporation and carbon dioxide evaporation. The latter includes a safety-related duty for the provision of carbon dioxide for fuelling machine cooling during fault conditions and pressure support in the event of some reactor depressurisation faults.

20. CW SYSTEM (FIG 5.28)

The CW system is a once through syphonic system using seawater drawn through static coarse screens and rotating drum screens in sequence from the seaward end of Heysham harbour. Its major duty is that of providing the cooling flows necessary to maintain the vacuum in the condensers of the two 660MW turbines.

Four other seawater systems are associated with the main system. Each reactor has its own, albeit interconnected, reactor seawater system which draws its supplies from the drum screen outlet chambers. Each discharges, through an isolating valve, into the surge shaft of the associated main outlet culvert on the seaward side of its isolating penstock. Each generating unit also has its own auxiliary CW system which provides cooling water for turbine auxiliaries and other services. These two systems, which are not interconnected, draw water from the inlet culverts upstream of the condenser inlet valves and discharge it into the outlet culverts downstream of the condenser outlet valves.

The total station requirement for cooling water is $50m^3/s$ of which $46m^3/s$ is for the turbine condenser duty, $3m^3/s$ for the two auxiliary CW systems and $1m^3/s$ for the two reactor seawater systems.

Equipment is provided, in a building alongside the CW pumphouse, for the production of sodium hypochlorite solution, by the electrolysis of sea water, and for its injection into the cooling water at the drum screen inlets in order to limit the growth of marine organisms in the culverts and the fouling of the condenser and other heat exchanger tubes.

20.1 Civil Engineering Works (Fig 5.29)

The CW pumphouse was constructed within a 73m diameter sheet piled cofferdam some 24.5m deep. As internal excavation progressed, the external overburden pressure on the cofferdam was taken by circular concrete beams cast in-situ. Four such beams were cast and their vertical separation was maintained by means of a system of steel columns and rods. The excavation was kept dry by a network of deep dewatering wells which passed through the sand overburden into the rock below.

The walls of the intake channel to the pumphouse were formed of mass concrete. It was convenient to employ tunnellers, so a layer of concrete was first cast in each cofferdam at ground level. Holes were left in the concrete through which sand or rock below could be mined and a further layer of concrete cast. This process continued until the requisite depth had been attained.

The length of harbour wall which formerly occupied a position at what is now the mouth of the intake was demolished by using explosives. The blasting techniques were carefully controlled to avoid damage to the nearby pumphouse which serves Heysham 1.

Each generating unit has two CW pumps from which a single inlet culvert of 3.66m diameter tunnels and shafts leads to the turbine house. An interconnecting tunnel, complete with isolating gates, connects the individual culverts of the two units.

Tunnelling between the pumphouse and the turbine house was carried out in the Triassic sandstone, by means of a rotary cutter, at invert levels approximately 20m below ground level. The tunnel walls were lined with pre-cast concrete segments bolted together. The spaces between the lining and the rock face were pressure grouted through holes in the segments, the internal pans of which were then in-filled with curved slabs to form a smooth hydraulic surface. Hydraulic bends at the top and bottom of shafts were formed by precast concrete segments set in reinforced concrete cast in-situ.

The two 3.66m diameter outlet culverts, which are not interconnected, have surge shafts which are open to atmosphere and contain motorised isolating penstocks. From the surge shafts, the tunnels pass under the sea wall to vertical marine shafts which rise to beach level on the foreshore.

In order to construct the marine shafts, a barge capable of jacking itself up on the foreshore was used to provide a platform from which 5m diameter steel liner tubes were vibrated through the beach sand to the upper level of the weathered rock. Vibration of the tubes through the weathered rock to penetrate the sound rock-head was aided by chiselling and grabbing out the debris inside the tubes. Both the weathered rock and the sound rock were then pressure grouted to establish a seal at the bottom of the liner tubes after which the tubes were pumped dry and the sinking of the shafts through the rock was completed in a conventional manner.

From the upper ends of the marine shafts, the cooling water is discharged along a 40m wide rocklined outfall channel which runs for some 720m across the foreshore to a deep water channel in Morecambe Bay. Jack-up barges were also used as the platforms from which to drive the two rows of sheet steel piles that form the entraining walls of this outfall channel.

20.2 CW Pumps and Valves

The main CW pumps are situated 17.5m below ground level in a pumphouse structure some 74m long by 25m wide which also accommodate the pumps for the reactor seawater systems.

Each of the main CW pumps is capable of pumping 12.25m^3/s at a generated head of 11m. Nominally this constitutes four 25% duty pumps for the station but the electrical output obtainable from the two turbine generators with fewer than four CW pumps in operation depends upon the seawater temperature. Normally, a minimum of two pumps would be in operation at all times that electricity is being generated. They are single-speed, vertical-spindle, mixed-flow pumps with concrete volutes. They have bearings with oil bath lubrication and face seals to prevent the egress of untreated seawater along the shafts. The face seals are cooled and flushed by seawater which is drawn from the pump outlets and first passed through cyclone filters to remove suspended particulate matter. Air seals are provided for startup and standby purposes.

The main CW pumps are driven by 11kV motors through epicyclic speed-reduction gearboxes. Forced lubrication is provided for the gearboxes and the drive motor bearings by means of shaft-driven oil pumps incorporated in the gearboxes. Electrically-driven oil pumps are provided for startup and standby purposes.

The main CW pump discharge valves are of the butterfly type, 2.4m in diameter and hydraulically-actuated. Each valve has its own hydraulic actuation system with a motor-driven pump for normal operation and a manual pump for use in the event of failure of the electric pump. If the hydraulic system fails, the valve will remain open as long as its associated CW pump is in service and be closed by the action of a weighted

HEYSHAM 2 – AGR NUCLEAR POWER STATION

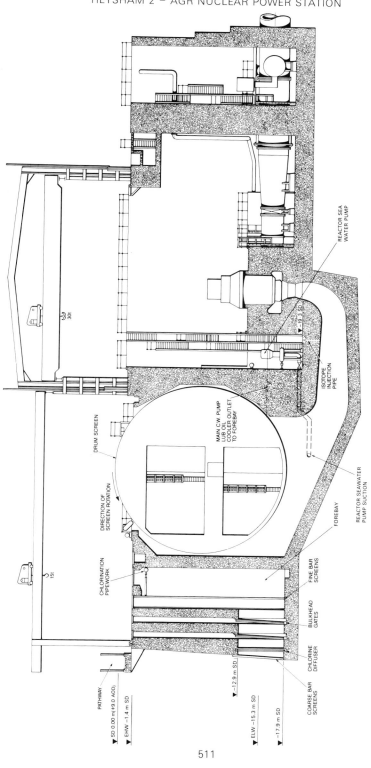

DRUM SCREEN

DIRECTION OF
SCREEN ROTATION

CHLORINATION
PIPEWORK

MAIN CW PUMP
LUB OIL
COOLER OUTLET
TO FOREBAY

REACTOR SEA
WATER PUMP

ISOTOPE
INJECTION
PIPE

REACTOR SEAWATER
PUMP SUCTION

FOREBAY

FINE BAR
SCREENS

BULKHEAD
GATES

CHLORINE
DIFFUSER

COARSE BAR
SCREENS

PATHWAY

▼ SD 0.00 m (+9.0 AOD)

▼ EHW −1.4 m SD

▼ −12.9 m SD

▼ ELW −15.3 m SD

▼ −17.9 m SD

FIG. 5.29   CW PUMPHOUSE ELEVATION

arm when the pump is shutdown. For a normal closure under hydraulic actuation, the valve disc will move rapidly to 50% closure and then complete the operation more slowly; the actual closure rates, however, are adjustable so that they can be set during the commissioning surge tests.

The startup and shutdown of CW plant is normally initiated in the central control room and effected and monitored by sequence control equipment but local manual control is also available. Condenser protection is afforded by air admission valves on the condenser waterboxes which open, after a CW pump trip, when the pressure in the waterboxes falls below 0.3 bar.

20.3 Electro-Chlorination Plant

The electro-chlorination plant serves both Heysham 1 and Heysham 2 and has an output capability of 374kg/h chlorine equivalent. This is sufficient to satisfy the seawater demand and leave a residual chlorine content of 0.2mg/litre at the condenser inlets. The power consumption of the plant decreases as the temperature of the seawater feed increases and for this reason, the feed is normally drawn from a Heysham 1 cooling water discharge point.

21. STATION ELECTRICAL SYSTEM (FIG 5.30)

The station electrical system provides the means of exporting power generated at Heysham 2 to the grid system and of controlling and distributing the power supplies needed for the operation of the station plant and equipment.

The station electrical system operates normally as a pair of independent systems. Each generating unit is provided with its own power system and the common station equipment is shared between the two. Normally-open interconnectors provide the means of ensuring continuity when a normal supply route is taken out of service for any reason.

Under normal circumstances and when the unit is in service, the power required by the plant and equipment connected to the power system associated with that unit is derived partly at 23.5kV directly from the main generator itself and partly from the 132kV busbars of the Heysham grid substation. If the unit is shutdown, the power supplies required for the reactor post trip equipment and the connected common station services are derived wholly, in normal circumstances, from the grid, partly from the 132kV busbars and partly from the 400kV busbars of the Heysham substation.

If the grid connection is lost, diesel generators are started up to provide power supplies to those power station auxiliaries whose operation can be temporarily interrupted. Those items of equipment which must be assured of a supply at all times, are connected to the uninterruptible power supply boards which are backed by batteries and, when necessary, equipped with inverters.

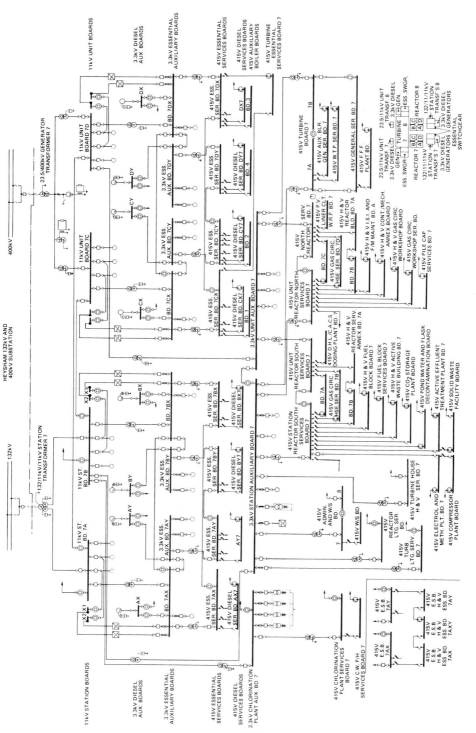

FIG. 5.30   ELECTRICAL POWER SYSTEM UNIT 7

21.1 Generator Main Connections

Each 660MW generator is connected through a 130kA 23.5kV generator switch and an 800MVA 23.5kV/400kV generator transformer to the 400kV busbars of the Heysham grid substation.

The generator switch provides the means of isolating the generator from the generator transformer and the station electrical system so that power supplies for the station can also be drawn from the 400kV busbars when the unit is shutdown. It opens automatically for all reactor trips and turbine generator faults other than short-circuits in the generator protection zone. If the latter occur, the 400kV circuit breaker on the grid side of the generator transformer opens to prevent a fault infeed from the grid.

21.2 Unit Electrical Power System

The electrical power system for each generator unit is based upon several voltage levels to which the various auxiliaries are connected according to their individual ratings. The highest level, to which the supplies from the grid substation and the main generator are converted by stepdown transformers, is 11kV.

21.2.1 11kV Supplies

The division of the boiler annulus into four quadrants and the provision of independent trains of post trip heat removal equipment for each quadrant led to a corresponding division of the 11kV busbars into four sections. Two of these are termed 'station' boards and the other two are termed 'unit' boards.

Those associated with the A and B quadrants are supplied from 132kV busbars of the Heysham substation through a 3-winding 90MVA/60MVA/60MVA 132kV/11kV/11kV 'station' transformer and constitute the 11kV 'station' boards. Those associated with the C and D quadrants are supplied from the main generator through individual 60MVA 23.5kV/11kV 'unit' transformers and constitute the 11kV 'unit' boards.

Although the traditional terms of 'station board' and 'unit board' are used here, they do not serve as adequate descriptions of the functional arrangements at Heysham 2 and should be regarded solely as a convenient means of distinguishing between the 11kV switchboards according to the source from which their supplies are normally derived.

Normally-open interconnections are provided between the A station board and the D unit board and between the B station board and C unit board. With a unit transformer rating of 60MVA, this permits full unit output to be maintained if the station transformer is out of service. In addition, each of the two station boards is interconnected, also on a normally-open basis, with the corresponding station board of the other generating unit.

The auxiliaries connected at 11kV are the standby feedpumps, the main CW pumps, the gas circulators and the test rig in the gas circulator workshop. The two standby feed pumps and the two CW pumps are allocated on the basis of one to each 11kV board.

The other circuits from the 11kV boards supply the 3.3kV essential auxiliaries boards and various 3.3kV boards which supply non-essential auxiliaries and 415V boards.

Gas Circulator Supplies

The two gas circulators of one quadrant are connected to the 11kV board associated with that quadrant. Each circulator circuit breaker has a double cable-box, one cable from which connects with the circulator whilst the other can receive a variable frequency supply through another circuit breaker. The two circuit breakers are interlocked so that the breaker on the 50Hz 11kV board must be opened before the breaker in the variable frequency circuit can be closed and vice versa. The variable frequency supply is derived from a solid state frequency converter which is supplied through a step-up transformer from the associated 3.3kV essential auxiliaries board.

21.2.2 3.3kV *Essential Supplies*

Each 11kV board feeds a 3.3kV essential auxiliaries board which is sectioned so that it can operate as two separate boards with the X trains for the associated quadrant connected to one section and the Y trains for that quadrant connected to the other. Normally the interconnector between the two sections is closed and the single 11kV/3.3kV transformer circuit feeds both sections. If, however, the 11kV supplies are lost, the interconnector is opened and each section is then supplied independently by separate diesel generators.

Normally-open interconnectors are provided between the A and B quadrant boards and between those for the C and D quadrants but no interconnection is possible between the essential and the non-essential 3.3kV boards.

The essential auxiliaries connected at 3.3kV are the decay-heat boiler feed pumps, the emergency feed pumps and the reactor seawater pumps. One of each category is connected to each 3.3kV essential auxiliaries board and is allocated to one or other of the sections according as to whether it is part of the X system or part of the Y system.

Each section of each 3.3kV essential auxiliaries board also supplies a 415V essential services board, allocated to the X or the Y system according to the section of the 3.3kV board from which it is fed.

21.2.3 415V *Essential Supplies*

At this voltage level, the boards allocated to the X and Y systems of each quadrant are separate and are supplied independently through 3.3kV/415V transformer circuits from the corresponding section of the associated 3.3kV board. Normally-open interconnectors are provided between the A and B quadrant boards and between those for the C and D quadrants but no interconnection is possible between the essential and the non-essential 415V boards.

In general, motors rated at 150kW or less are connected at 415V. The other supplies afforded by the 415V essential services boards are for the battery-backed uninterruptible power systems.

21.2.4 *Uninterruptible Power Supplies*

There is much equipment, of which instrumentation and circulator lubricating oil pumps are two examples, which must be assured of electrical supplies at all times. Various voltages are required; for example, computers require 415V ac, the reactor guardlines require 110V ac, emergency lighting requires 250V dc, switchgear closing solenoids require 220V dc and switchgear tripping requires 110V dc.

The uninterruptible power supplies (UPS) required by the plant and equipment in the X and Y trains of any quadrant are supplied from the 415V essential services boards associated with that quadrant. Each X system 415V essential services board supplies three separate UPS boards, each through its own rectifier and backed by its own lead-acid battery, to provide switchgear closing supplies at 220V dc, switchgear tripping supplies at 110V dc and, through an inverter, essential drives at 415V ac. The 110V ac instrument supplies are obtained through a 415V/110V transformer circuit fed from the 415V ac UPS board. Each Y system 415V essential services board provides 110V dc and 110V ac supplies through a common rectifier circuit and backed by a common lead acid battery.

For other equipment requiring uninterruptible supplies, separate battery-backed station and unit supplies are provided at 220V and 250V dc and at 415V and 110V ac. Although not part of the X and Y systems, the supplies are derived from the same 415V essential services boards.

Further diversity is introduced into the reactor protection system by arranging that the 110V ac supplies for the main guard lines are derived from X system 415V ac UPS boards whilst those for the diverse guard lines are taken from Y system 110V ac UPS boards.

The UPS incorporate various standby arrangements; for example, the main and diverse guardlines can also be supplied, respectively, from unit and station 415V ac UPS boards.

21.3 Earthing

The 23.5kV connections are normally earthed at the generator neutral point through a transformer/resistor arrangement which restricts the earth current to 10A in order to limit damage to the generator. All 23.5kV transformer windings are delta connected and to cover the situation when the generator switch is open, the 23.5kV connections are provided with a star-connected earthing transformer and a high resistance loading arrangement similar to that provided in the generator neutral.

The 11kV and 3.3kV systems are earthed at the neutral point of the unit, station and auxiliary transformers through resistors which limit the earth current to 1000A. The 415V system is solidly earthed.

21.4 Cabling

In addition to the wiring for such domestic facilities as the lighting and the public address system, there are about 2000km of cable installed at Heysham 2 in some 35,000 individual cable runs. The degree of security appropriate to individual cables is achieved both by appropriate choice of cable type and by the physical arrangement of the cableways within the station and the physical arrangement of different groups of cables relative to each other along any particular cableway.

The cableways which serve an individual reactor quadrant are segregated by fire barriers from those serving other quadrants. For the most part, this is achieved as a feature of the detailed design of the civil engineering structures but where this was not practicable, suitable barriers have been erected to maintain the integrity of the route.

In the area below the central control room, for example, special prefabricated 2.5 metre square fire-barrier tunnels are installed in order to provide the eight segregated cableways required.

Those cables which are not embraced within the quadrant concept are routed, where appropriate, through suitable quadrant cableways but are physically separated from the quadrant cabling. Along each segregated quadrant cableway, the X train cables are separated from those of the Y train by being placed on opposite sides of the tunnel. The different classes of cabling within each system are separated vertically with the power cables at the top, the control and instrumentation cables at the bottom and those associated with non-essential auxiliaries placed in-between.

The effects of a fire within any segregated cableway would moreover be restricted since specially developed PVC cables are employed which have been designed to minimise the spread of fire. In certain instances, cables with a short-time fire-proof rating are used.

Power cables are chosen according to the rating of the load supplied, the allowable voltage drop and the required short-circuit capability. For control and instrumentation circuits, twisted-pair and multicore cables are used. The needs of Heysham 2 are met by multi-core cables having up to 37 cores and by twisted-pair cables in the range from 2-pairs to 200-pairs.

21.4.1 *Arrangement of Control and Instrumentation Cabling*

The cabling for control and instrumentation purposes involves approximately 1.5 million separate cores. For the most part, these are organised as a matrix made up of cable runs and marshalling frames on which jumpers are used to transfer individual circuits from one cable to another. The marshalling frames are situated at appropriate points throughout the station. Below the central control room, for example, there are eight marshalling rooms, one for each of four quadrants of two reactors. As already stated, cables which are not embraced within the quadrant concept may be routed through suitable quadrant cableways and each of these marshalling rooms contains three frames which accommodate separately the cores for the X system, the Y system and the non-essential auxiliaries. Cores from individual items of equipment enter the matrix at appropriately located marshalling frames.

In certain cases, the added security of an uninterrupted solid connection is considered to be warranted. For this reason, multicore cables are employed to provide direct connections between items of equipment for such functions as electrical system protection and the automatic sequencing of reactor post-trip heat removal systems.

22. STATION CONTROL AND INSTRUMENTATION

The provisions made at Heysham 2 for control and instrumentation are based upon the centralised control and monitoring philosophy adopted for earlier AGR stations. In normal conditions, each generating unit can be operated by a single operator in the central control room with a minimum of staff required at positions local to the plant.

As a result of operating experience at Hinkley Point B, the extent of the instrumentation provided for monitoring the operating conditions within the reactor vault has been greatly increased. At Heysham 2, there are, for example, three thermocouples per fuel channel, about 1,000 on the boilers and another 1,000 or so on the reactor pressure vessel. The developments in post-trip heat removal further increase the amount of control equipment and instrumentation and to assist the operator in his task of supervising the plant, extensive use is made of computer-based control, display and monitoring systems.

Modulating control facilities are provided to allow the unit performance to be optimised and to keep plant parameters within working limits without the need for continuous operator supervision.

Sequence control is provided to assist the operator with the startup and shutdown of some plant systems.

Automatic protection systems of high reliability are provided to terminate all fault sequences for which prompt action is required. Automatic sequencing equipment controls the post-trip heat removal systems without the need for any operator action within the first 30min after the trip.

22.1 Control Rooms (Fig 5.31)

The design of the central control room (CCR) is such that the station can be operated from the CCR during startup, normal power operation and shutdown by one operator per unit and one control room supervisor.

As far as possible, all unit controls and indications relating to startup, normal power operation, shutdown and post-trip operation are mounted on the unit operators' desks. Operational philosophy and space requirements have dictated, however, that some controls and indications be mounted on vertical panels. All the other data required for maintenance, efficiency, administration or record purposes is available on demand in a data centre which is readily accessible from the CCR.

Each generating unit has its own unit operator's desk and vertical panel on which are mounted visual display units (VDU) and conventional equipment for communication, display and control. The supervisor has a separate desk which accommodates communications facilities and VDUs to enable him to monitor unit performance, station services and electrical auxiliaries supply alarms. There are vertical panels for the cooling water system station and reactor services, fire alarms and the station electrical system.

The unit operators' desks and panels are of a CEGB modular design which permits the desk layout to be optimised late in the construction programme to suit operating procedures. The layout is, however, based on the use of computer-driven VDUs as the primary source of information for both data and alarms and the principle that normal power operation is monitored and controlled from the centre section of the desk. The controls and indications for the less frequent operating regimes of startup and shutdown are located on the wings of the desk.

Hardwired indications and alarms are also provided since it must be possible to maintain the plant at steady load, to shut it down and to monitor post-trip heat removal if the computer system ceases to be available. There are, moreover, instances where it is ergonomically advantageous to have displays directly associated with controls or where it is necessary to provide indications by a diverse means in order to satisfy a safety requirement.

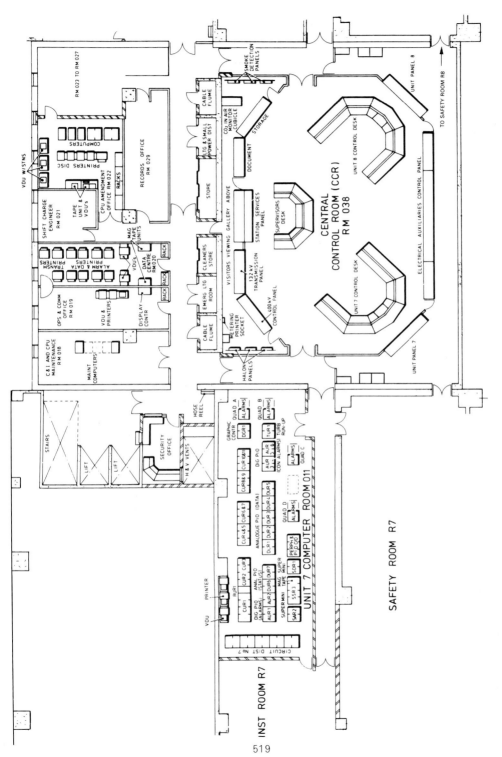

FIG. 5.31   CONTROL ROOM LAYOUT

An important facility, provided on each unit panel, is a post-trip monitoring mimic. Because of the extent of the post-trip heat removal systems and the large number of changes of plant configuration which can be initiated by the eight sets of post-trip sequencing equipment on each reactor, the mimic was designed to provide the unit operator with a functional overview of the post-trip heat removal situation. Thus, in the short term, the operator is relieved of the necessity to monitor the detailed actions of the post-trip sequencing equipment; it is sufficient for him to establish, with the aid of the mimic, that an adequate number of post-trip heat removal trains are operating.

To meet a particular safety criterion, an emergency indication centre (EIC) is provided remote from the CCR. The EIC is equipped with instrumentation to monitor the shutdown of the reactors in the event of the loss of the CCR. It is provided with post-trip monitoring mimics and a limited number of additional indications. The only controls available in the EIC are for tripping the reactors; all other control actions have to be taken local to the plant, for which purpose direct-wire telephone systems are provided.

22.2 Alarms

The computer driven VDUs are the primary means of presenting alarms in the CCR. A hierarchical system of alarm formats is available to the operator, from a total unit alarm overview through plant area alarm lists to detailed plant alarms. All alarms are available for display on the VDUs and in addition, some are also displayed on conventional annunciators.

Where alarm indications are required local to the plant, so-called 'intelligent' local alarm systems are provided which also transmit their alarms to the central data processing system. This enables the CCR operator to monitor local alarms through the central system.

22.3 Modulating Control

The operational characteristics of AGR stations differ from those of coal and oil-fired plants in that the nuclear stations are not as flexible in their ability to respond to changes in grid system frequency. The Heysham 2 control scheme is, therefore, of the turbine-follows-reactor type. The heat output required of the reactor is determined by the setting of the inlet guide vanes of the gas circulators and the electrical output of the turbine generator is adjusted to match that heat generation.

Some of the ancillary control loops on the turbine plant and the boiler system are implemented by electronic analogue controllers but the main unit control loops are implemented by 11 microprocessors which utilise the CEGB's high level engineer oriented computer language named CUTLASS (see Chapter 1 Section 12). The various control loops are distributed amongst the microprocessors in order to limit the consequences of failures. In general, redundancy is provided in respect of both plant inputs and processors so that a single failure does not cause a significant reduction in capability.

22.4 Sequence Control

Operational sequences are provided by the control microprocessors referred to for running up the 660MW turbine, for running up the boiler feed pump turbine, for the turbine gland steam sealing system and for starting and stopping the starting/standby boiler feed pumps. The main CW system is normally started up and shutdown by a programmable logic controller (PLC). PLCs are also used for the heating and ventilation systems.

The automatic sequencing of the post trip heat removal systems is provided by eight sets of equipment for each reactor. The four sets which control the X trains use microprocessors to implement the logic whilst the four sets which control the Y trains are based on relay logic in order to secure a further degree of diversity. They all send outputs to plant according to a predetermined logic and time sequence. Both X and Y systems function on a two-out-of-three basis in order to achieve the required reliability and the microprocessors continuously carry out internal checking of their own operation.

In addition to controlling the plant systems and the associated electrical equipment, the post trip sequencing equipments also isolate all CCR controls, the maloperation of which could prejudice post trip heat removal. Control from the CCR can however be restored by the CCR operator whenever required.

22.5 Protection Systems

In addition to the reactor protection system there are four other systems for protecting plant:

22.5.1 *Quadrant Protection System*

A high integrity protection system using hardwired electronic equipment and two-out-of-three relay logic is provided on each quadrant in order to:
(1) Protect the gas baffle against over pressurisation.
(2) Protect the gas circulators against lubrication system faults.
(3) Protect the plant generally against the damage which would ensue if reactor gas flow and boiler feed flow became mismatched.

On detection of a fault condition, the equipment removes the affected quadrant from service. Operation with more than one quadrant out of service is not permitted; thus, output signals from the quadrant protection system are fed into the guardline logic networks of the reactor protection system which will generate a reactor trip output on receipt of trip signals on a two-out-of-four basis.

The modulating control system copes with the transient caused by a quadrant trip and maintains the reactor on load at reduced output.

22.5.2 *Essential Plant Protection Equipment*

A set of relay-based plant protection equipment is also provided which is separate from and additional to the quadrant protection system. This equipment is, in general, intended for protection of the plant against harmful conditions as distinct from protection against situations which could give rise to a nuclear hazard.

22.5.3 *Boiler Protection System*

This system protects the boiler tubing from the deleterious effects which would follow from the ingress into the boiler feedwater, of chlorides from the main CW system through leaks in the turbine condenser or of chemical solutions carried over from the condensate polishing plant. A high level of protection is provided, using two-out-of-four logic and, so far as practicable, diverse sets of equipment. However at the time of writing the only form of sensor having the requisite reliability is one which measures the electrical conductivity of the feedwater. A degree of diversity is attained, nevertheless, by using different designs of conductivity instrument and different designs of relay-based logic systems.

22.5.4 *Turbine Supervisory and Protection Equipment*

The turbine supervisory and protection equipment is referred to in Section 19.2.3.

22.6 Control and Instrumentation for Heating and Ventilation Systems

There are, at Heysham 2, over 50 separate heating and ventilating systems, of which some 40 or so are safety-related. A number of these systems are controlled by PLCs which implement both sequential and modulating control. These PLCs are linked to a microprocessor, programmed in CUTLASS, which provides centralised logging and display.

22.7 Instrumentation

The instrumentation is made up either of designs which have been type tested to meet the requirements of the CEGB Instrument Approvals Scheme or which have been subjected to equivalent testing for the specific purpose of being accepted for use at Heysham 2.

Two new forms of instrument system have been introduced. One provides the CCR operator with measurements of reactor power over the whole of the power range. These measurements are derived from the pulses received from a single fission chamber and the instrument system changes over automatically from a form of pulse counting circuitry appropriate to low flux levels to one suited to high levels.

The other new system, introduced following successful trials at Hinkley Point B, is a method of detecting boiler leaks which has a considerably improved capability compared with that of earlier systems. It is based upon the changes in oscillation frequency which occur when a crystal adsorbs water upon its surface and it uses a differential technique in which readings from the four reactor quadrants are compared with each other.

22.8 On-Line Computer System

The station on-line computer system provides the main channel by which the CCR operators receive plant data and alarms. The system is unitised for the most part but does have some common station-based facilities. It consists, in total, of some 69 computers, located for the most part in two computer rooms on either side of the CCR, amongst which the various functional requirements are distributed. Redundancy and diversity are incorporated in order to ensure a high availability.

Thermocouple outputs and other analogue signals from the plant are scanned in blocks of about 500 by intelligent analogue multiplexers which deal with 2500 inputs per unit. Digital on/off inputs are scanned by intelligent digital multiplexers which deal with 4000 inputs per unit. Both types have local 16-bit microprocessors with random-access memories of 128 kilobytes. These microprocessors pass the signals through data links, which are subject to high level data link control (HDLC), to the central processing system which consists of three 32-bit super mini-computers with random-access memories of 2 megabytes.

The control loop microprocessors have their own input multiplexers and are connected together and to the central data processing computers by a duplicated coaxial cable local area network. This enables the VDUs to be used for displaying data and the numerical values of control terms in use and for diagnostic purposes after downline loading.

Each generating unit has its own central data processing system consisting of three super mini-computers, one dealing with data, one with alarms and the other acting as standby to either. This arrangement meets the requirement that a single fault shall not affect the service to the operators. Bulk storage of 80 megabytes is provided on rotating discs with moving heads. The output of the two systems serves seven colour-VDUs on each unit operator's desk, two on the CCR supervisor's desk and one on the CCR auxiliary electrical panel. Mobile VDUs are also available for use at other locations as required. Permanent records are provided in the data centre in the form of printouts and magnetic tapes which can be analysed on separate computers.

Changes to software are strictly controlled. After authorisation, they are effected, by means of downline loading, through a station amendment super minicomputer located in a separate software amendment office.

23. STATION TELECOMMUNICATIONS

A diversity of telecommunications systems is provided at Heysham 2 in order to meet various operational requirements. To some extent, they overlap so that a failure of any one system can be overcome, at least in part, by appropriate use of the others.

Under normal circumstances, the CCR constitutes the communications centre from which operational activities are coordinated. If the CCR is lost, local plant control actions are supervised from the EIC whilst the coordination of communications is effected from an emergency communications centre (ECC) located in the administration, welfare and workshops building which is situated to the east of the turbine house.

23.1 Telephone Systems

For general purposes, a private telephone exchange (PAX) with about 600 extensions is provided and connected by tie-lines to the Electricity Supply Industry corporate telephone network. There is however an emergency scheme associated with the PAX system which is centred upon a so-called 'non busy' handset on the CCR supervisor's desk. This handset can accept calls from up to five PAX extensions at any one time and the extensions which have thus obtained access to the supervisor's handset are identified on a digital display on the supervisor's desk. An extension to the supervisor's handset is provided in the gatehouse.

For operational purposes, highly reliable links between telephones located at appropriate points around the plant and key panels on the CCR desks, are provided by five independent direct-wire telephone systems. Each system can accommodate up to four telephones on each of 30 lines and in noisy areas, calling beacons and acoustic hoods are provided. For use in emergencies, key panels are also located in the emergency indication centre.

Each unit has its own maintenance and commissioning telephone jack system. These have passive circuitry which consists of a central patching cubicle and wires to telephone jack sockets located around the plant. In addition to portable telephones, headsets and portable amplifiers are provided for use in noisy areas.

23.2 Radio Systems

Five UHF channels are provided for communications with personnel working within the power station buildings or within the site boundaries. Two of these channels, those allocated to operational/emergency uses and to maintenance work, use a common aerial network of wall-mounted aerials which is aimed primarily at coverage within the station buildings. Two further channels, those allocated to common services, also use a common aerial network which, in this case, is designed for external coverage up to the site boundaries as well as within the buildings. The networks are fed from four distributed base stations and are partly run in so-called 'leaky' coaxial cable so as to augment the signal strength available within the complex, thick-walled structures of the reactor and fuel handling building. A total of nine controllers, having short-time fire-proof cabling, are provided. The fifth channel constitutes a separate system for use when personnel enter the reactor vault. It consists of a self-contained trolley-mounted transmitter/receiver with a battery-backed power supply and eight hand portables.

For the internal system, UHF permitted a number of channels within the allocated band and the use of a smaller aerial than that required for VHF transmission. However, a VHF system which is more suitable for longer range communication is provided for coordinating the activities of health physics personnel beyond the station boundaries. This is a single frequency amplitude modulated system with a single base station and an end-fed unipole aerial mounted on the roof of the reactor and fuel handling building. Controllers again with short-time fireproof cabling are located on the CCR supervisor's desk, in the emergency communications centre, in the health physics monitor room and in the district survey laboratory. Three switchable frequencies are available which permit three separate channels of communication to be used.

A VHF paging system is also provided with nine transmitters, most of which are in the reactor and fuel handling building because of its complexity and wall thicknesses. Paging receivers can be sent messages, usually PAX extension numbers, which are read from a digital display. Of the 99 receivers, 40 are suitable for receiving group calls which can be initiated from the supervisor's desk in the CCR. An interface with the PAX system allows individuals to be paged from any PAX extension and for that extension number to be transmitted as a message to the paged receiver.

23.3 Station Siren System

The station siren system is controlled from a panel mounted on the CCR supervisor's desk or from a standby panel in the EIC. Two equipment cubicles, situated at widely separated locations, operate sirens in nine fire zones and each zone contains sirens connected to each cubicle. For additional assurance, repeat tones are broadcast on the public address system. Operation can be either a wailing or a steady mode to indicate, respectively, a fire alert or a nuclear incident warning and can be checked, zone by zone, by means of a microphone in each zone which is connected to the control panel on the supervisor's desk.

Noisy areas also have beacons to give a visual warning.

23.4 Public Address Systems

Duplicate public address systems are provided with the two sets of central equipment situated at widely separated locations. The loudspeaker networks overlap and are so designed that in the event of failure of either system, the coverage provided by the other will still be above the ambient noise level. To achieve this, some 1200 loudspeakers and 58 amplifiers of 200 watts output are installed in each system. Noise monitoring microphones are provided in the turbine house, the gas circulator areas and the CW pumphouse so that the central equipment can compensate for the varying noise levels experienced.

The controllers, which can select the zones to which broadcasts shall be addressed, are situated on the CCR supervisor's desk, in the EEC, the EIC and on the reception desk in the administration, welfare and workshops building.

Short-time fireproof cabling is used to ensure the security of the system.

24. HEATING AND VENTILATION SYSTEMS

Heating and ventilation systems are provided in all areas of the nuclear island and ancillary buildings to:
(1) Control environmental conditions for personnel and plant.
(2) Control and limit the spread of radioactive contamination.
(3) Control the venting of hot gas or steam release to atmosphere thereby relieving pressure on walls and seals.

There are approximately 50 separate and largely independent systems comprising air handling units, fans, filters, fire dampers and grilles. Systems are generally of the balanced plenum type, ie each has its own supply and extract fans. Most of the heating and ventilation plant is housed in individual system plant rooms together with its control panels.

Air flow rates are designed to provide fresh air and ensure distribution, maintain temperatures within limits and remove excess heat and fumes.

Clean air zones include a recirculation facility of extract air to economise on heating requirements. Systems serving contaminated air zones include high efficiency particulate air filters and do not have a recirculation facility.

Contaminated air zones are maintained at a negative pressure relative to clean areas and the outside atmosphere, in order to prevent leakage of untreated air. Flow rates are arranged to promote flow from clean areas towards potentially lightly-contaminated areas and from these, to the more heavily-contaminated areas. Non-return dampers guard against reverse air flow from contaminated areas. This hazard is minimised by an arrangement of gaseous blowdown into an independent auxiliaries blowdown system.

The design air temperature range is 10°C to 35°C for an external ambient range of −1°C to 25°C.

Each hot gas release zone is sealed to prevent the gas from damaging plant in other areas of the building and has a venting route to the outside atmosphere to relieve pressure on the walls and seals.

Potential hot gas release vents are located in the charge hall, reactor quadrants, pile cap, bypass gas plant and secondary shutdown system room.

To protect the heating and ventilation systems, hot gas release dampers are located in the ductwork at the entry and exit points of the supply and extraction ventilation ducts which penetrate hot gas barriers.

25. FIRE DETECTION AND PROTECTION SYSTEMS

The fire fighting systems which detect, alarm and protect the various plant areas consist of the following:
(1) Waterspray and sprinkler systems.
(2) Fixed foam pipework system.
(3) Smoke detection system.
(4) Fixed Halon 1301 system.
(5) External fire hydrant system.
(6) Dry riser system.
(7) Fire break-glass system.

Waterspray and sprinkler systems are fed from a fire fighting trunk main which is constantly charged with water from a pressure tank. The tank water level is maintained by diesel engine-driven pumps located in the fire fighting pumphouse to one third of its total capacity of 45.45m^3. Pressure is maintained by compressed air in the remaining two thirds.

The waterspray systems comprise high and medium velocity systems which are automatically initiated, and a high velocity system which is manually initiated.

The sprinkler system consists of networks of fast-acting sprinkler heads which operate in groups on receipt of an alarm signal generated by heat detecting cables. In addition, each sprinkler head within the group provides a secondary auto release back-up action offered by conventional pattern sprinkers. Sprinklers operate automatically following rupture of the quartzoid frangible glass bulb which is sealed into the sprinkler head.

Fixed foam pipework with nozzles is fitted to diesel and auxiliary boiler fuel oil tanks. Each tank has a base injection arrangement from a foam generator installed in the foam line to the tank. The bund areas around each tank are similarly protected with foam pourers mounted on the bund walls.

Smoke detection equipment is fitted in selected areas of the station to give audible and visual alarms in the case of normal initiation and failure of the detection equipment.

Halon 1301 gas systems are used extensively throughout the station in termination and marshalling rooms and also false floor areas below safety equipment, instrumentation and computer rooms. The systems are automatically initiated by heat detecting cable in unmanned areas and manually initiated in manned areas.

The external fire hydrant system provides water to the conventional fire fighting equipment by means of electric motor-driven fire hydrant pumps or a diesel-driven fire hydrant pump. The water source is normally the raw water storage tanks, but in the event of this not being available, water may be taken from take-off points at the townswater reservoir and CW system.

The dry riser system comprises vertical risers in staircases of the control and mechanical annexes, reactor buildings and fuel handling building, with double branch connection points at the various floor levels. The risers are charged, when necessary, by mobile pumping equipment at ground level.

The station is divided into a number of zones in each of which the manually-operated fire break-glass switches are connected to control and indicating units located within the escape route for that particular zone. Audible alarms sound within the zone and remote alarm signals alert the station fire alarm system and central control room.

25.1 Lightning Protection System

Lightning protection is provided on the following basis:
(1) All main buildings irrespective of height.
(2) All buildings and structures over 15m in height which are not within the protective zones of the main buildings.
(3) Buildings and structures containing explosive or inflammable materials.

Each individual lightning protection system is interconnected to the station main earth network.

Building structures are protected by aluminium strip air terminations on the roofs, together with aluminium strip down-conductors and earth terminations. Metal roof decking, facing cladding and reinforcement in concrete columns are not relied on for protection.

Buildings or structures containing explosives or inflammable materials are protected against lightning by catenary type air termination networks.

26. COMPRESSED AIR SYSTEMS

Separate compressed air systems are provided to supply the needs for:
(1) General service air.
(2) Instrument air.

(3) Breathing air — low pressure and high pressure.

(4) Reactor purge air and turbine forced cooling.

General service air meets the needs of maintenance work for the turbine, reactor and workshops. Two 50% duty compressors provide the total supply requirements of $36m^3/min$ at 7.2 bar.

The instrument air systems provide dry, oil-free air for the continuous operation of pneumatically-controlled valves and instruments throughout the station. There are two separate and self contained systems each supplying one turbine-reactor unit; and each system comprises three 100% duty compressors to provide $10m^3/min$ at 8 bar.

The breathing air systems comprise a low pressure system and a high pressure system. Low pressure in the range 0 to 8.5 bar is used for face masks and half suit activities which are undertaken in contaminated zones that are subject to ambient air temperatures. High pressure air at up to 13 bar supplies full (hot) suits for both breathing and cooling when used for reactor entry. Each system has two 50% duty compressors to generate the normal duty requirements and emergency back-up from storage bottles.

Reactor purge air is provided from the turbine forced air cooling system compressor to purge the reactor vessel and other plant areas of the nuclear island of CO_2. It is also used for boiler dry out and leak checking of the pressure vessel.

27. MAINTENANCE FACILITIES

Many items which require maintenance become radioactive as a result of neutron absorption and/or become contaminated by radioactive material during use. Workshop facilities for such items are provided within, or contiguous with, the nuclear island. Conventional workshops and stores for other items are provided in the separate administration, welfare and workshops block which is situated to the east of the turbine house.

27.1 Active Workshops

Many active or contaminated items which require maintenance are associated with the fuel route and are dealt with in the irradiated fuel disposal (IFD) cells or the active maintenance facility or in other IFD facilities such as the fuelling machine maintenance bay and the pond equipment maintenance room.

The largest contaminated items which require regular maintenance are the gas circulators. Because of their size, number and test requirements, a separate maintenance workshop with its own decontamination facility is provided, with the reception bay at the same level as the gas circulator withdrawal positions. It adjoins the western end of the nuclear island and is connected to the circulator withdrawal positions by easily-decontaminated internal routes along which the machines are moved on air-flotation transporters. The workshop has a 45t crane and is designed to service four circulators a year. The circulators arrive during the 6 weeks annual shutdown and two of them are given a minor overhaul during that period and returned to service. The other two are replaced by overhauled spares and are subjected to a major overhaul prior to the succeeding annual shutdown.

Small components are dealt with in a decontamination centre and active workshop which are situated alongside one another on one of the upper levels of the central fuel handling building. Items which must pass through clean areas are suitably wrapped and/or enclosed in shielded containers before despatch to this facility. Separate dirty and clean handover bays are provided.

All active workshops are arranged and provided with facilities to minimise the radiation dose received by those carrying out maintenance activities. Features which lead to the attainment of this objective are proper ventilation, the layout of machine tools for ease of decontamination, the provision of portable shielding and protective clothing and the provision of special storage bins for swarf.

27.2   Decontamination

The function of the decontamination facilities is to reduce the contamination present on items requiring maintenance in order to minimise the radiation dose received by maintenance personnel from this cause.

Some items are not amenable to decontamination because of the materials of which they are made or because of the complexity of their construction. However, a large range of methods is available from vacuum cleaning, through scrubbing with a solvent, to abrasive processes. The choice is, in principle, decided on a cost/benefit basis; the alternative is disposal of the item as radioactive waste and its replacement.

The processes of equipping the decontamination facilities follows the establishment of detailed plant maintenance procedures but the kinds of equipment likely to be provided include:

(1) Portable vacuum cleaners with discharge of the exhaust to the contaminated ventilation system.

(2) Enclosed washdown areas.

(3) Portable units for spraying on latex which, after hardening, can be peeled off bringing any loose contaminants with it.

(4) Tanks in which items can be steeped in hot and other solvents.

(5) Drying cabinets.

(6) Low pressure blasting cabinets in which a mixture of an abrasive and water is used.

(7) Fume cupboards with sinks in which items can be scrubbed.

Subchange rooms are provided by means of which the control of personnel entering and leaving the decontamination facilities is effected.

27.3   Conventional Workshops

The conventional workshops occupy some 2200m^2 of floor space and have stores areas occupying a further 4000m^2. The main mechanical and electrical shop is provided with a wide variety of machine tools and a 40t crane capable of lifting a complete boiler feed pump turbine. A welding/blacksmiths' shop, a carpenters' shop and an instrument workshop are also provided.

28. ARCHITECTURAL TREATMENT AND LANDSCAPING

The architectural and landscaping proposals for the Heysham 1 twin reactor station were accepted by the Royal Fine Art Commission in February 1968. In the 11 years which elapsed before the proposals for Heysham 2 were submitted, the changes which resulted from enhanced safety requirements and operational experiences at Hinkley Point B led to the need for buildings much larger in volume than those of Heysham 1.

The proposals for Heysham 2 were accepted by the Royal Fine Art Commission in March 1979 with the observation that "the designs showed promise of a handsome building".

28.1 Architectural Treatment

The elevations of the main building block at Heysham 1, which comprises the nuclear island and the turbine house, are dominated by two materials, fair-faced concrete and mill-finished aluminium. The much smaller ancillary buildings have mill-finished aluminium sheeting above a red brick plinth.

The corresponding building block at Heysham 2 is larger though not higher than that at Heysham 1. Its impact is reduced by clearly expressing significant individual elements. For the reactor buildings, these consist of three horizontal layers. At the heart of the bottom layer lie the massive pre-stressed concrete pressure vessels which surround the reactor cores. The middle layer consists of exposed mechanical equipment such as pipes, exhausts and ventilators which lie along the south elevation of the lower part of the charge hall. The topmost layer is the upper part of the charge hall which is a comparatively light and simple structure.

In the past, the potentially forbidding character of power stations was sometimes relieved by vast areas of patent glazing which, however, created practical problems. More recently, the omission of most of the glass combined with the use of what, especially on exposed marine sites, is one of the most durable finishes, mill-finished aluminium sheeting, have made it difficult to avoid the drab effect of very large sheds.

However, the sensitive use of colour, combined with the expression of the three layers mentioned above, make it possible to express the character of the reactor buildings clearly, powerfully and, at the same time, elegantly.

The upper part of the charge hall is, especially when seen from a distance, the most prominent feature of Heysham 2. Its cladding consists of profiled aluminium sheeting with a white organic coating, echoing a nautical tradition. It has slightly curved corners to avoid a chopped-off look and is capped at about 70m above ground level by a shallow glazed fascia which lights the charge hall. The middle layer is a series of recesses flanked by reinforced concrete buttresses, which form a strong framework into which the many and varied pieces of exhaust and other equipment fit. At the end of the buildings, this layer is clad in aluminium sheeting with a black organic coating. The 37m high base, an expression of the massive pressure vessels within the buildings, is of concrete cast in situ with a slightly textured surface, further broken up by emphasising the horizontal construction joints.

FIG. 5.32   VIEW OF MAIN BUILDING LOOKING SOUTH EAST

FIG. 5.33   AERIAL VIEW LOOKING NORTH WEST OF THE STATION SITE

The turbine house, low by comparison with the reactor buildings, is also clad with the black aluminium sheeting and the upper levels of its north elevation contain one of the few large expanses of glazing in the station. In order to form a lively foreground at the human level, some small areas of more positive colours, mainly green and stone, are used for the minor structures which cluster around the main buildings.

Fig 5.32 shows a view of the main building looking south east.

28.2 Landscaping

It is CEGB policy to retain or improve the existing landscape at new power station sites both to improve the working environment and to soften the impact which the large power station structures have on the general public.

In order to hide the foreground and lower elements of the station, an existing ridge has been retained on the east side of the approach road and an artificial mound has been created on the other side. Both have been grassed and planted with bushes and ground cover to give a broken skyline. The station car park is provided with planted areas to break up and reduce the impact of a large mass of parked cars.

An earth bank, some 4m high, suitably grassed and planted with shrubs, has been formed on the eastern and south eastern boundaries of the CEGB land in order to provide a screen between the station and a golf course to the east and between the station and a holiday camp to the south east.

On completion of the station, the contractors' storage and temporary works areas which lie to the south east of the station buildings will be suitably levelled and grassed to provide a grazing area in keeping with the general character of the local landscape.

Local conservationists have drawn attention to the desirability of preserving and promoting the natural flora and migrant bird life in the marshy area which occupies the north east corner of the CEGB land. This area is therefore being left in its natural state and, together with an adjacent landscaped area, will be made available for educational and field-study purposes and be opened to the public as a picnic and recreational area.

Fig 5.33 shows an aerial view of the station construction site and also the existing Heysham 1 station.

29. MAIN CONTRACTORS, MANUFACTURERS AND PLANT SUPPLIERS

Nuclear Island

National Nuclear Corporation Ltd
Whessoe Heavy Engineering Ltd
Union Carbide (UK) Ltd
Emerson Electric Industrial Controls Ltd
Balfour Kirkpatrick Ltd
Babcock Bristol Ltd
Aiton & Co Ltd
NEI Nuclear Systems Ltd

Strachan and Henshaw
Fairey Engineering Ltd
GEC Reactor Equipment Ltd
Darchem Engineering
James Howden
Burnett & Lewis Ltd
Jordan Engineering (Bristol) Ltd
Babcock Power Ltd
Pipework Engineering (PED) Ltd

Davy MacKee (Minerals & Metals) Ltd
J. Kenyon (Swansea) Ltd
Flight Refuelling Ltd
G & N Construction (Engineers) Ltd
NEI Cranes Ltd
NEI Reyrolle Ltd
Motherwell Bridge Engineering Ltd
NEI Thompson Nuclear Engineering
NEI Thompson Ltd
GEC Energy Systems Ltd
NEI Electronics Ltd
Mather & Platt Ltd
Head Wrightson Teesdale
Pall Process Filtration Ltd
Centronic Ltd

Norsk Hydro Electrolysers
D. Robinson & Co Ltd
Permutit Boby Ltd
Robert Jenkins & Co Ltd
Condor Northern Ltd
Vickers Ltd
GEC Measurements
Protech Instruments & Systems Ltd
Cunnington & Cooper Ltd
Matterson Ltd
Fairey Microflex Ltd
B.O.C. Ltd
Plessey Controls Ltd
Hopkinson Ltd
T.I. Chesterfield Ltd
Contropanels Ltd

Mechanical

Turbine generators and associated plant
CW pumps and associated plant
CW screens and associated plant
Cranes
Structural steelwork
Water treatment plant
Diesel generators

Fire fighting equipment

Auxiliary boilers
CW electro-chlorination plant

NEI Parsons Ltd
APE-Allen Ltd
J. Blakeborough & Sons Ltd
Herbert Morris Ltd
Octavius Atkinson & Sons Ltd
NEI Thompson Ltd
Mirrlees Blackstone (Stockport) Ltd
Merrol Fire Protection Engineers Ltd
Babcock Power Ltd
Engelhard Industries Ltd

Electrical

11kV switchgear
Station transformers

415V switchgear
Auxiliary transformers

Generator transformers and aux transformers
3.3kV switchgear

Cabling

Telecommunications equipment

GEC Distribution Equipment Ltd
Hawker Siddeley Power Transformers Ltd
GEC Industrial Controls
GEC Distribution Transformers Ltd
NEI Peebles Ltd
NEI Reyrolle Ltd
Whipp & Bourne (1975) Ltd
Matthew Hall Mechanical & Electrical Engineers Ltd
Protowire Telecommunications Ltd

Batteries and charging equipment	Chloride Industrial Batteries Ltd
Radio equipment	Pye Telecommunications Ltd
Telephones, paging and PA systems	Plessey Communications Systems Ltd
Generator main connections	Balfour Beatty Power Const Ltd
Unit transformers	GEC Power Transformers Ltd
Generator switchgear	British Brown Boveri Ltd
Inverter systems	AEG-Telefunken (UK) Ltd

Control and Instrumentation

Control and instrumentation equipment	NEI Electronics Ltd
Conventional alarm annunciator equipment	D. Robinson & Co Ltd
Computer systems	Babcock Bristol Ltd

Civil

Civil works and buildings	Taylor Woodrow Construction Ltd
Heating and ventilation	Hayden Young Ltd
Cladding and roofing	Briggs Amasco Ltd

Consultants

Station layout	Powell Moya & Partners
Civil engineering	Allott and Lomax
	Nuclear Design Associates
Civil engineering contracts	E.C. Harris & Partners
Heating and ventilation and electrical works	Steenson, Varming, Mulcahy & Partners
Landscaping	Derek Lovejoy & Partners

Aerial Photography

Aerial Photography	Peter Joslin

6.

Sizewell B – PWR Nuclear Power Station

1. INTRODUCTION

Sizewell B power station site lies immediately north of the existing Sizewell A power station on land already owned by the CEGB.

In 1973 Sizewell was selected by the CEGB as the site for multiple consent and planning applications for the various additional reactor systems then under consideration. These applications were for a total capacity of 2500MW to be produced using one of the following:

Steam Generating Heavy Water Reactor (SGHWR)
High Temperature Reactor (HTR)
Light Water Reactor (LWR)

The latter is a generic name covering the Pressurised Water Reactor (PWR) and the Boiling Water Reactor (BWR).

In 1974 the Government selected the SGHWR system, and in February 1975 consent was received for a Sizewell B station using reactors of this type (2500MW in total). Following a review of the SGHWR system the project was cancelled in February 1978, when revised Government policy was announced which no longer included the SGHWR system and supported the power industry in its preference for a PWR in order "...to establish a flexible strategy for the UK nuclear power programme in the light of developing circumstances".

In December 1979 the Government agreed to the development of a PWR based on the Westinghouse design and in April 1980 the CEGB issued a letter of intent to the National Nuclear Corporation (NNC) to authorise the design and, subject to approval, the manufacture of a PWR. On 1 October 1980 the CEGB made a public announcement of its proposal to construct the first PWR at Sizewell. Applications for Section 2 consent and for deemed planning permission were submitted to the Secretary of State for Energy on 30 January 1981. On this latter date, application was also made for a revision of the existing nuclear site licence. It proposed that this new PWR station, to be known as Sizewell B should have a total generating capacity of 1110MWso.

1.1 Selection of Site and Fuel

In addition to the requirements given in Chapter 1 for a nuclear power station the population distribution around the site must enable the safety of the population to be assured in the event of an emergency at the station.

Previous power station development in England has resulted in a preponderance of oil-fired plant in the south (located particularly in the south east) while in the midlands and north coal-fired plant predominates. In view of the substantial rise in the cost of oil since the oil-fired stations were planned, they have become uneconomic to operate within the merit order unless the availability and costs of other fuels are adversely affected. Under normal circumstances, whilst providing a flexibility of capability, oil-fired stations would not be operated in preference to nuclear and coal-fired which would provide the bulk of the CEGB's requirements.

Operations of the CEGB's system in this manner results in a requirement for large power transfers from the midlands and north to the south of the country and this dependence upon power transfers will increase with time.

Fig 6.1 shows the location of Sizewell in relation to the supergrid system in the south of the country planned for 1987/88. Studies of the expected operation of this system in the early 1990s show that some 70 – 80% of the optimum economic power transfers to the south are likely to be required in the south east where there is a heavy demand for electricity. The studies also indicate that, under certain conditions, there is a possibility of the transmission capability to the south being inadequate and acting as a constraint on optimum economic operation. Therefore there are clear benefits to be derived from the location of new high merit generation in the south, and the south east is the preferred area for the first PWR. It would then contribute towards satisfying the heavy power demands in the area and thereby reduce the power transfers required to the south and south east.

By using land in the CEGB's ownership at partially-developed power station sites and by developing such sites to their full capacity as determined by environmental and technical limitations, advantage can be taken of facilities already provided for the existing station such as transmission system connections and improved roads. Furthermore the overall number of new greenfield sites required is reduced. Thus it is general practice to develop existing sites where this serves the electricity system requirements and is practicable.

In the south east there are three power station sites which have potential for further nuclear development — Bradwell, Dungeness and Sizewell. At Bradwell, where a 245MWso Magnox station is operating, the CEGB owns a substantial area of land. If another station was located there, the existing Bradwell-Rayleigh 275kV transmission line would require uprating to 400kV, a new 132kV line from the existing 132kV system in the Maldon vicinity would be required to Bradwell to provide a back-up for auxiliary and essential services plant, and, depending on timing, further transmission reinforcement might be necessary in the London area. In the case of Dungeness there is scope for considerable further development beyond the existing 410MWso Magnox station and the 1200MWso AGR station nearing completion there, but a new 400kV transmission line to London would be required for further new capacity.

At Sizewell, where a 420MWso Magnox station is already operating, there is sufficient additional land within the CEGB's ownership to accommodate the proposed B station and a further C station of about 1200MW each. The transmission lines built for the existing station, when uprated from 275kV to their design voltage of 400kV and with relatively minor modifications to the transmission system in East Anglia, would be capable of supporting a further 2500MW output from the site. No new transmission lines would be needed. The proposed B station would necessitate the uprating to 400kV of the existing 275kV transmission lines from Sizewell to Bramford and their extension within the site boundary, to a new 400kV substation included in the B station supply. Minor modifications would be needed to the connection arrangements at the junction point of these lines with the transmission system at Bramford and at the Norwich 400kV substation. For the full 2500MW development it would be necessary to provide additional conductors, on existing towers, on the transmission line from Bramford to Norwich and, depending upon the timing of the full development, similar

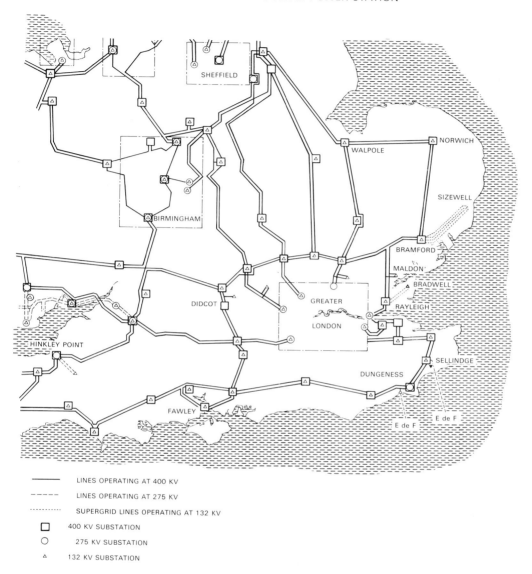

FIG. 6.1   **SUPERGRID TRANSMISSION SYSTEM**

work might also be needed on the transmission line between Norwich and Walpole. However, it may be necessary, for operational reasons, for part of this work to be carried out with the development of the Sizewell B station.

The Sizewell site meets all the desired criteria and has the further advantages that substantial preparatory work for a proposed station was carried out in 1971 and planning approval had already been obtained from the Local Planning Authority for a light water reactor in 1974. It was concluded that Sizewell has advantages over Bradwell and Dungeness for early development of the next nuclear station.

541

2. SITE LAYOUT

The Sizewell site of some 97 hectares once comprised heathland and conifer plantations, with low grade agricultural land and marshland at the north. Following levelling, the A station was constructed on a plateau formed at a level of about +9m OD at the south end of the site. The central part of the site, where Sizewell B will be built, was levelled to +6.4m OD during the preparatory work in 1971 whilst the north end comprises marshland at about +2m OD.

The level of the B station site is sufficient to protect it from seawater and surface water flooding.

Extreme high water level (EHWL) is +3.9m OD and the margin between this and the ground level of +6.4m OD gives adequate margin against surges above EHWL due to extreme combinations of wind and tide.

The proposed B station incorporates a 3425MW (thermal) nuclear steam supply system (NSSS) which provides steam to two turbine generators each of 622.5MW gross output. The design of the NSSS is based on a Westinghouse design incorporating a PWR as its heat producer. The layout and many of the auxiliary plant systems are based on the 'standardised nuclear unit power plant system' (SNUPPS) design produced by Bechtel in association with Nuclear Projects Inc., which is an association of several US electrical supply companies.

The relative arrangement of the main power block elements is thus fixed and only the position and orientation of the power block is variable. Fig 6.2 shows the layout adopted which has been constrained by the desire to keep both cooling water (CW) culverts and buried 400kV cables as short as possible. The former because culverts require the deepest excavations and extended construction time during which access around the site is severely restricted; the latter because of cost. With the sea and transmission lines at opposite extremities of the site the two requirements are incompatible, hence the location of the turbine hall on the south side of the reactor represents the best compromise between cable cost and culvert cost.

The station administration and welfare buildings, the workshops and stores are all positioned along the southern boundary of the site adjacent to the A station; they are built late-on in the construction programme allowing this part of the site to be used as temporary works areas by plant contractors during the earlier stages of the programme.

The layout of the station buildings on the site is shown in Fig 6.2. The site occupies an area of some 16 hectares to the north of the existing Sizewell A power station. The main power block houses the main power generation plant, shown in Fig 6.3, and comprises the following buildings:

Reactor building (with secondary containment).
Auxiliary building including decontamination shop.
Fuel building.
Control building and control building extension.
Diesel building.
Auxiliary boiler house.
Turbine house including mechanical and access control annexe.

SIZEWELL B — PWR NUCLEAR POWER STATION

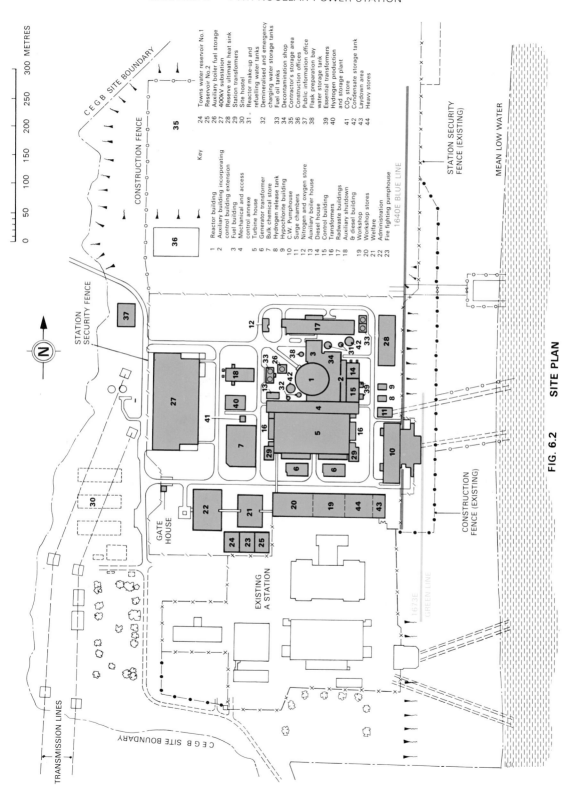

Key

1. Reactor building
2. Auxiliary building incorporating control building extension
3. Fuel building
4. Mechanical and access control annexe
5. Turbine house
6. Generator transformer
7. Bulk chemical store
8. Hydrogen release tank
9. Hypochlorite building
10. C.W. Pumphouse
11. Surge chambers
12. Nitrogen and oxygen store
13. Auxiliary boiler house
14. Diesel house
15. Control building
16. Transformers
17. Radwaste buildings
18. Auxiliary shutdown & diesel building
19. Workshop
20. Workshop stores
21. Welfare
22. Administration
23. Fire fighting pumphouse
24. Towns water reservoir No.1
25. Reservoir No.2
26. Auxiliary boiler fuel storage
27. 400kV substation
28. Reserve ultimate heat sink
29. Station transformers
30. Site hostel
31. Reactor make-up and refuelling water tanks
32. Demineralised and emergency charging water storage tanks
33. Fuel oil tanks
34. Decontamination shop
35. Contractor's storage area
36. Construction offices
37. Public information office
38. Flask preparation bay water storage tank
39. Essential transformers
40. Hydrogen production and storage plant
41. CO_2 store
42. Condensate storage tank
43. Laydown area
44. Heavy stores

FIG. 6.2   SITE PLAN

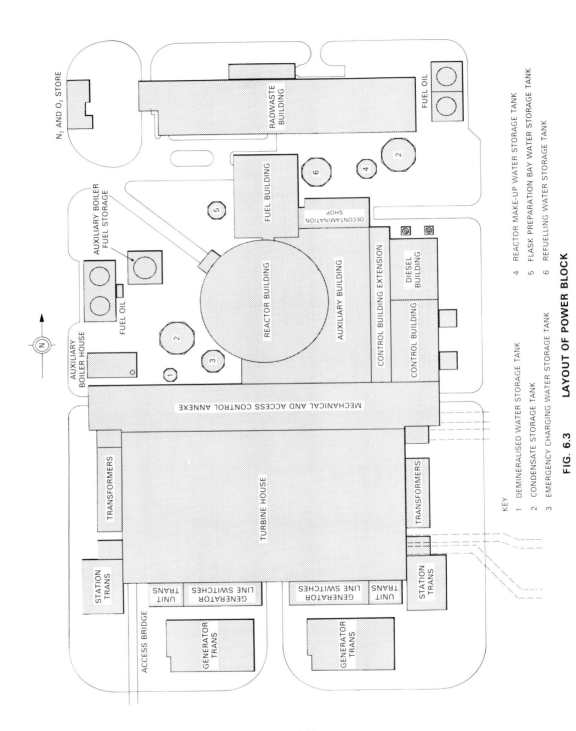

FIG. 6.3   LAYOUT OF POWER BLOCK

KEY

1 DEMINERALISED WATER STORAGE TANK
2 CONDENSATE STORAGE TANK
3 EMERGENCY CHARGING WATER STORAGE TANK
4 REACTOR MAKE-UP WATER STORAGE TANK
5 FLASK PREPARATION BAY WATER STORAGE TANK
6 REFUELLING WATER STORAGE TANK

The land to the east of the station buildings, comprising the Bent Hills and a beach, is owned by CEGB but has been leased to the Local Authority for public access. This arrangement requires that land enclosed within the permanent station fence should not in general extend beyond co-ordinate 1640E (this is known as Blue Line, shown on Fig 6.2). Prior to construction of the A station it was agreed that structures for the cooling water pumphouse may project beyond the Blue Line up to a position of 1673E (known as Green Line). This line also defines the eastern limit for construction of permanent works of the B station pumphouse.

Important buildings which are separate from the power block are the cooling water (CW) pumphouse, the radioactive waste (radwaste) processing and storage buildings, the auxiliary shutdown and diesel building, together with workshops, offices, storage tanks, towns water reservoirs, etc. A service tunnel for pipe, cable and personnel access joins the radioactive waste processing building to the auxiliary building. West of the power block is the 400kV substation.

For reasons of safety, where there are two mutually redundant sets of safety plant external to the main buildings, then they are generally sited well apart. Thus the auxiliary shutdown and diesel building and its associated essential auxiliary transformers and fuel tanks are situated about 140m from the control building and diesel house on the opposite side of the reactor building. Station, unit and auxiliary transformers are separated into two groups, one on either side of the turbine house.

Groups of water storage tanks of similar duty are situated on opposite sides of the main buildings and remote from the towns water reservoirs.

Cable routes from both the control building and the auxiliary shutdown and diesel building follow segregated routes. The same applies to cable routes between the station and generator transformers and the 400kV substation.

Essential service water pipework from the CW pumphouse to the auxiliary building and component cooling water pipework from the auxiliary building to the reserve ultimate heat sink is segregated from other services.

The main building foundations are constructed of reinforced concrete. In the power block, the reactor building, auxiliary and control building, the fuel building, the diesel building and the turbine house have separate foundations. Differential settlements subsequent to building completion are calculated to be in the range 15mm to 30mm.

The buildings of the power block are constructed of reinforced concrete with some internal steel framing. The reactor building comprises a cylindrical steel-lined post-tensioned concrete structure and internal structures supporting the primary coolant system components.

The auxiliary building houses the majority of reactor auxiliary and safety systems including:

chemical and volume control system (CVCS),
residual heat removal system (RHRS),
safety injection system (SIS),
containment spray system (CSS),
heating, ventilating and air conditioning plant.

The fuel building contains the spent fuel storage pond. It is the reception point for new fuel, and contains the spent fuel flasking facility and the loading bay for transport of spent fuel from the site.

The turbine hall follows the layout used for nuclear stations in the UK since 1970, with the main steam end of the set towards the reactor and the machine axis transverse to the longitudinal axis of the building. A central loading bay separates the two sets and the raw water treatment plant is located below it.

The mechanical and access control annexe separates the turbine hall from the reactor complex and contains the steam generator feed pumps, the steam distribution mains and the access control area for the reactor auxiliary building.

The annexe is a reinforced concrete structure designed to withstand the design earthquake conditions, albeit with some structural damage, without collapsing onto the auxiliary building. Most of the strength is derived from the four reinforced concrete shear towers, one at each end and two in the centre incorporated into the annexe structure. The central shear towers also supply the necessary restraint to the main steam headers against pipewhip in the event of a break in the steam pipes.

2.1 Seismic Design

The design of plant within the station is influenced by the need of some of the systems to remain operative and ensure safe shutdown of the plant following extreme disruptive events. Of such events, an earthquake is probably the most widespread in its effect, and in the design process very specific provision is made for this occurrence.

For design purposes an earthquake is characterised by the peak accelerations experienced by the surface of the ground. For the design of Sizewell B, the specified peak accelerations are 0.25 times that due to gravity (0.25g) in the horizontal direction and 0.17g in the vertical direction. The probability of these accelerations being exceeded is estimated to be no more than one in 10,000 per year. The seismic event specified for design purposes is known as 'safe shutdown earthquake' (SSE).

When a seismic event occurs, the acceleration of the ground varies continuously. The extent to which the station structures and components are affected by this vibratory motion depends on whether their natural frequencies of vibration coincide with the predominant frequencies of the motion of the ground. Analyses are carried out to calculate the movement of the ground under and around the station, to provide response spectra at the various critical points in the building structures.

2.1.1 *Design of the Plant*

As it is not reasonably practicable to design the whole station to withstand a seismic event, it is first necessary to decide which parts of the plant are required to withstand it. This is done in four steps:
(1) Selection of those plant faults which must be considered in conjunction with the SSE. Such faults are chosen either because they are sufficiently probable that their coincidental occurrence must be considered credible, or because they could be caused by the event.
(2) Specification of the safety functions which are required to achieve safe shutdown following these faults.

(3) Identification of the equipment and structures which are required to satisfy these functions or whose failure could prevent satisfaction of these functions.
(4) Specification of the design requirements for this equipment and structures.
   Plant which must be designed to withstand a seismic event are generally within the following categories:
   (a) Plant needed to trip the reactor and initiate the operation of safeguard equipment if a frequent fault occurs. Such faults include inadvertent reactor trip, loss of the grid and loss of main feed.
   (b) Plant needed to achieve a hot shutdown condition after frequent faults.
   (c) Plant needed to achieve a cold shutdown condition and to maintain it there safely.
   (d) All equipment and structures whose failures could directly or indirectly cause a fault whose consequences cannot be mitigated by equipment capable of surviving the event.
   Structures and equipment are classified according to their role following a seismic event using a system of seismic classification. This system has four classes of plant which indicate the importance of the plant to safety in the event of an earthquake. The classification assigned to a plant item indicates whether or not it has to function during and after an earthquake and, where functioning is not required, whether failure could damage safety-related plant or lead to activity release from the radwaste system. The classification provides a systematic means of identifying which plant needs to be designed and qualified to withstand a seismic event.

2.1.2 Design Limits

   The limits on seismically-induced stresses are in line with current practice in Europe and the US. These limits are chosen to ensure adequate functioning of the plant during and after a seismic event. Adequate functioning varies from plant item to item. For example, a fluid system pipe may be required to retain pressure; a structure not to collapse; a valve to open; and another to remain closed. Because it is not intended that the station is designed to be operated at power after a major seismic event without the possibility that repair will be needed, the limits chosen allow equipment or structures to distort where this does not affect their functioning. Generally where motion is involved, such as valve movement or pump operation, undue distortion would affect functioning and hence the stress limits chosen are lower than when the component or structure performs a purely passive role.

2.1.3 Seismic Qualification

   Demonstration of the adequacy of the design is known as the 'qualification', and is achieved in different ways depending on the nature of the plant.
   The adequacy of buildings is demonstrated by analysis. The stresses in the structure are calculated and shown to be lower than design code limits.
   The reactor coolant system (RCS) is qualified by analysis using methods similar to those used for building analysis. RCS components have already been seismically-qualified for a number of different station designs on sites having a range of soil properties and input motions. Preliminary calculations for Sizewell B indicate that qualification of these components is not a problem.

Either calculation or testing may be used for qualifying equipment. Most electrical equipment and some mechanical equipment for which adequate functioning involves moving parts are qualified by testing according to standards laid down by the US Institute of Electrical and Electronic Engineers (IEEE 344 - 1975). Qualification of equipment by testing is accomplished by mounting the equipment on a shaker table and applying a motion representative of that to which it would be subjected during an SEE.

3. MAIN PLANT DATA

Station Output

Rated thermal power of reactor (inc 14MW from pumps)	3425MW
Gross electrical output	1245MW
Station internal power consumption	70MW
Net electrical output	1175MW

Fuel Assembly (per reactor)

Array	17 × 17
Number of fuel rods	264
Number of guide tubes:	
for absorber	24
for in-core instrumentation	1
Full length (without control spider)	4.058m
Width	214mm
Rod pitch	12.6mm
Number of grids	8
Mass of UO_2	523.4kg
Fuel rod: length	3.851m
outside diameter	9.5mm
cladding material	zircaloy 4
cladding thickness	0.57mm
initial internal pressure (He)	24.1 bar
Fuel pellet: material	UO_2
density (% of theor. density)	95%
length	13.5mm
diameter	8.19mm

Reactor Core

Number of fuel assemblies	193
Active height	66m
Equivalent diameter	3.37m
Rod cluster control assemblies: absorber	Ag-In-Cd

Number of assemblies	53
Absorber rods per assembly	24
Fuel rod linear heat rating: average	17.8kW/m
maximum	41.3kW/m
Fuel rod surface heat flux: average	$600kW/m^2$
maximum	$1389kW/m^2$
Enrichments: first core	2.1/2.6/3.1%U235
reload	3.1%U235

Coolant

Design conditions	
pressure	172.4 bar
temperature	345°C
Operating conditions:	
pressure: vessel inlet	158.3 bar
vessel outlet	155.1 bar
temperature: vessel inlet	293.4°C
vessel outlet	324.9°C
Flow rate	18,749kg/s

Reactor Vessel

Overall height: with the head	13.55m
without the head	10.08m
Inside diameter	4.39m
Total thickness (opposite the core)	215mm
Minimum stainless cladding thickness	3mm
Inlet nozzle inside diameter	698mm
Outlet nozzle inside diameter	736mm
Mass (including head)	385t

Steam Generator

Overall height	20.6m
Upper part diameter	4.47m
Lower part diameter	3.44m
Tubesheet thickness	534mm
U-tubes: number	5626
outside diameter	17.48mm
thickness	1.02mm
total heat transfer area	$5110m^2$
Masses:	
total (empty)	325t
total (for normal operation)	393t

Reactor coolant characteristics:
    inlet temperature   324.9°C
    outlet temperature   293.3°C
    inlet pressure   155 bar
    pressure drop   2.6 bar
    flowrate   4687kg/s
Feedwater temperature   227°C
Steam temperature   285°C
Steam pressure   69 bar
Steam flow rate   477kg/s

Reactor coolant pump

Speed   1485rev/min
Developed head   88m
Flow rate   6.33m^3/s
Motor power rating   6MW

Pressuriser

Overall height   16.1m
Inside diameter   2.12m
Design conditions: pressure   172.4 bar
                   temperature   360°C
Volumes: total   51.0m^3
         water (at full power)   30.6m^3
Heaters: number   78
         total heater power   1.8MW
Relief valves: number   2
               unit capacity   26.5kg/s
Safety valves: number   3
               opening pressure   172.4 bar
               unit capacity   52.9kg/s
Pressuriser relief tank:
    total volume   51.0m^3
    normal liquid volume   38.2m^3
    design pressure   7.7 bar
    design temperature   171°C

Reactor Coolant Pipes

Inside diameters:
    hot leg   737mm
    intermediate leg   787mm
    cold leg   699mm
    pressuriser surge line   292mm

Accumulators

Number	4
Total volume per accumulator	57.3m^3
Water volume per accumulator	36.1m^3

Turbine Conditions

Pressure at inlet	66.6 bar
Temperature at inlet	282°C
Flow rate at inlet	955kg/s
Condenser pressure	53mbar

Condensate Extraction Pumps

Number per turbine	$3 \times 50\%$
Axis	vertical

Main Feedwater Pumps

Number per turbine	$3 \times 50\%$
Drive	induction motor
Axis	horizontal
Design flow per pump	$0.5\text{m}^3/\text{s}$
Developed head	973m
Temperature	160°C
Speed	variable through fluid couplings

Cooling Water Pumps

Number per turbine generator	$2 \times 50\%$
Axis	vertical
Design flow per pump	$12\text{m}^3/\text{s}$
Developed head	9m
Motor power rating	1.4MW

3.1 Operational Characteristics

Although intended primarily as a base-load station, the PWR system is capable of considerable operational flexibility which makes it an attractive plant for both early and late-life operation.

Table 6.1 shows the control requirements specified for the plant to take advantage of the reactor system flexibility.

TABLE 6.1. CONTROL REQUIREMENTS

Operating Mode	Specified Requirement	Plant Capability
Continuous operation	Anywhere in the range 20% to 100% CMR	Design basis
Load changing	Up to 5% per min.	Design basis (See Note 1)
Frequency control	Up to 50% per Hz	Achievable but can be a factor limiting plant component life (See Note 2)
Spinning reserve		Up to 10% step increases in power is achievable
Load following	3400 cycles 100% to 80% CMR 4400 cycles 100% to 60% CMR 1100 cycles 100% to 50% CMR	Design basis for plant taken as single daily cycles
Load rejection	20% loss in unit load without unit trip. Total loss of external load without reactor trip.	Turbine generator bypass dump facility provided to meet these requirements
Single turbine trip	Reactor must run through without tripping.	Turbine generator bypass dump facility provided to meet this requirement
Variable frequency	49.5Hz to 51Hz. 47Hz to 49.5Hz	Full output available. CMR reduced in proportion to frequency.

Note 1
Initial raise to power requires the fuel elements to be 'conditioned' by very slow power increases of about 1% per hour. Similarly after prolonged low power operation, this slow rate of increase is required to avoid sudden thermal growth of the fuel pellet and subsequent splitting of the cladding.

Note 2
Prolonged operation in the frequency control mode can cause higher rates of wear on control rod drive mechanisms and their earlier replacement. Also, reassessment of the inspection frequency of pressure parts may be needed to reflect the additional high cycle fatigue loading.

3.2 Availability and Plant Life

The PWR is an off-load refuelled reactor system with the refuelling period normally in the summer months. Hence the 'settled down' average annual availability of about 70% reflects a winter peak availability of about 85%. In making the economic justification for the Sizewell B PWR, an average annual availability of 64% was used.

A life of 40 years is assumed for the figure of 70% availability. Whilst this assumes some major plant replacements, the limiting factor is the life of the reactor pressure vessel. The design life is based on an exhaustive analysis of the vessel operating regime. However, a life of 35 years has been used in the economic justification calculations.

4. PLANT SELECTION AND REFERENCE DESIGN

The main plant falls into three groupings:
(1) The nuclear steam supply system. This area of plant is specific to the reactor type chosen as well as to the equipment supplier.
(2) The power conversion systems (including heat rejection systems), which are similar in most respects to plant performing the same duties in fossil-fired power stations.
(3) Auxiliary systems, such as fuel handing, radioactive waste management and safeguard systems which are broadly similar to comparable systems in existing UK nuclear stations, although specific to the detailed requirements of the chosen nuclear steam supply system (NSSS).

4.1 Selection of Nuclear Steam Supply System Type and Licensor

The choice of NSSS to be adopted for a power station is absolutely central to its concept and the starting point for the design process. The choice of a NSSS design is closely associated with the choice of a licensor.
The NSSS comprises a number of components:
(1) Fuel assemblies.
(2) Control assemblies and their drives.
(3) Reactor internals.
(4) Reactor instrumentation.
(5) Reactor pressure vessel.
(6) Steam generators.
(7) Reactor coolant pumps.
(8) Pressuriser.
(9) Interconnecting pipework.
(10) Auxiliary systems such as: chemical and volume control, residual heat removal, tools for handling fuel and reactor internal components.
There were three possible approaches to the choice of the NSSS design:
(1) Choose an established existing design.
(2) Develop a UK design from scratch.
(3) Put together a design incorporating the best features of existing designs.
Of these, only the first was considered practical. To develop a design of NSSS from scratch would invalidate one of the main reasons for considering a PWR, which is the wealth of existing experience. To develop a design incorporating the best features of the various existing designs would be no better: it would require the UK designers to put the pieces together and ensure that all interactions between them were identified and adequately catered for. Furthermore this latter approach could involve a number of different NSSS vendors, thereby causing significant commercial difficulties related

to confidentiality, guarantees and responsibilities. However, having decided to adopt an existing design, there is every reason to examine it in the light of established UK practices and experience to see whether improvements should be introduced.

The CEGB confirmed the choice of the NSSS design and hence of the licensor in 1978/79. At this time there were four companies with developed and operating commercial units which were considered to be suitably proven. Having selected Westinghouse as the supplier of a proven design it was necessary to consider which of the Westinghouse size options should be adopted.

4.2 Choice of Nuclear Steam Supply System Size

The current Westinghouse model range includes units from 1800MW(Th) to 3800MW(Th). Changes of unit size within this range are achieved by providing units having two, three or four standard heat transfer loops each comprising a steam generator, a reactor coolant pump and the associated connecting pipework. 2-loop units have thermal outputs of about 1800MW; 3-loop units, about 2700MW; and 4-loop units, either 3425MW or 3800MW.

The main considerations in the choice of the number of loops were:
(1) Economics.
(2) Compatibility with the use of turbine generators whose design is based on that of existing standard UK designs.
(3) Technical maturity of the unit.
(4) Trends in the choice of unit size in the USA and Europe.

There are two standard sizes of turbine generator which are used in the CEGB's most recent power stations. These have electrical outputs of 500MW and 660MW. Use of turbine generators of these sizes leads to consideration of four potential NSSS/turbine generator combinations:

1800MW(Th) 2-loop NSSS with 1 × 660MW(e) turbine.
2700MW(Th) 3-loop NSSS with 2 × 500MW(e) turbines.
3425MW(Th) 4-loop NSSS with 2 × 660 MW(e) turbines.
3800MW(Th) 4-loop NSSS with 2 × 660MW(e) turbines.

The economics of electricity generation are such that the larger NSSS units have lower generating costs per unit of output. Consideration of this economy of scale led to the judgement that detailed assessment of the first of these options was not warranted, particularly as relatively few 2-loop plants had been ordered or constructed and future trends favoured three and, more particularly, 4-loop plants.

The use of a 3-loop unit to supply two 500MW(e) standard generators is practicable and, at the time the decision was made, there were more 3-loop Westinghouse units in service and being constructed than 4-loop units. Despite this, it was judged that it was preferable to use the more modern 660 MW(e) turbine generator components. It was also more economic to use the larger NSSS. For these reasons the 4-loop unit was chosen. See Fig 6.4.

There are two existing Westinghouse 4-loop designs from which to choose: a 3425MW(Th) unit and 3800MW(Th). Although the 3425MW(Th) reactor unit would not utilise the full capacity of the twin 660MW(e) turbine generators, it was recognised

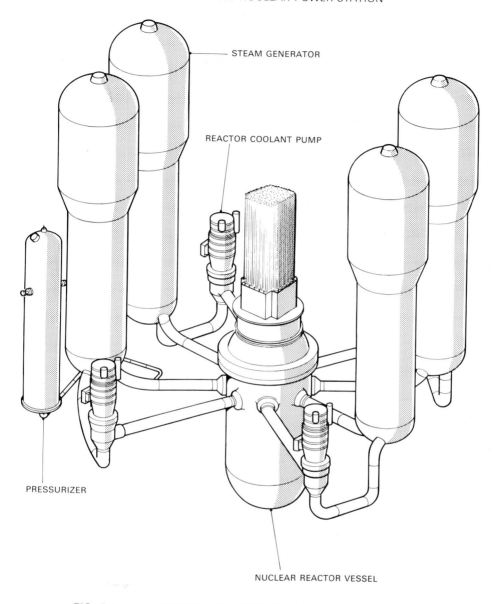

STEAM GENERATOR

REACTOR COOLANT PUMP

PRESSURIZER

NUCLEAR REACTOR VESSEL

FIG. 6.4   GENERAL ARRANGEMENT OF REACTOR SYSTEM

that of the two 4-loop Westinghouse designs the 3425MW(Th) unit was better established. Six of these units were in service in 1979 when the decision was made and others were under construction.

On the other hand the 3800MW(Th) unit, whilst preferable from an economic point of view, was a recent development and, although several units were under construction or on order, none was operating. The 3425MW(Th) unit was therefore chosen as the basis of a UK PWR because of the more extensive design, manufacturing and operational experience available at the time of the decision, and because of the substantially greater volume of operating experience which will have accumulated by the time Sizewell B is scheduled to enter service. At some time in the future, when more experience is available with the 3800MW units, the choice of unit will be re-assessed to decide whether advantage should be taken of the economic benefits of the larger unit in later PWRs.

4.3 Selection of Power Conversion System

The majority of PWRs operating and under construction use a single turbine generator: a few use two. The CEGB had the choice, therefore, between the provision of a single turbine generator of 1200-1300MW capacity, which would require either that a new UK turbine generator be developed or that a UK manufacturer obtain a licence to build a turbine generator to an overseas design; or the provision of two turbine generators each of 600 to 660MW capacity.

Despite being the more expensive in terms of capital cost, the second of these alternatives was chosen for the following reasons:

(1) The turbine generators could be based on and would make extensive use of standard components and technology established in 660MW units in current use in the CEGB's fossil-fired and AGR power stations. The generator would physically remain unchanged and the turbine would require limited changes, which were well within the scope of existing UK technology, to allow it to utilise steam at the lower pressure and temperature obtained from a PWR.

(2) The use of turbine generators based on the well-established 660MW UK design would reduce the possibility of delays to the project and would contribute to the achievement of high station availability, in that national spares exist for turbine, generator and generator transformer components.

(3) Sizewell B is being built to examine the PWR option. To develop a new turbine design in parallel might well prove a limiting factor to that examination.

(4) Whilst of lower capital cost, it is not clear that development of a new larger turbine would prove to be more economic for Sizewell B. It is not obvious that the larger unit would be as reliable as the current 660MW units. Judging by experience with the 660MW units, it might be some years before the larger units achieve the levels of availability which are currently achieved by the 660MW units. Additionally there are the development costs of such a turbine which could only be justified if a series of orders was proposed.

4.4 Selection of Reference Design

Having chosen a specific NSSS plant there are many ways of proceeding to an overall station design.

In the USA the overall design of the station and the specification of that plant which is not within the scope of the NSSS is normally the responsibility of an Architect-Engineer. For Sizewell B this task was initially undertaken by NNC under contract to the CEGB. NNC studied a number of basic station layouts including:

(1) The standardised nuclear unit power plant system (SNUPPS) layout.
(2) A rearrangement of the SNUPPS layout.

(3) A layout based on a spherical containment with an annular auxiliary building having some similarities to the PWR station offered by the German firm Kraftwerk Union.

(4) A layout based on a cylindrical containment having an annular auxiliary building.

In all, some 10 different layouts were examined. Throughout this examination of alternative layouts, the NSSS remained the Westinghouse 3425MW(Th) model, although the design of the safeguard features was refined progressively.

Whilst NNC has a long experience of designing and constructing nuclear power stations in the UK, its experience with PWRs is more limited. In view of this and to follow effectively its overall strategy of utilising proven designs, components and techniques wherever possible, the CEGB specified that the services of an experienced Architect-Engineer should be retained to advise and assist NNC in its task. Bechtel was chosen as consultant Architect-Engineer by NNC in consultation with the CEGB.

After considerable detailed consideration and evaluation by the CEGB and NNC, in consultation with Westinghouse and Bechtel, the SNUPPS design was finally adopted as the basic reference design for Sizewell B power station.

SNUPPS is being built on two sites, in the USA, Calloway in Missouri and Wolf Creek in Kansas. It embodies the Bechtel generic design developed by Bechtel from experience gained on over 30 nuclear power projects.

Although the SNUPPS design was closely followed, certain modifications have been identified as necessary for a UK nuclear plant being built in the 1980s. The modifications are well-understood by the UK designers and have been thoroughly reviewed by the Bechtel and Westinghouse consultants to ensure that they were necessary and did not affect adversely the basic SNUPPS design.

The changes which were made resulted from seven main considerations:

(1) The need to comply as far as reasonably practicable with the CEGB's PWR Design Safety Guidelines and to take account of the Nuclear Installations Inspectorate's (NII) Safety Assessment Principles.

(2) The requirement for additional provisions to limit the exposure of station operating staff to radiation,

(3) The specification by the CEGB of two turbine generators.

(4) The differences that exist between Sizewell and the SNUPPS design sites. Because Sizewell is a sea coast site, sea water is used for cooling purposes whereas the SNUPPS sites obtain their cooling water from a lake or river. Because of its corrosive nature, it is not desirable for sea water to be used for a number of cooling duties whereas lake or river water is acceptable in the SNUPPS design. Examples of these are the coolers used for containment and safeguards equipment room air cooling. These duties have had to be transferred to the component cooling water system which uses purer water as the cooling medium.

(5) The frequency of the electricity supply in the UK is 50Hz compared with 60Hz in the USA. This means that each electric motor must rotate at a different speed from its USA counterpart and, therefore, there must be some changes in the design of rotating machinery. Because the UK motors run at a lower speed, rotating machinery is generally of larger dimensions to perform the same duty.

(6) The requirement to design a British PWR to allow for any differences in engineering practices in the UK, particularly to allow equipment of British manufacture to be used extensively.

(7) The need to incorporate experience gained in the construction of SNUPPS and from PWR design and operating experience since the SNUPPS plant was designed.

A 'Revised Reference Design' was issued in April 1982 and this chapter describes the plant and systems conforming with this 'Revised Reference Design'.

5. GENERAL DESCRIPTION OF MAIN PLANT

Like all major thermal power stations, be they fossil-fuelled or nuclear, the PWR has three major sections.
(1) Steam supply system.
(2) Power conversion system.
(3) Heat rejection system.

The power conversion and heat rejection systems are essentially the same processes using similar plant whatever type of fuel is used.

Fig 6.5 shows diagrammatically the general arrangement of the three heat transfer cycles in the PWR power station. The NSSS consists of the reactor, the reactor coolant system, and a number of auxiliary and safety systems.

The reactor coolant system (RCS) (also referred to as the 'primary circuit') consists of the reactor within its pressure vessel and four cooling loops connecting the reactor vessel to heat exchangers (steam generators). The reactor vessel and the cooling loops are filled with demineralized natural water which is pumped through the reactor and the steam generators. The water takes heat from the reactor and gives it up in the steam generators. This is a closed cycle during normal operation at power though some water is bled from the system for purification and returned to the system. In order to produce steam at 285°C in the steam generators the reactor coolant water has to be at a higher temperature, so that to prevent boiling the reactor coolant is held at a pressure of some 155 bar. The pressure is controlled by a pressure control system and a surge tank known as the pressuriser vessel connected to one of the cooling loops. The coolant temperature is raised from 293°C to 325°C as it passes through the reactor core.

The reactor, the reactor coolant system and parts of its auxiliary systems are enclosed within a cylindrical prestressed concrete containment vessel. The purpose of this is to provide a further barrier against leakage to atmosphere of radioactive products should any escape from the primary circuit.

The power conversion system (also referred to as the 'secondary circuit') consists of the condensate, feedwater and steam systems and the steam turbines. Condensate is pumped from the turbine condensers by condensate extraction pumps and feed pumps through the steam generators where it takes up heat from the reactor coolant, and boils, producing steam. The steam passes out from the top of the steam generators, and flows through the main steam pipes to the two steam turbines. The steam gives up about one third of its acquired energy in rotating the turbines and electrical generators. The steam is then condensed in the turbine condensers by sea water, to which the steam gives up the remaining two thirds of its energy. This system is also a closed cycle in normal operation, apart from minor losses. The condensate is below atmospheric pressure, but is raised in pressure to about 70 bar by the pumps before entering the steam generators. In the two turbine generators, work is performed by expansion of

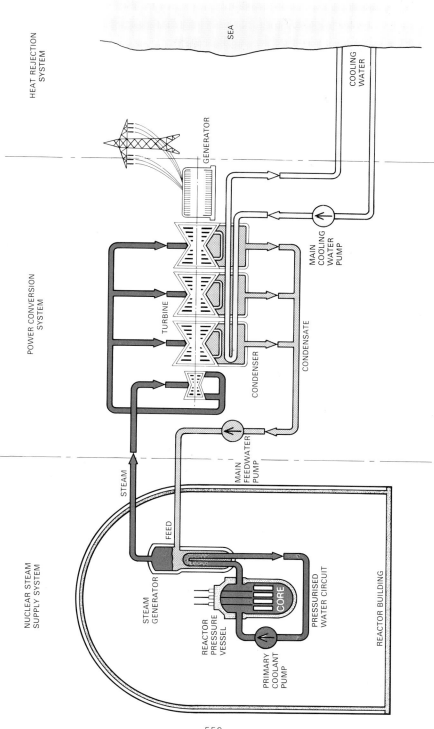

FIG. 6.5   PWR SYSTEM SCHEMATIC

the steam produced in the steam generators. Each turbine is coupled directly to a 660MW electrical generator on the same line of shafting which produces alternating current at 23.5kV. This is raised to 400kV by the generator transformers and transmitted from the station through the National Grid.

The third heat transfer cycle is the cooling water (or circulating water) system. This takes in large quantities of sea water through off-shore tunnels, pumps it through the tube banks in the turbine condensers and returns it to the sea. The flow quantity is about 48m³/s and the heat it takes up in the condensers raises its temperature by about 11°C. This is an open cycle system.

The steam generator tubing forms the boundary between the reactor coolant and the secondary system water and steam. The condenser tubing forms the boundary between the secondary fluid and the sea water.

The main plant referred to above is supported by many auxiliary and control systems. These systems may be categorised in accordance with the plant they support, eg reactor auxiliaries, general station services, etc, or they may be categorised according as to whether they are or are not provided primarily for safety purposes. If the latter categorisation is used the systems provided primarily for safety purposes are variously termed safety systems, safeguard systems, engineered safety features, essential systems, according to usage. Many systems have a role both in normal circumstances and as safeguards in the event of faults or accidents.

In many instances where plant is provided for safety purposes more items of plant than are required for the duty are included in case some of the plant should fail or be under maintenance when called upon. This is termed 'redundancy of plant'. In some cases separate sets of plants of different design are supplied to perform essentially the same function. This is to provide a safeguard against failures of systems due to a common cause affecting the whole system despite the provision of redundant plant. This is termed 'diversity of plant'. When redundant or diverse plant is provided this is usually segregated so that hazardous occurrences such as fires or missiles would be less likely to cause failure of the whole function for which the redundant or diverse plant is provided. Such segregation is achieved either by installing appropriate physical barriers between the units of plant or by separating the units by appropriate distances.

6. REACTOR COOLANT SYSTEM (FIG 6.6)

The main functions of the reactor coolant system (RCS) and its pressure retaining boundary are to:
(1) Transfer heat from the reactor core to the steam generators.
(2) Contain the reactor coolant under operating temperature and pressure and limit leakage of coolant and radioactivity into the containment atmosphere and the secondary system.
(3) Control the pressure of the reactor coolant and accommodate coolant volume changes generated by temperature changes.

The reactor vessel, pressuriser and steam generators are constructed of carbon steel of controlled composition and manufacture. The surfaces in contact with reactor coolant are lined with stainless steel to provide corrosion protection. The coolant pump bowls and interconnecting pipes are constructed of stainless steel. The pressure

SIZEWELL B — PWR NUCLEAR POWER STATION

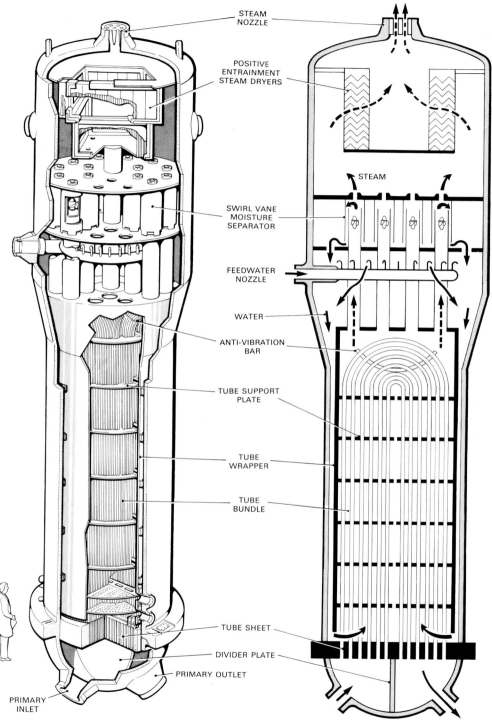

STEAM
NOZZLE

POSITIVE
ENTRAINMENT
STEAM DRYERS

SWIRL VANE
MOISTURE
SEPARATOR

FEEDWATER
NOZZLE

WATER

ANTI-VIBRATION
BAR

TUBE SUPPORT
PLATE

TUBE
WRAPPER

TUBE
BUNDLE

TUBE SHEET

DIVIDER PLATE

PRIMARY OUTLET

PRIMARY
INLET

STEAM

FIG. 6.8   **STEAM GENERATOR**

entrained water, which drops down to mix with the incoming feedwater. Steam driers above the moisture separators increase the steam quality to a minimum of 99.75% (ie the steam contains not more than 0.25% of water). Steam off-take is through a nozzle in the upper dome, to one of the four main steam pipes.

Access openings are provided in the bottom head to permit steam generator tube inspection. Access openings are also provided for inspection of the moisture separators and for water lances to clean away deposits which may settle on the upper surface of the tube plate.

The steam generator is vertically supported by four articulated columns fixed to the concrete floor and the bottom head. Lateral restraint is provided by lower and upper lateral supports which are designed to permit thermal expansion but withstand seismic loads and loads which would arise if a reactor coolant loop or main steam pipe should fail.

A steam generator blowdown system is provided to draw a small water flow continuously from each steam generator just above the tube-sheet to remove any concentration of non-volatile impurities.

This water is flashed into a high pressure (HP) blowdown vessel; the flashed steam is passed into the HP feed heating system of either unit while the water passes to a low pressure (LP) heater which can be connected into either unit LP feed train. The cooled blowdown liquid is then passed to the secondary liquid waste system.

6.3   Reactor Coolant Pumps and Motors (Fig 6.9)

The reactor coolant pumps circulate coolant through the reactor vessel to the steam generator for heat removal and return it to the reactor vessel. There is one pump in each coolant loop, located in the cold leg between the steam generator and the reactor vessel giving 5.9m^3/s flow against 88m head. The pump is a single-stage centrifugal unit driven by a 6MW single-speed electric motor at 1485rev/min. The vertical pump and motor unit is designed as an integral assembly with the motor mounted above the pump. Attached to the lower end of the vertical pump shaft is a single stage 'mixed flow' impeller. The reactor coolant is drawn up through the impeller, discharged through passages in a diffuser and out through the discharge nozzle in the side of the casing. Above the impeller are a water-lubricated bearing and a shaft seal assembly.

The point at which the reactor coolant pump shaft enters the reactor coolant is one of potential leakage. The seal assembly is provided to maintain the integrity of the reactor coolant pressure boundary whilst allowing passage of the rotating shaft through the boundary. Three face-type seals are provided in series to seal against the full pressure in the pump. The first is a controlled leakage, film riding face seal and the second and third are rubbing face seals. The first seal is nearest to the impeller. Water from the chemical and volume control system (CVCS) is injected into the seals at high pressure, so that about 60% of the injected water passes through the first seal into the primary circuit, so preventing outward leakage from the primary circuit. The remainder is either returned to the CVCS or passed to the reactor drains system. Between the impeller and the bearing is a thermal barrier and heat exchanger, which limits heat transfer between hot reactor coolant water and the seals, so that the seals do not fail due to overheating if the seal water supply fails.

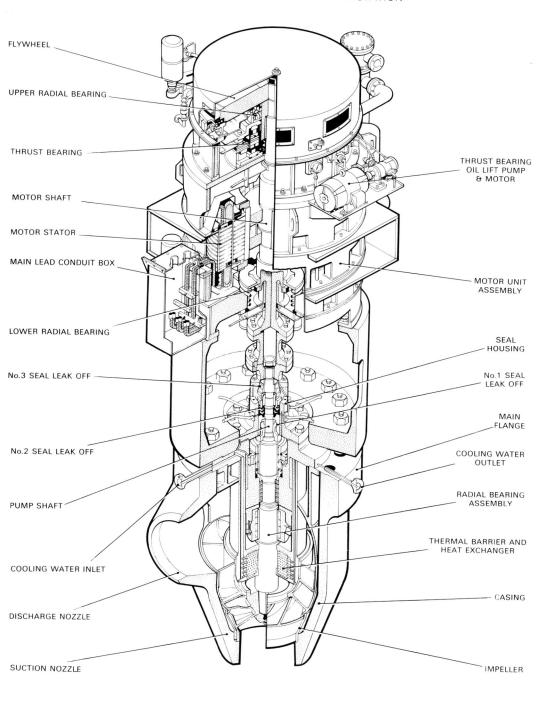

FLYWHEEL

UPPER RADIAL BEARING

THRUST BEARING

MOTOR SHAFT

MOTOR STATOR

MAIN LEAD CONDUIT BOX

LOWER RADIAL BEARING

No.3 SEAL LEAK OFF

No.2 SEAL LEAK OFF

PUMP SHAFT

COOLING WATER INLET

DISCHARGE NOZZLE

SUCTION NOZZLE

THRUST BEARING
OIL LIFT PUMP
& MOTOR

MOTOR UNIT
ASSEMBLY

SEAL
HOUSING

No.1 SEAL
LEAK OFF

MAIN
FLANGE

COOLING WATER
OUTLET

RADIAL BEARING
ASSEMBLY

THERMAL BARRIER AND
HEAT EXCHANGER

CASING

IMPELLER

FIG. 6.9   REACTOR COOLANT PUMP

The motor is a single-speed, air-cooled, vertical, squirrel cage induction type. The motor bearings include an oil lubricated double thrust bearing and two oil lubricated radial bearings. Each motor is equipped with an anti-reverse rotation device. A flywheel situated above the motor ensures that if the pumps trip they will have sufficient rotational inertia to run down slowly, so providing some core cooling in the initial transient state.

The pump is vertically supported by three columns fixed to the concrete and to the casing. Lateral dampers are provided between the pump casing and the motor supports and the concrete wall.

6.4 Pressuriser and Pressure Control System (Fig 6.10)

The pressuriser is a vertical, cylindrical vessel with hemispherical top and bottom heads, constructed of carbon steel with austenitic stainless steel cladding on all surfaces exposed to the reactor coolant. It has a volume of $51m^3$ which is about one eighth of the total reactor coolant system volume. A surge line connects the pressuriser to one reactor hot leg. The pressuriser is supported by a skirt welded to its bottom head, bolted to supporting steelwork which is anchored to the concrete floor. Lateral support is provided against seismic or pipe rupture loads.

During normal operation the pressuriser is partly filled with water and partly with steam at a pressure of about 155 bar. The steam space in the pressuriser acts as a buffer which limits the magnitude of the pressure change due to a coolant volume change and so facilitates control of the pressure which is effected by the pressuriser heaters and sprays. If the reactor coolant pressure needs to be increased, electrical immersion heaters within the pressuriser are automatically switched on increasing the quantity of steam and consequently increasing the pressure. These heaters, 78 in number, are located in the pressuriser by penetration sleeves, seal welded on the internal surface of the bottom head in three concentric circles around the surge nozzle. A small quantity of cool water is normally sprayed through the spray nozzle, fed by two spray lines connected to two of the cold legs of the loops. The cooling effect is offset by a small amount of immersion heating. If the pressure needs to be decreased this spray flow is increased. This additional cool water spray causes a net condensation of the steam and consequently decreases the pressure.

The pressuriser has sufficient steam and water volumes to prevent uncovering the heaters or discharging water through the safety valves during normal operation or minor fault conditions. If there is an increase in the reactor coolant pressure which is beyond the capacity of the pressuriser spray to control, two solenoid-actuated power-operated relief valves open automatically. These valves are provided with electrically-operated isolation valves in case they should fail to reseal after use. Additional protection against overpressure is provided by three spring-loaded safety valves which open by direct fluid pressure action.

Steam discharged through the safety and power-operated relief valves is received in the pressuriser relief tank. This tank is located on the floor of the reactor containment. It is normally part-filled with water with nitrogen gas in the space above. Use of nitrogen, which is a relatively inert gas, reduces the potential for sudden pressure changes and chemical corrosion. Steam from the safety and relief valves is introduced

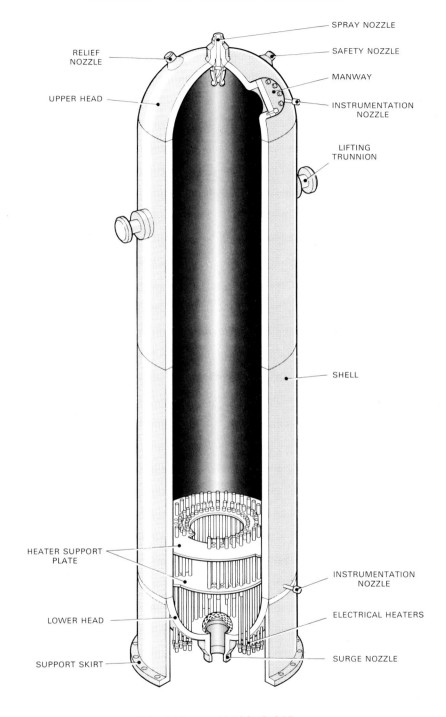

FIG. 6.10   PRESSURISER

through a sparge pipe below water level in order to achieve effective condensation and cooling. Should the relief tank become overpressurised it is relieved by bursting discs discharging into the containment.

6.5 Reactor Core, Fuel Assemblies and Control Rods (Fig 6.11)

The nuclear fuel for the reactor is contained in fuel rods 3.85m long and 9.5mm in diameter. Each fuel rod consists of a column of cylindrical pellets of slightly enriched uranium dioxide contained in cladding of cold-worked Zircaloy-4 tubing which is plugged and seal welded at the ends to encapsulate the fuel. All fuel rods are pressurised with helium during fabrication. This reduces stresses and strains in the cladding during operation when the reactor power is changed. 264 fuel rods are held together in a square section assembly. 193 of these fuel assemblies comprise the reactor core.

In each fuel assembly the 264 fuel rods are mechanically located in a 17 × 17 square array (289 positions). The fuel rods are supported at intervals along their length by eight grid assemblies which maintain the lateral spacing between the rods. The grid assembly consists of interlocked straps which contain spring fingers and dimples for fuel rod support as well as coolant mixing vanes in six of the grids. The centre position in the fuel assembly is reserved for the in-core instrumentation, while the remaining 24 positions in the array are equipped with Zircaloy guide thimbles joined to the grids and the top and bottom nozzles. Depending upon the position of the fuel assembly in the core, the guide thimbles are used as core locations for rod cluster control assemblies, burnable poison rods, neutron source assemblies, or are simply plugged. The bottom nozzle is a box-like structure which serves as a bottom structural element of the fuel assembly and directs the coolant flow through the assembly. The top nozzle functions as an upper structural element and a partial protective housing for the rod cluster control assembly or other components.

Control of the nuclear chain reaction in the core is achieved by neutron absorbing rods arranged in clusters, called rod cluster control assemblies, and by varying the concentration of boric acid in the reactor coolant. The control rods allow for rapid control of the chain reaction including its rapid termination (reactor tripping). Variation of the boric acid concentration is used for relatively slow trimming of the nuclear processes to allow for fuel depletion which requires the boron concentration to be progressively reduced over the periods between successive refuellings, or prolonged operation at a changed load, which results in changes in the neutron absorption in the fission products contained within the fuel.

A rod cluster control (RCC) assembly comprises a group of 24 individual absorber rods fastened at the top end to a common hub and forming a spider assembly. The neutron absorber, silver-indium-cadmium alloy rod, is contained in cold-worked stainless steel tubing sealed by welded end plugs. There are 53 rod cluster control assemblies arranged in four control banks and five shutdown banks. These banks contain 4, 8, 8, 5, 8, 8, 4, 4, 4 RCC assemblies respectively. Within any bank all control rods are maintained at the same level and the whole bank is moved as a unit. The shutdown banks remain at the full withdrawn position except when inserted to shut the reactor down. The control banks are used in normal operation to make short term adjustments to reactor power. They may be inserted into the core only in a prescribed

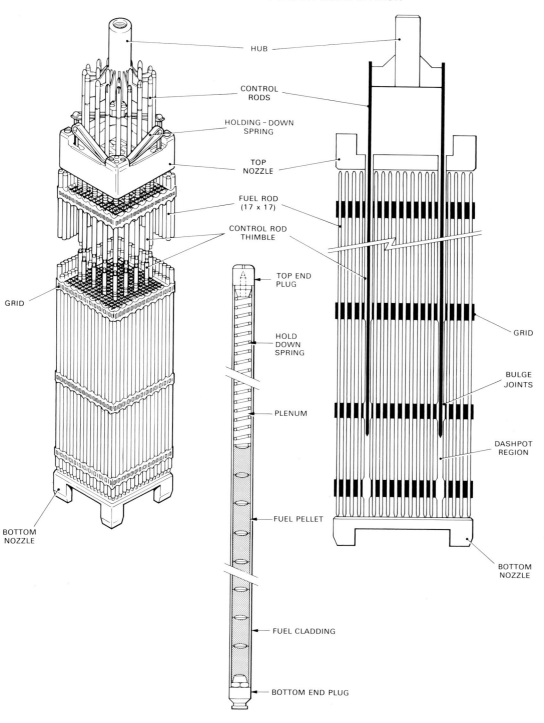

HUB

CONTROL RODS

HOLDING - DOWN SPRING

TOP NOZZLE

FUEL ROD (17 x 17)

CONTROL ROD THIMBLE

GRID

BOTTOM NOZZLE

TOP END PLUG

HOLD DOWN SPRING

PLENUM

FUEL PELLET

FUEL CLADDING

BOTTOM END PLUG

GRID

BULGE JOINTS

DASHPOT REGION

BOTTOM NOZZLE

FIG. 6.11   FUEL ASSEMBLY

571

sequence with a specified overlap between partially inserted banks. These arrangements are designed to give a good neutron flux and hence power distribution in the reactor core. The neutron flux is a measure of the flow rate of neutrons and hence of reactor power.

When the reactor is started up for the first time with a complete core of new fuel there is a surplus of reactivity. To absorb this, burnable poison rods are used. These are neutron absorbers which have the characteristic of becoming progressively less effective as they are irradiated in the reactor. The rods are grouped into assemblies which can be inserted into the guide thimbles of a fuel assembly. With the first charge of fuel, 68 such assemblies are inserted, but are not needed thereafter.

In order to provide a measurable base neutron flux level to ensure that the neutron flux detectors are operational when the reactor is shutdown, primary and secondary neutron source rods are used. The primary source rods, containing a radioactive material (californium 252), spontaneously emits neutrons during initial core loading and reactor startup. The secondary source rod contains a stable material (antimony and beryllium) which is activated by neutron bombardment during reactor operation.

6.6 Control Rod Drive Mechanisms (Fig 6.12)

Each rod cluster control assembly is actuated by a control rod drive mechanism. The drive mechanism is an electromechanically-operated jacking device which is designed to move the rod clusters up or down in a series of 15.9mm steps. It is operated by energising, in sequence, electrical coils which are outside the primary circuit pressure boundary. It requires a continuous electrical current to hold the control rods out of the core. If the supply is cut off, whether by accident or design, the rods will fall rapidly into the core by force of gravity, thus tripping the reactor.

7. CONTAINMENT (FIG 6.13)

A large containment building is provided around the reactor coolant system and some of the associated auxiliaries. The containment performs a number of functions such as:

(1) It serves to house the reactor coolant system and other equipment during normal operation, containing any minor leakages from the enclosed systems, protecting them from external incidents and shielding operators and the public from their emitted radiation.

(2) Following events involving large energetic releases of steam and water (ie loss of coolant accident (LOCA) and steam or feed pipe breaks) it contains the energy release so that general damage to the station and, in particular to safeguards equipment, does not occur.

(3) Following events such as LOCA it contains radioactivity released from the fuel or the reactor coolant system to the containment atmosphere and thereby prevents its release to the general environment.

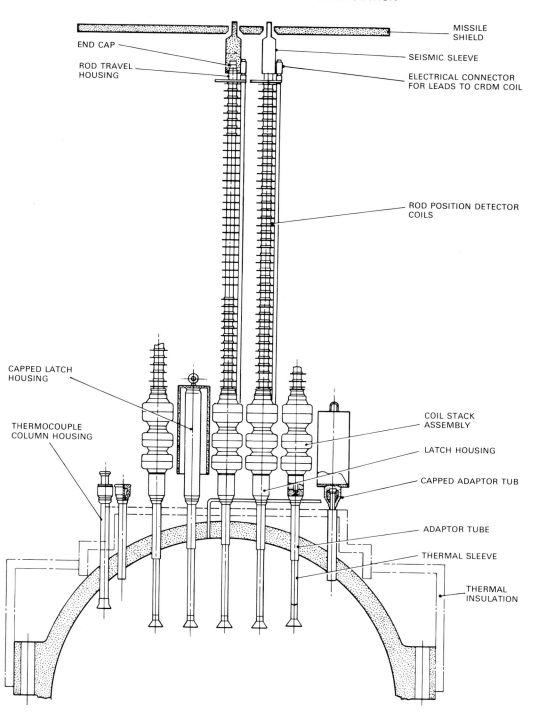

MISSILE
SHIELD

END CAP

SEISMIC SLEEVE

ROD TRAVEL
HOUSING

ELECTRICAL CONNECTOR
FOR LEADS TO CRDM COIL

ROD POSITION DETECTOR
COILS

CAPPED LATCH
HOUSING

THERMOCOUPLE
COLUMN HOUSING

COIL STACK
ASSEMBLY

LATCH HOUSING

CAPPED ADAPTOR TUB

ADAPTOR TUBE

THERMAL SLEEVE

THERMAL
INSULATION

FIG. 6.12   CONTROL ROD DRIVE MECHANISMS

573

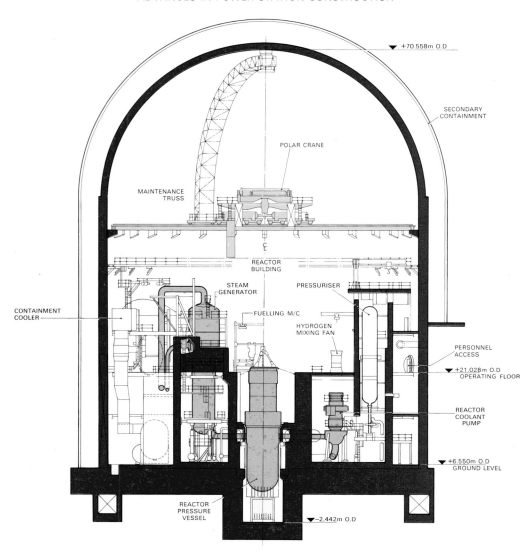

+70.558m O.D

SECONDARY
CONTAINMENT

POLAR CRANE

MAINTENANCE
TRUSS

REACTOR
BUILDING

STEAM
GENERATOR

PRESSURISER

CONTAINMENT
COOLER

FUELLING M/C

HYDROGEN
MIXING FAN

PERSONNEL
ACCESS

+21.028m O.D
OPERATING FLOOR

REACTOR
COOLANT
PUMP

+6.550m O.D
GROUND LEVEL

REACTOR
PRESSURE
VESSEL

-2.442m O.D

FIG. 6.13   CONTAINMENT BUILDING

The containment is a prestressed concrete cylindrical structure with a hemispherical prestressed concrete dome roof. The containment base is of reinforced concrete containing a keyhole-shaped vault to accommodate the reactor pressure vessel and its instrument guide tubes at low level. For leaktightness the concrete structure has a welded 6mm thick steel liner attached to its internal surfaces.

The containment also serves to protect the reactor primary coolant plant. Massive reinforced concrete internal structures support the primary coolant plant within the containment. Structures embedded in these are designed to restrain the plant and prevent damage in the event of an earthquake and limit damage due to postulated

574

failure of major high pressure pipes. The concrete internal structures also function as radiation shields to shield operating and maintenance staff from excessive radiation exposure from the reactor coolant system components. The containment internal height is 64m, its internal diameter is 45.7m, and its wall thickness is 1.3m. It supports within it at high level a polar crane rated at 200t for operations and 450t for construction.

Penetrations are provided for pipes and cables to enter the containment and for man access. The two man access penetrations each incorporate two interlocked doors with an air lock between. There is an equipment hatch for bringing large plant components out of the containment; this is provided for plant construction, repair and replacement purposes. All these penetrations are firmly anchored and sealed into the containment wall. A containment isolation system is provided to ensure that if radioactivity is released inside the containment all penetrations of the containment shell are closed, except those essential for the safety of the reactor. All pipes and ducts passing through the containment wall have at least two isolation valves or barriers. On receipt of a 'safety injection' signal or a manual signal from the control room all the appropriate penetration isolation devices are automatically closed.

As well as the containment isolation system there are other systems which are provided to ensure effectiveness of the containment after an accident. These are the containment spray system, the containment fan coolers and the combustible gas control system.

To provide further control of leakages from the containment, a secondary containment is provided. This consists of the auxiliary building and a steel-framed cylindrical enclosure building, with a hemispherical dome and sealed cladding. The enclosure building is sealed to the auxiliary building and fuel building. These buildings surround the whole of the primary containment, except the immediate locality of the steam and feed pipe penetrations, and are arranged to act as a leakage collector. On receipt of a signal indicating that a LOCA has occurred the normal building ventilating systems are closed down and an emergency exhaust system started. The emergency exhaust system draws a slight vacuum in the auxiliary and enclosure buildings to control out-leakage and it discharges through high efficiency filters and a stack to atmosphere.

8. AUXILIARY SYSTEMS

There are a number of auxiliary systems needed to support the reactor coolant system and the containment systems. These in turn are dependent on the station service systems.

8.1 Reactor Auxiliary Systems

8.1.1 *Chemical and Volume Control System (Fig 6.14)*

The chemical and volume control system (CVCS) is one of the principal auxiliary systems supporting the reactor coolant system during normal operation. It controls the chemistry of any reactor coolant as well as the volume (ie the inventory) of coolant in the reactor coolant system.

During normal operation a small quantity of reactor coolant (4.7kg/s) is continuously discharged from a reactor coolant cold leg to the CVCS where it is chemically processed before being returned to the RCS. The flow taken from the RCS is known as a letdown flow and that returned as a charging flow. Because the CVCS processes require relatively low temperatures and pressures it is necessary first to cool and depressurise the letdown flow.

The first stage of cooling is achieved inside the containment using the regenerative heat exchanger, in which some of its heat is given up to the charging flow, and depressurisation by the letdown orifice. The second stage is achieved outside the containment using a heat exchanger, cooled by component cooling water, and the low pressure letdown valve.

An important function of the system is the control of radioactive contaminants and other impurities in the reactor coolant within specified limits. If chemical clean-up is required the letdown flow is passed through a mixed bed demineraliser and, if further purification is required, through a cation bed demineraliser. The purification process removes corrosion products and contaminants in solution and suspension in the reactor coolant and so reduces the radioactivity deposited within the reactor coolant circuit. After clean-up, or directly if no clean-up is required, the letdown flow is passed to the volume control tank through a spray nozzle in the top. Fission product gases are removed from the reactor coolant in this tank. A partial pressure of hydrogen is maintained above the water in the tank so that dissolved hydrogen in the coolant limits the level of radiolytic oxygen.

The letdown flow is returned from the volume control tank to the RCS by two electrically-driven centrifugal charging pumps in parallel, taking suction from the volume control tank. Normal charging flow can be provided by one of the pumps. The charging flow splits into two paths. The bulk of the flow is directed to the reactor coolant system through the tube side of the regenerative heat exchanger, where it receives heat from the letdown flow which passes through the shell side of the heat exchanger. The remaining flow (2kg/s) is directed, through a filter, to the reactor coolant pumps for seal water injection where it either passes into the RCS where it contributes to the charging flow or it leaks back through the seals and is returned to the inlet of the charging pumps through a component coolant water-cooled heat exchanger.

The processes of letdown and charging are used to control the inventory of coolant in the RCS. A control valve is provided downstream from the charging pumps to control the charging flow so that the correct water level is maintained in the pressuriser. The processes of letdown and charging are also used for emptying and filling the RCS for refuelling, inspection or repair and when the plant is being commissioned.

occasionally be required to operate in order to achieve optimum water conditions if there is a large heat load in the pond, for example when the core is completely removed for vessel inspection, or if ambient temperatures are high. Separate water purification systems are provided for the water in the fuel storage pond and the refuelling cavity.

10. POWER CONVERSION SYSTEM

The power conversion system takes the thermal energy supplied by the reactor system and converts it to electrical energy.

The reactor system and its support systems described in the previous sections are known collectively as the nuclear steam supply system (NSSS).

The NSSS delivers its thermal energy in the form of saturated steam (0.25% wet) at 69 bar, 285°C and at a flow rate of 1908kg/s to the power conversion system.

This system is made up of the following major sub-systems:
(1) Main steam piping system.
(2) Turbine generator system.
(3) Condenser system.
(4) Condensate and feed water supply.

Systems 2, 3 and 4 are housed within the turbine hall and mechanical annexe.

10.1 Main Steam Piping System

This system takes steam from the four steam generators and delivers it to the two turbines. Four 700mm diameter pipes, one connected to each steam generator pass through the reactor containment wall into an adjacent steam and feed valve cell and from thence into the mechanical annexe and the turbine house.

The four pipes then divide into two pairs, each pair feeding one turbine generator steam header. From this header four 500mm pipes take the steam to the HP cylinder of the turbine; two pipes feed the moisture separator reheaters and two pipes feed the turbine bypass system together with a take-off for the shaft seal glands for each turbine. The arrangement of the piping is governed by two criteria:
(1) The arrangement of piping and the associated crossovers between the pipes must be such that the steam generators are equally-loaded over the entire operating range of the turbine generator systems.
(2) The piping design and restraint must be such as to ensure no damage to any safety system in the event of a pipe breaking.

At the outlet of each steam generator there is a flow restrictor to limit the flow in the event of a steam pipe breaking.

In the steam and feed valve cell, which lies between the containment and the mechanical annexe, each steam pipe carries branch connections for six safety relief valves, a branch supplying the auxiliary feed and emergency charging pump turbines and the main steam pipe isolating valve. The latter is a quick-acting valve whose actuator is provided with dual power systems controlled from two separate control sources. These valves are designed to close within 5s against the flows associated with steam pipe breaks either upstream or downstream, thus limiting the escape of steam and the associated transient disturbances.

The steam pipes are firmly anchored and sealed to the containment wall where they pass through, using forged junction components in the steam pipes to form the connection with the penetration sleeves of the containment steel liner. The latter are also anchored within the concrete wall of the containment.

Within the containment, steam pipes are provided with restraint steelwork to prevent pipe whip which could cause escalation of the accident in the event of a pipe failure.

Within the mechanical annexe, pipe whip design is catered for by locating the first series of bends of the piping system within concrete shear towers which are able to absorb the energy of any pipe break impact without affecting the structural integrity of the mechanical annexe.

Manually-operated isolating valves are provided in the pipes leading to the turbine header to enable double isolation to be achieved for maintenance of either turbine.

10.2 Main Turbine Generators

The two identical turbine generators are arranged side-by-side in the turbine house with a central maintenance and loading bay between them. The orientation of the turbine axes has been chosen to reduce the probability of missiles striking the reactor and auxiliary buildings in the event of a turbine disintegration. The turbine generators are 3000rev/min machines each with a normal full load gross electrical output of 622.5MW. Each turbine has a double-flow HP cylinder and three identical double-flow LP cylinders. The turbine generators and exciters are carried on steel frame foundations mounted on the reinforced concrete foundations of the turbine house.

The actual steam cycle chosen has been based on optimising plant and auxiliary power costs against the marginal benefits of increased output. A major difference from conventional steam turbine cycles arises from the saturated state of the inlet steam. On expansion the steam rapidly becomes wetter as it passes through the turbine thus requiring careful detail design to remove water from the steam flow and make provision to counter possible corrosion/erosion effects in the bled steam piping and other plant. Interstage moisture extraction is incorporated into the HP cylinder design and, although steam temperatures are low compared to fossil-fired stations, the high moisture content of the steam requires high strength steels similar to the high superheat machines.

To protect the LP cylinders two moisture separators and reheaters are interposed between the HP and the LP cylinders. These are located one on either side of the turbine and are enclosed within pressure vessels. Wet steam from the HP cylinder exhaust enters the separator where the moisture is removed. The dry steam then passes over the reheater tubing, becomes superheated and exits to the LP cylinders. The reheater tubing is supplied with live steam from the main steam manifold which is fed from the steam generators. The water from the separators passes to the feed and condensate system. Employment of moisture separators and reheaters enables the steam conditions and the design of the LP cylinders to be similar to those of the 660MW sets in AGR and fossil-fuelled stations; this arrangement permits the use of LP rotors from other 660MW plants in an emergency, albeit with modified blades in the first two stages.

Steam from the main steam manifold is brought into the HP cylinder through four steam inlet chests each containing one stop valve and one governing valve. The LP cylinders are connected to the two moisture separator-reheater outlets by two pipes each with one LP emergency stop valve and one LP governing valve. These governing and stop valves all act as quick-closing isolating valves to stop steam flow through the turbine if the generator load is lost and so prevent excessive turbine overspeed. In addition the speed governing system is designed to limit the turbine speed increase in the event of complete loss of off-site electrical load so that tripping on overspeed is avoided. Duplicate overspeed trip devices are provided to give back-up protection in case of failure of the governing system.

The generator is of established design used for AGR and fossil-fuelled stations. The rotor is cooled with hydrogen and the stator windings are water-cooled. The hydrogen coolers, stator water coolers and exciter air coolers are supplied with water from a closed loop demineralised water system, from which heat is removed by a heat exchanger to a sea water system — the auxiliary cooling water system. (See Section 11).

10.3 Condensers

Each of the three LP cylinders has an underslung, spring mounted, transverse condenser unit attached to it.

Each unit contains two independent tube banks, each bank comprising 5432 seam-welded titanium tubes 25.4mm od and nearly 17m long.

The tube banks are arranged side-by-side and are independently connected to the cooling water supply and discharge culverts via isolating valves, thus permitting the condenser to remain in service on one tube bank, whilst the other is undergoing maintenance.

Titanium was chosen because of its well-proven resistance to corrosion and erosion in sea water systems. The tubes are supported by mild steel sagging plates, so spaced as to prevent vibration of the tubes in service. Their vertical location is arranged such that the tubes hog slightly to ease drainage from them on shutdown.

At each end the tubes are roller expanded into double tube plates which are diaphragm mounted to permit differential expansion between the tubes and the condenser shell.

Double tube plates are employed with an interspace sealing system designed to prevent sea water entering the steam side of the condenser. The connections into the interspace for vent and drainage also permit rapid location of the level of any leaking tube.

The steam side tube plate is of mild steel, the seawater side tube plate is of aluminium bronze alloy E.

Air extraction is drawn from the cold end of each tube bank to ensure the lowest possible ratio of steam to incondensible gases at discharge.

The tube banks are fitted with condensate collecting trays so arranged that all condensate passes through a deaerator tray section before entering the condenser hot well. The steam flow pattern within the condenser ensures that the trays are well supplied with steam to scrub the condensate, thus avoiding condensate undercooling and the retention of dissolved incondensible gases.

Each condenser shell is fitted with two steam dump diffuser tubes set in the condenser neck. These receive the steam from the turbine bypass system and distribute it to follow the same path as the turbine exhaust steam. The distribution holes are small so that the energy in each jet is low and soon lost as the steam diffuses. Care is taken to ensure that no jet impinges on any structural component of the condenser.

10.4 Turbine Bypass System

This system is designed to bypass main steam into the turbine condensers and to atmosphere in a controlled manner, thus allowing the reactor to remain in service despite sudden large turbine load transients. It allows the reactor to operate independently of the turbine generators provided the condensers are still in service.

Typically the system is used as follows:

(1) At reactor startup until there is sufficient steam to load the   turbines.
(2) At shutdown to remove decay heat generated steam until the residual heat removal system can take over.
(3) Following turbine trip (one or both machines) when the condenser dump system can accept 59.5% main steam flow and the atmospheric dump valves a further 10.5%. The remaining 30% will be absorbed within the thermal capacity of the RCS and the steam generators while reactor power is reduced at the maximum controlled rate consistent with not tripping the reactor.

The system comprises six, pneumatically-operated, high speed, pressure reducing valves per turbine, drawing steam from a 700mm diameter bus main connected to the main steam supply header.

Each valve discharges its steam through a multi-plate, multi-orifice diffuser and spray cooling section connected to one of the dump tubes in the condenser.

The two atmospheric dump valves are connected into the main steam system upstream of the turbine main steam inlet manifold isolating valves.

Because of its emergency duty the turbine bypass system is kept warm by warming drains whenever the reactor is on-load.

10.5 Condensate System

Each turbine is provided with three 50% duty condensate extraction pumps. These draw water from the condenser hotwell and pump it forward through the condensate polishing plant and LP feed heating system to the deaerator located at high level in the mechanical annexe.

Immediately after leaving the extraction pumps the water passes through the gland steam condenser before entering the condensate polishing plant.

This plant comprises four 33% capacity mixed bed ion exchange demineralisers, three in service while the fourth is being regenerated. This plant ensures that the water going forward in the feed train contains none of the corrosion inducing impurities that could affect the steam generator tubing or lead to the build-up of sludge within the steam generator.

Water leaving the condensate polishing plant is continually monitored for quality. Hydrazine and ammonia are injected at this point to remove oxygen and control the pH. Should the water quality deteriorate below the specified limits the associated unit is tripped.

The cleaned, chemically-dosed water then passes through the four stages of LP feed heating before reaching the deaerator. All the heaters are of the horizontal tube and tube plate type. The deaerator is of the spray type whereby heating and deaeration is achieved by a fine spray of the incoming condensate through the heating steam space onto a system of splash baffles. The heating steam is taken from the HP cylinder discharge and is fed through the stored water in the tank before reaching the steam space and spray zone. The entire contents of the deaerator tank are thus kept at saturation temperature.

Electric heaters and steam heating coils are fitted in the deaerator tank to provide adequate heating under no-flow and startup conditions.

The deaerators are located at the 26.35m OD level in the mechanical annexe; three downcomers drop from each deaerator, one to each feed pump.

An automatic backwash full-flow filter is provided in each feedpump suction line to prevent any particulate debris greater than 200 microns reaching the pumps which are positioned on the –0.9m OD level in the mechanical annexe basement.

Each unit has three feed pumps which are electric motor-driven through fluid couplings for speed control. Two pumps per unit are required for full-load duty; the third pump on each unit being for standby. The pumps deliver to a common unit header from which two streams flow through the three stages of HP heating. The HP feed flows from each unit then join into a common feedwater header, from which four pipes lead to the four steam generators. The pipes pass through the containment wall in the same manner as the main steam pipes. Each pipe carries in order, going towards the steam generator, a main feedwater regulating valve with a bypass regulating valve, a main feed isolating valve and a non-return valve. The bypass regulating valve is for use during startup and after reactor trip when small feedwater flows are required. The feedwater flow is automatically regulated to maintain the water level within specified limits in the steam generator.

11. MECHANICAL SYSTEMS

The major mechanical systems comprise:
(1) Sea water cooling systems — condenser main cooling water
                               — auxiliary cooling water
                               — essential service water.
(2) General service water system.
(3) Compressed air.
(4) Auxiliary steam system.
(5) Raw water treatment and chemical storage.
(6) Fire protection systems.

11.1   Sea Water Cooling Systems

There are three sea water cooling systems:

(1) The condenser main cooling water system requiring approximately $48m^3/s$.

(2) The auxiliary cooling water system requiring approximately $2.15m^3/s$.

(3) The essential service water system requiring approximately $1.9m^3/s$.

All three systems draw their water from a common cooling water (CW) pumphouse forebay fed by a single, submerged, supply tunnel of $21m^2$ cross section from an intake 750m off-shore. (See Fig 6.17).

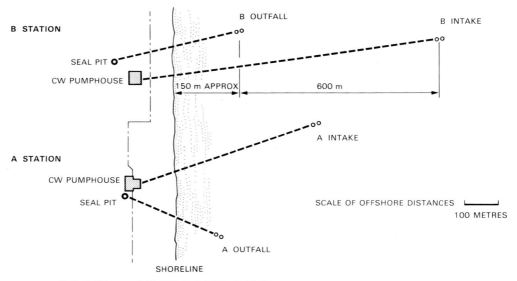

FIG. 6.17   RELATIVE POSITIONS OF A AND B STATION CW INTAKES

The original Sizewell B station design had two separate tunnels each serving one half of a split forebay and the tunnels were driven in the conventional manner using compressed air equipment.

Two tunnels were employed to allow one tunnel to be shutdown and drained for cleaning while the units could still operate at reduced load using the other tunnel for water supply.

Similarly, the CW discharge was via two tunnels from a split surge chamber. The offshore ends of the tunnels had superstructures carrying the closure plugs that allowed the tunnels to be drained.

Following a review of culvert performance it was decided that isolation and draining for cleaning was not necessary where an adequate sodium hypochlorite dosing system can be engineered to discourage marine growth. Moreover, Sizewell B is a single reactor station with a full 40 days shutdown every year for refuelling and an extended shutdown of 50 days every 10 years, which would allow remote-operated, submerged, cleaning machines to be usedwithout affecting plant availability if marine growth did occur.

Consequently, a single tunnel scheme was chosen and, as it will never be dewatered, it is possible to adopt a submerged tube concept using pre-cast concrete sections sunk in a trench in the sea bed. Had dewatering been required the empty tunnel would have been too buoyant to use this technique economically.

A trial section of trench was excavated to determine the rate of refilling due to tidal drift and this confirmed the feasibility of the sunken tube method.

The relative location of outfalls and intakes is extremely critical. The situation at Sizewell is complex in that the sea is shallow with little increase in depth with distance offshore.

The tidal flow is also complex with interaction between flows into the North Sea from the English Channel and from the North Atlantic as well as the local effects arising from the Sizewell Bank.

The dispersal of warm water plumes and the need to limit recirculation make the design of the system very difficult to assess and requires extensive studies to be done very early on in the station design process. Model studies are undertaken to determine the extent of warm plume cover to the intake structure and the intake itself is designed to minimise drawdown of the warmer surface water. (Fig 6.18).

Detail design of the intake also has to consider the effect of tidal flow on the intake flow to ensure that intake velocities never reach a level at which fish are entrained; typically 0.5m/s is taken for the maximum velocity through the coarse screens that surround the intake areas.

For a site like Sizewell, where an existing power station is in operation, the tidal flows and dispersion can be studied by aerial surveys using infra-red techniques; additionally, weather satellites can give trends over large areas and for long timescales. Fig 6.19 shows the decrease in discharge water temperature away from the discharge point, in relation to the ambient temperature.

The Sizewell B intake comprises two vertical shafts to the inflow tunnel, each shaft being capped by pre-cast intake structures designed to enhance horizontal flow into the structure. The sill level of the intake is set at 2m above the sea bed to allow for a predicted possible accretion rate of 40mm/annum as well as reducing uplift of sand by the intake flow. (Fig 6.20).

The intake tunnel discharges into the forebay of the pumphouse through a diffusing section developed to give even flow to the four rotating drum screens. Drum screens are standard for sea water cooled stations because of their ability to withstand full pump suction head in the event of severe blockage by seaweed, fish or other debris.

The drum screens rotate in the direction of water flow. Pressure jets, set above the screen deck level, wash off debris adhering to the screens into trash channels which convey the debris into trash collecting baskets. Alternatively, it can be diverted into the surge chamber on the system discharge side. Isolating gates give double isolation of screens for maintenance.

The size of the forebay and the level of the surrounding walls is set to contain any surge in water level arising from a pump trip.

The pumphouse requires a very deep excavation; consequently, there is considerable economic advantage in keeping the size down, particularly in the deepest area between the screens and the pumps themselves. The Sizewell B pumphouse is a development from the Littlebrook D design adapted to accommodate the additional pumps for the essential service water system.

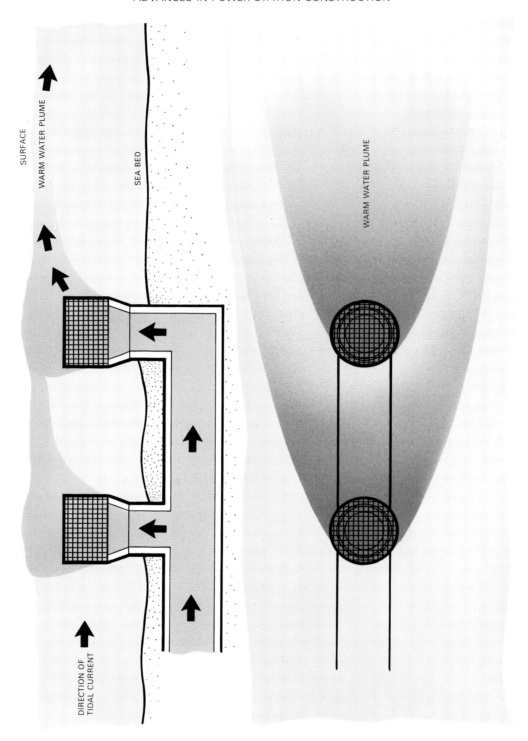

SURFACE

WARM WATER PLUME

SEA BED

DIRECTION OF
TIDAL CURRENT

WARM WATER PLUME

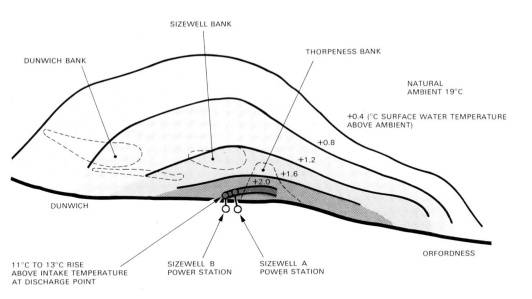

FIG. 6.19   TYPICAL HEAT FIELD AT END OF FLOOD TIDE

The pumphouse well is divided into two chambers by a reinforced concrete division wall to restrict flooding to one chamber in the event of serious leakage. A single superstructure spans both parts and carries the pumphouse crane.

Each pump chamber contains two CW pumps, two auxiliary CW pumps and two essential service water pumps together with all their isolating valves, non-return valves and pipework.

A third deep well contains the cross-over isolation valves between the two halves of the CW system on the pump discharge side.

The individual systems are described in the following sections.

11.1.1   *Condenser Cooling Water System*

This system supplies the condenser cooling water for the turbine generator. Two CW pumps in parallel serve each machine (four pumps total per station) each pump passing 11.8m³/s against 9m head. The pump is single speed running at 150rev/min driven through a reduction gearbox by a 1000rev/min 1.4MW motor. The pump itself is a mixed flow design in a specific speed range, vertical spindle with a concrete volute. The discharge pipe from each pump contains a hydraulically-operated butterfly valve which isolates the pump when shutdown as well as tripping-closed on reverse flow to protect an individual pump from reverse rotation on loss of drive power.

The pump discharges are then commoned in pairs into a single octagonal culvert leading to each turbine. A crossover duct links the two systems which allows both machines to run at full-load from three CW pumps for much of the year. (Fig 6.21).

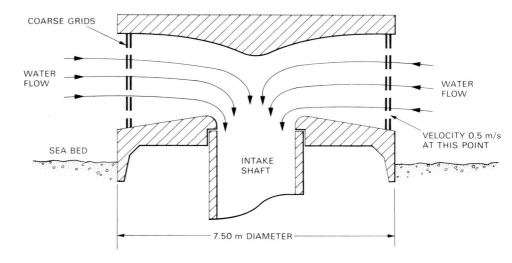

FIG. 6.20   TYPICAL ARRANGEMENT OF INTAKE STRUCTURE WITH VELOCITY CAP

The inlet culvert to each turbine terminates in a manifold with six valved connections into the waterboxes of the condenser tube banks. The six outlets from the condenser tube water boxes are fitted with butterfly valves that permit throttling of the CW flow to even the flow through the condenser sections as well as to isolate the water box to allow for tube inspection and repair with the unit on load. The outlet culverts discharge into a common surge chamber fitted with gates to give single isolation to an individual machine culvert. The surge chamber decouples the onshore culverts from the outfall tunnel and absorbs surges resulting from pump trip.

The outfall tunnel is similar to the intake tunnel, but terminates 180m offshore.

11.1.2 Auxiliary Cooling Water System (Fig 6.22)

The auxiliary cooling water (ACW) system removes heat from the auxiliary plant within the turbine house and mechanical annexe; it is not a safety classified system but, as it shares the same suction manifold as the essential services system, the isolating valves on the ACW pump suction and their dismantling joints are classified as safety-related and constructed to the higher standard with their actuators positioned above flood level.

There are four auxiliary CW pumps each sized to deliver up to $1.2m^3/s$ during station operation over the full anticipated tidal and temperature range. The system is split into two halves with two pumps serving each half. Normally one pump runs in each half of the system delivering water to its associated turbine auxiliary coolers. In addition either half of the system serves the station coolers for general service water and steam generator blowdown cooling.

The turbine auxiliary coolers discharge into the associated unit outlet culvert whilst the station cooler discharges can be directed to the discharge culvert of either unit.

Full flow, continuously self-cleaning strainers are fitted on the discharge of each pump to protect plate type coolers from blockages.

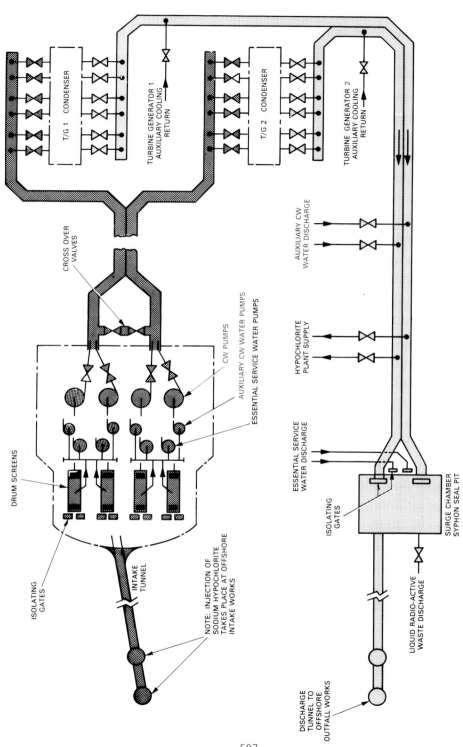

FIG. 6.21   PROCESS FLOW DIAGRAM OF MAIN COOLING WATER SYSTEM

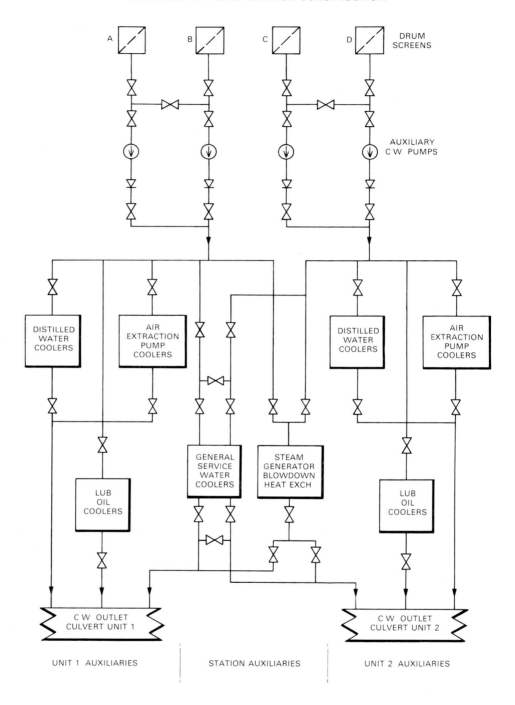

FIG. 6.22   AUXILIARY COOLING WATER SYSTEM

The piping used throughout the system is either glass-reinforced plastic or lined low-carbon steel depending on its location and duty.

11.1.3 *Essential Service Water System (Fig 6.23)*

The essential service water system is a high integrity seawater cooling system for removing heat from the reactor component cooling water system.

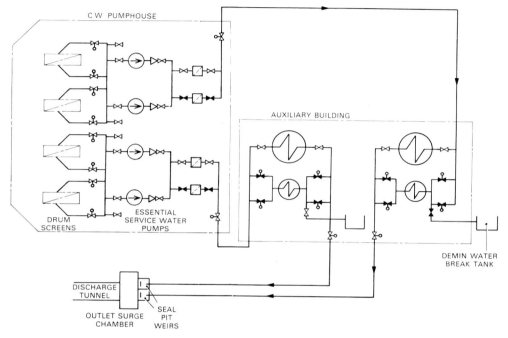

FIG. 6.23   ESSENTIAL SERVICE WATER SYSTEM

Its design caters for loss of a single active component coincident with loss of off-site power. All its motor drives in the CW pumphouse are located above extreme high tide level to guard against flooding. In the event that the CW drum screens become blocked, all other take-offs are tripped to give the essential service water pumps priority.

The system is not designed to withstand the safe shutdown earthquake condition; following that event the reserve ultimate heat sink maintains the reactor auxiliary plant in a safe condition.

Seawater is taken from the clean side of the drum screens to supply a suction header in each half of the CW pumphouse. Each header is fed by valved connections from two screens with a dividing valve to give double isolation for maintenance on a shutdown pump whilst the other remains in service. Each pump is followed by a continuous back-wash self-cleaning strainer.

Each pair of pumps supplies water for one of the two component cooling water system trains located in the basement of the control building. Currently two main and one auxiliary cooler are provided for each train.

Because the system performs a safety function the piping is run either in tunnels or in trenches to allow ready in-service inspection. Similarly, frost protection is provided to guard against extreme low temperature.

The discharges from each cooler train are run separately to the CW discharge surge chamber and discharge through a weir trap that maintains syphonic recovery in the system for all states of tide.

11.2 General Service Water System

The general service water system uses demineralised water and removes heat from those plant auxiliaries within the turbine house and mechanical annexe where sea water cooling is undesirable due to the risks inherent from leakage of sea water into the cooled medium.

The components served are:
(1) Main feed pump coolers.
(2) Compressed air system coolers.
(3) Drain pump oil coolers.
(4) Extraction pump bearing oil cooler.
(5) Chilled water plant coolers.
(6) Sample coolers for water and steam systems.
(7) Generator switch cooler.

Three 50% duty pumps deliver into a common header feeding three 50% duty heat exchangers. A bypass controls the exit temperature from the cooler banks. At times of low sea water temperature as much as 90% of the system flow is through this bypass.

Demineralised water is supplied to the system surge tank for filling and make-up while a small tank in the pump bleed-off bypass is used to add potassium chromate to the water to inhibit corrosion. Carbon steel is used throughout the system except for the cooler tubes which are titanium to ensure long life in the sea water environment.

11.3 Compressed Air System

There are five independent compressed air systems:
(1) General service air system.
(2) Instrument air system.
(3) Clean air system — X.
(4) Clean air system — Y.
(5) Breathing air system.

11.3.1 General Service Air System

The system comprises two 50% duty asymetric screw compressors discharging air through two 50% aftercoolers to a pair of air receivers connected in parallel to a manifold. The receivers store a limited volume of air for use following a compressor failure.

The maximum demand is expected to occur during a shutdown period and the 7 bar, 600 litres/s installed capacity is estimated to be sufficient to meet this demand.

No standby plant is provided and the compressors are connected to separate non essential 415V electrical supply boards and cooled by the general service water system. Maintenance is based on the replacement of a compressor.

11.3.2 *Instrument Air System*

This is a station system designed to provide a continuous supply of filtered, dry and oil free air for instrumentation and non-nuclear safety valves. There are three, skid-mounted, 100% duty air compressing trains each comprising a motor-driven, reciprocating compressor, an aftercooler and a receiver connected in parallel to a manifold. From this manifold the air is processed to the required cleanliness and dew point in one of two 100% filter/dryer trains prior to distribution.

The capacity of the system is based on 75% of all the valves operating following a seismic event, LOCA or loss of off-site electrical supplies causing a unit trip, coincident with a single failure.

The system is provided with nitrogen back-up from the station nitrogen system to maintain system pressure to mitigate against failure of all three compressors.

The compressors are connected to separate non-essential 415V electrical supply boards and are cooled by the general service water system.

Maintenance is based on the changeover of a complete compressor to reduce downtime to a minimum. The system normally operates on demand under pressure switch control but manual overrides are provided for commissioning and maintenance.

11.3.3 *Clean Air System*

The clean air system is divided into two trains, X and Y. Each train is an independent safety system providing filtered, dry and oil-free-air to valves designated 'essential to safety' and valves considered to be significant to the 'probabilistic risk assessment'. Approximately half of these valves are allocated to each train X and Y to prevent common mode failure of the valves. Both trains are substantially the same as the instrument air system except for minor differences. Train X derives its nitrogen back-up from a bank of dedicated bottles while train Y draws its intake air from outside the building to ensure that both trains do not take-in air from the same fire area. In addition each train has two seismically qualified high pressure nitrogen storage vessels providing a dedicated back-up to the pilot operated relief valves.

11.3.4 *Breathing Air System*

This is a station system designed to provide a continuous supply of breathing-quality air to a limited number of areas where entry may be required into a possible hostile environment for inspection, maintenance or repair.

The system is based on the requirements for six men in full work suits.

The three 50% duty compressors are motor-driven, reciprocating units fitted with dry carbon piston rings. Air intake is drawn from outside the building at roof level to prevent the ingestion of vehicle exhaust fumes and other harmful gasses. The compressed air is initially cooled by an after cooler and then processed using a refrigerant dryer and a series of filters to produce clean, odour free and vapourless air. An emergency supply of hospital-quality air is provided from a dedicated bank of bottles which together with the air stored in the single air receiver is sufficient to provide a total escape time of 100 man minutes. Because the system will only be required to operate periodically no special provisions have been made for maintenance.

The compressors are supplied with secure electrical supplies and cooling water.

11.4 Auxiliary Steam System

This system supplies saturated steam at 9.6 bar to the domestic loads and plant loads shown in Fig 6.24.

The steam is supplied by two oil-fired shell boilers each rated at 22,500kg/h. The maximum demand is 44,000kg/h but essential loads can be met by one boiler only.

An alternative supply of steam is available via a pressure reducing valve from the main steam system but this is restricted to plant loads to avoid any contamination of the domestic loads by possible activity from the live steam supply.

When the live steam supply is in operation, condensate flows from the 'plant' side to the turbine condenser with the exception of that from the radwaste evaporators which is always returned to the secondary liquid radwaste system.

When the live steam supply is isolated both domestic and plant clean condensate returns to the auxiliary boiler feed system.

This system is a major departure from that on SNUPPS which employed a large reboiler to produce the steam using bled or live steam as the heating medium. When adapted to a twin turbine arrangement the bled steam supply system becomes so complex and the piping so extensive that the overall cost of the system is such that the oil-fired alternative system is considerably cheaper, much of the cost of the oil being offset by the increased value of the turbine generator output.

11.5 Raw Water Treatment Plant

This plant provides high quality demineralised water using town's water for feed supply.

Two 100% duty process streams are provided, each capable of a net output of $136m^3/h$. The run time of each stream is 8 hours minimum and together they provide a total of $3264m^3/day$. Increased output can be obtained for a period by running both streams in parallel. Each stream comprises raw water filters, cation demineralisers, CO_2 degasser, anion demineralisers, mixed bed demineralisers and a final vacuum degasser.

The equipment is located in the basement of the turbine house below the central loading bay as is the plant for resin regeneration.

Bulk storage of caustic soda and sulphuric acid used for resin regeneration is provided external to the turbine house and comprises bunded tanks with capacity for 26 weeks supply. The same area also stores the ammonia need for condensate pH control.

The water treatment plant is designed for unmanned operation. Instruments detect that a demineraliser bed is approaching exhaustion and signals an alarm in the control room; the operator then brings the other stream on line and initiates the regeneration cycle on the exhausted stream. Detail control of the plant operation and regeneration cycle is by an integral control microprocessor. Local manual controls can be used if necessary.

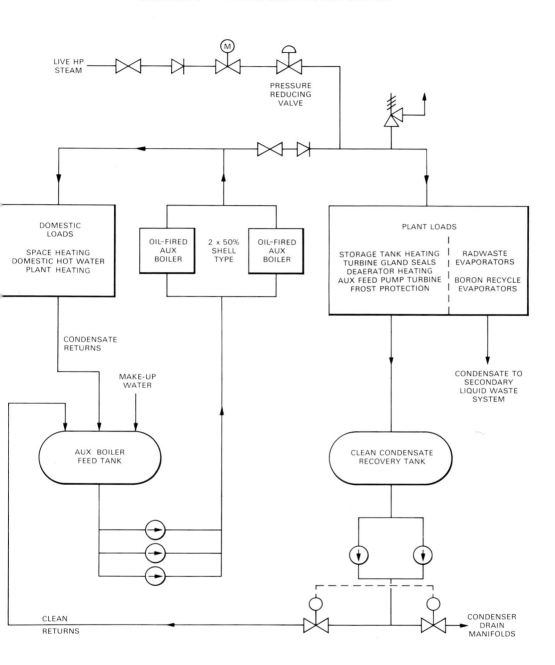

FIG. 6.24   AUXILIARY STEAM SUPPLIES

11.6 Fire Protection System

11.6.1 *General*

The fire protection system is designed to minimize the effects of fires by providing the ability to detect and fight any fires encountered in all plant areas. The system is designed so that in the event of the failure of the primary suppression system a back-up system is available.

In areas containing safety-related equipment the fire protection system is designed to minimise damage to such equipment. Inadvertent operation does not impair the safe shutdown ability of the plant.

Where the fire protection system requires electrical supplies for correct operation the supply security is designed to the appropriate level of reliability.

11.6.2 *Fixed Protection Equipment*

The following types of fixed fire protection are installed:
(1) Automatic Water Sprinklers
   Automatic water sprinklers consist of a network of piping distributing water to closed-head sprinklers, arranged to open automatically when the temperature rises above a preset figure. The network stands full of water under pressure.
(2) Automatic Preaction Sprinkler System
   Automatic preaction sprinklers are similar to (1) above but the distribution pipework stands empty. Heat or smoke detectors in the area of the closed-head sprinklers open the alarm valve to charge the distribution system with water and simultaneously initate a remote alarm. Water is applied to the risk as and when the frangible bulbs shatter.
(3) Manual Preaction Sprinkler System
   Manual preaction sprinklers are similar to (2) above but the control valve is manually-operated. Heat or smoke detectors in the area of the closed-head sprinklers initiate a remote alarm to call for the operator to open the valve. Water is applied to the risk as and when the frangible bulbs shatter.
(4) Automatic Water Spray Systems
   Automatic water spray systems use open nozzles arranged strategically to spray the whole of the equipment being protected. Each nozzle is designed to give a solid cone spray pattern. The nozzles and their distribution pipework normally stand empty and are controlled by a quick opening valve (deluge valve) which is operated by heat detectors in the same area as the nozzles. These consist of detector sprinklers with a frangible bulb closure on a network of piping pressurised with air via a small compressor. Thermal rupture of any bulb releases the air pressure causing the deluge valve to open admitting water to the nozzles protecting the plant at risk.
   Local and remote fire alarms are initiated from a pressure switch on the deluge valve.
   Low air pressure in the detector piping raises an alarm in the main control room.

(5) Manual Water Spray Systems

Manual water spray systems use nozzles as described in (4) but the water is controlled by normally-closed manually-operated ball or butterfly valves for each protection zone. Line isolators are provided to allow for regular exercising of the controller.

Heat detectors initiate a fire alarm in the main control room to call for operator action.

(6) Internal Hydrant/Hose Points

Additionally, hose/hydrant points are positioned in all areas of plant containing combustible materials located, in general, adjacent to stairways and at interior columns so that no more than 30m separates adjacent hose racks. The points are fed from the wet pressurised distribution system.

Inside the containment the hose racks are supplied from normally dry standpipes. Break-glass pushbutton stations create an alarm in the control room, whereupon the control room operator can open the containment isolation valve to charge the standpipes.

11.6.3 *Pumping Plant*

The largest water demand associated with the fixed fire protection would normally result either from a fire on one of the two turbine generator sets including the associated lubrication system and the protection of the turbine generator support legs, or from one generator transformer unit and its associated conservator tank and coolers. In either case an additional allowance of 230m³/h is included for fire hoses.

The pumping plant comprises three pumps, two of which are adequate for the above demand. There is thus reserve capacity to allow for one pump failure or maintenance.

Each pump is self contained, ie independent of external supplies (except for battery charging) and consists of a diesel engine and pump with its own controls and duplicate starting batteries for automatic and for manual starting. The pumps start in sequence, automatically, according to water demand. Associated with the pumps is a pressurised storage tank (for initial water feed while engines start) with its own charging system. The distribution pipework employs internal and external pressurised ring mains with control and isolating valves for each fire risk area/zone. The system contains section isolating valves enabling sections to be repaired or maintained without impairing the rest of the system.

The inlet manifold for the pumps has two independent feeds from separate reservoir sections each containing a supply of 455m³ dedicated for fire fighting purposes.

Frost protection is afforded for the external mains and pipework by use of dry systems, trace heating and burying as appropriate.

11.6.4 *External Hydrant System*

This system is entirely separate from the fixed protection equipment. A ring main positioned around the station buildings has sufficient hydrants for all buildings to be readily protected. The ring main normally stands full of water but unpressurised.

Located in the fire pumphouse, two pumps of 100% duty are connected to the ring main and started manually as required. Each is self-contained, driven by a compression ignition engine and has its own controls for remote and local starting.

In general the hydrant points are positioned between 15m and 30m from the station buildings and spaced not more than 45m apart. Each hydrant point has a wooden ventilated weatherproof cabinet on legs containing branch pipes, nozzles, key and 45m of 65mm bore lined hose with instantaneous couplings.

A number of section isolating valves not more than 150m apart are installed in the ring main to enable the system to remain usable while any part of the system is under repair.

Two emergency pumping-in points, widely separated, each with mobile pump hose discharge connections, are provided for use by the local fire authority.

The inlet manifold for the pumps has access to a maximum water quantity of $3640m^3$ from the reservoir. This is in addition to the two reservoir sections reserved for the main fire fighting pumps.

Frost protection is by burying or trace heating and lagging as appropriate.

Connections to the CW culverts are provided to enable water to be drawn and pumped into the hydrant ring main.

11.6.5 *Halon* 1301 *Gas Extinguishing System*

Halon 1301 (bromotrifluoromethane gas) is the extinguishing agent used to protect rooms containing essential control, alarm and processing equipment such as computer suites, data processors, cable termination cubicles, motor generator sets, essential switchgear, containment electrical penetrations, etc.

The gas is applied in concentrations not greater than 6% by volume within the protected area or by local cubicle flooding.

The gas will be stored in pressurised cylinders local to the risk area or within separate containment areas associated with the risk. It will be released automatically on receipt of the detection of either smoke or heat within the protected area. Manual override facilities are provided.

11.6.6 *Mobile and Portable Equipment*

Portable fire extinguishers for manual suppression of fires are provided throughout areas of the plant where access is unlikely to be prevented by fire conditions. They are located generally at access points, entrance doors or corridors. These will be carbon dioxide, bromochlorofluoromethane (BCF), pressurised water and dry chemical as appropriate for the class of combustible located in the hazard area. In addition a wheeled dry chemical extinguisher of 136kg capacity is provided for the turbine building.

11.6.7 *Smoke Detection*

An automatic alarm system based on the use of a smoke detection system is provided for the main control room, cable flats and risers and other areas of fire risk which may be uninhabited for long periods during normal operation.

The station is arranged into specific zones dictated by normal access routes to the area and installed fire fighting capacity. Each defined zone raises an individual alarm in the main control room. Each zone of protection consists of a number of approved smoke detector heads connected in cascade to a control cubicle, the control cubicle nominally serving 20 to 30 zones as required. The allocation and grouping of zones to control cubicles is on the basis that the cubicle is reasonably near the access route to the zones it contains. The system is provided with self monitoring circuits to raise alarms on equipment failure, supply failure or zones disabled.

Table 6.2 gives the type of protection fitted to each major plant area.

TABLE 6.2. AREAS COVERED BY THE FIRE PROTECTION SYSTEM

System	Plant
Automatic water sprinkler	Auxiliary boiler room Dry waste compactor room (radwaste building) Access control area Hydrogen store Main cable routes in the auxiliary and control buildings Upper and lower cable spreading rooms Vertical cable risers Cable area above access control area
Automatic preaction sprinkler	Fuel building loading bay Diesel generator rooms Areas below turbine generator operating floor
Manual preaction sprinkler	North and south cable penetration inside the containment
Automatic water spray	Turbine generator lubricating oil tanks and plant and associated bunded floor areas Clean and dirty oil tanks Main feedpumps, bearings and associated oil systems Hydrogen seal oil unit Generator, station, unit and other auxiliary transformers and harmonic compressors Main fire fighting pumps External hydrant pumps Other inflammable fuel and oil risks
Manual water spray	Turbine generator lubricating oil pipework and bearings and associated floor areas
Halon 1301 gas extinguishing systems	Essential switchgear rooms in control and communication building Non-essential switchgear and transformer rooms in control and communication building Control cabinet. Load centre and rod motor control motor generator sets room Computer room proper and below floor Control room cable trenches and associated wall chases

12. ELECTRICAL SYSTEMS

The electrical systems may conveniently be divided into:

(1) The connection of the station to the national grid. This is the interface between the power station and the transmission network for power import and export.

(2) The main electrical system within the station, which provides electrical power to the station auxiliaries during plant startup, shutdown and normal plant operation. The main electrical system is also required to provide bulk power supplies to the essential electrical system during abnormal or emergency situations.

(3) Low voltage non-safety ac and dc systems. These systems provide ac and dc supplies of a better integrity than the grid.

(4) The essential electrical system, so called because it provides distribution of grid supplies or emergency diesel generator supplies to components essential to reactor safety.

(5) The essential instrument and dc systems. These systems are provided to ensure that sufficient equipment is operable at all times to prevent an unacceptable release of radioactivity to the environment.

Fig 6.25 illustrates the principal electrical systems in simplified form.

12.1   Connections to the Grid

The main generators produce electrical energy at 23.5kV. This is transmitted from each generator to an associated main generator transformer through phase-isolated bus bars. The generator transformers are standard units each rated at 800MVA which raise the voltage to 400kV. Standard single-phase units are used to permit replacement of a failed unit by a national spare. These transformers are located outside the turbine house in line with the turbine axes. Power is transmitted by underground cables to the 400kV substation on the site from which connections are made to the 400kV national grid at the Norwich, Pelham and Bramford grid switching stations. Each of the main generators can also supply power to the station internal distribution system via a unit transformer which reduces the voltage to 11kV.

The 400kV substation also provides a source of power from the grid to the Sizewell B station. This power is imported directly via two 400kV/11kV station transformers feeding the internal distribution system at 11kV. When the main generators are not running, power can also be imported via the same paths that are normally used for export, that is, via the generator transformers and thence to the unit transformers, which in turn feed the internal distribution system at 11kV. Switch disconnectors, located in the main connections between each generator and the generator transformers, enable the main generators to be individually coupled to or decoupled from the rest of the electrical system without affecting these routes. This enables post-trip supplies to be immediately available to all 11kV boards. Each path may also be individually isolated at the substation or at the 11kV level by means of suitable breakers. The use of generator switch disconnectors and 400kV/11kV station transformers improves the reliability and safety of the electrical system by a factor of approximately 400.

The use of 'open' type cable routes for Sizewell B initiated the rapid development of cable insulating and sheathing compounds to produce cables which not only have reduced propagation characteristics but also low smoke, acid and toxic emissions. In all other respects the cables for Sizewell B are identical to those of other power stations.

Cables installed in the reactor containment area must, in addition to having low propagation of smoke, acid and toxic emissions, be able to withstand the effects of postulated accidents such as loss of coolant accident (LOCA) or main steam line break (MSLB). LOCA and MSLB accidents exert high temperatures, high pressures and high radiation levels on the cables. The cables are designed and tested to be able to withstand and, where necessary, operate through these accidents.

To maintain the integrity of the primary containment, electrical penetrations are cast into the containment walls.

Sets of penetrations into the primary containment are provided, as necessary to meet segregation requirements, the outside end of each set being enclosed in a fire-rated cell, equipped with a Halon gas fire suppression system, the inboard end being exposed to the containment environment. Enclosures are provided at both ends of the penetration to protect terminals or cable joints.

Each essential cable will be allocated to one of four essential cable segregation groups which will generally be consistent with the segregation requirements of the terminal equipments. Each group of reactor protection cables and each pair of groups of essential cables will be segregated from all other cables.

Generally two categories of separation and segregation are applied to plant cables commensurate with their safety role and the hazard from which they are to be protected. These are defined as 'electrical segregation' and 'fire segregation'.

Electrical segregation is spatial separation between specified electrical separation groups to protect the cables of one separation group from being affected by electrically-generated fires occurring in another separation group.

Fire segregation is separation by fire barriers, or their equivalent, for safe shutdown cabling, to ensure that the required protection exists against postulated exposure type fires (including those caused by transient combustibles) such that the redundant safe shutdown capability is not affected by the fire. Fire segregation is defined in terms of the number of hours of fire resistance rating which exists on a fire barrier placed between two redundant separation groups. One-hour and three-hour fire resistance ratings are established on barriers in accordance with accepted fire test standards.

In plant areas outside the containment, fire segregation between redundant safe shutdown cabling is provided by three-hour rated fire barriers ensuring that redundant trains are free of fire damage.

For cabling inside the containment, fire segregation between redundant safe shutdown cabling is provided by one of the following fire protection means:

(1) Greater than 6 metres spatial separation with no intervening combustibles.

(2) Automatic fire detection and suppression.

(3) Provision for a non-combustible radiant energy shield between redundant trains.

(4) Have an alternative shutdown capability.

Non-essential cabling associated with the two main turbine generators and normal reactor operation is allocated to two non-essential segregation groups which are separated from each other such that a single fault (including a fire) in one group will not affect the operation of the other group.

13. RADIOACTIVE WASTE MANAGEMENT

13.1 General Principles

Radioactivity associated with the nuclear fuel cycle is almost entirely created within the reactor, mostly within the fuel but also from neutron activation of other materials within the reactor circuit.

At a nuclear power station, small quantities of fission products escape from the fuel and these, together with some neutron activation products, appear in wastes within the plant.

The radioactive wastes arise in gaseous, liquid and solid forms. Their activity ranges from the insignificant to levels that could be deadly without proper precautions in their handling.

The active waste management plant's function is to handle all such arisings in a manner that is safe for the operators and the general public and dispose of the waste into the environment in a safe and acceptable manner.

Discharge of radioactive effluents has been limited by statute (Radioactive Substances Act 1960, etc) but the CEGB practice has been to reduce discharges to far below these levels and the design of the radioactive waste (radwaste) systems at Sizewell B is in line with the ALARP (as low as reasonably practicable) principles.

The radioactive waste management of the plant will comply with the following general principles:

(1)   All radioactive discharges to the environment will be carried out in a controlled manner via well-defined routes which are continuously monitored.

(2)   Radioactive source terms for discharge streams will be calculated using a continuous design failed fuel rate of 0.02%. Provision will be made in the plant installed to cope with shorter periods of reactor operation with a higher failed fuel rate.

(3)   It is not intended that the installed plant should cater for a complete primary circuit decontamination. Special equipment would have to be installed for such a procedure.

(4)   Radioactive process equipment will be installed in shielded cells and where liquid spillages can take place these cells will be adequately bunded and drained to contain the maximum liquid spill.

(5)   All process plant will be housed in a building with its own heating and ventilating system. Adequate personnel access will be included and contamination control implemented.

(6)   Radioactive solid wastes (including sludges and concentrates) will be encapsulated in a suitable solid medium and placed in drums before storage and subsequent off-site disposal.

(7)   The turbine building drains will be monitored and discharged directly to sea via the CW outfall. No significant radioactivity is expected in this waste stream during normal operation. An interconnection to the radioactive waste building caters for abnormal conditions, eg large leakages from the primary to secondary side of the steam generator.

(8)   The laundry liquid waste will be collected, filtered, monitored in the radioactive waste building and discharged.

(9)   It is proposed to operate the liquid waste plant, under normal conditions, on a collection, decay, filtration, sampling and discharge philosophy. This is based on circumstances in which the failed fuel rate is less than 0.02%. This reduces quantities of solid waste and therefore costs of processing and storage. It also reduces the operator man-rem budget and equipment dose rates.

(10) It is expected that with failed fuel rates based on current operating experience (0.005%) active discharges to the environment will be well within the required limits. Should failed fuel rates rise, then the liquid processing equipment provided, which includes demineralisation and evaporation plant, may be brought into service in order to reduce discharge levels.

(11) Tritium discharges from the plant will be made intermittently in order to reduce the levels in the primary circuit and therefore working environment levels. This will be carried out with the condensate stream of the boron re-cycle evaporator discharging to the liquid waste system and then to the sea via the CW outfall. Discharge of any other active isotopes via this route will be negligible.

Fig 6.26 shows schematically the systems involved.

13.2   Volume Arisings

Normally all radioactive waste occurs in systems associated with the primary circuit. All potentially radioactive waste from the secondary circuit is referred to as 'secondary waste'.

The plant will be designed to cater for the following volume arisings.

13.2.1   *Gaseous*

The primary source of radioactive gas is via the continuous hydrogen purge of the volume control tank at a rate of $0.02m^3$/min. Smaller quantities are also received, via vent connections, from other process equipment throughout the plant.

13.2.2   *Liquid*

The liquid waste volume arisings are detailed in Table 6.3.

13.2.3   *Solid*

The estimated solid waste volume arisings are detailed in Table 6.4 together with an approximate estimate of the specific activity based on a failed fuel rate of 0.02%.

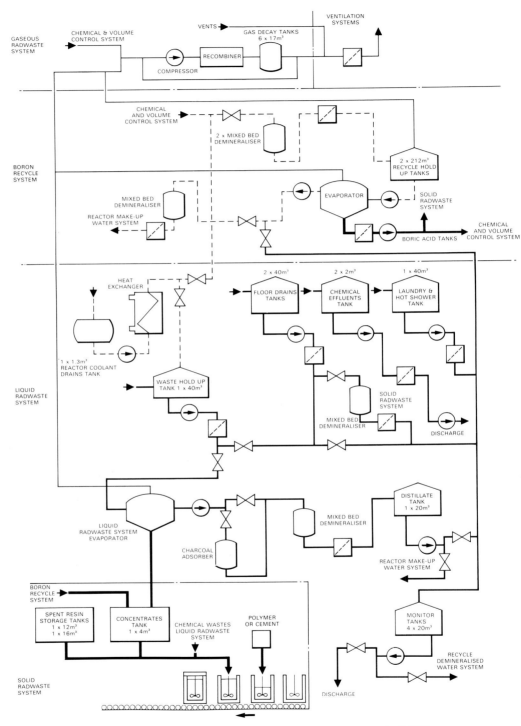

FIG. 6.26 **RADIOACTIVE WASTE TREATMENT SYSTEM**

TABLE 6.3. LIQUID WASTE VOLUME ARISINGS

Waste Stream	Volume of Waste	Activity Level
Aerated reactor grade waste (equipment drains)	$500m^3/y$	Medium
Non-reactor grade floor drains	$1425m^3/y$	Low/Medium
Turbine building drains	$18,000m^3/y$	Very low
Condensate polishing plant regeneration waste	$60,000m^3/y$	Very low
Laundry waste	$13,200m^3/y$	Low
Chemical drains	$100m^3/y$	Low
Steam generator blowdown	$40,000m^3/y$	Low
Tritium control via boron recycle evaporator condensate	$2400m^3/y$	Low

TABLE 6.4. SOLID WASTE VOLUME ARISINGS

Type of Waste	Volume of Waste	Activity Level	Remarks
Primary spent ion exchange resins	$15.33m^3/y$	High	
Secondary spent ion exchange resins	$8.48m^3/y$	Low	This figure is based on the blowdown demineralisers being on standby duty during normal operation
Primary system spent filter cartridges	18/y	High	
Secondary system spent filter cartridges	24/y	Low	Could be treated normally as low-active waste
Evaporator concentrates	$4m^3/y$	Medium	
Dry waste (clothing, paper air-filters etc) after compaction and/or incineration.	$180m^3/y$	Low	
Active chemicals	$3.0m^3/y$	—	Small batches of waste defouling and decontamination chemicals.

617

13.3 Liquid Waste Management System

13.3.1 *Plant Role*

The liquid waste management system collects, processes, stores and safely disposes of radioactive and potentially radioactive wastes generated as a result of normal operation, including anticipated operational occurrences, of the boron recycle system, the floor and equipment drains system and the secondary liquid system.

The system can also process and dispose of liquid wastes collected from the secondary liquid system if the radioactivity in that system is too high for direct discharge.

The liquid waste management system performs no function related to the safe shutdown of the plant and its failure does not adversely effect any safety-related system or component. Consequently the system is not safety-related except where it interfaces with safety-related systems.

All discharge paths are monitored for radioactivity, and should the radioactive content exceed specified limits, automatic valves terminate the discharge.

13.3.2 *System Description (Fig 6.27)*

The liquid waste management system is divided into sub-systems so that the different waste streams can be processed separately depending on the nature and activity of the various influents. The system is manually-initiated, except for some functions of the reactor coolant drains sub-system which is in continual use when the reactor is at power. The system includes adequate control equipment to protect the components, together with instrument and alarm functions to provide operator information to ensure proper operation.

The bulk of the radioactive liquid discharged by the reactor coolant system is processed in the boron recycle system and then either discharged or returned to the reactor make-up tank for re-use. This limits the input to the liquid waste management system.

All the liquid waste management system components are located in the radwaste building except for the reactor coolant drains tank sub-system.

The Reactor Coolant Drains Tank Sub-System (RCDT)(Fig 6.28)

The RCDT sub-system consists of a tank, two pumps and a heat exchanger, together with valves and controls. The RCDT collects reactor coolant from leakoff type drains inside the containment at a central collection point for further disposition through a single penetration via the pumps.

Only water which can be directed to the recycle hold-up tanks enters the reactor coolant drains tank. The tank is provided with a hydrogen cover gas because the water must be compatible with reactor coolant and it must not contain dissolved air or nitrogen to minimize pressure build-up in the gaseous radwaste system.

Sources of water entering the reactor coolant drains tank include the reactor vessel flange leakoff, valve leakoffs, reactor coolant pump seal number 2 leakoffs and the excess letdown heat exchanger flow. (No continuous leakage is expected from the reactor vessel flange during operation).

In normal operation the reactor coolant drains tank sub-system is automatic and requires no operator action. The system can be put in the manual mode, if desired.

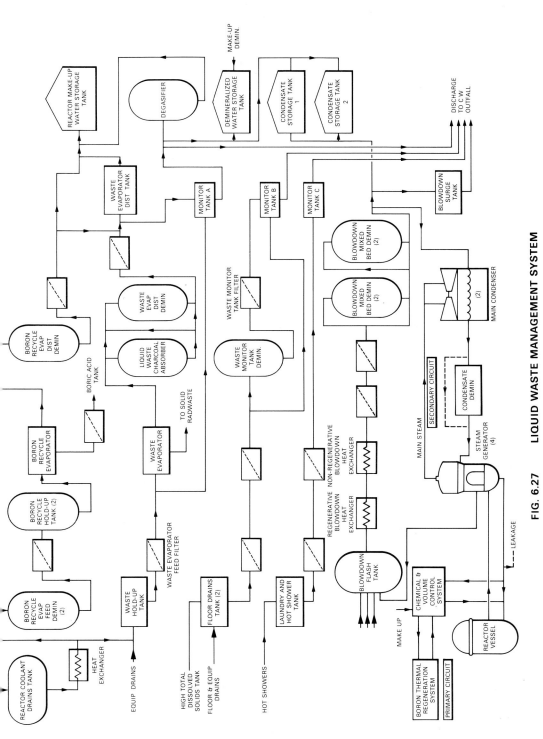

FIG. 6.27   LIQUID WASTE MANAGEMENT SYSTEM

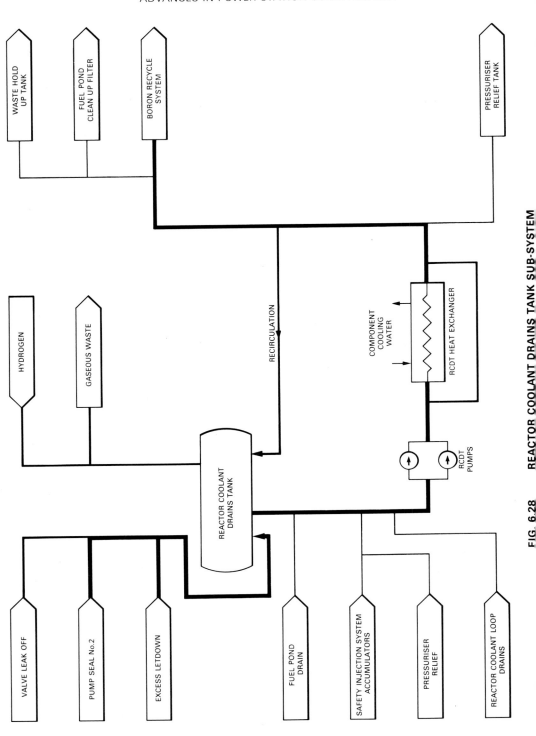

FIG. 6.28 REACTOR COOLANT DRAINS TANK SUB-SYSTEM

All influents to the tank are of reactor grade and are therefore suitable for re-use after processing. If however, the radioactivity is low or there is sufficient tank capacity to delay the final discharge, the effluent may be diverted to drain channel A, see Fig 6.29. This discharge may be used for tritium control.

One of the reactor coolant drains tank pumps normally discharges to the waste hold-up or boron recycle system. The level in the tank is automatically maintained by running one pump continuously and using a proportional control valve in the discharge line. This valve operates on a signal from the tank level controller to limit the flow out of the sub-system. The remainder of the flow is recirculated to the tank. The tank heat exchanger is sized to maintain the tank contents at or below 77°C assuming an in-leakage of 37.85 litre/s at 316°C.

A cover gas system is provided to prevent wide pressure variations in the tank. Hydrogen cover gas is supplied from the service gas system and is automatically maintained between 1.14 bar and 1.41 bar by pressure regulating valves. One valve maintains a minimum tank pressure by admitting hydrogen, while another maintains maximum tank pressure by venting the tank to the gaseous radwaste system. The hydrogen is supplied from no more than two small bottles, to limit the amount of hydrogen gas which might be accidently released to the containment atmosphere. The tank vents to the gaseous radwaste system to limit any releases of radioactive gases.

The reactor coolant drains tank sub-system may also be used in the pressuriser relief tank cooling mode of operation. In this mode, the level control valve at the discharge of the reactor coolant drain tank is closed and the pressure relief tank contents are circulated through the reactor coolant drain tank heat exchanger, by the reactor coolant drain tank pumps, prior to returning to the pressure relief tank. In this mode of operation, the drains tank heat exchanger is capable of cooling the relief tank contents from 90°C to 49°C in less than 8 hours. In all cases of relief tank cooling, the tank is vented to less than 4.5 bar to prevent overpressurisation of the drains tank sub-system.

The reactor coolant drains tank sub-system may be used to drain the reactor coolant loops by first venting the reactor coolant system, then connecting the spoolpiece in the pump suction piping. The design objective of this mode of operation is to drain the reactor coolant system to the midpoint of the reactor vessel nozzles in less than 8 hours with both drains tank pumps running.

The reactor coolant drains tank sub-system may be used to drain down portions of the refuelling pond which cannot be drained by the residual heat removal pumps. In this mode of operation, the heat exchanger is bypassed and the level control valve may be bypassed to maximize flow through the fuel pond cooling and clean-up system to the refuelling water storage tank. An alternative drain line is provided from the refuelling pond to the containment sump to route decontamination chemicals away from the drains tank sub-system and minimize the possibility of contaminating the clean systems downstream of the pumps. The containment sump is drained by the dirty radioactive waste (DRW) sub-system to drain channel B, see Fig 6.30.

Drain Channel A Sub-System (Fig 6.29)

Drain channel A normally collects clean tritiated wastes from the reactor coolant system via the equipment drains system (clean radioactive wastes sub system), the gaseous waste system (its compressors, recombiners and waste gas drain traps) and the

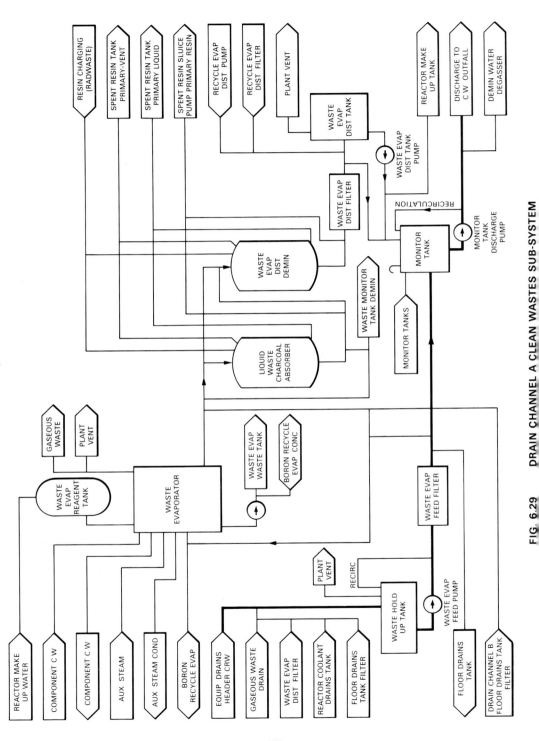

FIG. 6.29   DRAIN CHANNEL A CLEAN WASTES SUB-SYSTEM

reactor coolant drains tank. It can also collect effluents from the waste evaporator condensate filter for re-processing and from drain channel B via the floor drains tank filter when water from drain channel B requires processing.

All influents to drain channel A are suitable for re-use after processing in the waste evaporator and the waste evaporator condensate demineraliser. (The liquid waste charcoal absorber is not normally used unless organics have been detected in the influents). When the condensate is to be recovered, the discharge is directed to the waste evaporator condensate tank, which is fitted with a diaphragm to exclude oxygen, and pumped from there to the reactor make-up tank by the waste evaporator condensate pump. However, in order to control tritium and to minimize processing costs, these wastes are normally held up for decay and then discharged directly via monitor tank and monitor tank discharge pump.

If however, radioactivity is higher than acceptable or there is insufficient time for delay, the influents will be processed as described.

The waste evaporator is of the same type as the boron recycle evaporator and each acts as standby to the other to cover for abnormal conditions.

When the evaporator is used as standby to the boron recycle evaporator and the influent is of high quality, the evaporator residuals (bottoms) are concentrated to 4% by weight of boric acid when they are sampled and may be transferred to the boric acid tank if they are shown to be suitable. If the quality is unsatisfactory, evaporation is continued until the concentration reaches 12% by weight boric acid when the concentrates are discharged to the waste evaporator bottoms tank for disposal in steel drums at the drumming station.

When the evaporator is used to process waste without boric acid the evaporator bottoms may be concentrated above 12% dissolved solids by weight. Evaporation is then terminated and the concentrates are discharged to the evaporator bottoms tank.

Prior to transferring the bottoms out of the evaporator, it is necessary to ensure that the temperature of the pipe transferring the concentrate is above 24°C when 4% by weight of boric acid is being transferred. When 12% by weight of boric acid bottoms are transferred it is essential to ensure that all piping trace heating is on. This prevents solidification of the bottoms in the piping.

The transfer of bottoms is followed immediately by a timed flush of hot reactor make-up water.

Appropriate connections are provided to allow the operator to boil out the waste evaporator and the primary bottoms tank. A chemical reagent tank is provided for chemical addition to promote the cleaning activity.

Drain Channel B Sub-System (Fig 6.30)

The influents to this system are dirty and potentially radioactive wastes from the radioactive area floor and equipment drains sub-system which consist of miscellaneous leakage from systems within the controlled area of the auxiliary building. Generally, the amount of highly radioactive reactor coolant leakage in these areas is very small, the bulk of the leakage being from non-radioactive or only slightly radioactive systems. This system also collects decontaminated water derived from washdown areas, spent fuel flask washdowns and laboratory equipment decontamination rinses via the floor drains system. Highly chemical-contaminated solutions are not allowed to enter drain

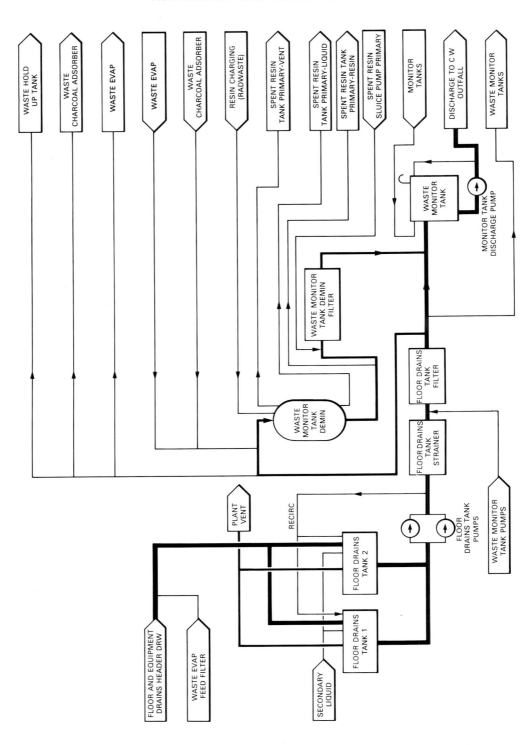

FIG. 6.30   DRAIN CHANNEL B DIRTY WASTES SUB-SYSTEM

channel B nor are large volumes of component cooling water for which separate disposal arrangements must be made. The chemical drains are collected directly by the chemical equipment drains tank.

The second floor drains tank is provided to allow one tank to be isolated and sampled prior to discharge or processing, or to provide a greater surge volume during periods of abnormal waste generation.

Chemically, the influents to drain channel B system are variable and may at times be high in dissolved solids. However, the radioactivity is normally low so these wastes are held up for a short period and then discharged via the monitor tank and the associated discharge pump.

If the radioactivity is too high for direct discharge and the total dissolved solids are below 25mg/litre, the effluent may be treated in the waste monitor tank demineraliser before disposal.

If the radioactivity is high and the total dissolved solids are also high, the effluent may be transferred to drain channel A for processing either directly by the waste evaporator or to the waste hold-up and subsequent processing by the waste evaporator if necessary. Wastes from drain channel A may also be processed in the waste monitor tank demineraliser to effect additional clean-up.

In unusual circumstances, when the secondary liquid is too radioactive for direct discharge, the secondary liquid waste system will pump radioactive effluents to the floor drains tanks where it may be held up for decay or processed in the waste monitor tank demineraliser of the waste evaporator as necessary. When the waste evaporator is used in this mode, chemical limitations must be maintained to safeguard the waste evaporator. Systems carrying primary water must be flushed before re-use, locked valves in conjunction with non-return valves prevent cross contamination between primary and secondary systems.

Laundry and Hot Shower Waste Sub-System (Fig 6.31)

This sub-system collects waste from the washing machines, the washers used for reusable personnel gear and waste from personnel decontamination showers, via a drain tank in the basement of the administrative building. This drain stream contains detergent, but has low radioactivity and does not normally require treatment.

Chemical Drains Tank Sub-System (Fig 6.32)

This sub-system receives spent samples composed of chemically-contaminated tritiated water from the plant sample stations. Chemically-contaminated decontamination wastes from the decontamination centre, or decontamination wastes from local decontamination operations may also be received by the chemical drains tank if they are not suitable for processing. Evaporator defouling chemicals may also be received in this tank. The contents of the tank are sampled, pH adjusted and then discharged. Alternatively these chemical wastes may be directed to the solid radwaste system for solidification. Operation is intermittent and manually-controlled. A high level alarm is provided from the chemical drains tank for operator information.

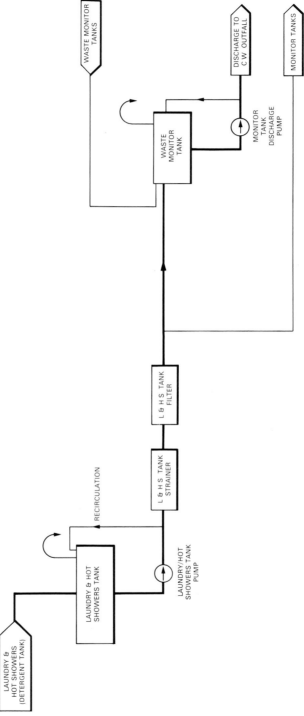

FIG. 6.31   LAUNDRY AND HOT SHOWERS WASTE SYSTEM

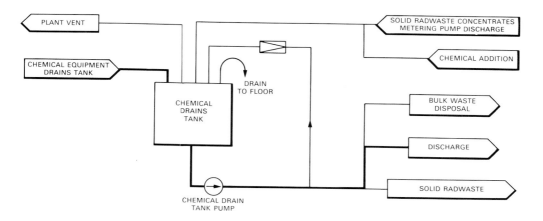

FIG. 6.32   CHEMICAL DRAINS TANK SUB-SYSTEM

General

All tanks are fitted with a recirculation line to allow the tank contents to be mixed before sampling. The monitor tanks and pumps are equipped with sufficient piping inter-connections to allow the tanks to operate together and to return the effluent for further processing in the event of higher than acceptable radioactivity being detected.

All systems (except the chemical drains tank) discharge to the CW outfall culvert where the effluent is well diluted before ultimate discharge offsite.

13.4   Gaseous Waste Management System

The gaseous waste management system controls, collects, processes, stores and disposes of gaseous radioactive wastes generated as a result of normal operation, including anticipated operational occurrences.

The system performs no function related to the safe shutdown of the plant and its failure does not adversely affect any safety-related system or component. Although the system is not safety-related, it controls gaseous release to the environment.

The system receives active gas from the volume control tank, the boron recycle hold-up tank, the boron recycle evaporators and the pressuriser relief tank. The gases are processed and stored for a specified decay period, sampled for activity and then released as appropriate.

13.4.1   *System Description (Fig 6.33)*

The main flow path in the system is a closed loop comprising waste gas compressor, catalytic hydrogen recombiner, four gas decay tanks for service at shutdown and startup. A second waste gas compressor and recombiner are provided as back-up. The system also includes a gas decay tank, drain collection tank, drain pump, gas traps to handle normal operating drains from the system, and a waste gas drain filter to permit maintenance and handle normal operating drains from the system. All of the equipment is located in the radwaste building.

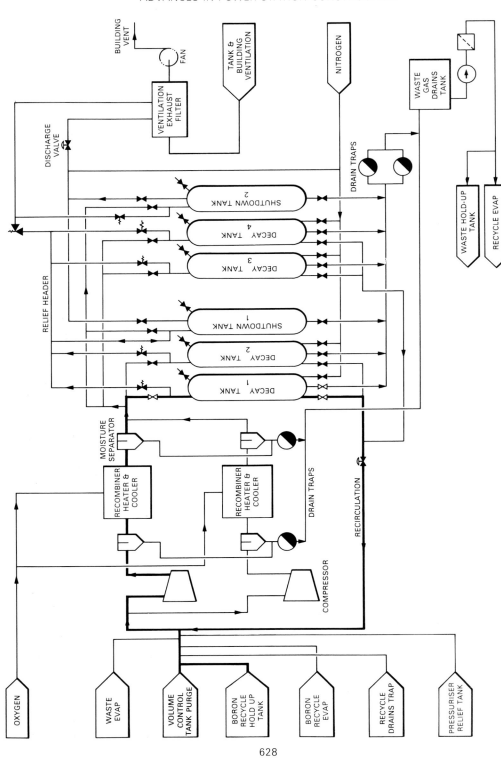

FIG. 6.33   GASEOUS WASTE MANAGEMENT SYSTEM

The closed loop has nitrogen for a carrier gas. The primary influents to the system are combined with hydrogen as the stripping or carrier gas. The hydrogen that is introduced to the system is recombined with oxygen, and the resulting water is removed from the system. As a result, the bulk of all influent gases is removed, leaving trace amounts of inert gases, such as helium and radioactive noble gases to build up.

The primary source of the radioactive gas is via the continuous purge of the volume control tank with hydrogen. The continuous operation of the system serves to reduce the fission gas concentration in the reactor coolant system which, in turn, reduces the escape of fission gases from the reactor coolant system during maintenance operations or through equipment leakage. Smaller quantities are received, via the vent connections, from the recycle evaporator gas stripper, the reactor coolant drain tank, the pressuriser relief tank, and the recycle holdup tanks. The waste evaporator gas stripper is normally vented to the radwaste building exhaust, but is vented to the gas system when it is used as a substitute for the recycle evaporator.

Since hydrogen is continuously removed in the recombiner, this gas does not build up within the system. The largest contributor to the non-radioactive gas accumulation is helium generated in the reactor core. The second largest contributors are impurities in the bulk hydrogen and oxygen supplies. Stable and long-lived isotopes of fission gases also contribute small quantities to the system gas volume accumulation.

When operating at power, fission gases are normally delivered to one of the four gas decay tanks, causing the system pressure to rise gradually. When this decay tank is full or partly full the tank is isolated and the contents are allowed to decay for the order of 45 days before they are discharged to atmosphere in a controlled manner.

The second decay tank is then brought into service and the procedure is repeated. By the time the second tank is used, the first tank is empty and ready for service. If necessary the third and fourth tanks are similarly brought into service.

During startup and shutdown nitrogen is admitted to the CVCS volume control tank (rather than hydrogen). During a startup, one of the startup tanks, which contains nitrogen from the previous shutdown, is used to collect the fission gases. The startup tank is normally sufficient to hold the fission gases generated during startup indefinitely but if required the tank contents can be discharged after a delay in a similar manner to the decay tanks.

The second startup tank is provided as standby but is normally used to accept relief valve discharges from the normal operation of the gas decay tanks.

13.5 Solid Waste Management System

13.5.1 *System Role*

The solid waste management system collects, processes and packages radioactive wastes generated as a result of normal plant operation, including anticipated operational occurrences and stores this packaged waste for disposal in a safe manner. The system is designed for the following functions:

(1) Provide a remote transfer and hold-up for spent radioactive resins from the chemical and volume control system, fuel storage pond cooling and clean-up system, boron recycle system, liquid radwaste system, steam generator blowdown system and spent radioactive activated charcoal from the liquid radwaste system.

629

(2) Solidify and package concentrated waste solutions from the boron recycle, liquid waste evaporators, spent radioactive resin from the demineralisers, spent activated charcoal from the absorber, spent filter cartridges and laboratory and decontamination chemical wastes for shipment off-site.

(3) Provide a means to semi-remotely remove and transfer the spent filter cartridges from the filter vessels to the solid waste processing system in a manner which minimizes radiation exposure to operating personnel and the spread of contamination.

(4) Provide for collecting, storing, compacting and packaging miscellaneous dry radioactive materials, such as paper, rags and contaminated clothing.

(5) Provide means of collecting, storing and incinerating low active combustible waste and waste oil. The waste oil may also be filtered if required.

(6) Provide storage for the low, the medium and the more highly active wastes.

13.5.2 *System Description*

The system is located in the radwaste building with the active stores forming an annexe and is shown diagrammatically in Fig 6.34. It consists of the following sub-systems:

(1) Dry waste system.
(2) Demineraliser resin charging system.
(3) Primary spent resin system.
(4) Secondary spent resin system.
(5) Evaporator bottoms tank system.
(6) Filter handling system.
(7) Solidification system.

The activity of the influents to the system depends mainly on the activity in the reactor coolant system and the decontamination factors achieved in the chemical and the volume control system, the boron recycle system and the fuel pond clean-up system. The activity of the influents is also dependent on the activity of the floor drains system, the steam generator blowdown system and the decontamination factors achieved in the liquid waste management system.

(1) Dry Waste System

The dry waste system consists of one compactor (hydraulic power mechanical ram device) and incinerators. The compactor is used to reduce the volume of compressible dry waste, such as paper, rags, filters and contaminated clothing. The incinerators are used to reduce the volume of low activity combustible waste and possibly waste oil. Both the compactor and the incinerators are designed with exhaust arrangements to control the spread of airborne dust.

(2) Demineraliser Resin Charging System

The demineraliser resin charging system consists of the resin charging tanks of the radwaste system and the CVCS. These tanks are transportable and can be located over one of a series of standpipes, each of which is connected to an individual demineraliser (or charcoal absorber) vessel. When a connection is made to an individual standpipe, fresh resin can be sluiced to the appropriate demineraliser by reactor make-up water moving under gravity. The filling operation is remotely controlled from the vessel being filled.

SIZEWELL B — PWR NUCLEAR POWER STATION

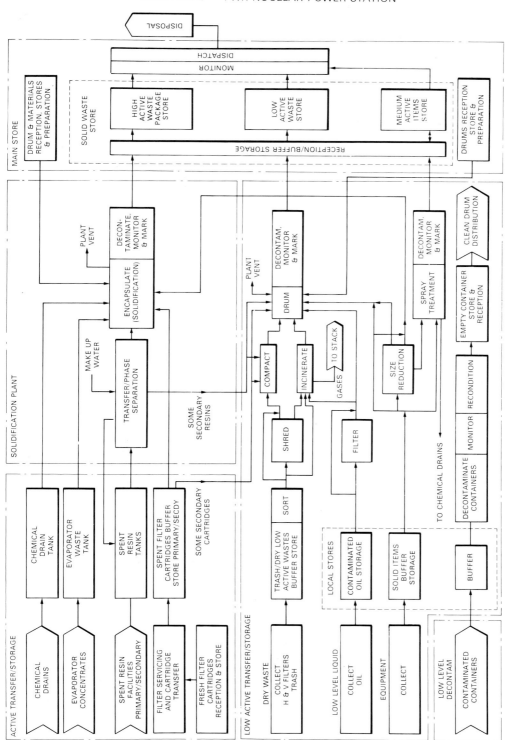

FIG. 6.34   SOLID WASTE PROCESSING SYSTEM

(3) Primary Spent Resin System

The primary spent resin system consists of a spent resin storage tank, a sluice pump and a sluice filter. This sub-system provides remote removal of spent radioactive resin from the demineralisers.

In the resin transfer mode, the spent resin sluice pump takes suction from the storage tank via a screened connection on the tank and pumps water through the respective vessel to first back-flush the resin and then sluice the resin to the storage tank. Positive indication that the resin has been sluiced to the storage tank is provided by an ultrasonic density element located in the sluice header.

Spent resins are transferred from the storage tanks to the solid radwaste decant station by pressurising the storage tank with nitrogen and supplying sluice water at the outlet nozzle on the tank. Positive indication that resin has been transferred is provided by the instrumentation in the solid radwaste decant station. Upon completion of the resin transfer, the tank is vented to the radwaste building ventilation system.

(4) Secondary Spent Resin System

The secondary spent resin system is similar to the primary system in design and operation except that it can also transfer and store spent radioactive charcoal.

Secondary spent resins may be transferred by gas pressure to the bulk waste disposal point for use when the radioactivity of the spent resins does not justify solidication. (Primary resins may be similarly disposed of when appropriate).

(5) Evaporator Bottoms Tank System

The evaporator bottoms tank system consists of a tank and a tank pump. The tank is provided with a mixer and the pump provides a relatively high flow for recirculation of the tank contents to maintain the concentrates in a homogeneous state. The tank is fitted with external strip heaters and all piping which can contain the concentrated waste is heat traced to prevent crystalization.

The evaporator bottoms tank collects the concentrates from the boron recycle evaporator and the waste evaporator. The contents are then mixed and sampled and if necessary caustic soda is added for pH control. The concentrates are then pumped to the drumming plant metering pump for solidification.

The tank pump is also arranged to deliver concentrates to the bulk disposal point for use in circumstances when the radioactivity of the concentrates does not justify solidification.

(6) The Filter Handling System

The filter handling system consists of a lead shielded cask, transport arrangements, a working plug and a set of tools which are designed to unbolt filter heads and manoeuvre the filter cartridges.

The filter handling system is a semi-remote system which removes spent radioactive cartridge filters from their filter housings and transports them to the solid radwaste processing area in the radwaste building.

(7) The Solidification System

The solidification system consists of a decant station, a solid radwaste drumming station, a cement filling station, a polymer filling station, a metering pump, a new drum store, the active waste store and means of handling both new and filled drums.

The solidification system operates on a batch basis to solidify evaporator concentrates, chemical wastes, spent resins and spent activated charcoal. It can also be used to immobilise spent radioactive cartridge filters. The system is designed to process the designated wastes resulting from normal plant operation, while operating for less than one shift per day, and is capable of continuous operation to meet anticipated transients which place a greater-than-normal load on the solidification system.

Solid inputs to the solidification system, such as spent resins and charcoal, are sluiced to the solid radwaste decant station. In the decant tank, the solids-to-water ratio is adjusted to provide the water content necessary to assure complete solidification when the slurry is mixed with the solidification agent. This is accomplished by removing only excess water, with the aid of the decant pump and ultrasonic level sensors. The decanted water is returned via the decant pump to either the primary spent resin storage tank or secondary spent resin storage tank, depending upon the tank which supplied the waste to the decant tank.

A positive displacement metering pump is provided for accurate transfer of the slurry to the solid radwaste drumming station for solidification.

In the drumming station, empty drums are transferred from the new drum store to the injection station where a measured quantity of cement or polymer is transferred to the drum. The partly-filled drums are passed to the next station where a metered amount of concentrates, resin slurry or chemicals are delivered to the drum. The drum is then mixed and transferred to a curing station and finally the drum is sealed and capped. Radioactive spent filter cartridges can also be loaded into the drums. Each section of the drumming station is fitted with sprays which are used when it is necessary to decontaminate the equipment. All drain water is recovered in the dirty radioactive waste sub-system.

Upon completion of the drumming process, the drum is weighed, the radioactivity content of the drum is checked and the drum surface is monitored for contamination. Radiation levels are recorded (scintillation counters with remote readout are provided) and the drum is transferred remotely to the solid radwaste storage area.

The solid radwaste decant station and drumming station are automatically flushed, following each batch, with reactor make-up water. Evaporator concentrate lines and spent resin slurry lines on the drumming station are automatically flushed into the drum with heated reactor make-up water to avoid potential crystallisation when contacted with cold flush water.

The active storage area is divided into a high active waste package section and medium and low active areas. Each drum in store is identified, and its radiation level recorded. The activity of each drum is measured immediately before despatch and after activity has decayed. The more highly active areas are shielded and means are provided for remote handling of the packages. Local buffer stores are provided for low active waste.

The active store also includes non-radioactive sections in which the clean containers and materials are received, stored, prepared and issued when required. Provisions are incorporated to decontaminate used containers which are only slightly contaminated.

14. CONTROL AND INSTRUMENTATION

The control and instrumentation (C & I)provided on the PWR follows the natural development of facilities and equipment in CEGB stations together with the proven Westinghouse Integrated Protection System.

14.1 Operational Facilities

C & I facilities are provided as necessary for the safe and efficient automatic or manual control of the station in all operating and shutdown modes. These modes include:

(1) Startup from a pre-defined plant state with off-site power supplies available.

(2) Normal operation with the power at any level within defined limits.

(3) Shutdown to defined states and maintenance in a safe shutdown state.

(4) Operations necessary to maintain reactor safety following any fault or hazard.

The C & I systems required for reactor safety provide manual and automatic operational facilities associated with the identification, detection and limitation of faults and abnormal station conditions. To enable safe conditions to be maintained at all times, the systems provide for control and monitoring of:

Reactivity.

Reactor core cooling and heat removal.

Reactor coolant system integrity and inventory.

Containment integrity and radioactivity releases.

14.2 Principles

The main control room is the main centre for control of startup, operations at power and shutdown. Additional controls, indications, etc not needed in the control room are provided elsewhere for plant preparation, maintenance, etc as appropriate.

The auxiliary shutdown room is separate from the main control room and provides emergency control facilities for use if the main control room is damaged or not habitable.

The reactor protection system monitors the state of the reactor and initiates reactor trip and engineered safety features plant when appropriate. This system also provides signals derived from its sensors for control and indication. In such cases the signals are provided via isolation amplifiers, isolation data links or both, such that any single fault on the output circuit will not prejudice the protection function.

The reactor protection system provides protection by means of a primary protection system and a separate secondary protection system, the equipments of which are diverse and independent so far as practicable.

The primary protection system is implemented by means of a development of the Westinghouse integrated protection system and provides reactor trip and engineered safety features actuations for all faults within the design basis. The system also includes necessary interlock and veto facilities to ensure safe and proper operation.

The secondary protection system provides an adequate degree of protection for the more frequent faults. When a fault is detected, a reactor trip and double turbine trip with appropriate engineered safety features are initiated, separate from the primary protection system.

The engineered safety features have all necessary manual controls and associated indications in the main control room.

The reactor protection system, the engineered safety features equipment and the essential main control room equipment are located in seismically-qualified buildings and are themselves seismically-qualified.

14.3   Control Rooms and Associated Support Facilities

14.3.1   *Main Control Room*

The main control room is situated on the 21m level of the control building and has an area of approximately 19m × 11m. The equipment is allocated and laid out to allow established CEGB operating procedures to be followed with a normal manning of a single operator and a supervisor. The principal facilities are:

(1) Operator's desk for normal operation of the reactor and both turbines.
(2) Supervisor's desk for general supervision and off-site communication.
(3) Steam cycle auxiliaries desk for pre-startup and manual back-up control of auxiliary plant.
(4) Post fault and reactor plant auxiliaries panel for manual control of auxiliary plant and engineered safety features.
(5) Electrical panel for control of the main and essential power systems and connections to the grid system.
(6) General services panel for general services plant including storage and transfer, heating and ventilation and meteorological systems.

Display information is provided in the main control room to enable the operator to:

(a) Operate the plant within defined safety limits.
(b) Assess reactor state.
(c) Detect the onset and assess the severity of fault conditions and significant plant faults.
(d) Assess the status of engineered safety features.
(e) Perform manual actions such as to achieve shutdown, or initiation of engineered safety features, where appropriate.
(f) Assess the plant state after reactor trip.
(g) Monitor the progress of faults.
(h) Detect and provide warning of internal and external hazards where appropriate.

The main method of displaying information is via computer-driven cathode ray tube visual display units (VDU). Conventional instruments and controls are arranged on desks and panels using the CEGB designed 72mm modules, as currently practiced on other stations.

Adjacent to the main control room are the computer-based data processing system, reactor protection system and miscellaneous support system equipment rooms together with main control room personnel messing facilities.

The design of the main control room and identified C & I facilities is such as to allow it to continue in operation and supervise the safe state of the plant following any credible seismic disturbance.

14.3.2 *Auxiliary Shutdown Room*

The auxiliary shutdown room contains the auxiliary shutdown panel which can be used to bring the reactor to a hot shutdown state and maintain it in that condition should the main control room become uninhabitable for any reason. In addition to the displays required for reactor safety, other instrumentation of value to the operator is provided and alternative on and off-site communications are available.

Where possible, the C & I in the auxiliary shutdown room mimics that in the main control room to avoid operator confusion. Circuit design is arranged so that no fault can invalidate any facility in both of the rooms.

The auxiliary shutdown room is in the auxiliary shutdown and diesel building situated at the opposite side of the reactor building from the main control room, and safe evacuation routes are defined between the main control room and auxiliary shutdown room; the latter is not normally manned and entrance is supervised.

14.3.3 *Control Priorities*

Controls in the main control and auxiliary shutdown rooms can be enabled or disabled to give control priority and avoid possible spurious operation from faulty controls.

14.3.4 *Technical Support Centre*

To allow assistance to be provided to the operator during fault conditions a technical support centre is provided in the administration building. Following an incident this centre will be manned by experts who will be able to communicate with the operator in the main control room and will have full access to all operational information via the data processing system. In this way expert assistance will be available to the operator without overcrowding of the main control room.

The technical support centre will also serve as an information gathering and plant assessment centre for use by the station's technical experts during normal operation and will thus contain historical information which may be relevant to the development of a fault condition.

14.3.5 *Simulators*

Normal operator training will be on a full replication of the main control room with a computerised plant simulator located at the CEGB training centre.

A similar plant simulator with replication of a selection of main control room facilities and the auxiliary shutdown room panels is also provided at the station adjacent to the technical support centre. This will be on site some 2 years before fuel loading in order that operational procedures can be developed. Its main use during station life is to back up the formal training at the Nuclear Power Training Centre with regular on-site practice facilities.

14.4 Station Control System

14.4.1 *Function*

The station control systems are designed to maintain operating conditions within defined limits automatically in the primary circuit and the secondary circuit during steady state operation and following normal transients such as load changing. The control systems are provided to reduce operator work load whilst maximizing efficiency and availability. Another important purpose of the control systems is their role in reacting to plant disturbances in a way which limits departures from defined conditions and returns the plant to these conditions thereby reducing the demands on the reactor protection system.

The functions of the station control systems include those needed to carry out the following:

(1) Adjust control rod positions to maintain reactor coolant average temperature within limits defined by turbine power demand.

(2) Provide the facility to set the power output of the station and, as a selected facility, to enable the station output to be varied in response to grid frequency variations within defined limits.

(3) Control the reactor pressuriser pressure and maintain the pressuriser level within defined limits.

(4) Control an alternative heat sink for the reactor steam in the event of large load rejections, eg trip of one turbine and loss of grid.

(5) Control the feedwater flow, in order to maintain steam generator level within satisfactory limits.

(6) Control the steam/feed header differential pressure, to optimise feedwater pump operation.

(7) Maintain the water inventory in the feed system of each of the two turbines by the avoidance of an excessive mismatch of the levels of water in the deaerators.

(8) Protect the station against loss of generation by avoiding situations likely to cause abnormal power conditions.

To facilitate these requirements, the following control loops are included in the station control system:

  Coolant temperature.
  Pressuriser pressure.
  Pressuriser level.
  Steam generator water level (4).
  Steam dump.
  Steam dump interlocks.
  Steam generator atmospheric relief valves (4).
  Feed pump speed.
  Turbine load.

14.4.2 *Equipment*

The foregoing control systems form part of a functionally and physically distributed system. The control systems are microprocessor based with incremental actuators where possible. The control system network is a sub-system of the overall station data processing system (DPS) but does not require any information from the rest of the DPS to function. The DPS normally provides the operator with the relevant information to enable him to operate the control loops. Each control loop is autonomous and will continue to function on failure of any other loop or the main data processing system. Manual facilities are provided for each loop in the main control room for the operator to take control in the event of failure of any loop.

There is a data link from each train of the protection system to the automatic control systems. These data links are isolated wired serial links. The data transmitted along these links is received by four interfacing computers, one per data link. The interfacing computers are connected to the automatic control systems via data links and transmit to the relevant control computers the information received by the interfacing computers from the protection system plus any other information generated within the interfacing computers. The control computers are also linked to the station DPS through separate links.

All data used by the station control system is validated and each control loop microprocessor is self checking and, in the event of any error or failure, the particular control loop automatically reverts to manual with no resultant movement of any control device such as an actuator. A high integrity of control is thus achieved.

14.5 Reactor Control System

The PWR is virtually self regulating due to its inherent feedback characteristic, ie its negative temperature coefficient of reactivity. The coolant temperature is programmed and regulated according to load demand.

14.5.1 *Coolant Temperature Control*

The reactor control system is designed to restore the coolant temperature of the reactor to its demanded value during normal operational transients by regulation of the insertion of the appropriate bank of control rods. The 53 control rods are moved in a predefined sequence under the control of the control rod drive equipment.

The coolant temperature is obtained from four measurements in each of the four coolant loops. Rod movement is caused by a difference between the measured and programmed coolant temperatures.

A mismatch between the total station load and the reactor nuclear power as measured by four ex-core nuclear detectors, is also used to move the rods as an anticipatory signal. The boron concentration can be manually adjusted to compensate for fuel element life or long term load changes in order to maintain the normal rod positions at their most effective position.

14.5.2 *Pressuriser Pressure Control System*

The coolant pressure is maintained at a set value to obtain efficient heat transfer.

The pressuriser pressure is compared with its demanded value to form the pressure difference signal which is then conditioned to form the actuation signals for the heater and spray controls. Increasing the heater power increases the pressuriser pressure and increasing the spray rate decreases the pressuriser pressure.

14.5.3 *Pressuriser Level Control System*

The pressuriser level is controlled in order to maintain a steam bubble above the coolant and allow effective pressure control. The volume of coolant, and hence pressuriser level, changes with temperature and so the demanded level is programmed as a function of coolant temperature to maintain a relatively constant mass inventory. The programmed demanded level is maintained by the chemical and volume control system which takes a constant recycle sample from the main coolant for treatment purposes but has a storage capacity for let-down and make-up control.

14.5.4 *Steam Generator Water Level Control System*

The water level in the steam generator is controlled at a level to avoid water carry over to the turbine and to obtain maximum heat exchange from the primary coolant. The difference between demanded level and measured level is used to control the feed water regulation valve for each steam generator. The steam flow and feedwater flow is also measured and a difference signal also feeds to the feedwater regulation valves as an anticipatory signal.

14.5.5 *Feed Pump Speed Control System*

There are six electric feed pumps each rated at 25% of total capacity. These are divided between the two turbines so that in each turbine feed train, two pumps are normally running and one on standby. The feed pumps are driven by fixed speed motors with a hydraulic scoop coupling to give a variable speed pump.

The feed pump speeds are controlled to maintain a main steam header to main feed header set differential pressure, ie pressure drop across the steam generators, and are scheduled according to the individual turbine loads to maintain the individual deaerator levels.

14.5.6 *Steam Dump Control System*

The steam dump system is arranged to dump steam on a load rejection or reactor trip, from the main steam header into the condenser of the two main turbines via the dump valves. The dump valves for each turbine are arranged into three banks which operate sequentially, the first two banks dumping steam into the condensers while the third bank dumps steam directly to atmosphere. One power operated atmospheric relief valve is installed on the outlet piping from each steam generator. The dump valves are operated by an error signal generated by the difference between the measured steam header pressure and the set pressure.

Operation of the steam dump system is inhibited during small load fluctuations below 10% total or 5% individual turbine load reductions.

14.5.7 *Load Control System*

The main station load demand is scheduled between two turbine load controllers. These controllers allow a maximum change of 10% per minute and have compensation from grid frequency, ie load droop.

The load controllers give load demand signals to the two turbine electronic governors.

14.6 Turbine Control System

14.6.1 *Function*

Each turbine governor controls the speed and load of the turbine in a consistent manner to provide long term security against thermal damage to turbine components. It runs the turbine up to speed when initiated, and after automatic synchronisation applies a block load (approximately 5%). It then accepts load increase or decrease signals from the station control system. Metal temperature measurements are used to override any changes in load demand which would cause thermal stress. Two levels of speed droop can be selected to allow the turbine throttle valve to respond to changes in grid frequency as required.

14.6.2 *Equipment*

The governor equipment comprises an hierarchial multi-microprocessor system of high integrity and availability.

Internally, the system is organised into two levels of control. At the lower level a number of distributed microprocessors perform the basic speed error and valve control functions. At the upper level a single microprocessor performs the more complex supervisory functions.

The lower of these two control levels is termed the 'base level governor'. It provides the fundamental means of speed and load control combined with acceleration protection, with one valve controller for each of the 10 valves per turbine. The safety and availability requirements of the turbine control system are met at this level. Once initiated and running the governor is capable of operating as a stand-alone narrow range governor. Individual manual control of each valve is provided.

The upper control level containing the unit processor provides all the other functions required in an advanced governing system. All interfaces to the operator, plant, etc, except the basic signals hardwired directly to the governor, are part of the upper control level system. Upper control level output requests, to alter speed, load or operating mode of the turbine, take effect through the governor, which accepts and acts on them only if they meet the acceptability criteria stored at the base level.

14.7 Reactor Protection System

14.7.1 *Function*

Plant parameters are continuously monitored to detect the occurrence (or potential occurrence) of a fault condition. When a departure from defined conditions is detected the reactor is automatically tripped and safeguards equipment operation (opening or closing valves, starting or stopping pumps, etc) initiated as appropriate. Fault detection and safeguard equipment initiation is provided by automatic operation of the reactor protection system during at least the first 30 minutes following fault detection. The operator may take control within 30 minutes and will need to do so some time after 30 minutes, the exact time after depending on the nature of the initiating fault and the subsequent operation of the safeguards systems.

The reactor protection system (RPS) is sub-divided into:

(1) Primary protection system (PPS).

(2) Secondary protection system (SPS).

Each of these sub-systems operates a separate set of circuit breakers in the control rod power supply to trip the reactor and a separate engineered safety features actuation system (ESFAS). The SPS is provided as a system largely independent and diverse from the PPS, for the more probable faults. The PPS central equipment is in the control building and the SPS is centred in the auxiliary shutdown and diesel building.

The occurrence of unsafe reactor conditions is determined by measuring relevant operating parameters. Each parameter is measured by four sensors and one is associated with each of four logic groups of equipment. Reactor protection is provided by means of these four logic groups of equipment, each of which independently detects potential faults in the reactor. Each logic group is commonly referred to as a 'guard line'. If two or more of the four guard lines indicate a faulty condition then a decision logic system gives a reactor trip, and activates the engineered safety features provided to deal with the fault detected.

The ESFAS is provided to initiate automatically the operation of the appropriate safety and support plant and take other appropriate action following the detection of a fault on the basis of the logic system described above. The primary ESFAS actuations include, as required:

Trip of both turbines.

Diesel generator start.

Main steam line isolation.

Main feed pump trip and feed line isolation.

Auxiliary feed pump initiation.

Containment isolation.

Emergency boration.

Safety injection.

Steam generator blowdown isolation.

Control room ventilation operation.

The secondary ESFAS actuations include, as required:

Trip of both turbines.

Main feed line isolation.

Auxiliary feed pump initiation.

Provision of the automatic systems described for reactor protection does not prevent manual action by the operator to trip the reactor or initiate safeguards should he consider it necessary. The reactor protection system automatic actions are mainly concerned with the short term safety requirements, partly to avoid the need for operator action in a very short time. Subsequently the operator is expected to exercise control, but specific actions are not required of the operator until at least 30 minutes after the reactor trip.

14.7.2 *Equipment*

There are four trains of equipment in the PPS and four in the SPS. Each parameter measured has four sensors and one is physically associated with each guard line train of equipment. The measurement is taken from the sensor to the guard line cubicle of the PPS or the guard line cubicle of the SPS where the partial trip status of the signal is detected and sent by data link or otherwise to the other three guard lines.

Each guard line equipment and associated parameter detectors and cabling is given complete electrical and physical segregation to avoid any common fault caused by power failure or fire for instance.

The detected trips are communicated between each guard line to enable each guard line to perform two-out-of-four parameter voting logic. Each guard line controls a pair of circuit breakers, arranged that any two guard lines must trip their circuit breaker to give a reactor trip, by removal of control rod supplies. Both trip systems incorporate this four-way redundancy and two-out-of-four voting, but using different equipment for the logic.

The control rod tripping circuit breakers are mechanically latched 3-phase breakers of well-proven design. Either of the PPS or SPS trip circuit breaker systems will alone totally interrupt the power supplies to the control rod system.

The circuit breakers on the PPS are of a diverse type to those of the SPS.

14.7.3 *Primary Protection System*

The PPS is a Westinghouse system specifically developed for application to the Westinghouse PWR. It provides reactor trip, ESFAS and safety interlock functions.

The sensors for the PPS are connected into the integrated protection cubicle which detects individual parameter trips by means of computers. Trips detected are transmitted via optical data links to the cubicles of the other three guard lines.

In each guard line, computers are used to combine the trip requirement data received from the other three guard lines with that data derived directly within its own guard line two-out-of-four logic to initiate a circuit breaker trip.

A large proportion of the PPS computer logic is self-testing and an automatic injection tester is available to vary all input signals to allow a comprehensive check of all trip functions. It is envisaged so that this test, which is under control of a master locked keyswitch, will be carried out every 30 days and take about 10 hours to run.

The PPS provides digitally-measured values of reactor parameters used for protection directly to other systems with full isolation by means of optical data links. This is used by the station control system and the station data processing system.

14.7.4 *Secondary Protection System*

The SPS is based on the use of the laddic module which is well tried and proven on the British AGR stations and also some of the Magnox stations.

A comprehensive lamp indication facility is provided at each local guard line cubicle giving trip and alarm status of all parameters. All parameter partial trip and trip approach alarms are connected into the station data processor and made available to the main control room.

14.8   Reactor Instrumentation

14.8.1   *Flux Measurement*

Instrumentation is provided to monitor the neutron flux and generate any necessary alarms and trip signals during various phases of reactor operating and shutdown conditions. It also provides signals for a secondary control function and indicates reactor power during startup and power operation.

To cover the very large range of neutron flux between shutdown and full power there are three separate ranges of instrumentation channels. Three types of neutron detectors, with appropriate solid state electronic circuitry, are used to monitor the neutron flux, covering the range from sub-critical conditions to 120% of full power and capable of recording brief overpower excursions up to 200% of full power. Four detectors of each type are provided in housings external to the reactor pressure vessel. One detector of each type is provided at each of four equally spaced circumferential positions.

The lowest range (the 'source' range) covers six decades of neutron flux on a logarithmic scale. This embraces the flux levels induced by the neutron sources. The next range (the 'intermediate' range) covers eight decades, also on a logarithmic scale. The highest range of instrumentation (the 'power' range) covers approximately two decades on a linear scale. The intermediate range overlaps the upper end of the source range and the lower end of the power range.

Instrumentation within the reactor pressure vessel consists of thermocouples and movable neutron flux detectors. The thermocouples are positioned in preselected locations to measure fuel assembly coolant outlet temperature. They are removable through the vessel upper head. The movable flux detectors can traverse the entire length of selected fuel assemblies from below, thus providing a three dimensional map of the neutron flux distribution in the core. They are removable from the lower head of the vessel through curved guide tubes terminating in an accessible seal plate within the containment. The temperature and flux distribution can be used to determine the power distribution in the core. This information can be used in the calibration of the neutron flux detectors external to the reactor vessel.

The nitrogen N-16 power monitoring instrumentation provides an input into the reactor control system and the PPS. It provides a measure of the thermal power of the reactor by detecting the level of N-16 present in the coolant system. N-16 is a short-lived isotope of nitrogen generated by neutron activation of oxygen contained in the water. The rate at which N-16 is produced in the primary coolant is directly proportional to the fission rate in the core. Decay of the N-16 isotope produces high energy gamma rays which penetrate the wall of the high pressure piping and are monitored by two ion chambers located adjacent to the hot leg piping of each coolant loop.

14.8.2 *Control Rod Position Measurement*

A position detector is supplied for each control rod cluster drive rod. Each standard detector is in essence a tube with 42 coil assemblies spaced along its length so that the ac impedance of each coil changes as the rod moves through it. To obtain high reliability, alternate coils are used by two separate measurement systems and an overall accuracy of ± 4 steps is obtained, equivalent to ± 64mm.

15. CONSTRUCTION METHODS

15.1 General

The site construction phase of a power station is crucial in determining the outturn cost.

Overall, the construction period is planned to last for some 7 years, and has a peak labour force in excess of 3000 and a total labour content of some 24 million man hours.

Delays in construction, particularly in the later parts of the programme, lead to additional costs in interest charges, site continuation costs and loss of net effective cost benefit. These penalties create a keen incentive to achieve a short construction period and the site management needs to be meticulously planned.

The construction falls into three distinct phases:
(1) Site preparation and preliminary works comprising:
    Land clearance, access roads, dewatering, foundation excavation and construction of temporary facilities:
    Target duration 11 months.
(2) Construction of plant:
    Runs from placing first permanent concrete up to the point where the plant is ready to receive nuclear fuel. Target duration 66 months.
(3) Raise power and setting to work:
    Target duration 6 months.

Planning for each stage requires precise knowledge of the activities of the succeeding stages to limit rework to a minimum and to ensure that proper access and logistical support by way of craneage, lighting, electric power, compressed air and gas services, water and weather protection are available. Other factors considered are the levels of manpower to support the planned rate of work, the desire that individual contractors achieve smooth labour curves in the interests of economic resourcing and reducing the potential for industrial relations problems associated with fluctuating labour forces.

15.2 Construction Planning

From the earliest stages of the station conceptual design the requirements of the construction phase are kept under review and by the time the design phase opens the nucleus of the construction management team is set up to co-ordinate all construction aspects.

A major tool is the use of both engineering and construction models to develop a full understanding of erection philosophy as well as detailed examination of problem areas. The construction model of 1:75 scale covers the area of the whole power block. This

FIG. 6.35   CONSTRUCTION MODEL SHOWING MAIN BUILDINGS LAYOUT AND ACCESS

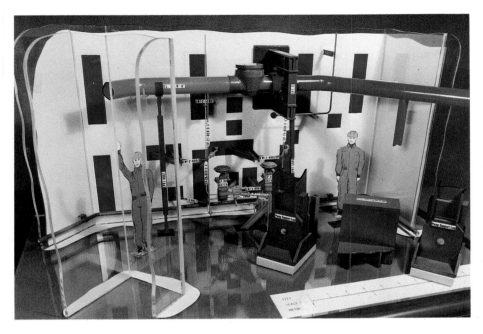

FIG. 6.36   A MODEL SECTION OF THE SNUPPS HIGH PRESSURE COOLANT
INJECTION TEST LINE

FIG. 6.37   A MODEL SECTION OF THE SIZEWELL B HIGH PRESSURE COOLANT
INJECTION TEST LINE

model is built up in stages from the basic foundation profile to completion of each stage showing progress. The model includes temporary access for plant and personnel, construction craneage, temporary works and storage areas. Photographs were taken for inclusion in the work plan for issue with the PWR Contract Enquiry documents to brief potential contractors on the overall target construction programme.

Such a model is a live construction tool and can be used to evaluate the impact of changes to the construction sequence and to minimise the effects of any delays. Fig 6.35 shows typical details taken from the model.

The detailed engineering models covering the power block are at the following scales:

Reactor building	1:20 and 1:10
Auxiliary building	1:20
Turbine house and mechanical annexe	1:33

These models can be used to study areas in very real detail, particularly those areas which are most congested. The best aid to construction is to simplify the design. The reactor building is a very fruitful area for this approach and Fig 6.36 and 6.37 show what can be achieved here.

In all areas the maximum use is made of modularised construction where fully complete modules of plant are installed thus limiting site work to setting, securing and making-off to the external terminals rather than installing the components piecemeal and then interconnecting them. The increased first cost of the modular approach is more than offset by the simplification in site activity.

The PWR lends itself to this approach in that all the major primary circuit items are delivered as complete entities and the site work mainly involves the piping systems to and from the primary circuit.

Having established the overall site plan and developed the critical paths for construction activities through the major plant items, the key activities are put on network programmes which are in themselves supported by subsidiary networks in very great detail.

The word 'network' implies a graphical approach but nowadays these are mainly computer-based networks capable of handling vastly greater numbers of activities than the earlier graphical methods. In especially sensitive areas these computer programmes (eg SMART) can optimise the workload against available resources to achieve the most effective use of the workforce.

The construction programme is reflected right back down the station design chain and the engineering and procurement programmes are developed to service that programme.

7.

Project Management

1. INTRODUCTION

This Chapter reviews the development of the project management function from the beginning of the 500MW unit conventional and AGR station construction programmes up to the most recent current projects. Project management procedures and practices, particularly with reference to design, contract strategy, financial control and site industrial relations, were subject to development throughout this period as experience was gained from the earlier projects. The success of this development has resulted in good construction progress on both the Drax Completion and Heysham 2 projects which are currently (early 1985) meeting their construction programmes and are within their financial sanctions.

During the 1960s it became clear that future power station projects would be fewer in number and considerably more complex in the scope of supply and in their managerial and technical demands. This was particularly true of the commercial AGR programme which had begun in 1965 at Dungeness B. Against this background the CEGB determined to concentrate its three dispersed power station project group resources in one location which led to the formation of the Generation Development & Construction Division (GDCD) at the onset of the 1970s.

Prior to the formation of GDCD at Barnwood, the allocation of projects to individual Project Groups was determined by geography and the relative workload of the CEGB's three Project Groups when new projects were initiated. The work of the three Project Groups was supported by a Generation Design Department at CEGB Headquarters, which was responsible for specialist plant engineering and design assessment.

In 1984 a separate CEGB organisation, known as the Project Management Board, was formed at Knutsford, in Cheshire, in association with the National Nuclear Corporation (NNC) with specific responsibility for project management of the proposed Sizewell B PWR power station, subject to the required consents and sanction being granted.

2. DEVELOPMENT OF A PROJECT MANAGEMENT ORGANISATION

2.1 The Client's Role and Responsibility

Through GDCD, the CEGB (as client) specifies, co-ordinates and controls the activities of contractors involved in power station construction. As part of this client responsibility, GDCD maintains a supervisory management team at each construction site, but undertakes no direct manufacture or construction work itself. This approach is adopted as the CEGB remains accountable for expenditure, the quality of the design and for plant safety and to assure itself that the electricity consumer obtains value for money in constructing new plant.

The CEGB as client plays a dominant role in defining the contract strategy, maintaining the programme, establishing compatible programmes, interface coordinating of design and programming, promoting the timely clearance of safety issues, and endeavouring to create conditions in which contractors can most effectively fulfil their performance obligations. To secure this last objective, GDCD has in recent years adopted a series of policies designed to assist in remedying the problems previously experienced in power station construction. These include the establishment of all significant design features before the construction phase commences, and the creation of management groups at each site to promote stable and cohesive site industrial relations.

Whilst the client's objectives in managing all types of power station projects are identical, the methods adopted to achieve the objectives differ between conventional and nuclear projects and can also differ between different nuclear projects.

2.2 Project Management of Conventional Power Stations

GDCD's role in the project management of conventional power stations encompasses all aspects of station design, architect-engineering, specification of plant, contract management, control of site construction and commissioning.

On conventional power stations (fossil-fired thermal stations and hydro-electric or pumped storage schemes) the work is executed by a single GDCD project team who places a series of separate contracts for items such as boilers, turbines, civil works and cabling. These separate contracts are then let against an overall project programme which is subsequently used to monitor performance in erecting the station.

For each contract the project manager's team supported by GDCD specialist branches draws up specifications incorporating user requirements, issues enquiries, assesses tenders, negotiates and finally places the contract. Each contract is then administered by the project team, which includes approving design submissions, attending to payments, variations, claims and extensions of time. The team also ensures the contract conditions are observed and progresses co-ordination with that on other contracts. The project team monitors the performance of each contractor by acting as a design, technical, quality and financial auditor both within the manufacturers works and on site. The team also monitors a contractor's progress against programme, the productivity of the work-force and industrial relations on site.

During the 1960s the CEGB's strategy called for the placing of lump sum or remeasured contracts for separate items of plant and civil works, each embracing total design, supply and erection requirements. This resulted in placing approximately 100 direct contracts for a typical conventional project, with some 50 to 60 different main contractors engaged on site at the same time together with a substantial number of sub-contractors.

A particular feature of the 1960s was the high rate of ordering 500MW units. During the decade orders for 49 units were placed, 47 of which were ordered in the first four years. The high ordering rate over-strained the capability of the plant manufacturing and construction industry, whose resources were dispersed across a variety of different designs. The difficulties were exacerbated by other factors, such as rapid technological advance and gave rise to serious delays in completing power station projects. The underlying reasons for these shortfalls were examined by a series of Government appointed inquiries, notably the Wilson Committee in 1969 (Cmnd 3960), the National Economic Development Office (NEDO) in 1970 and 1976, the Price Commission in 1979 and the Monopolies and Mergers Commission (MMC) in 1981. The recommendations emanating from the 1969 Wilson and 1970 NEDO Reports were taken into account by the CEGB in formulating its project management strategy for the early 1970s. It was recorded by the MMC in 1981 that, when reviewing prescriptions for improvement proffered by inquiries over the previous decade, a particular feature of the recommendations was the advocacy of a more active role in project management by the client. This required the development of stronger project management by the client, with higher status and management skills to complement engineering expertise. It also required the client to ensure that contractors had adequate site industrial relations expertise, and to co-ordinate contractors' labour management policies. NEDO also recommended that the client should monitor performance and audit incentive bonus schemes in a more adequate manner.

Following the 1969 and 1970 recommendations the project management strategy implemented on two major projects at Grain and Ince B restricted the number of main contractors on site to a minimum and employed reimbursable forms of contract. For Grain, only five main erection contractors together with a service contractor were appointed, with plant suppliers providing technical supervision of the erection. The mechanical erection contractors were employed on cost-plus reimbursable contracts which allowed for a lump sum management fee subject to contract price adjustment and reimbursement of all other direct costs. Approximately 122 'supply-only' contracts were let.

The outcome of this strategy was unsatisfactory. Restricting the number of erection contracts greatly increased their value and scope. Such large scale contracts overstretched contractors' managerial ability with adverse consequences for both technical performance and industrial relations. Moreover, the use of reimbursable contracts transferred the financial risk to the client and weakened the resolve of contractors' management to contain costs. It also encouraged site work people to believe that their productivity was no longer relevant to the viability of their employers, the project or their future employment prospects.

As a consequence, for the Drax Completion project begun later in the 1970s, the CEGB developed a revised project management strategy, the principal features of which are as follows:

(1) Completion of as much design work as possible before commencement of construction, using design-phase contracts, which include options to continue into the manufacturing phase at pre-established contract prices.

(2) Replication and standardisation of plant.

(3) More rigorous vendor assessment procedures to ensure that contracts placed are within the performance capabilities of contractors invited to tender.

(4) The use of a revised form of lump sum contract in preference to reimbursable arrangements.

(5) A 'key-date' procedure which provides for contractual payments to be withheld if specified work is not completed by the defined dates, thereby stimulating contractors' senior management into prompt corrective action. The 'key-date' procedure is more fully described in Section 6.5.3.

(6) A requirement that contractors participate in a management group which determines policies and fosters a co-ordinated approach to site labour relations for the project.

(7) The implementation (from November 1981) of the new National Agreement for the Engineering Construction Industry (NAECI), described in Section 9.4, with the associated establishment of the Drax Project Joint Council.

(8) The introduction of double dayshifting to secure the required rates of production on site.

The Drax project has 11 major erection contracts (compared with five for Grain), and more than 100 other direct contracts of relatively modest value. The major contracts include a number of work areas which in the 1960s would generally have been let as separate direct contracts. Thus the Drax strategy represents a compromise between the arrangements of the 1960s and those of the 1970s. Its success is evidenced by the project's successful progress within its original programme and budget, leading to the commissioning of the first unit in April 1984 7 months ahead of the original commitment.

2.3 Project Management of Nuclear Power Stations

2.3.1 *The Magnox Programme*

CEGB nuclear power station construction over the past 25 years has been characterised by the same difficulties and shortfalls outlined for conventional stations, including site industrial relations problems, poor productivity (particularly in the 1970s) and fluctuations in workload. The project management problems arising from rapid technological advance have been considerably greater in the case of nuclear as opposed to conventional power stations.

The project management of the CEGB's first nuclear programme of Magnox stations, which began in the late 1950s, was undertaken by several nuclear consortia under contract to the CEGB. The responsibilities of these consortia were similar to those discharged directly by the CEGB for conventional stations. They included total station design, an architect-engineering role, plant design, procurement, management of the site and commissioning in conjunction with the CEGB. The consortia were not manufacturers of plant, but this activity was undertaken by some of their parent companies. Turnkey contractual arrangements with the consortia worked satisfactorily for the Magnox programme, which ultimately comprised eight twin-reactor stations. Construction time and cost overruns, where they occurred, were small and the arrangements were highly successful.

2.3.2 *The First AGR Programme*

During the mid to late 1960s and early 1970s work began on five AGR stations, four for the CEGB at Dungeness B, Hinkley Point B, Heysham 1 and Hartlepool, and one for the South of Scotland Electricity Board at Hunterston B. The AGRs were thus largely built in the poor industrial relations climate described and which had adversely affected the construction performance of fossil fuel stations during the same period. Each AGR built at this time was ordered under a comprehensive contract with one of the three consortia then in existence. Each consortium developed its own AGR design.

In retrospect it is apparent that the design resources available were inadequate to develop three different station designs, involving substantial extrapolation from experience with the experimental 30MW Windscale AGR and the successful Magnox stations.

The most important causes of delay were the large amount of design innovation and the failure to ensure that all prototype features were properly proven before being committed to production. These delays were revealed during the construction, testing and commissioning phases, and consequently took an appreciable time to resolve as all modifications had to be approved and agreed on safety-related grounds.

The disruptions in site erection work caused by these factors often led to reductions in bonus earnings and low morale amongst contractors' work forces. The endemic industrial relations problems associated with large industrial sites were thus exacerbated by low labour productivity attributable to the impact of repeated design changes.

2.3.3 *The Second AGR Programme*

The problems and uncertainties associated with the first AGR programme and the experience gained led the CEGB to its strategy for the project management of the subsequent Heysham 2 AGR station. The CEGB adopted arrangements designed to make the best use of the respective strengths of GDCD and the National Nuclear Corporation (NNC), the nuclear design and construction contractor, in a complementary manner. The Heysham 2 strategy also employed the series of initiatives previously described for Drax, including design-phase contracts, a management group for Heysham 2 and extensive double dayshifting of the site workforce.

All contracts, including those for the Heysham 2 nuclear island, were drawn up between the CEGB and the contractor concerned. NNC was retained to act as an agent of the CEGB, thereby ensuring that performance guarantees remain directly available to the CEGB. This arrangement was formalised in the Heysham 2 Agency Agreement between the parties which requires NNC to manage the nuclear island work as the CEGB's agent. In addition NNC are responsible for the overall outline design of the station and for the systems engineering of the nuclear island. GDCD provides overall project management and is responsible for all construction outside the nuclear island on a basis similar to that adopted for conventional stations. Thus GDCD undertakes the systems engineering for this area of plant, placing and managing a series of separate contracts. The joint CEGB/NNC arrangements for the Heysham 2 project have operated successfully and the project has maintained its original programme and is within the sanctioned budget.

2.3.4 *The Pressurised Water Reactor - Sizewell B*

Other arrangements were implemented during 1984 for construction (subject to consents and financial sanction) of the Sizewell B PWR station which place the management of the project under direct CEGB control. These arrangements take account of the principles advocated by the Monopolies and Mergers Commission in 1981 for the construction of future nuclear power stations, including their observations on the bearing of financial risks and the need for unified control over the execution of projects.

The new Sizewell B project structure therefore reflects the following key considerations, which are judged to be particularly important when introducing a reactor system new to the UK:

(1) The CEGB has absolute responsibility for safety.
(2) The CEGB is accountable for the quality of the design.
(3) The CEGB is responsible for funding the project and must recognise that it cannot place significant financial risk on its contractors.
(4) The procurement strategy, including judgements about the extent of UK supply, is the responsibility of the CEGB.

To implement this strategy for the Sizewell B project the CEGB has established a Project Management Board (PMB) which acts as a subsidiary company of the CEGB and reports directly to the Generating Board. It is chaired by a CEGB Board Member and its Directors are drawn from both CEGB and NNC. While the PMB remains responsible for and retains control over the project, it has delegated executive responsibility to a Project Management Team (PMT), staffed in approximately equal numbers by CEGB and NNC personnel. The team is responsible for managing the project as a single integrated organisation. In this task it will adapt and apply the project management strategies which have proved successful at Drax Completion and Heysham 2. In particular it is the aim of the CEGB and the PMB to maximise the opportunities for British suppliers to participate fully in Sizewell B and to secure the successful transfer of PWR technology to the UK.

2.4 Appraisal of Past Experience

It has been stated that the CEGB's experience in constructing large conventional and nuclear stations during the 1960s and 1970s led to a number of policy changes being adopted for the Drax Completion and Heysham 2 projects. These changes encompassed plant replication, design control and contract strategy. A policy to replicate 'standard' plant of proven quality was introduced although in practice there are a number of potential constraints on full replication of an earlier power station project. Full replication is only feasible or desirable for a limited tranche of stations programmed for sequential construction without a significant gap in the plant manufacturing programme. Despite these constraints considerable effort was devoted to ensuring that the designs of the Drax Completion and Heysham 2 projects followed as closely as practicable the proven design basis of the earlier Drax and Hinkley Point B Stations.

The potential for delay and increased costs resulting from inadequate or late design effort was clear from experience in constructing both the 500MW unit conventional stations and the earlier AGRs. To reduce the risk of costly design changes during construction (including design changes required for nuclear safety), the CEGB now ensures that sufficient design work is completed before the start of construction, to minimise the risk of changes arising during construction which have a significant impact on cost or programme. Every effort is made to ensure that safety-related design changes to meet the requirements of the Nuclear Installations Inspectorate (NII) are accommodated prior to construction. The advantages of increased attention to early design work will not, however, materialise unless there is an adequate mechanism for controlling subsequent design changes. To ensure that proposed design changes can be effectively controlled during the design process, the design is progressively 'frozen'at successive stages in its development. All changes need to be fully justified and are subjected to rigorous review.

The CEGB's contract strategy for recent projects has also changed and a return to a larger number of lump sum contracts covering the majority of erection activities has been implemented. Competitive tendering is employed wherever possible. In the past, the ultimate sanction on a defaulting contractor was the threat of extracting liquidated damages, where the contractor could be held responsible for delay. This sanction was found to have limitations, as it could only be applied in practice at the end of the contract and the damages recovered were insignificant compared to the cost incurred by the CEGB as a result of the delay. The CEGB currently uses a key date procedure on all major plant contracts. This provides, if necessary, for the withholding of payments due to a contractor at any stage of the contract from design to completion of erection should the agreed programme not be met. The application of the keydate procedure is described further in Section 6.5.3.

The nature of industrial disputation on CEGB projects required a different series of policy initiatives if improvements were to occur. Such disputation was principally due to the multiplicity of national labour agreements with differing terms of employment in use on site and the effect of poorly constructed and badly managed incentive payment schemes. The resulting anomalies between different groups of employees working side-by-side on the same construction site and employed on broadly similar erection activities proved to be a source of continuing labour unrest. Accordingly, the CEGB at Drax and Heysham 2 introduced means whereby a consistent approach to the harmonisation of employment conditions on its construction sites could be achieved. The CEGB together with other parties involved in large site construction joined in promoting one comprehensive national agreement for all mechanical and thermal insulation contractors. In 1981 a new National Agreement for the Engineering Construction Industry (NAECI), was introduced covering the mechanical construction industry. The NAECI agreement provides for the control of bonus payments on a national basis and for increasing the proportion of basic pay to incentive bonus, thus reducing the adverse impact of poorly devised bonus schemes. At Heysham 2 and Drax all contractors are required to participate in a management group which also includes representatives of employers' federations and the CEGB. Contractors must abide by the site industrial relations policies determined in that group, which seeks to harmonise employment conditions on the site and allow contractors from different industrial sectors to manage their own labour forces in a compatible manner. The development of the NAECI and the establishment of management groups are fully described in Section 9.

3. PROJECT MANAGEMENT TEAMS - AIMS, STRUCTURE AND TECHNIQUES

3.1 Aims of Project Management

The overall planning control and co-ordination of a project from inception to completion is aimed at meeting a design intent and ensuring completion on time, within cost and to required quality standards.

Project management skill lies in marshalling valuable resources including manpower, materials, machinery and finance to this end. It demands the ability to:
(1) Plan and programme all the work and resources required to complete the project.
(2) Commit all those involved with the project to its aims and objectives.
(3) Design the plant, the plant support systems and structures.
(4) Place contracts with manufacturers, suppliers and constructors to procure plant and services at the right time, of the right quality and at an economic price.
(5) Organise the construction of the plant.
(6) Commission the plant by progressively integrating individual systems into a reliable power station.
(7) Monitor and control progress to achieve the programme.
(8) Control the financing of the project and ensure that costs are contained within the planned budget.

The techniques and processes available to aid a Project Manager in the achievement of these aims are detailed in Section 3.3.

3.2   Structure of a Project Team

3.2.1   *Directors of Projects*

The CEGB carries out and controls its operations on the principle that authority is delegated to the point in the organisation where it can be most effectively exercised. The Director-General of GDCD is personally accountable for the satisfactory development and construction of the CEGB's power stations. He is directly responsible to the CEGB Executive for performance within his delegated authority. Responsibility for individual power station projects is further delegated to Directors of Projects, who are charged by the Executive with ensuring the execution and completion of the scheme to programme, within budget and to the required performance.

A Director of Projects, in addition to being the 'responsible officer', is also the 'engineer' for the project. As such he administers the contracts for plant and civil engineering works acting as the agent of the CEGB. He is also required to exercise his powers independently on certain matters, for example when a dispute occurs between the purchaser and the contractor.

3.2.2   *The Project Manager*

A Director of Projects delegates certain of his responsibilities to a project manager, who reports directly to him. His remit is to control and progress all project activities to ensure that the station is completed satisfactorily to time and cost. The project manager heads a project team of engineers, draughtsmen and planners together with commercial, financial and administrative staff who are seconded from functional departments within GDCD or transferred directly from other project teams. Appointments to the project team take place progressively as the station progresses from design to commissioning. The project manager's responsibilities cover two main areas:

(1) Design and procurement for which he is assisted by a project engineer (Fig 7.1) supported by a team of engineers and draughtsmen located at Barnwood, and by a contracts officer concerned with the contractual aspects of procurement.

(2) Site management for which he is assisted by a site manager whose staff are located at the power station site (Fig 7.2).

In fulfilling these responsibilities the project manager:

(a) Establishes time and cost parameters for all aspects of work on the project and ensures that the commitment to time and cost is clearly disseminated to all participants in the project.

(b) Undertakes the duties of 'responsible officer' with respect to tenders and recommendations for contract approvals. He places contracts subject to his delegated financial authority and is empowered to authorise variations to these in accordance with the CEGB's contract procedures.

(c) Ensures that there is effective continuity between the design and construction phases of the contract and that the construction skills of GDCD are channelled into the design phases of the project. Similarly he is responsible for ensuring that there is continuity in the design and engineering approach during the construction stages of the project.

In addition to his own staff the project manager utilises the expertise of specialists within GDCD functional departments, who are accountable to him for their own contributions to design, procurement and construction of the station.

3.2.3 *The Project Engineer*

The project engineer is responsible for the engineering of the power station construction project both in its design and manufacturing phases. His responsibilities are delegated to him by the project manager and in the latter's absence he acts as his deputy.

In the 'design phase' the project engineer's responsibilities include:

(1) The establishment of the overall station design to meet the requirements and functions which have been identified in the Station Development Particulars issued by the CEGB Headquarters Corporate Strategy Department.

(2) Progressively to develop this design in more detail and to prepare procurement specifications and drawings either for individual components or for whole systems. These procurement specifications set out the requirements of plant and services to be ordered and the design parameters with which they must conform. In setting the specifications he pays particular attention to the interfaces between systems and plant items.

(3) To develop programmes for all aspects of the work on the project and to maintain and update these programmes as the project progresses.

(4) Supplier assessment. A formal assessment is carried out of the commercial and technical competence of possible contractors to supply the plant and equipment specified. This includes an analysis of their financial status, design, manufacturing and erection capabilities, quality assurance arrangements, production control systems and their capability to carry out the specific task for which they are being considered.

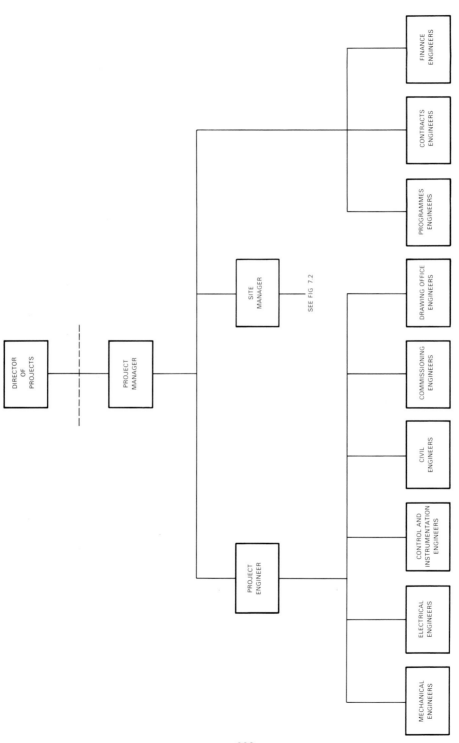

FIG 7.1   TYPICAL GDCD PROJECT TEAM ORGANISATION TREE

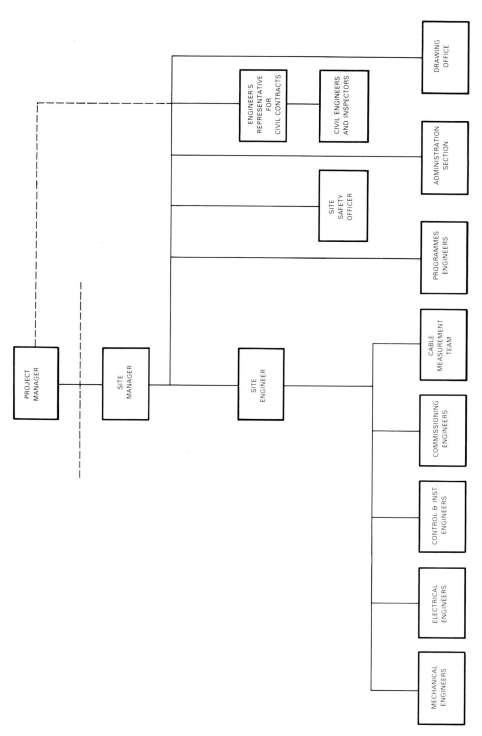

FIG 7.2   TYPICAL GDCD SITE ORGANISATION TREE

(5) Once potential manufacturers have been identified a normal tender and assessment process follows after which a manufacturer is selected. The project engineer ensures that individual tenders are assessed for compliance with the specifications and that these assessments are carried out on a consistent and impartial basis.

(6) Following these assessments he makes recommendations to the project manager on the choice of contractor. The assessment of the contractual aspects of tenders is carried out by the contracts officer. The project engineer and his staff are also responsible for assessing design phase submissions where these are specifically called for by the contract.

During the project construction/manufacturing phases the project engineer is responsible for ensuring that all contract work from design through manufacture up to delivery on site meets specification and is carried out to programme. This involves day to day monitoring of each contract and, where necessary, the initiation of action to remedy deficiencies. He is also responsible for ensuring that quality assurance requirements are met. During site construction and commissioning phases he provides support information with respect to the design and performance of plant items for which he has procurement responsibility.

Within the project engineer's team, responsibility for design, specification, assessment of tenders and the monitoring of progress on individual contracts, is normally carried out by the engineer responsible for the discipline appropriate to the contract, who reports directly to the project engineer as shown in Fig 7.1.

3.2.4 *The Site Manager*

The site manager of the project has responsibility for ensuring that power station site construction, commissioning and the handing over of the station to the generating region is carried out to time and cost. He is thus responsible for the overall management and co-ordination of all activities on site and in addition has responsibility delegated from the project manager for the implementation of the site quality assurance programmes. The site manager's specific responsibilities include:

(1) Engineering and contract control of all permanent works on site. In the case of the civil contracts, these are directed by the CEGB site civil engineer who can act as the 'engineer's representative' in certain circumstances. He is accountable to the site manager for ensuring that the civil works are carried out in accordance with the project programme, but reports directly to the project manager in respect of the engineering of the civil works (Fig 7.2).

(2) Ensuring that contractors conform with the CEGB's industrial relations policies and the implementation of these policies on site.

(3) Monitoring performance of all site related contracts forming part of the project to ensure that the overall project master programme is maintained.

(4) Management of site services, eg canteen and medical services and site security.

(5) Overall responsibility for site safety.

(6) The promotion of effective communication between all major contractors on site. Where necessary, eg on matters such as site safety, he may issue working procedures to which all on site must conform.

(7) Initiating the commissioning activities on site; co-ordinating all the resources required of the contractors to ensure that the commissioning procedures are carried out satisfactorily and safely within the programme dates.

(8) Acting as the focus of all relations with the general public with regard to activities on site and for ensuring that these are carried out with the minimum practicable inconvenience to members of the local community. It is the site manager's responsibility to ensure that the agreements which have been reached between the CEGB and, for example, the local authorities on matters such as control of off-site traffic are implemented.

The site manager is assisted by a site engineer who leads teams of engineers from the main engineering disciplines on site. They in turn have delegated responsibilities for the engineering and progress of site contracts within their own particular disciplines.

3.3 Project Organisation Interfaces

In discharging their responsibilities the project manager and his team must contend with the legitimate if sometimes competing interests of many different organisations whose spheres of activity impact upon the project. Fig 7.3 diagramatically represents these varied interests whose inter-action one with the other may create additional demands and challenges for the project team to manage. While the team itself is multi-disciplined it still cannot contain in its ranks all the specialist expertise required to execute the project. Within the CEGB the advice of specialists in finance, legal affairs, health and safety, operations, corporate strategy, public relations, contracts and industrial relations must all be accommodated. On site several thousand workpeople employed by one or two hundred different contracting companies and represented by some ten or twelve trades unions in several separate bargaining arrangements poses a variety of managerial challenges. Several thousand manufacturing company employees may be employed off-site at one time or another in supplying materials and components to the project, all subject to very different managerial, employment and representational arrangements.

All these companies and their employees will in turn maintain links with national trade associations, government, union organisations and the relevant statutory authorities. Government interests extend from the local authorities to the central ministries for Energy, Environment, Employment and others. Nor should the public relations aspects of a project and the media ever be ignored. The emergence of local interest groups concerned with the environment, regional infra-structure and economic development also properly demand the attention of the project team. There are clearly no completely comprehensive 'text book' analyses and solutions that can be developed in advance to aid project management in its task. There are however techniques and procedures that can be adopted to amass material and so order it for retrieval and application that the tasks of project management are to some degree simplified. Some of the main procedural aids to project management activity are outlined in the following sections of the Chapter.

3.4 Project Management Control Techniques

3.4.1 *Introduction*

Computing systems are widely used in design and in handling information relating to the requirements of project management. The systems have been developed over many years, employing some proprietary software but mostly utilising systems developed by the CEGB's Computing Information Systems Department to GDCD's specification. Most of the systems have evolved as the technology of computing facilities has developed. Systems that used to employ punched cards for input to batch processing with overnight or weekend running are now wholly or partly inter-active from desk terminals. The increasing integration of information is now encapsulated in a large data base known as the Total Project Information system. The main areas contained within this data base system are:

(1) Engineering related data (which is mainly plant and system related).
(2) Free text information (generally covering reports, articles, correspondence, etc).
(3) Graphical information in which computer aided drawing and computer assisted graphical calculations predominate.
(4) Management information and management reports derived from the above data.
(5) An index and repository of information related to the total life of a power station, including manufacture construction and commissioning and usually known as 'lifetime records'.

The concept of a total project approach to computer aided engineering already in use on current projects results in the progressive integration of plant related information into a single system which includes:

(1) Structured information on the plant items of the station, eg electrical plant, cables, control and instrumentation, valve schedules.
(2) Unstructured or free text descriptions of plant systems and components, eg safety reports.
(3) Graphical/topological information for which the main source is the computer aided draughting (CAD) system, including 2-dimensional representations and 3-dimensional models.

The following Sections (3.4.2 to 3.4.6) provide brief descriptions of the principal programs which have been implemented as aids for project management over the past few years. Program development and enhancement is continuing in many of these areas within the general concept of integration of individual programs into comprehensive systems.

As stated, programs are usually written by the CEGB's Computing Information Systems Department (CISD) to GDCD specifications, for operation on the CEGB's integrated computing network. As a general principle, it is normal to introduce the large and more complex type of new program on a new project and, in order to avoid any interruption, not to make a major change during the period of that project. However, because of the duration of construction projects, some modifications to current programs can prove worthwhile and new developments may be implemented on a project when feasible and beneficial.

3.4.2 Programs for Time, Cost and Resource Control

Several programs are available to assist the project manager in the tasks of work and resource planning and monitoring.

(1) SNAP (System of Network Analysis Programs)
The SNAP 3 program has been used for many years for time and financial control using network analysis. The program will carry out standard calculations for critical path networks calculating earliest and latest start and finish dates and float values to all activities and events in the network.

Financial values may be allocated to activities and events so that planned and actual incidence of expenditure may be calculated and analysed into a series of financial groups. There is a wide range of standard printouts covering input data, critical path bar charts, progress reports, reports at various management levels and financial analyses. The program is designed to manipulate data on a single network with a maximum of 5000 activities.

Information from the SNAP 3 program files can be accessed by the CEGB's INFSYS program to provide additional analyses and output flexibility. INFSYS can also be used to extract data from the SNAP files for further processing, eg for graphical plots or from combination with other support data files.

Because of the GDCD's increasing interest in the planning and monitoring of resource availability and usage, there is now a tendency to use the more comprehensive features of SMART 2 described in item 2.

It is also possible to carry out a machine transfer of networks from the SNAP to the SMART programs.

(2) SMART 2 (Scheduling Manpower and Resource Techniques)

Although the SMART 2 program is intended for resourcing studies, it incorporates network time analysis and financial analysis modules which provide similar facilities to those of the SNAP 3 program (although SMART can handle larger networks).

In addition, it provides comprehensive routines for the establishment of resource data banks, the allocation of resources to network activities and scheduling of activities within the user variable constraints of time, network logic and resource availability. Multi-scheduling may be carried out using alternative criteria such as early or late dates and several networks may be scheduled against a common resource bank. Achievements and history can be recorded but the emphasis of the program is toward the planning, modelling and scheduling of current and future work activities. Procedures based on the INFSYS program, described in item (4) and Section 3.4.3, are available to access the SMART 2 program files to provide a more flexible output format and to extract data for further processing and combination with other support data files.

(3) Programme Completion Forecasting

Conventional (deterministic) network analysis systems provide a single forecast of project completion with no indication of the likelihood of achievement. A program is available which accepts probability distributions of durations for activities in a network and, then, by repeated project time analysis (using SMART) establishes a range of predictions for the project completion data (probabilistic forecasting).

(4) Internal Resource Planning & Monitoring

The INFSYS program has been used for various aspects of planning internal resource allocation and monitoring resource usage including:

(a) Modelling direct and indirect human resource requirements for a single project or for a series of projects and providing tabular analyses by time, discipline and grade.

(b) An application which provides facilities for scheduling work items and their planned dates and resource requirements. The system produces questionnaire type outputs which are used for weekly returns of time spent and monthly return of progress. A variety of reports can be produced showing progress against elapsed time, analysis of manpower costs, actual against planned resource, usage, etc, sorted and sub-totalled in various modes.

(c) An application known as 'TIMER' which was operated to create an historical record of the usage of drawing office manpower resources in a detailed form suitable for flexible analyses. The system stores several libraries of resources, projects and contracts/jobs and activity categories, against which time allocations can be input. Analyses may then be made on an incremental or cumulative basis, in tabular or graphical form, which are suitable for allocation of resource costs and planning for future resource requirements.

(5) Divisional Resource Management and Information System

The computer processing of the system introduced through GDCD in April 1984 is a development of a previous system providing schedules of planned work, resources available and resources employed on the planned work. Preformatted questionnaires are provided for weekly returns of time spent.

Outputs are available in schedule and graphical form comparing actual work with plan, broken down by task type, work location, plant area, rechargeable work, etc. These analyses, which may be on an incremental or cumulative basis, are suitable for determining allocation of resource costs and for planning future resource requirements.

3.4.3 *Information Storage and Retrieval*

Considerable quantities of information are now held in text or numerical form in computer based files or are indexed by means of a computer program. Computer programs provide a powerful means of retrieving information. There are two basic approaches, either to hold the whole information in an accessible form or to hold only an index or key.

Computer accessible magnetic storage has become steadily cheaper over the last 10 years but it is still relatively costly to input data just to store it. As a general rule, unless information is already input to the computer for another purpose it is more economic to hold only an index. Many computer programs are useful for storage or retrieval and three are currently used by project managers namely INFSYS STATUS and ASSASSIN which are described as follows. A number of INFSYS applications are also described.

(1) ASSASSIN Information Retrieval Program

The name ASSASSIN is an acronym for Agricultural System for the Storage and Subsequent Selection of Information, and was orginally developed by Imperial Chemical Industries, Agricultural Division. Under this program abstracts of textual information are held in computer files and can be searched as required on the basis of a number of indexing terms.

A distinguishing feature of this program is its ability to provide a current awareness facility. A user's interests are defined by means of key words which are used as indexing terms. The indexing terms from the text of each new entry to the file are compared with the user's defined interest and a list is produced of those new entries which are relevant. It is possible to set up separate sub files covering selected topics, eg containing documents relevant to a specific power station, contractor's name or a series of contract meeting minutes. The program will then provide an up-to-date index. GDCD holds files of unpublished information although it is possible to add references to published data where this is required by the project manager.

(2) STATUS Information Storage & Retrieval System

Under this program, complete documents or details of mainframe databases and any text can be searched for and retrieved. STATUS includes facilities for maintaining the databases and has security arrangements which can be adjusted to suit any particular application.

This program is the basis of systems for the PWR project. It is also employed for TEXT which has been applied successfully to some GDCD requirements.

(3) Many of the programs described are written in languages such as PL1, FORTRAN, and COBOL. Modification and maintenance of these systems require the attention of trained programming staff. However, frequent reference is made to INFSYS (the CEGB Information System) which has been developed in order to enable someone with little or no computer programming experience to perform simple enquiries and analyses on data held in a computer in a standardised form. The main features of the program are:

(a) An enquiry language consisting of ordinary English words with a few technical terms.

(b) No special alignment of works or items in columns, continuation indicators or similar restrictions.

(c) Automatic facility for producing acceptable printed reports with the minimum effort on the user's part.

(d) Extensive options to allow the experienced user to determine more precisely the results required.

The program can also be used to maintain up-to-date information by adding, changing or deleting data in files. In addition users with programming knowledge can add extra facilities by attaching their own programs to be run as part of the system.

GDCD makes extensive use of INFSYS:

(1) To set up and manipulate relatively simple data scheduling systems.

(2) As an interface with other particular application programs so as to provide more flexible output facilities.

(3) As an aid to the development of major new systems.

Examples of some of the applications for which project managers have used the INFSYS program are as follows:

(1) CONAID: The file is designed to hold technical and descriptive data on C&I hardware and provide printouts of subsets of this data.

(2) CONVAD: The file is designed to hold data on the conventional alarm system and provide printouts of subsets of this data.

(3) COMINOD: The file is designed to hold data on the inputs to and outputs from, the station computer and provide printouts of subsets of this data. The site-based process control computer used to monitor progress of achievement against various levels of responsibility and incorporates the facility for reporting trends of forecast completion dates, etc.

(4) DESIGN SCHEDULES: This is a file of design events used to monitor progress of achievement against various levels of responsibility and incorporates the facility for reporting trends of forecast completion dates, etc.

(5) CORRESPONDENCE MONITORING: This system provides for the details of correspondence sent or received to be recorded under unique identification numbers together with a cross reference to the associated reply. Specially selective/sorted listings are available to assist in monitoring the status of correspondence.

3.4.4 *Project Cost Evaluation - PROVAL*

The PROVAL program was developed to aid project management in the monitoring and control of project cost estimates and expenditure. It provides a standard method of processing and reporting project cost information which is compatible with GDCD procedures and enables both regular and ad hoc analyses to be produced as required. The system can accept all known data relating to a particular project. Data which can be accommodated ranges from firm committed costs to the most tentative estimates of future work, both in respect of conventional contracts and 'non-contracts'. Included in the 'non-contracts' are consultants' fees, engineering costs, agreement for services or materials and miscellaneous costs.

A breakdown of the financial estimates and authorisations is established and transactions which determine or re-allocate expenditure can be input into the system. Such transactions are recorded against the appropriate financial provisions and a variety of reports may be generated displaying summations of suitably classified potential spendings and current balances for a complete project or part of a project.

Data is normally entered by the project team accounts staff, and it is imperative that the project team engineers, site staff and contracts department ensure that all relevant data is channelled through the team accountant. Collation of data and appropriate calculations are then carried out within the computer to provide a selection of reports on request.

The program has been further developed (PROVAL 2) to provide additional features including more comprehensive handling of changes in price levels and the closer correlation of project/contract cost data and capital estimates.

An interactive computer system is available to assist in the calculation of contract price adjustment (CPA) values for power station construction contracts. CPA calculations involve lengthy formulae which use cost index values. Cost indices can be stored, displayed, forecast, modified (eg rebased on a new date, or combined together to form fresh indices). CPA formulae are also stored and calculations can be performed using input parameter values prescribed by the user to define precisely the calculation required. Certain input parameters can be varied automatically to investigate the effects of such variations of CPA values.

3.4.5 *Drawing Office and Design Aids*

Considerable use is now made of computing systems to assist drawing office and design activities. A number of these are described in Sections 5.9 and 5.10.

4. PLANNING AND PROGRAMMING

4.1 Introduction

A project involves establishing and eventually disbanding a large multi-disciplined team, which interacts with many other organisations with interests in the project including contractors, consultants, designers and local and national government bodies (Fig 7.3). The flow of accurate information between all interested bodies is a vital ingredient in achieving overall success. It is as important that the machinist at the manufacturers works has the correct detailed component drawing as it is for the project manager to know the project's current status.

It is essential that fully detailed programmes are prepared for every stage in the execution of a project and that they are properly related. These programmes will provide the detail for tender inquiry documents and will cover the investment appraisal programme, design programme, manufacturing programme, construction programme and commissioning programme.

Many interrelated programmes are used to plan, organise and control the design, procurement and construction. Some programmes are in bar-line form, others are networks, and many are schedules of requirements and dates. The more important programmes normally used for a single power station project are indicated in simplified form in Fig 7.4. Three of these programmes are used to establish overall project timescales and are referred to as:

(1) The project investment programme.
(2) The executive's target programme.
(3) The project manager's target programme, which is often referred to as the project master target programme and is usually in network form.

The project investment programme is drawn up in the formative stages of a project and its purpose is to indicate the construction duration on which capital investment comparisons are calculated and investment decisions made by the CEGB executive.

It is in the consumer's interest that construction duration should be as short as is economically attainable. Thus, when a project has been selected for construction, has received the necessary consents and is presented for formal financial sanction and release for construction, the overall project programme is established by the project manager and agreed with the Executive. This is referred to as the 'executive's target programme' and is invariably shorter in duration than the programme used in the derivation of estimates for investment decision.

The project manager sets himself a target programme a few months shorter in duration than the executive target, and it is to his target programme that the project manager places contracts and to which each contractor is committed by his own contract programme which will fall within the project manager's overall target. The project manager's programme also has to make provision for co-ordination between contract programmes. This target programme dictates the schedule for the production of design intent memoranda and the processing of design information to specifications, tenders and contracts for design, manufacture, erection and commissioning.

A variety of programming techniques have been used. The most common are networks, bar charts and schedules, all of which are amenable to modern computer analysis and updating, as described in Section 3.4.2. Different contributors require

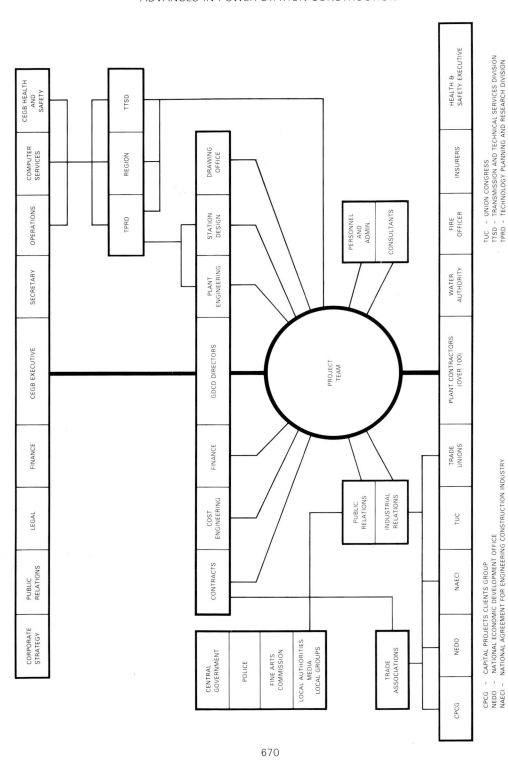

CPCG – CAPITAL PROJECTS CLIENTS GROUP
NEDO – NATIONAL ECONOMIC DEVELOPMENT OFFICE
NAECI – NATIONAL AGREEMENT FOR ENGINEERING CONSTRUCTION INDUSTRY

TUC – UNION CONGRESS
TTSD – TRANSMISSION AND TECHNICAL SERVICES DIVISION
TPRD – TECHNOLOGY PLANNING AND RESEARCH DIVISION

FIG 7.3   PROJECT ORGANISATION INTERFACES

PROJECT MANAGEMENT

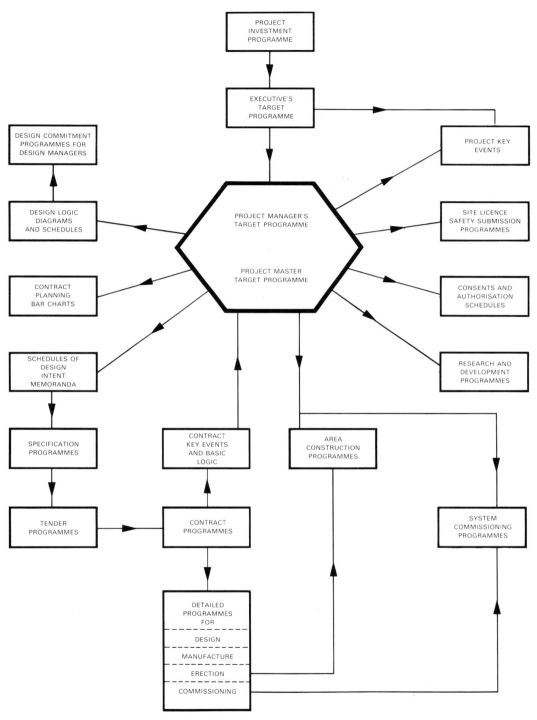

FIG 7.4   PROJECT CONTROL PROGRAMMES

different techniques and they must all be accommodated in the central system, as the best technique in any particular application is often that in which the individual user has most confidence. The aim is to ensure that all satellite programmes yield accurate information to facilitate sound decision-making at project master programme level.

For a conventional fossil-fuel station, a typical programme suite would consist of the programmes outlined in Sections 4.2 to 4.8.

4.2 Principal Project Programmes

These programmes contain three important statements of management objectives:
(1) CEGB Executive Target Programme
   This sets out the key objectives of the project programme providing top level decision and control guidelines. The programme format is usually that of a simple interlinked bar chart.
(2) The Project Master Programme (PMP)
   As the project develops the Executive Target Programme is expanded to include further important activities. This senior management tool ties together the main activities and identifies the key events in the project chain and sets the parameters for the lower level programmes. It covers the principal sequence logic of design, manufacture, erection and commissioning and shows the interdependencies between the various station systems, together with their relationship to the main plant programmes.
   The PMP's influence increases as major project decisions are made and though in its infancy it merely 'opens-out' the outline programme to about 300 activities, it ultimately embraces between 1500 and 3000 activities. It is the principal source of data for the management information system and it conditions every decision taken by the project team.
(3) Contract Planning Chart
   This sets out the contract strategy and logic, showing how the initial design activities lead-in to the issue of enquiries to contractors, when each tender is due for return and all the major contract key dates. Manufacture and delivery periods are also shown related to programme dates for work on site.

4.3 Contract Specification Programme

This informs tenderers of the programme commitment to the project required for each contract package. It details the requirements not only in relation to hardware manufacture, supply, erection and commissioning, but also the information and data flows required to meet the total design, and to facilitate effective monitoring and control of the contract. Together with specific clauses in the standardised tender documents, the specification programme provides the time bases on which the tender should be prepared.

4.4 Contract Tender Programmes

These are the tenderer's response to the contract specification programmes. His programme submission is examined by the project team to ensure that it is consistent with other related tender data and compatible with the specification programme as part of the tender assessment procedure. When a preferred tender emerges, deviations from specification and qualifications attached to the tender are discussed and resolved before a contract is placed. The resolution of differences provides the basis for the contract programme, because benchmark dates important to the project will be settled at this stage.

4.5 Contract Programmes

These are further developments of the tender programme in detail but without alteration to the agreed benchmark dates and are agreed between the contractor and the project manager after the contract is placed. They show precisely how and when each part of the contract will reach completion and provide the means to monitor the contractor's performance. As a general rule, contract programmes are finalised and agreed four weeks after the contract is placed but, for larger plant contracts, a six month period may be allowed for the development and agreement of the erection and commissioning sections of the programmes.

The contractor is required to regard the contract programme as his firm commitment to the project and the information source of his own management control. He must understand the importance of the quality of this data because, ultimately, it affects the updating to the PMP. On the larger plant contracts, where on past projects serious delays have occurred, the contract programme also provides the basis for the key dates system of payments. Notwithstanding the key dates system, programmes are designed in such a way that at any time they can be updated and reviewed over their full duration.

Manufacturing progress and quality are monitored by the CEGB specialist department Engineering Services Department (ESD) and consequently the manufacturing programmes must stand alone and contain their own progress milestones. ESD for its part is also aware of the criticality of each manufacturing programme in relation to the project master.

4.6 Design Logic Diagrams

The project's key design logic is embodied in the early stages of the PMP but amplification is required in each area of work to ensure that key events are clarified and that each design discipline works to compatible objectives. This amplification is achieved through design information schedules, the scheduling technique having proved the most effective means of accommodating the flexibility inherently necessary in design work.

The majority of the early design work is delegated to GDCD specialist departments such as:

> Electrical Engineering
> Nuclear Design
> Systems Engineering
> Station Design
> Boiler Design
> Turbine Generator Design
> Civil Design

In the case of certain aspects of civil engineering, civil design consultants are employed. At this stage, the work is controlled by engineering discipline rather than by contract package. However, as the contract package is the ultimate destination of the design data, it has to be accommodated in the programme. Otherwise the specialist departments might work to disparate priority objectives to the general detriment of the project.

The approach which delegates to specialists the design work advances the design on a wide, fast moving front and ensures that the most up-to-date information and best practices are incorporated. But, from the project management viewpoint, it also increases the span of control to be exercised over a most critical phase of the project. The regular monthly design review meeting chaired by the project manager introduces accountability in the delegated process and serves other critical purposes such as:

(1) Allows the project manager to review design progress compared with programme.
(2) Identifies current priorities for the departments to which design work had been delegated.
(3) Provides the input to update the programme to include completed work, accommodate and recover delays and re-direct resources to handle new priorities.

A typical design commitment programme is illustrated in Fig 7.5 It shows how heavily the critical path of a project depends upon civil design, and how the civil designer is dependent on plant designers for his basic information. Though only the key activities and events are shown on the programme, it would be normal for hundreds of significant design interface decisions to have been co-ordinated by the project team in the first two years of the programme and for this level of activity to be maintained and increased as the plant contracts become active.

4.7 Site Construction Programmes

The contractor's construction intentions are incorporated within the agreed contract programme. This data is extracted by the site planning team and built into the overall site construction programme where inter-contractor dependencies and access restrictions are given particular emphasis. The main interfaces are referred back to be incorporated in the PMP.

4.8 Commissioning Programmes

The commissioning content of the agreed contract programme is refined to accommodate interfaces with other plant items. Interlinking provides comprehensive programmes for commissioning complete systems and through a hierarchical programme family a firm plan for the whole station is developed. The major milestones are referenced back to and incorporated into the PMP.

4.9 Recovery Programmes

In the event of the site construction work falling behind programme, it is sometimes necessary to institute a 'recovery programme'. This would necessitate an examination of existing programmes with the view to re-scheduling particular items to enable the principal milestones in the PMP to be maintained. Measures taken by the recovery programme might include acceleration and reallocation of works manufacture and

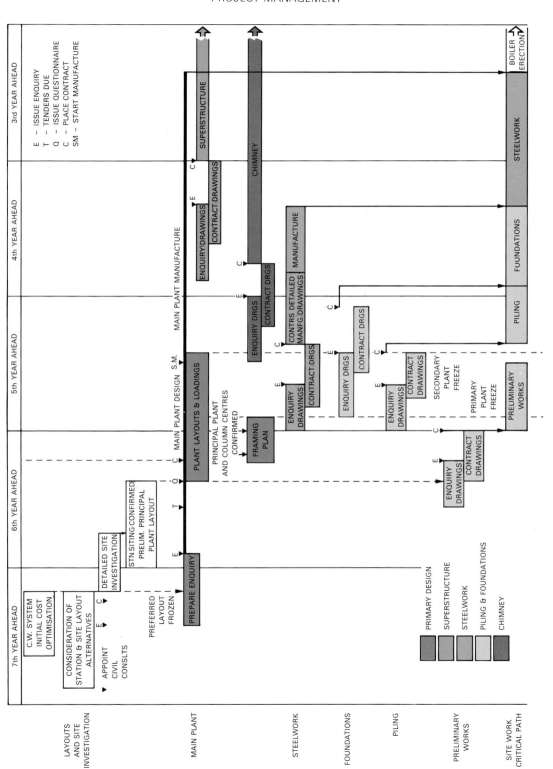

FIG 7.5   CIVIL DESIGN COMMITMENT PROGRAMME

delivery, preparing additional access and storage facilities, rescheduling of erection and commissioning activities, increasing dependent work resources by double dayshifting, etc.

5. DESIGN AND ENGINEERING

5.1 Introduction

The design phase of any major project can be the key to success or the path to delays, overspending and frustration. This phase of the project however is the most difficult to manage and involves the project manager and his senior staff, throughout the whole of the period, in a series of vigorous actions to ensure that the final design is satisfactory in all respects.

Design always involves conflicting needs and interests and therefore generates continuous requirements for decisions in order to maintain design impetus. Considerations like performance efficiency, constructability, accessibility, maintainability, reliability and cost may conflict with one another and need to be carefully balanced in reaching optimum solutions. Experience shows that the key areas requiring managerial control are:

(1) Clear definition of responsibilities and scope of work for sections and individuals.
(2) Prompt transfer of design information between interested parties.
(3) Disciplined, prompt decision making.
(4) Avoidance of late design changes.
(5) Early reduction in numbers of design alternatives to avoid confusion and eliminate unproductive effort.
(6) Incorporation of the whole design phase into the design programme in order that priorities can be established logically and progress accurately monitored.

The engineering development of a power station from its conceptual stage to the establishment of a detailed contract design is a very complex set of activities which take place over many years. Civil, mechanical and electrical engineering disciplines are involved and the process is very interactive, calling for careful co-ordination of the numerous parallel design activities in addition to the overall design process requiring detailed programming which must be well regulated and controlled. The design of a power station is progressive and cannot be split into discrete sections although three design phases can be identified. The first phase comprises 'conceptual design' which should be based on an existing design if it is to be consistent with the policy of design evolution. The second 'system design and layout' or 'reference design' phase develops the conceptual design to satisfy the functions and requirements set in the Station Development Particulars. In the case of a nuclear plant, the reference design is the basis of the Pre-Construction Safety Report (PCSR). The third 'contract design' phase results from the progressive refinement of the reference design and its detailing such that it embodies all the information needed for manufacture, construction, testing and operation.

5.2 Conceptual Design

5.2.1 *Site Investigations*

The first contracts to be placed for a proposed new power station are those for site investigation. These involve both on-shore and off-shore/river or lake investigations to confirm the already known national data that is available and also to establish the total viability of the project on the particular site in question.

The information from ground investigations are utilised to optimise the location of power station buildings and plant, establish foundation and piling requirements and allow reasonable cost estimates to be made. The off-shore/river bed investigations provide some of the essential data for establishing the basic designs for cooling water systems, jetties, berthing arrangements for vessels, off-shore structures, etc and the costs associated with such provisions.

All this data and basic station design are required in order to establish the proposed power station viability on the chosen site and the estimated cost of the project. This information is then utilised in the decision of whether to proceed with the proposed project with a particular type of power station at this site or at another possible location.

5.2.2 *Station Development and Technical Particulars*

The initial documents which form the basis for the overall design of a new power station are the 'Station Development Particulars' and the 'Station Technical Particulars'. These documents provide the following information.

Station Development Particulars

The main parameters of a new power station and the practices to be followed arising from statutory and other requirements, are eventually brought together in a comprehensive document - the Station Development Particulars. Although the development of the document is the responsibility of the CEGB HQ System Planning Department, GDCD specialists provide much of the technical input. When the statutory consents have been granted and the project is sanctioned by The Board for construction the Station Development Particulars receive the approval of The Board and are handed over formally to GDCD with a directive to execute the design, construction and commissioning of the station.

Station Technical Particulars

In parallel with the production of the Station Development Particulars, the technical requirements of the new power station are set out in the Station Technical Particulars with particular attention to the design features that will be required to enable the plant to integrate with the national grid.

The document's prime aims are divided into separate sections covering:

(1) Main plant, which details the intrinsic and output characteristics of the turbine generators.

(2) Operating flexibility, which gives the expected loading regime and the plant performance parameters required. Conventional plant is now generally required to run at a wide range of loads on continuous or 2-shift operation. The document provides details of the most rigorous characteristics, overload capacity and reliability standards.

(3) Plant control arrangements which reiterate the basic need to be able to control the CEGB's system within the statutory limits of frequency and voltage whilst meeting station operational requirements within specified emission limits.

(4) Auxiliary power supplies system which emphasises the tenacity required from these systems when the station output is under stress. For example, margins are built into auxiliary plant designs to ensure that continuous maximum rating can be maintained when the system frequency varies from 50Hz.

(5) Auxiliary gas turbine plant, stating the design aims of providing power for a dead station startup, station system frequency correction and emergency generation capacity as a national grid reserve.

(6) Transmission reinforcement provisions, which detail all the grid switchgear and transmission line modifications that ensure the delivery of the stations output to the correct locations to suit grid loading.

5.3 System Design and Layout

5.3.1 *General*

The development and technical particulars outline the project's essential design requirements. The early design work therefore involves optimising the layout of the major plant items whilst accommodating decisions already reached (like CW system data and site levels), in order to achieve the first major decision — the first stage layout freeze. This in turn, releases major areas of work including the main plant design which lies on the critical programme path.

At this stage the proposed plant is laid out in as much general detail as is possible from the information available. At the same time the station CW system, HP pipework system, electric power system, basic control system, fuel systems, etc are all developed to the point where they can be translated into design intent memoranda and subsequently specifications.

The CEGB's past experience has demonstrated that it is important to control the degree of innovation introduced into a project. Hence, a policy of replication of proven designs and general component standardisation has been adopted wherever possible. Further, where innovation is necessary, the principle of introducing new designs for major plant items in a cautious manner is followed.

This policy of design evolution is illustrated by the case of the Drax Completion project which is based on the original Drax design, and by the Heysham 2 project which follows the Hinkley Point B design. However, for replication to achieve the greatest benefits the gap between orders should not be greater than about two years. In the case of both Drax and Heysham 2 the orders were separated by some 12 years from their earlier stations. Changes had therefore to be made to the designs to take account of developments in technology, enhanced safety requirements and to take advantage of lessons learned during the construction and operation of the earlier stations.

5.3.2 *Design Intent Memoranda*

The purpose of a design intent memorandum (DIM) is:

(1) To provide definitive information at an early stage in the development of a project on station and plant performance requirements and design intentions so as to enable station development and plant specifications and design to proceed on a firm and consistent basis.

(2) To identify areas of uncertainty in the design where further work is required to enable a definitive statement to be made.

DIMs are brief, factual and informative documents. Whilst a detailed technical analysis of design intentions is not required a summary of the major calculations and cost factors together with the influence of planning and safety constraints are included to justify the design intention. Where this is not possible an orderly assembly of facts is given to support any judgements made. A brief statement on the rejection of alternative proposals, where relevant, is given. Emphasis is placed on identifying those aspects of the design which are likely to affect the decisions of others involved in the project.

Where the DIM is based on preliminary or assumed information this is stated and attention drawn to the degree of freedom permissible in proceeding with the detailed design.

DIMs make reference to but do not duplicate agreed uniform specification standards, generation design memoranda and any other information from technical branches and major design correspondence or submissions from consultants, contractors and other CEGB departments. Where departures from previous practice are intended the reasons are stated.

5.4 Contract Design

Following on from the DIMs, the contract packages identified in the contract strategy document (see Section 6.3) are built up, terminal points are established and technical specifications prepared. As the design proceeds, data of increasing authentication is incorporated into the design drawings. Consequently, by the time a specification is issued for tender it can be complemented with the relevant design drawings. The post-contract phase adds to the activity as the location and orientation of the major plant items are established and civil and structural details finalised. Auxiliary plant items, are fitted into the station whilst the connection of the plant by pipework and cables is properly co-ordinated. When novel plant arrangements are envisaged, scale models have been constructed to assist in these processes.

The major plant design work is undertaken by the plant contractors. GDCD however vet all design and drawings prior to any manufacture commencing to ensure that they:

(1) Comply with the specification.
(2) Match-in with the overall design of the station.
(3) Conform with the necessary standards.
(4) Give all the information required for civil works, cabling, pipework, etc.

In order to ensure the timely provision of the information GDCD now frequently awards design contracts to major plant contractors, prior to the release of the associated hardware contracts.

5.5 Design Programming

The design programme provides part of the project manager's means of achieving the desired control. He ensures that the programme comes to be regarded as an integral part of the design process for the project. This programme shows:

(1) The work of the specialist departments and details the activities each must complete and the required timescale.
(2) The source of initiating information.
(3) Major design aspects to be considered at each stage.

The programme is computer processed to facilitate easy monitoring.

Problems can arise when design information is required from contractors many months ahead of the date by which they themselves need it for manufacturing purposes. In all such cases, the consequences for the project programme must be known and firm action taken if the project manager's commitments are to be safeguarded.

5.6 Design Contracts

The award of design contracts prior to manufacture and construction ensures the timely provision of vendor information, which is essential to the development of the overall station design and in particular the civil design. The civil design information is required very early in the construction of the station. To develop this design the civil engineers require information from the mechanical and electrical contractors with regard to plant location, floor loadings and support requirements. The design contracts placed with the manufacturers require them to provide this information against a programme of submissions which satisfies the requirements of the overall station design.

These design contracts, coupled with the development of the safety case for nuclear projects and its acceptance in principle by the NII, form an important part of the CEGB's strategy to enter the construction phase with a secure design and minimum risk of substantial subsequent design changes.

Design contracts consist of design schedules in which GDCD list the submissions required from the contractor. The submissions are linked to a design submission programme to which the contractor must comply. Payment to the contractor is similarly linked to the design submission programme and should a contractor fail to provide all the submissions required by a programme date then GDCD can normally withhold the payment otherwise due. This gives an incentive to the contractor to provide the information needed in accordance with the programme.

Considerable effort has been applied in the field of 'graded information'. Graded information gradually raises the accuracy of information provided by a contractor. For example the information required from a main plant contractor for civil works may be graded C, B and A. The first information required is the C class information and the date for this will be clearly indicated in the design submission programme. C class information will provide sufficient detail to enable civil construction tender documents and drawings to be produced and initial design undertaken. The next grade of information, B class, then follows at a later programmed date and this will provide information to enable detailed design of piling, foundations and other aspects of civil works. The final A class information contains and expands the information given at the B class stage such that detailed design work can be completed.

5.7 Design Quality

The design work as it proceeds is subjected to design reviews to reveal whether the design objectives are being met properly and that the design programme status is being accurately assessed and reported.

A second check on design quality is provided by feedback from plant in operation. The process of manufacture, erection, commissioning and operation provides the greatest test of a design. An example of this aspect is provided in the cooling water system DIM for Drax Completion. After assessing the cost effectiveness of possible improvements the designer's conclusion was:

"The study of the existing system (ie the first three units in operation at Drax 1) has shown that the basic principle of replication can be adopted. Operational experience has however, shown that the cooling tower ponds should be capable of individual isolation without affecting the operation of other towers and this memorandum supports this change. The purge pipework needs increasing in size on account of underestimates of system throughput and hydraulic losses during the original station design investigations. The pump isolation provisions on the first station are not adequate and this memorandum describes the additional valves required."

Operational feedback is thus used to improve station availability through modifications to the original detail.

Finally, it is the policy of the CEGB that for all items of power generating and transmission plant and associated systems there shall be in force appropriate arrangements for providing assurance of quality at all stages from design to decommissioning. Within the GDCD this is implemented by the establishment of quality assurance arrangements, based upon relevant British Standards, whose scope encompasses design control.

5.8 Control of Design Changes

A major objective in the design of a project is to minimise design changes. To this effect, early on in the project programme 'design freezes' are imposed on layout, foundations, steelwork and subsequently plant. After these freezes every proposed design change is evaluated against the criteria of technical necessity, constructability, effect on programme, effect on cost, personnel or plant safety, operational desirability and maintenance. The procedure calls for all relevant parties to be consulted and all design changes to be approved by the project manager.

In the case of major plant a standard procedure for 'notification and securing approval of changes in design' has been established. In this procedure the plant contractor must submit in standard form detailing the design change he is requesting, the reasons for proposing the change together with any cost and programme implications. Should the design change be agreed upon after considering all the criteria referred to in the previous paragraph the form submitted by the contractor will be signed by the project manager. Only when this signed form is received by the contractor may he proceed with the change proposed.

Frequently requests for design changes are received from the CEGB Region who will subsequently operate the plant. These requests are normally initially discussed at a project liaison committee where the changes are again assessed against the same criteria as that for a design change from a contractor. Following this discussion the Region involved must then make a formal request to the project manager for the required change and agreement or rejection is then routed back in writing under the signature of the project manager.

5.9 Drawings

A power station project can involve up to 80,000 drawings and it is therefore essential to have a reliable system which facilitates retrieval and permits progressing and updating of information on all these drawings. This is achieved at GDCD by utilising a drawing office records information system (DORIS) which is a computerised system.

5.9.1 *Drawing Office Records Information Systems (DORIS)*

This is basically a drawing office record and cross reference system, allowing seven standard printout types to be extracted. The program also has the facility to progress the internal approval of drawings and the response of outside interests to copies of drawings sent to them. This latter facility provided by the program requires feedback from the project teams to the drawing office. This feedback has been achieved by extracting drawing codes and update types from outgoing mail.

In summary, this program:
(1) Indexes all drawings on a project.
(2) Cross refers GDCD and contractors' drawing numbers.
(3) Progresses drawings submitted to the GDCD.
(4) Progresses drawings issued by the GDCD.

INFSYS procedures have also been developed to access information held in DORIS for use in other programs. An example is the Dinorwic system of monitoring the issue of drawings against the design schedule planned dates with exception reporting showing lateness.

An enhanced system (DORIS 3) incorporating on-line updating and enquiry facilities is now available for all new projects.

5.9.2 *Computer Aided Draughting*

GDCD employs computer aided draughting for some of the drawings it has to produce. The draughting system, called MEDUSA, is used for:
(1) Preparation of mechanical and electrical diagrams.
(2) Extraction of data from the diagrams and transfer to other CEGB systems eg CADMEC. (See Section 5.10.1).
(3) 2- and 3-dimensional detail and assembly designs.
(4) 3-dimensional solid modelling of building designs.

5.10 Cable Design and Management

The management of the electrical installation on a power station project can be divided into three parts:

(1) Electrical Equipment
  - transformers
  - switchgear
  - marshalling cabinets
  - junction boxes
  - cable steelwork
  - lighting and small power
(2) Cable Installation
  - power cables
  - control cables
  - safety cables
  - special cables
(3) Cable Terminations and Jumpers
  - switchgear
  - plant equipment
  - marshalling cabinets
  - junction boxes

The magnitude and complexity of such electrical contracts combined with the stringent segregation and quality assurance requirements gives rise to the need for a comprehensive system for design and contract control and hence the following computer aided system is used:

5.10.1 *Computer Aided Design and Management of Electrical Contracts*

CADMEC is an acronym for 'computer aided design and management of electrical contracts' and is a comprehensive computer based design and management control system, covering the activities both in the design office and at site, including contract planning aids, work face instruction and evaluation. CADMEC comprises seven programs which for a single project interface with a common TOTAL Data Base. Each program maintains specific areas of the data of which there are seven as follows:

  Basic project parameters
  Equipment scheduling
  Cable scheduling
  Extended equipment scheduling
  Cable routing
  Core allocation
  Measurement & stores control
Outputs are produced using two 'report generator' packages
  SOCRATES - batch outputs
  MANTIS - online outputs

Scheduling in the CADMEC Program
The following basic design data is recorded within the system and forms the nucleus of the database:

(1) Plant Scheduling

The areas of plant are split into separately erectable/commissionable areas based on a hierarchical breakdown of systems, sub-systems and groups. The concept being that a sub-system comprises a major commissionable entity, eg systems are boiler, reactor systems, etc, sub-systems are a gas circulator, turbine barring gear, etc, and groups are a single drive, eg a jacking oil pump.

(2) Equipment Scheduling

All equipment items that are to be either separately erected or cabled are scheduled together with the appropriate plant group code.

The location of the equipment, the supplier of the equipment and other design data about the equipment may also be given as desired.

(3) Cable Scheduling

All cables are scheduled in their appropriate plant group as going from one equipment item to another.

(4) Designing

There are two design packages within the CADMEC 2 System, cable routing and sizing, and core allocation.

(a) Cable Routing & Sizing

Initially, cableways are identified for the main cable routes, making provision for the necessary segregation and separation of cables according to unit, circuit function and cable duty.

The cable routing and sizing details are then input. These define where the cables join and leave the route matrix, the segregation and separation classes applicable for the cable and the cable size or circuit load.

The program will then on request determine the shortest possible route compatible with the above restraints and the segment capacities. The program will also size the cable in accordance with the sizing details specified and the route length and finally record the segment capacities used.

(b) Core Allocation

The core allocation section of the system will assist the design engineer in allocating and recording the control cable cores that are required together with the interconnections at marshalling boxes etc, for each control circuit within a pre-designed cable network.

In order that this can be done, the user first has to establish the cables, marshalling boxes and junction boxes etc, which form the network and are considered to be available to the core allocation system. These may be multi-pair cables hard wired through marshalling boxes and jumpering fields, multicore cables connected via junction boxes, or any combination of the above. The user then specifies the equipment terminals at which each wire with a different identity enters the network.

The segregation and separation classes are also specified for each circuit and cable cores are allocated either automatically or manually.

In the case of automatic allocation the cables/cores chosen are those that minimise a weighted combination of circuit length and number of connections compatible with segregation and separation requirements.

This allocation is then recorded in the data-base together with the jumpers that are determined by the system in order to complete the circuitry.

Organising

Key target dates based on the project master commissioning programmes may be either generated automatically from SNAP or SMART critical path network programs or explicitly input.

The organisation depends upon the control of resources both material and labour to match the current work programme and access.

All material orders and deliveries are fed into the system and in conjunction with the figures derived by the system from progress data form the basis of a stock control system which compares these quantities with the requirement, which is computed from the scheduled quantities.

The labour resources required for the current work programme are computed from the previously entered labour content of each task to be performed by the contractor.

Work face instructions for all of these tasks are computer produced in the form of work cards bearing all the necessary information to carry out the task. Work cards are produced for:
- cable installation
- equipment installation
- termination instructions
- jumper instructions

Completed work is recorded on the work cards which are then used as computer input documents.

The site manager is able to co-ordinate cabling contractors' activities with regard to both his other activities and the activities of other contractors. To this end both the current and imminent site availability are input to the computer.

Cable laying access is determined by the user maintaining the access status for each route matrix segment, whereupon the computer reports on the availability of access for each cable.

Equipment erection access and termination access is determined by the filling-in of questionnaires reporting on an equipment, group or sub-system basis for each erection contractor.

Reporting

An important aspect of the aid to management afforded by CADMEC is the quality of reporting that is possible. In addition to special ad-hoc enquiries, there are routine reports that can be produced showing the progress and status of the contract works against programme.

Management summaries are produced showing the total work done on each section of the contract. These may also be produced in graphical form so that current trends can be analysed.

6.   COMMERCIAL ASPECTS

6.1   Introduction

In its simplest form a contract is an agreement between two parties defining respective obligations, the basis for discharging them and remedies for default.

In major power plant contracts, the agreement is conditioned by so many factors that the parties need to specify these conditions in considerable detail. To this end the GDCD has assembled comprehensive standard conditions of contract that are adaptable to the needs of particular professional bodies and different engineering disciplines. Each piece of plant or structure which is the subject of a tender (eventually a contract) requires its own technical specification to complement the contract conditions.

Before contracts can be let a number of project management decisions and actions are necessary.

6.2 Project Strategy

As already described during the 1970s GDCD's contract strategy was heavily influenced by the NEDO report on large industrial sites. This strategy proved unsuccessful and led to the re-adoption of lump sum contracts with greater attention to commercial incentives and penalties than previously was the case.

Policy decisions on the outline contract strategy are taken early in project development. The strategy document lays down general guide-lines within which the project team will operate. It covers for example:

(1) Whether separate design contracts will be let and how a design contract will be linked to subsequent hardware options.
(2) Whether supply and erection contracts will be split or combined.
(3) The type of contract for each extent of supply, ie lump sum, remeasured or some form of reimbursable contract.
(4) Whether key date procedures will apply.
(5) Level of risk to be carried by the parties.
(6) Special management aspects such as management groups, which need to be established within the contract.
(7) The number of major contracts to be placed and the nomination of sub-contractors.

6.3 Contract Strategy

The objective of a project's contract strategy is to place contracts that the CEGB is satisfied are within the capacity of the contractors to discharge competently and to time under terms and conditions which are equitable to both parties and will encourage a high standard of performance. The cost to the CEGB of failure by the contractor is many times greater than can be expected to be recovered from the contractor. Therefore the optimum form of contract is that which provides such safeguards and incentives as will harmonise to the maximum practical extent the conflicting objectives of the two parties and accords with CEGB policies.

The project team's first task is to develop an appropriate contract strategy by deciding the number and type of contracts to be let. The project is divided into recognisably discrete plant areas with specific terminal points that can stand alone for engineering and contract purposes and each is assessed for its contract risk factors. Both experience and 'hard data' provides guidance on expected contract prices and this data assists in the preparation of the project budget and its inclusive risk margin values.

The parameters used here also condition decisions on the type of contract most appropriate to each area. The greater the uncertainty the greater the financial risk that must be borne by the client. For example a contract for an exact replica of a piece of plant supplied by the same contractor would contain little or no uncertainty at all, no risk would accrue to the client and therefore a firm price lump sum contract could easily be achieved. If on the other hand, a completely novel feature was to be incorporated in a prototype station, it would be most unlikely that a realistic lump sum offer would be bid because of the high risk to the contractor. In these circumstances it would be appropriate to consider a form of reimbursable contract.

However, simple choices like this rarely occur and many other factors have to be considered before the contract strategy can be concluded. The most important of these are:

(1) The state of design development.
(2) The number of interfaces between contractors that must be controlled by the CEGB.
(3) The size of the contract relative to the financial resources of probable contractors and whether further sub-division would usefully open the work to wider competition.
(4) The risks associated with the industrial relations climate.
(5) The current and anticipated state of the commercial climate.
(6) The CEGB's resources available to manage the anticipated number of contracts.
(7) The projected workload and available capability of the contractors.
(8) The management ability, technical, labour, plant and commercial resources of potential contractors.

These deliberations result in a primary contract strategy for up to 15 principal task areas. The definition of the scope of these principal contracts generates a secondary contract listing either from strategic splitting of the principal works or from identification of the smaller remaining work packages. Wherever possible, the secondary list adopts contract forms that are consistent with the principal list and provides a starting point for the preparation of specifications by identifying the scope of the individual contracts.

Following agreement of the scope of each contract, their strategy has to be developed. This raises the following considerations:

(1) Contract type and form.
(2) Programme requirements.
(3) Manufacturing and supply/delivery sequences and programme consequences.
(4) Sequence of taking-over the plant and the consequential payment regime including the effect on the CEGB incidence of capital expenditure.
(5) Whether contract price adjustment will be necessary and what formula will be used.
(6) Whether 'key date' penalty arrangements are appropriate.
(7) Management of sub-contracts.
(8) Special features such as shiftwork arrangements at site and common site agreements that might be enforced by contract.

6.4 Specifications

6.4.1 *Aims*

The specification contains the technical requirements of the contract, programme management, and quality assurance requirements, the conditions of contract and the special commercial aspects of the contract strategy. Its purpose is to describe precisely and in detail an acceptable outcome. For example, the specification for a design contract would have as its aim a firm design which can be costed, programmed and its performance forecast. Performance is defined as output, sustained output (availability) and output on demand (reliability).

This key document has two other prime functions:
(1) To define the terms of the competition under which tenders are submitted.
(2) To be used as a management reference guide throughout the life of the contract and as a contractual document; consequently it requires great care in the preparation if ambiguity is to be avoided.

A specification is concerned with ends, it is for the contractor to explore the best means and it must indicate how successful designs will be identified by objective assessment. It aims also to delegate responsibility for design, manufacture and construction which allows the contractor the maximum freedom to deploy his particular expertise and experience in meeting the client's needs. To allow contractors to update their design parameters as the balance between capital and running costs change, critical cost data is included. This takes the form of a capitalised value for each of the key performance criteria relating to the contract.

It is important that the parties to the contract accept that the specification is attainable.

6.4.2 *Content*

Specifications can be divided into three areas:
(1) Areas where the client is expert include explicit requirements for plant which is to be matched to the generation/transmission system (eg cost, time, performance and Station Technical Particulars); definition of functions to be performed as set out in agreed design intent memoranda; use of standards, codes and uniform documentation; choice of particular equipment based on operational experience; station layout; control and contract strategy and definitions of interfaces. This part contains unequivocal statements of requirements and criteria for acceptability. The criteria must be stated in such a way that intangible objectives and exhortations can be turned into a form capable of being qualitatively and quantitatively measured.
(2) In areas where the designer/tenderer/contractor is expert the specification should describe acceptable outcomes such that viable solutions are seen to be possible. Only plant which has passed through the research and development programme of feasibility or has already been proved in operation would normally be considered. Constraints must be identified. The constraints might be environmental (cooling water, amenity, transmission, chimney emissions) or man-made (rules, regulations, laws, etc) or the natural limits of technology.

(3) Areas of uncertainty where creative dialogue and mutual consent between client and contractor/tenderer must be achieved. These might include: specification of safety requirements; the use of scarce or developing data; the use of scarce or strategic resources; the interpretation and application of new data; ranking or priority of requirements which can only be expressed in qualitative terms.

6.4.3 *Component Specification*

Deficiencies identified on some of the plants built in the 1960s have led to a closer definition of the functional requirements of the components of main plant and less reliance on overall guarantees and final acceptance tests. Thus the performance and margins required of each component is defined and, where appropriate, tests are specified for components to demonstrate, prior to setting-to-work, that the specification has been met. While, for example, the life of pressure parts was always specified, this has been extended to a number of other plant components. For electrical and control components this is defined as 'mean time before failure' (MTBF) and, for mechanical equipment which is subject to wear, minimum lives often in multiples of the intervals between statutory outages, are defined.

At the contract-placing stage the plant is not defined in all respects. Some components have to be developed, design details have to be evolved and the plant has to be integrated into the station. To ensure that this process is carried out in a systematic manner the relevant design schedule is included in the enquiry document. This schedule defines both the design submissions required and the information that has to be provided by the contractor and the CEGB using dates obtained from design network and the PMP.

The margins in hand for each component of any system must be known so that margins are similar throughout the plant, rather than having some components which are conservatively rated whilst others are working near their maximum limits.

Where the plant has to be compatible with services provided under different contracts, terminal points are accurately defined in the specification, preferably by reference to relevant standards. Non-destructive testing (NDT) and fabrication requirements are defined in appropriate international, national and CEGB standards. CEGB standards are used to ensure that control and instrumentation and electrical equipment included with the main plant complies with the station control and instrumentation and electrical specifications. A uniform format is used for each plant specification so as to ensure consistency and optimum standardisation across the whole of the station.

In writing a specification, full advantage is taken of previous work and experience and this is achieved by standardising all those parts of the specification where design has become optimised over the years and the use of the uniform specifications developed by GDCD.

6.5 Contractor Selection

6.5.1 *Tenders*

The project team prepares invitations to tender for each contract ensuring that each set of documents is identical and that there are no internal contradictions, for example between the technical specification and the conditions of contract. The principal documents are:
(1) Instructions to tenderers.
(2) Form of tender.
(3) Conditions of contract.
(4) Specifications and drawings.
(5) Contract specification programme and design schedule.
(6) Commercial requirements.

As a general rule, selective tendering is used with tender lists compiled from the objective examination of contractors' commercial and technical competence. In this regard, GDCD accepts as a duty the need to ensure that prospective tenderers can meet its stringent specifications, for neither party benefits from a contract with weak suppliers. The supplier can suffer considerable commercial damage and a project substantial critical delay and additional cost, when there is a failure to meet the required standards. History shows that the cost to the CEGB and hence the electricity consumer of such a failure is many times the loss borne by the defaulting supplier. Consequently, GDCD takes great care to assess prospective suppliers before they are admitted to tender lists and also checks that established supplier's standards are being maintained. The procedure involves an examination by interview and questionnaire (and sometimes also by audit) of the companies' abilities in regard to such matters as:
(1) Corporate structure.
(2) Financial stability and funding for major projects.
(3) Design control procedure.
(4) Manufacturing capacity, ability and quality assurance systems.
(5) Past performance record and reference in relevant type of work.
(6) Availability of suitably qualified personnel to handle prospective CEGB work.
(7) The ability to commit resources.
(8) Industrial relations practices and past record at works and at site.

6.5.2 *Contract Award*

In order to ensure fairness throughout the tendering process, mandatory procedures are applied to ensure propriety of practice both within the GDCD and in its relationship with others. Considerable formality surrounds the despatch, receipt, opening, confidentiality of contents and assessment of the documents. Administrative control procedures ensure independence in the processing of documentation from invitation-to-tender to contract acceptance.

The authority for final decision on the award of contracts is vested at different levels of the CEGB hierarchy. The level at which authorisation is necessary depends on the type of tender invited, the contract value and the magnitude of the perceived financial risks.

6.5.3 *Incentives and Key Date Sanctions*

The significant difficulties experienced in constructing most of the capital projects let on reimbursable contract arrangements in the last decade has resulted in a return to lump sum contracting policies. The main provisions of this strategy are:

(1) The use of lump sum contracts (with contract price adjustment) embracing both supply and erection requirements.

(2) The introduction of financial incentives for effective and timely completion through a key date procedure which ties progress payments to satisfactory completion of programmed work by certain specified dates. Where progress is behind programme, the client has the right to withhold payments due at a key date until all the outstanding work has been completed. The aim is to encourage the contractor to improve the quality and effectiveness of his management, both in the manufacturing and erection phases by stimulating a positive response at the time when it can be most effective.

The key date procedure has played an important role in the management of the Drax Completion project in ensuring the progress of the contracts for supply and erection of the station's two major plant items, the turbine generators and the boilers. The key date procedure provides for a joint review at six monthly intervals by the CEGB and the relevant contractor at Director level, of achieved performance relative to contract programme. Where performance has been unsatisfactory the contracts provide for the withholding from the contractor of the progress payment due at that key date. At each contract review meeting the participants carry out a detailed review of the status of the contract, covering as necessary the progress achieved since the preceding key date in relation to the contract programme. Crucial decisions are also taken as to the extent of money to be withheld by the CEGB should the contractor default in the completion of major sections of work to programme. However, if GDCD becomes aware of unsatisfactory progress on the part of a contractor it does not wait for the key date meeting to expedite resolution of the problem. Instead steps are taken immediately to recover the delay. The contractor is always obliged to notify the CEGB as soon as problems are encountered. For those contracts utilising the key date procedure, the contractual arrangements provide for payment only on completion of agreed sections of work.

Key date procedures are also used for contract control on the Heysham 2 project for eight contracts critical to the overall project programme. Six of these relate to the nuclear island works managed by NNC and two relate to conventional plant managed by the CEGB. Progress monitoring and control are very similar to that at Drax, but differ in a number of detailed respects reflecting the role of NNC as manager of the nuclear works.

(3) The increase in the number of main direct supply-and-erect contracts in order to reduce the scope of each erection contract and thereby improve site labour management control.

(4) The regulation of site labour relations by establishing a management group comprising all major contractors on site and employers' federation representatives. All tenderers are required to confirm their willingness, as a contractual

commitment, to comply with the decisions of this management group. The objective is to achieve harmony in payment structures and site procedures across the whole project in order to control one of the contractors' major risk areas.

7. PROJECT COST CONTROL

7.1 Introduction

This Section describes the arrangements for cost control on a conventional power station project such as Drax. For the Heysham 2 project the principles described are unchanged, but their detailed implementation differs insofar as they reflect the financial authorities granted to the NNC for nuclear works under the Agency Agreement (Section 2.3 refers).

Cost control on individual power station projects takes place within the CEGB's overall system of financial control. It comprises four main stages:
(1) Budgetary approval.
(2) Sanction of schemes.
(3) Commitment and expenditure.
(4) Monitoring.

7.2 Budgetary Approval

The first step in cost control for a power station project is the establishment of a preliminary power station cost estimate as an input to the assessment of the need for the station by the CEGB's planning departments. This project estimate is included in the CEGB's investment programme which is submitted to the Secretary of State for his approval and begins the process of seeking formal investment authority for the station. Separate consent for the project itself must also be obtained from the Secretary of State.

7.2.1 *Estimates*

The basis of financial control is an approved estimate, but the quality of an estimate depends on the amount of information that can be established at the particular stage in the project's life when the estimate is being prepared.

Sections of an estimate which are concerned with quantities but which have no time dependencies can be based on past experience. This data can provide a relatively high degree of certainty even in a project's formative stages.

In contrast the other sections of the estimate which involve time dependent or undefined quantities can vary considerably. These uncertain estimates should then be assigned risk margins by the project manager (Section 7.2.2).

Recognising that unless the estimates are properly prepared, the monitoring of cost would be invalid, a detailed and standard procedure has been built-up by the CEGB. The power station is divided into major plant items which are then further subdivided into individual sections. In preparing the estimate, the standard procedure used for building-up costs ensures that the estimator covers all major items. Throughout the life of the project it is this detailed analysis which is up-dated regularly in such a way as to identify actual expenditure against the original estimate. Although this standard approach is particularly useful where schemes are of basically similar nature and comparisons can be studied, the principles can still be adapted to a unique type of project.

As the design and specification of a prospective station progress, the estimate for each section of work can be developed until a definitive estimated cost to completion of the project is established.

7.2.2 *Risk Margin*

Construction of power stations inevitably involves some risk and uncertainty. The inclusion of a risk margin in the overall cost estimate of a scheme being considered for sanction enables the CEGB to make realistic provision for these uncertainties. At the same time, procedures are instigated to ensure that the use of these additional resources is subject to strict overall control.

A proportion of the project's risk margins are allocated to the project manager and may be expended by him subject to the submission of periodic financial reports. During construction of a power station the project manager may apply for release of the retained elements of risk margin by submitting a detailed report on the current and prospective financial status of the project to the Project Director and the GDCD Finance Director. They in turn would call for a comprehensive review of the project as a whole should they not be satisfied with the financial position of the project and the measures being taken to contain costs. The use of risk margin is the subject of regular reports through to the CEGB Executive. One of the main purposes of risk margin is to provide funds to allow corrective action to be taken to overcome delays which inevitably arise during the course of a project.

7.2.3 *Contract Strategy*

On the Drax Completion project, for example, a contract strategy was drawn up prior to seeking financial sanction from the CEGB Executive. As well as having regard to matters such as the capability of suppliers and potential industrial relations problems arising from different contract structures, the contract strategy took into account a number of specific factors having a direct bearing on effective control of station cost. They included:

(1) The number of contracts and their form. For Drax, 11 main supply and erect contracts and approximately 30 other major contracts were placed compared with the five contracts placed for Grain in 1971 and more than 100 for previous projects. In light of the experiences with Grain and other stations commenced in the 1970s period, it was decided that none of the contracts should be reimbursable. Thus in numbers and type of contract the CEGB reverted to the contract strategy that was in general use prior to the Wilson and NEDO Report recommendations.

(2) The use of negotiated contracts in areas where there is limited competition of monopoly supply.

(3) The level of financial incentives for major contracts such as the use of stage or progress payments. The results of these assessments are included in a contract strategy document which is subject to the approval of GDCD's Directors.

7.3 Scheme Sanction

Prior to making commitments to spend on any large project the scheme is submitted to the CEGB Executive for sanction. It includes the estimated cost to completion and risk margin and also contains an economic appraisal. The cost estimate having been sanctioned then forms the reference for monitoring and reporting on any movements in the station's cost. The authority to permit commitments to proceed is vested in the 'responsible officer' who is normally the Project Director for major projects and he is accountable for the completion of the scheme to time and cost.

7.4 Commitment and Expenditure

Once the project is sanctioned, the placing of contracts proceeds in accordance with the contract strategy. The process of preparing tender lists, issuing enquiries, assessing tenders, and awarding contracts is carried out within established CEGB procedures using a uniform style of specification and conditions of contract which have been agreed with the main trade associations.

7.5 Contract Cost Monitoring and Control

Cost control on the Drax project utilises close monitoring of both the estimated cost to completion of the project and the financial progress of individual contracts.

In addition to analysing the costs over individual items of equipment, expenditure is also identified with individual contracts in order to provide consistency with other management control systems. There is a continuous updating of the estimated cost to completion of each individual contract, using the PROVAL program described in Section 3.4.4 and, where it appears that there is potentially an overspending on the contract, a detailed investigation is carried out making use of the information and analysis included in the detailed estimate.

The responsibility for monitoring the financial status of contracts is given to nominated individuals. For each contract, a continuous review and update is made of changes and potential changes in cost from those originally sanctioned, including variations and claims.

Use may be made of the PROVAL program to provide a continuous record of the latest cost position on each contract including potential and firm variations, outstanding claims and the utilisation of risk margin, together with an up-to-date estimated cost-to-completion of the contract. The information is aggregated to show any changes in the estimated cost-to-completion.

7.6 Monitoring and Reporting

The monitoring and financial management of contracts, forms the basis of a monthly statement prepared for the project manager showing the current station estimated cost-to-completion compared with the original sanctioned sum, together with an analysis of the use of risk margin.

Each month the project manager formally reports on the progress of the project, including its financial status, to his Project Director and this report is considered by GDCD Directors at their monthly meeting. The financial position of the main individual contracts is considered in more detail in the form of a quarterly report to the Directors.

Also every quarter, a report covering the estimated cost-to-completion of each power station project is submitted to the CEGB Executive. In this report, particular attention is paid to the monitoring of the use of the overall contract risk margin. Once a year, the Executive receives a full status report on each project covering all aspects of its progress including its financial status. In addition to this regular reporting, both the Executive and the GDCD Directors may call for specific reports on individual projects as and when they see fit.

8. THE CONSTRUCTION PHASE

8.1 Introduction

The project management team is a triumvirate of project manager, project engineer and site manager. Within this group the site manager has the responsibility to ensure that the site works are completed and commissioned to a predetermined programme, a controlled budget and a specified engineering standard. The site manager is also the legal occupier of the site.

His work starts on site when all the statutory requirements have been satisfactorily cleared (and this could include any Public Inquiry required by Government policy) and access to the site is fully available to start the preliminary site works. He will previously have been involved in the development of the project so he fully understands the project manager's contract strategy. Under the project engineer's authority the site layout will be developed and the necessary site services and temporary works will be provided. These will include site offices, security systems, canteens, site roads, electrical and water supplies and the layout of the contractors' working areas.

He will ensure that the various contractors have adequate facilities and are established on the site in time to meet their programmed commitments. He will ensure also that they understand the site labour agreements so that they can co-operate and play their full part in establishing satisfactory industrial relations. At this time he will start to build up his own site team and will ensure that each of his engineering staff can identify with their head office counterpart. In this way the design, manufacture and site erection activities can be fully co-ordinated.

8.2 Construction Programme

The overall construction programme is drawn up to be entirely compatible with the PMP but has to be in more detail and in a form that allows accurate monitoring of the site works. On recent contracts this has taken the form of a key date programme, the advantages of which are:
(1) Encourages the contractor to give the earliest possible warning of actual or potential difficulties.

(2) Ensure that senior management in the contractor's organisation are made aware of any serious problems at an early date.

(3) Provides a focus for early discussion of any potential problems and possible remedial measures, whilst clearly maintaining the contractor's responsibility for recovering delays.

(4) Helps to foster a climate amongst all parties that no extension of site deliveries and erection schedules are allowable.

It will be appreciated that delays incurred by one contractor, particularly where these are on the critical path, may in turn delay the work of other contractors. In these circumstances recovery of the project may involve:

(1) Reprogramming the works of the affected contractor.

(2) Expenditure of additional money from risk margins to accelerate part of the works.

Within these key events very detailed programmes are drawn up by the contractors to indicate resources and also to indicate access for other contractors ie for cabling etc. In this way the site team can adequately control the interface and inter-reactions between the different contractors and ensure that the requirements of the commissioning programmes are met.

8.3 Progress Monitoring and Control

The site manager's key task is to continuously monitor site progress against the agreed programmes and to initiate whatever corrective action is necessary to maintain satisfactory site progress. A monthly progress meeting is held with each contractor where formal reports are tabled giving an agreed progress statement. From these agreed progress statements, an accurate prediction of the state of the programme is available, which allows the site manager and his staff to adjust, if necessary, the activities of that contractor or indeed any affected contractor.

This also helps to establish the forward requirements of plant, materials and labour and to make adequate arrangements for their receipt on site.

To cover the manufacturing progress, the CEGB monitors the manufacturing works progress in great detail and ensures that adequate reports are issued reflecting the true state of manufacture. These allow early warning on security of delivery dates, particularly those which are crucial to the site construction programme.

8.4 Plant Commissioning

Commissioning a modern power station is a complex operation. It is the phase in the project when all the design, manufacturing, erection and quality assurance expertise are put to the test. The complexity of the task and the associated stringent safety controls, demand that formal procedures are adopted in its execution. The essential procedures are laid down by the CEGB and apply to all projects, but the site manager and his team play a central role in their implemetation and day-to-day operation.

In essence commissioning is controlled by the station commissioning panel whose chairman is the station manager and the deputy chairman is the site manager. From this committee individual commissioning teams are set up covering agreed plant areas. Each commissioning team has a CEGB chairman from either the Region commissioning staff or the site manager's staff.

Each team has appropriate representatives from the contractors, the CEGB commissioning engineers and a member of the site manager's staff, with design engineers or specialist engineers being called upon as necessary.

The terms of reference of each commissioning team is clearly defined by the commissioning panel and includes a detailed programme and statement as to the procedure for bringing into service each item of plant indicating, what inspections will be required, what tests will be necessary and naming the people authorised to carry out the tests. Once the commissioning panel has approved the prepared procedures, no alterations can be made without the approval of the commissioning panel.

When construction work is complete and the appropriate checklists signed off (this will include presentation and inspection of all statutory test certificates), the commissioning teams take over and work through the agreed procedure to prove that the plant is, in every respect, fit for service.

All test data is recorded including any departures from specification either under or over and an agreed report is prepared by the team for submission to the commissioning panel.

Additionally, the commissioning panel will set up a generating unit team consisting of the deputy station manager, site engineer, and the senior commissioning engineers of the contractors. This unit team monitors the detailed work of the individual commissioning teams as, in the end, they have the ultimate responsibility for commissioning of the unit.

Membership of the commissioning panel includes: senior generation, transmission and technical representatives from the operating Region; and senior project management and site representatives from GDCD.

Formal meetings of this panel are held monthly, at which the chairmen of the individual commissioning teams are required to submit a detailed report.

The terms of reference of the commissioning panel include approval of the proposed commissioning procedures and preparation of a commissioning programme. CEGB specialist engineers and contractors' specialist engineers are called into the meeting as necessary to deal with any detailed or specific engineering points.

8.5 Documentation and Certification

Those involved in commissioning need proper documentation before any commissioning commences to enable them to agree with colleagues and contractors what is to be done and what performance standards are required from each plant item. Checklists are designed to ensure that the plant has been properly installed and all the necessary services are available and that appropriate safety measures have been taken. Certification covering initial energising, operation, takeover of plant, reporting of defects and tests results provides the permanent record of plant status and performance at the commissioning stage. Finally, all of the completed documents are carefully preserved since they are the foundation of the plant records system for the station.

9. INDUSTRIAL RELATIONS

9.1 Introduction

During the 1970s the CEGB in common with other national clients experienced a decline in site industrial relations standards, as evidenced by falling levels of labour productivity on site erection contracts accompanied by increasing levels of disputation. These symptoms of labour unrest resulted in extended delays to power station construction programmes and made a significant contribution to the escalation of erection costs. Time lost in disputes on CEGB Construction Projects from 1970 to 1983 is shown in Fig 7.6 with detailed data in Table 7.1.

In the period 1970 to 1979 the annual average loss of planned productive effort due to disputation amounted to 872,000 manhours, whilst in 1972, 1976 and 1978 CEGB construction projects lost over one million manhours due to a number of causes. For example, in 1972 a national strike in the Building and Civil Engineering Industries severely affected progress on civil engineering operations at the Grain, Ince B and Heysham A Projects. In 1976 problems affecting the Grain site associated with general labour disciplinary matters, productivity performance and hygiene considerations brought the project to a standstill for several months. In 1978 bonus payment issues caused notable disruption at Dungeness B and Ince B Projects. Ongoing labour unrest at Grain continued until April 1980, when the CEGB suspended all work on two of the station's five units and some 600 employees were made redundant. The majority of strikes and other forms of industrial action that occurred over the period 1970 to   1979 can be attributed largely to disputes over wage rates and earnings levels, particularly problems associated with the operation of bonus incentive schemes. These problems were diverse in origin and required solutions affecting changes in attitudes and the development of expertise within the Industry. External threats to bonus scheme integrity derived from:

(1) Traditions of the industry.
(2) Labour shortage.
(3) Trades unions militancy.
(4) Monetary inflation.
(5) Government earnings controls.
(6) Lack of expertise in the design of bonus schemes.

Coincident with an increase in the level of industrial disputes, the productivity of site workpeople declined significantly and contributed to the lengthening of the period required to commission a power station and the escalation of site erection costs. Several years in the 1970s were characterised by hyper-inflation, during which site earning levels on power station sites were cushioned by increases in bonus payments made by contractors to their workpeople without a concomitant productivity benefit. The effect was exacerbated by secondary factors such as:

(1) Inexperienced operators.
(2) Ineffective bonus schemes.
(3) Union identification of critical areas.
(4) Demarcation.
(5) Inflexible working hours.
(6) Impact of other contracts.

TABLE 7.1 ANNUAL SITE DISPUTE STATISTICS ON
CEGB
PROJECTS 1963 TO 1983

Year	No. of stoppages on sites	Manhours lost	% Man-hours
1963	Not Available	164,716	0.73
1964	88	345,192	0.84
1965	124	507,944	1.04
1966	187	517,075	0.94
1967	215	576,574	1.17
1968	377	454,748	0.89
1969	402	748,685	1.89
1970	429	391,623	1.41
1971	574	325,776	1.32
1972	135	1,300,490	6.19
1973	216	454,512	2.49
1974	148	447,799	2.55
1975	229	635,882	3.37
1976	221	1,574,415	7.73
1977	287	655,176	2.95
1978	311	1,310,121	5.76
1979	303	751,768	3.33
1980	261	322,781	1.58
1981	197	319,065	1.60
1982	119	166,090	0.89
1983	86	105,826	0.58

*January to June 1983 only.

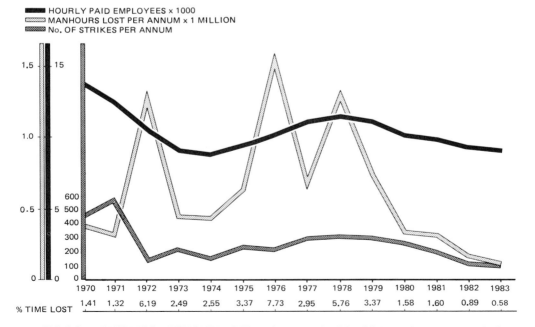

FIG 7.6   NUMBERS EMPLOYED, STRIKES AND HOURS LOST ON CEGB PROJECTS

These problems affected all UK clients' construction projects to a greater or lesser extent, so that the site industrial relations risk area became a significant investment decision factor, particularly as construction experiences in the Thamesside and Merseyside areas proved to be unfortunate owing to extreme labour difficulties. It is in these very difficult labour areas that the CEGB's initiatives (see Section 9.3) have proved successful in achieving a very significant reduction in serious labour problems.

9.2 Site Industrial Relations in the 1970s

The site industrial relations problems encountered in the 1970s, originated largely in the great variety and disparity in conditions of employment offered by different contractors to their workpeople. A multiplicity of national labour agreements, negotiated by many employer associations and trades unions, possessed very different wage structures. This led to employee dissatisfaction, which was expressed continually in the form of 'leapfrogging' wages and employment conditions claims. The demotivating effects of anomolies in national agreements interacting between groups of workpeople working side by side on the same site and engaged in broadly similar erection activities, were a continual restraint on the achievement of increased morale, improved productivity and acceptable standards of erection performance. The site industrial relations environment through the duration of a typical power station project in that era is graphically presented on Fig 7.7.

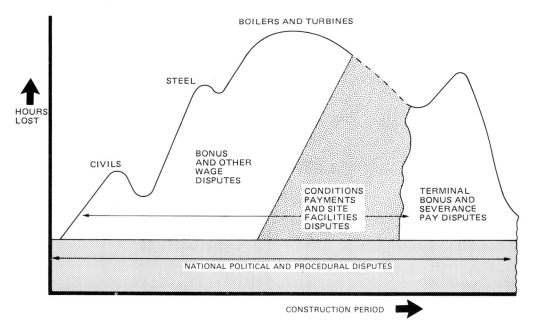

FIG 7.7   INCIDENCE OF INDUSTRIAL DISPUTES AND CAUSES

A number of contractors and trades unions (principally within the petro-chemical sector) embarked upon payment strategies involving high basic hourly rates and small, fixed bonuses (or other supplements). These strategies suffered from the impact of several national anti-inflation incomes control policies and consequently experienced extreme industrial relations pressure. This situation caused the original steelwork contractor at Grain Power Station to relinquish his contract.

As previously stated much industrial relations trauma was caused by poorly designed and badly administered incentive schemes. Some major labour agreements (such as those covering civil engineering and mechanical erection labour) allowed incentive payments to be negotiated by individual contractors with workpeople's representatives on each contract, the national base rate only being a minimum level and not an earning rate accepted by the labour as being applicable to large project work. Consequently, civil engineering and mechanical erection contractors had to provide additional payments to meet the earnings expectations of the operatives concerned on any particular large construction project. In some instances, the payment system was not rigorously controlled and a number of abuses occurred, resulting in these contractors and their site workpeople coming to regard incentive payments as a primary method of securing acceptable levels of income, and not as payments directly related to high levels of incentive effort obtained through good standards of site supervision and labour discipline. In addition, poorly designed bonus schemes gave rise, during periods of earnings restraint imposed by the Government, to increased second tier earnings over basic rates not justified by a corresponding improvement in performance, as shown in Fig 7.8.

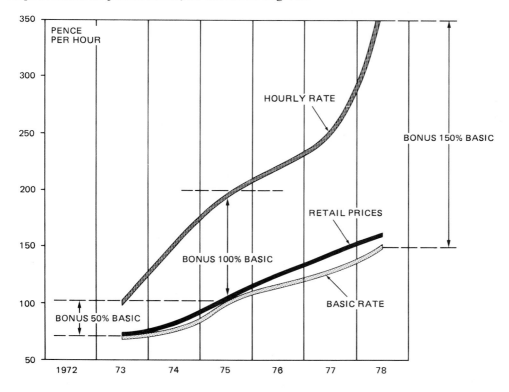

FIG 7.8   THAMESSIDE MECHANICAL LABOUR - ESCALATION OF SITE EARNINGS AND RETAIL PRICES

This general industrial relations problem was exacerbated by conflicting client strategies which impacted adversely on the projects managed by GDCD. During the last decade, the CEGB was continually investing in new plant and as a result, tended to adopt a different approach to contractors' site labour policies from some other national clients, who were only committed to one large project. In such situations, contractors were able to adopt expedient labour policies including severance and completion payments. Site labour employed on CEGB projects expected some of these provisions which, when resisted by CEGB contractors, caused industrial disputes.

The 1970s also demonstrated a hardening of employee attitudes to work, conditioned by a number of factors. Locally recruited labour forces offset the prospect of unemployment by the use of job protection practices leading, in turn, to lowered productivity. This strategy by locally residing workpeople to secure increased continuity of employment was implemented in the form of bans on overtime working, insistence on 'working to rule' and the overzealous adoption of safe working methods. Furthermore, the hostility of some local union members to the employment of highly experienced travelling men, was prompted by the same desire to secure the maximum employment benefit for workpeople resident in a particular project locality.

Government intervention to maintain employment in the plant manufacturing industry through the early ordering of generating capacity, fostered a belief among site workpeople at the Ince B oil-fired power station construction project, that large sites could be viewed (under the particular circumstances) as job creation schemes, which were intended primarily to relieve male unemployment in the particular locality. Under these circumstances, contractors experienced difficulty in convincing their employees that their earnings were not being funded directly by the consumer or the taxpayer. This in turn led to problems for contractors in upholding the integrity of their bonus payment agreements. The increasing complexity of employment legislation during the last decade coupled with the inadequacy of training activities, the lack of employment continuity within the 'casual' site erection industry and the erosion, and even reversal, of site supervisory salary differentials, demotivated contractors' management and supervision and intensified the decline in productivity and labour discipline.

These employee attitudes, coupled with the variety of national agreements and the site level negotiation of bonus payments, led to much of the industrial relations influence and negotiating power at site level being held by shop stewards rather than full-time union officials. The structure of the Construction Industry also facilitated this, since it normally consists of many small sites and a few large projects, all of which demand more or less equal attention from full-time trades union officials. These officials were sometimes unable to devote adequate time to the many problems that occurred with the large workforces employed on power station construction sites. This enabled the stewards and other part-time workpeople's representatives to assume the dominant role. High levels of inflation and consequent wage control measures during the 1970s, further enhanced the status of stewards, who were often able to force the payment of compensatory increases in site bonus payments.

Large sites came to provide optimum conditions in which industrial action could achieve the maximum disruptive effect. Certain CEGB construction projects in the last decade were continually subjected to unofficial strike action of all descriptions, including cross-picketing (or secondary picketing), protest marches, demonstrations, worker occupations and 'sit-ins'.

The organisation of site levies and other collections was easily made and the disruption of the working day of a large number of employees not involved in a particular dispute was quickly achieved. Such actions were usually unofficial and were often instigated without the knowledge or involvement of the full-time unions officials. They were usually successful in achieving the maximum attention from the media and public relations assumed a major importance in the public conduct of site industrial relations activities.

It is interesting to observe that many industrial relations and labour productivity problems originated off-site in other technical and managerial areas, their solution having become a major reason for improved site industrial relations in recent years. There are historically several problem areas that may be distinguished. Serious delays in the supply of materials to the workplace were due to many causes, eg labour disputes in a manufacturer's works or design changes resulting from component failure or ongoing improvements in technology. This factor also necessitated modifications to plant and equipment already erected on site. Such problems not only created difficulties in work scheduling but also led to loss of morale and demotivation of the site workforce. The cumulative effect of delays on one contract frequently resulted in a lack of access and work availability for succeeding contractors, who would have already built up their labour resources to meet their own contract requirements.

9.3 GDCD's response to the Site Industrial Relations Problem

During the latter part of the last decade the CEGB took a number of initiatives aimed at improving the site industrial relations and labour performance of its contractors. In retrospect, this client management strategy has proved to be a significant contributory factor in the considerable fall in manhours lost through recent labour disputes.

On sites in areas of traditional labour management difficulty, such as Thamesside and Merseyside, the CEGB was forced to intervene in a positive manner to assist contractors in establishing a more stable labour management environment. For example, labour productivity on projects was monitored closely using, where required, work measurement techniques including activity sampling. In addition, the CEGB was also forced to intervene in the handling of particular site labour problems, which had adversely affected construction work over an extended period. For example, at Ince B on Merseyside, some 700 contractors' employees were to be declared redundant in December 1979, following a failure to meet completion dates due to low labour productivity. Contractors' employees were warned that continued failure to achieve the revised (and delayed) completion dates would jeopardise the remaining construction programme and their site livelihood. The year 1980 saw a substantial improvement in the rate of completion of the revised construction programme.

Another example, the Grain construction project on Thamesside experienced continual labour unrest throughout the 1970s, chiefly in connection with problems caused by bonus levels and payment disparities. As outlined earlier, disputation on these grounds was effectively terminated in April 1980, when the CEGB suspended all work on two of the station's five generating units and some 600 contractors' employees were made redundant. The remaining workforce was warned that all work would cease at the end of June 1980, unless a solution was found to the current labour dispute involving thermal insulation operatives. As a result of these measures a novel, formal 'site understanding' was established between the project's contractors and trades unions, which imposed increased labour disciplines on the workforce and created a common maximum gross hourly earnings level over the whole project. Contractors' bonus schemes still continued to operate but were linked to the project's revised completion dates. A new form of procedure for resolving industrial disputes was also introduced, which allowed all parties with an interest in the project to contribute to speedy 'local' solutions of industrial relations grievances.

As already stated in Section 2.4, the CEGB also adopted a policy of establishing management groups on its most recent projects. Major contractors at the Drax Completion and Heysham 2 projects were required (as a formal contractual commitment) to participate in a federal manner within the groups and abide by their projectwide labour management policies. The management groups seek to create policies and procedures that harmonise employment conditions on their respective sites and allow contractors from different industrial sectors to manage their own labour forces in a consistent manner. Senior project management representatives of the CEGB act as chairmen to the groups and are supported by specialist site industrial relations staff. The promotion of a 'federal' approach by contractors on a particular project necessitates the management group performing a central co-ordinating (as well as industrial relations policy forming) role.

By introducing the management group concept, the CEGB has been able to support contractors in dealing with issues common to a power station construction site as a whole and which cannot, therefore, be dealt with effectively through the procedures of any single national agreement. Consultative steering groups or study groups have been introduced, involving both contractors and trades unions representatives from all sectors engaged in construction work on a project. As many of the issues discussed in these groups directly involve client interests, the CEGB also involves itself in the activities of these bodies. These joint consultative groups review (on a regular basis) site progress and monitor matters of common interest, including work scheduling and access, productivity, shiftworking, bonus schemes, demarcation and other disputes, safety, welfare and general labour questions. Their main objective is to identify developing problems and advise upon early corrective action.

9.4 The National Agreement for the Engineering Construction Industry (NAECI)

It has been stated that many of the industrial relations problems which have aggravated the difficulties of power station construction in the past can be attributed to uncontrolled bonus payment regimes and the multiplicity of national agreements with differing terms of employment. The CEGB encouraged over a number of years, other parties in large site construction to promote one comprehensive national agreement for all mechanical erection and thermal insulation contractors. The new NAECI became effective on 2 November 1981. Its objective was to improve the cost effectiveness of engineering construction in the UK and its introduction has aided the stabilisation of site industrial relations on the CEGB's construction projects since its inception.

A new National Joint Council (NJC) is responsible for the control and national conduct of the industry's industrial relations. The NJC comprises equal numbers from the site unions and construction employers concerned and is supported by its own permanent full-time staff. It is responsible for monitoring the application of the agreement on all mechanical engineering construction sites throughout the country. Project Joint Councils (PJC) have been established on all major 'nominated projects' (including the Drax Completion and Heysham 2 Projects) with responsibility for implementing the terms of the agreement through the negotiation of a Supplementary Project Agreement (SPA) relevant to the needs of that particular project. The PJC consists of representatives from employers and trades unions including a select number of shop stewards from the site workforce. An NJC staff member serves as secretary to the PJC and acts as the NJC's representative within the PJC.

The new agreement has proved to contain a number of provisions which have aided the eradication of certain major labour difficulties. Bonus payments are now controlled nationally by the establishment of fixed payment limits, within which incentive bonus payments may fluctuate. Gradual progress has been made in increasing the proportion of basic rate to bonus pay in the weekly wage packet, thereby partially stabilising earnings and reducing the impact of potentially problematic bonus schemes. The agreement has also created dispute procedures and investigatory arrangements, involving all parties interested in the project, as a means of securing a rapid and effective resolution of these problems.

Other developments to raise the standards of skill within the engineering construction industry's labour force and improve training methods have also been implemented in recent years and this should improve labour productivity performance on future CEGB projects.

9.5 Development of Shiftwork

The NAECI increased average allowances for working a 2-shift system from 17.3%, under the previous Mechanical Construction Engineering Agreement, to 25% of the higher NAECI basic hourly rate. At the same time the working week was reduced from 40 hours to 39 hours on 2 November 1981. This change occurred in a period of high and rising unemployment when trades unions antagonism to all overtime work started to limit effectively the manhours available from any given size of workforce. The CEGB and its contractors (with general site trades unions support) adopted a policy of increasing site manning levels by employing the great majority of mechanical and electrical erection workpeople on shift systems in recent years. This new approach to resourcing CEGB construction projects has promoted more man-hours working per week than was possible in the old 40 hour week utilising additional, uneconomic overtime. As a result of adopting double dayshift working arrangements at Drax Completion and Heysham 2 projects, faster programme completion rates than those experienced on many sites in the last decade, have been achieved. The shiftworking concept has provided a high ratio of staff to operatives and a lower than normal density of workpeople in attendance during each working period. These two factors have assisted the achievement of the required levels of productivity in addition to a greatly enhanced utilisation of capital plant.

9.6 Current Situation Review

Despite the severe industrial relations problems which have delayed programme completion rates on certain CEGB construction projects in the last decade, the situation has changed due to a number of factors. CEGB managerial initiatives, the introduction of shiftwork for large numbers of site workpeople and the establishment of a new national agreement have already improved the general site labour climate leading to faster programme completion rates.

10. PROJECT MANAGEMENT ON CURRENT PROJECTS - DRAX COMPLETION AND HEYSHAM 2

10.1 Introduction

Work on the Drax Completion site began in early 1978 with site preparation work. The main foundations contract commenced in March 1979, following the completion of piling for the first of the three new generating units. The full project management and architect/engineering role for the Drax Completion project was carried out by GDCD. The GDCD role covered system design, layout, preparation of specifications, evaluation of bids, contract management, supervision of erection at site, quality assurance and quality control, progress monitoring, time and cost management and commissioning, as described in the previous sections in this Chapter.

Government approval for Heysham 2 and a sister station for the South of Scotland Electricity Board at Torness, Lothian, Scotland was announced in January 1978. Later that year a joint design phase contract was placed with the National Nuclear Corporation (NNC) to enable a firm design basis to be established before permanent work began on site. Preliminary site work began in 1979 followed by the start of the station's main foundations in August 1980. The project management arrangements for Heysham 2 reflected the involvement of both CEGB and NNC in the design, engineering and site management of the project. Following the system design of the whole of the station and preparation of the pre-construction safety report, NNC undertook the engineering of the nuclear island. The CEGB was responsible for engineering the remainder of the station, managing the conventional plant contracts directly, and managing the overall project. NNC acted as agents for the administration and site management of the nuclear island contracts.

10.2 Drax and Heysham 2 Project Management Arrangements

Arrangements for the construction of Heysham 2 were established which would make the maximum effective use of the resources available in both NNC and GDCD, with NNC concentrating on the nuclear works and GDCD dealing with the conventional works.

GDCD assumed overall project management responsibility for both the conventional and nuclear works and responsibility for the detailed design and engineering of the conventional works. The CEGB also has three central client roles in relation to the project namely, responsibility for acceptance of the design, contractual responsibility as the purchaser, and responsibility for nuclear safety. The role of NNC was defined in an agency agreement between the CEGB and NNC which came into effect in February 1981. The agreement set out the role of NNC as designer and as the CEGB's agent and manager for contracts associated with the nuclear work areas. Under this agreement (which succeeded a design phase contract signed in 1978) NNC's responsibilities included:

(1) The overall station system design, the detailed design of the nuclear works and the detailed design of the works to be carried out under the CEGB's main civil engineering contract.

(2) Acting as agent and manager of the CEGB for the placing and administration of contracts for the nuclear works and parts of the common works (essentially plant and equipment items such as cabling which service both the nuclear and non-nuclear sections of the station), this role having been made subject to delegated authorities set out in schedules to the agreement.

(3) Managing the construction and administration of the nuclear works and parts of the common works, again subject to delegated authorities.

(4) The preparation of operating and maintenance manuals for the nuclear works to the CEGB Specification TP30.

The intent of both parties in drawing up the agency agreement was that the resources of NNC and CEGB should be fully co-ordinated and used in an effective manner to bring the project to a successful completion. Particular attention was paid to ensure that the responsibility and accountability of both parties was clearly defined. To achieve this objective, the main clauses of the agreement were supported by detailed schedules which defined areas of responsibility with particular reference to the interfaces between the two organisations' roles and laid down the working procedures to be followed in carrying out the project.

The project management arrangements for Heysham 2 differ from those for Drax Completion in that NNC is involved. However in common with the Drax project, the GDCD Heysham 2 Project Director is also the engineer and responsible officer for the project. In addition to the delegations which he makes to the GDCD project manager he also delegates part of his responsibilities to NNC for the management of nuclear island activities in accordance with the agency agreement. These delegations include the right, within defined authorities, to place contracts and issue contract variations. Notwithstanding these delegations, the Project Director remains accountable for completion of the whole project. The NNC project manager has responsibility for those aspects of the project which fall within the scope of NNC's supply. In addition to his overall delegated responsibilities for management of the Heysham 2 project, both on and off site, the GDCD project manager carries out the management of the contracts in the conventional and common works areas; the design and engineering of the non-nuclear island works; the management of the interfaces between the nuclear island and conventional work areas and the management of the conventional site works.

As in the case of Drax Completion, the GDCD project manager has a project engineer and site manager who report directly to him and the project engineer deputises for the project manager in his absence.

The Heysham 2 project engineer's responsibilities are (in outline) very similar to those of the Drax project engineer. They differed in detail to the extent that NNC also played a role in station design and procurement for the nuclear island. Formal design work on the Heysham 2 project commenced with the placement by the CEGB of a design phase contract with NNC. The design phase contract had two main elements, these being the production of a pre-construction safety report (PCSR) and the development of and outline design for the proposed station. The PCSR forms the framework of the CEGB's submission to the NII of the safety case for the Heysham 2 project. After endorsement by GDCD and the CEGB Health and Safety Department, the PCSR was formally submitted to the NII. Assessment of the NNC outline design for the station as a whole was the responsibility of the GDCD project engineer. The outline design formed the basis on which more detailed design development was carried out by the project team to enable design specifications to be prepared in respect of the contracts for plant outside the nuclear island. The project engineer was responsible for the various steps leading up to the award of these contracts.

At the manufacturing stages of the project, the project engineer has specific responsibility for ensuring the progress of contracts for plant and equipment outside the nuclear island. He also has responsibility for managing the engineering interfaces defined in the agency agreement and for monitoring progress on the project as a whole to ensure that this was in accordance with the overall programme requirements. As provided for in the agency agreement, design specification and tender assessment of plant within the nuclear island was carried out by the NNC. The project engineer also ensures the compatibility of the interfaces of items of plant and systems between the nuclear island and balance of plant areas respectively and participates in the management of the common works contracts as provided for in the agency agreement.

The role of the GDCD site manager for Heysham 2 differs from that for Drax in that the site management, supervision of erection, and commissioning of the nuclear island works are the responsibility of NNC. An NNC site manager was appointed to undertake this work and is responsible to the GDCD site manager for ensuring that the NNC role is carried out within the framework of the CEGB's overall management of the site. Of the common works on Heysham 2, the single largest contract was for the construction of the foundations and superstructures of the Heysham 2 nuclear island/turbine complex. The overall direction of this contract was undertaken by the Project Director as the Engineer, but within this framework NNC manages the nuclear island aspects and GDCD manages those works relating to the turbine complex. Other common works on the Heysham 2 site are managed in a similar fashion.

With these exceptions the role of the GDCD site manager of Heysham 2 and Drax are similar. Because Heysham 2 is a nuclear station, the GDCD site manager has the additional and important responsibility, delegated from his project manager, of ensuring that the conditions of the station's site licence from the Nuclear Installations Inspectorate (NII), which relate to site activities, are observed up to the time the station is taken over by the operating Region.

In order to ensure that NNC's activities in the design manufacture and construction of the nuclear island on both Heysham 2 and Torness are effectively co-ordinated between the two NNC project management teams, a single NNC Project Director is responsible to the NNC Board for both projects. On Heysham 2 he is accountable for the proper execution of the agency agreement to the GDCD Project Director for this station.

10.3 Design Standardisation and Design Change Control

10.3.1 *Introduction*

The policy of design standardisation but with some design evolution is illustrated by the case of the Drax Completion project which was based on the original Drax design, and the Heysham 2 design which follows that for Hinkley Point B. However, for the achievement of the maximum benefits from replication the gap between orders should not be greater than about two years. In the case of both Drax Completion and Heysham 2 the orders were separated by some 12 years from their earlier stations, and changes had to be made to the designs to take account of developments in the technology, enhanced safety requirements, etc, in addition to taking advantage of lessons learned during the construction and operation of the earlier stations. The following sections describe how design change was controlled through the progressive development of the design. The procedure was similar for Drax Completion and Heysham 2, but the emphasis is on the Heysham 2 procedure, as it also covered the safety and licensing requirements for nuclear power stations.

10.3.2 *Design Control - Heysham 2*

For the Heysham 2 project, the Hinkley Point B design was chosen as the basis for the new station design. This was because, of the three previous AGR station designs, Hinkley Point B had been the most successful in terms of its construction record and the station was operational. From the outset it was recognised that the operating experience updated safety criteria and changes in technology since the original design was produced would have to be taken into account. However, both the CEGB and NNC were determined that this reassessment should not result in significant changes from the proven design basis of the Hinkley Point B station. As a result, in the development of the conceptual design for Heysham 2, a rigorous assessment was undertaken of each change from the Hinkley Point B design. Every proposed design change had to be justified on at least one of a number of criteria. The criteria comprised the need to satisfy current safety requirements, the degree of improvement in plant operability or constructability and whether or not the proposed design change was necessitated by changed technology. The individual proposals were then subject to detailed review and approval by a joint CEGB/NNC committee which was chaired by the CEGB Project Director. In deciding whether to sanction a change the committee paid particular attention to the impact which it might have on the construction programme and overall station cost.

Despite some changes, the fundamental design concept of Hinkley Point B was successfully retained on Heysham 2, and the amount of innovative design was small. Following the review of the Hinkley Point B design an 'outline' or 'reference' design for Heysham 2 was produced and a pre-construction safety report (PCSR) based on this design was submitted to the NII as the basis for the issuing of the nuclear site licence. The development of the outline design and the preconstruction safety case were carried out by NNC under a design phase contract. The objective of the PCSR was to establish the framework within which the detailed safety case would be constructed and to present sufficient analysis and detailed arguments to enable high confidence to be gained that later design changes (which might have disrupted the programme) would not be necessary to satisfy the safety case. The PCSR was formally issued to the NII and some additional amplification was later provided by the CEGB. During this phase of the project a number of design contracts were placed with the intended suppliers of critical plant items.

The award of design contracts prior to manufacture and construction ensured the timely provision of vendor information, this being essential to the development of the overall station design and in particular to the civil design. The civil design information was required very early in the construction of the station, and to detail this design the civil engineers required information from the mechanical and electrical contractors with regard to plant location, floor loadings, and support requirements.

In October 1980 the NII confirmed that they were satisfied that the safety principles and design intent as set out in the PCSR were acceptable and that once manufacture and construction had started there would be only a small risk of significant modifications subsequently being required for safety reasons. Hence the NII granted consent to the commencement of construction of the pressure vessels, enabling the main construction activities to proceed. The CEGB considered that the statement by the NII confirmed its own judgement that the status of the design was now sufficiently advanced and that further substantiations of the safety case was very unlikely to lead to modifications which would threaten the project programme or cost. Accordingly the CEGB allowed work to proceed on the construction of the station. The subsequent five years of construction experience reinforced the judgement CEGB made at that time. Work on the substantiation of the safety case for Heysham 2 has not identified any new and unforeseen safety-related problems and no significant safety-related design changes have been required during this period by the NII.

The start of site construction in 1980 and the associated letting of hardware contracts were undertaken against an established design (the 'contract design') which incorporated all the previous design changes so constituting what is frequently termed a 'design freeze'. Further design changes arising during the construction of the station are the subject of a well-established procedure within GDCD, NNC, and the contractors' organisations, within which a design change is defined as any change possessing the potential to affect the safety case or having a significant cost impact, or affecting the programme. Design changes are controlled against the 'contract design' frozen at October 1980, this design also being the basis for the station estimated cost-to-completion which forms the financial control reference of the works. Responsibility for control of design changes for the conventional plant contracts is vested solely on the GDCD, but under the design change procedure NNC are consulted should there be any identifiable consequence for the nuclear island activities. Design changes to the nuclear works may be initiated by the CEGB or by NNC. The CEGB design change procedure for Heysham 2 reflects this in the following manner: All proposals by NNC for nuclear works design changes are sent to the CEGB project manager who consults with interested parties before commenting, giving his approval or rejecting the proposal. Changes to the nuclear works design initiated by CEGB, whether they originate from the project team or the CEGB specialist departments, must be approved by the project manager before being submitted to the NNC for their agreement.

Design changes initiated by NNC for common works contracts are developed by NNC up to a point where their implications for cost, programme and other nuclear works activities have been identified. Details of the design change are transmitted to CEGB for assessment and, if deemed appropriate, communicated to the contractor. The procedure for dealing with design changes initiated by CEGB is exactly as that for conventional plant contracts. Any differences of view over proposed changes to the design of nuclear works are resolved between NNC and CEGB. All design changes which have a potential impact on the nuclear safety case are subject to further review by a joint CEGB/NNC Safety Design Change Committee. Any change which has a bearing on the safety case is advised at the earliest opportunity to the NII.

Design changes are subject to close financial and programme control and, where a change affects project costs, a formal variation instruction is given to the contractor. The issue of variation instructions is subject to the CEGB's established delegation of financial authority procedure. Design changes may not be implemented until they have received financial sanction as these may require a call on the risk margin and be reflected in the estimated cost-to-completion. Proposed design changes which would lead to delays in contract completion are subject to additional control. In particular, any change which would result in a delay to the CEGB's target commissioning commitment date may only be approved by the CEGB Project Director.

10.4   Status of the Drax Completion and Heysham 2 Projects

The success of the various measures taken to control progress on the Drax Completion and Heysham 2 projects, including design and industrial relations initiatives is reflected in the progress which has been achieved at both sites. At Drax Completion the first unit was synchronised within programme in December 1983, 58 months after the start of main foundations. This was achieved despite a number of delays in the early part of the project programme which at one stage left the first unit running some 26 weeks late. Commissioning work on this unit met the target commissioning date of April 1984 whilst of the two remaining units one has already been synchronised and both are running to original programme.

Similar progress is being achieved at Heysham 2 with the first unit back on programme following the success of a recovery programme instituted at the end of 1982. Work on the second unit is also progressing well and the successful installation of the pressure vessel liner roof was some 10 weeks ahead of programme.

Most design and manufacturing risks have been eliminated on the project and attention is now concentrated on the site to anticipate and overcome problems which are likely to arise there. The major task is still to complete the large volume of detailed work necessary to achieve the timely installation, setting-to-work and commissioning of a large number of individual plant items rather than resolving individual technical problems. In addition, approval from the NII for the outstanding detailed safety submissions must be obtained.

Combined engineering tests of the first reactor are programmed to commence in October 1985, with the achievement of substantial load by early 1987.

Comprehensive Contents List

Chapter 1 Construction History and Development

List of Sections

Chapter 1   Construction History and Development (Cont'd)

Chapter 1   Construction History and Development   (Cont'd)

Chapter 1   Construction History and Development (Cont'd)

List of Illustrations

Chapter 1 Construction History and Development (Cont'd)

Chapter 2   Littlebrook D Oil-fired Power Station (Cont'd)

Chapter 2   Littlebrook D Oil-fired Power Station (Cont'd)

Chapter 2   Littlebrook D Oil-fired Power Station (Cont'd)

Chapter 2   Littlebrook D Oil-fired Power Station (Cont'd)

Chapter 2   Littlebrook D Oil-fired Power Station (Cont'd)

List of Illustrations

List of Tables

Chapter 3   Drax Coal-fired Power Station

List of Sections

Chapter 3 Drax Coal-fired Power Station (Cont'd)

Chapter 3   Drax Coal-fired Power Station (Cont'd)

Chapter 3 Drax Coal-fired Power Station (Cont'd)

Chapter 3   Drax Coal-fired Power Station (Cont'd)

List of Illustrations

Chapter 3   Drax Coal-fired Power Station (Cont'd)

List of Tables

Chapter 4   Dinorwig Pumped Storage Power Station

List of Sections

Chapter 4   Dinorwig Pumped Storage Power Station (Cont'd)

Chapter 4 Dinorwig Pumped Storage Power Station (Cont'd)

Chapter 4   Dinorwig Pumped Storage Power Station   (Cont'd)

Chapter 4   Dinorwig Pumped Storage Power Station (Cont'd)

Chapter 4   Dinorwig Pumped Storage Power Station (Cont'd)

List of Illustrations

Chapter 4   Dinorwig Pumped Storage Power Station (Cont'd)

List of Tables

Chapter 5   Heysham 2 - AGR Nuclear Power Station

List of Sections

Chapter 5   Heysham 2 - AGR Nuclear Power Station (Cont'd)

Chapter 5   Heysham 2 - AGR Nuclear Power Station (Cont'd)

Chapter 5   Heysham 2 - AGR Nuclear Power Station   (Cont'd)

Chapter 5   Heysham 2 - AGR Nuclear Power Station (Cont'd)

List of Illustrations

Chapter 5   Heysham 2 - AGR Nuclear Power Station   (Cont'd)

List of Tables

Chapter 6   Sizewell B - PWR Nuclear Power Station

List of Sections

Chapter 6   Sizewell B - PWR Nuclear Power Station (Cont'd)

Chapter 6   Sizewell B - PWR Nuclear Power Station (Cont'd)

Chapter 6   Sizewell B - PWR Nuclear Power Station (Cont'd)

List of Illustrations

Chapter 6   Sizewell B - PWR Nuclear Power Station (Cont'd)

Chapter 7   Project Management

List of Sections

Chapter 7 Project Management (Cont'd)

Chapter 7   Project Management (Cont'd)

List of Illustrations

List of Tables

Index

In addition to the comprehensive contents list for each chapter, which the reader will, in most cases, use to locate the information required, the following index is an additional facility for locating information from an alphabetical key title list related to chapter and section numbers.

It will assist the reader, in using the index, to remember that prefixes 2 to 6 represent the chapter numbers of power stations Littlebrook D, Drax, Dinorwig, Heysham 2 and Sizewell B respectively; that prefix 1 refers to Chapter 1 which is general information and not necessarily related to any particular station; and that prefix 7 refers to Chapter 7 Project Management information.